Introduction to Engineering Statistics and Lean Six Sigma

Theodore T. Allen

Introduction to Engineering Statistics and Lean Six Sigma

Statistical Quality Control and Design of Experiments and Systems

Third Edition

Springer

Theodore T. Allen
Industrial and Systems Engineering
The Ohio State University
Columbus, OH, USA

ISBN 978-1-4471-7419-6 ISBN 978-1-4471-7420-2 (eBook)
https://doi.org/10.1007/978-1-4471-7420-2

Library of Congress Control Number: 2018957066

This Springer imprint is published by the registered company Springer-Verlag London Ltd. part of Springer Nature
The registered company address is: The Campus, 4 Crinan Street, London, N1 9XW, United Kingdom

Dedicated to my wife and to my parents

Preface

There are four main reasons why I wrote this book. First, six sigma consultants have taught us that people do not need to be statistical experts to gain benefits from applying methods under such headings as "statistical quality control" (SQC) and "design of experiments" (DOE). Some college-level books intertwine the methods and the theory, potentially giving the mistaken impression that all the theory has to be understood to use the methods. As far as possible, I have attempted to separate the information necessary for competent application from the theory needed to understand and evaluate the methods.

Second, many books teach methods without sufficiently clarifying the context in which the method could help to solve a real-world problem. Six sigma, statistics and operations-research experts have little trouble making the connections with practice. However, many other people do have this difficulty. Therefore, I wanted to clarify better the roles of the methods in solving problems. To this end, I have re-organized the presentation of the techniques and included several complete case studies conducted by myself and former students.

Third, I feel that much of the "theory" in standard textbooks is rarely presented in a manner to answer directly the most pertinent questions, such as:

> Should I use this specific method or an alternative method?
> How do I use the results when making a decision?
> How much can I trust the results?

Admittedly, standard theory (e.g., analysis of variance decomposition, confidence intervals, and defining relations) does have a bearing on these questions. Yet the widely accepted view that the choice to apply a method is equivalent to purchasing a risky stock investment has not been sufficiently clarified. The theory in this book is mainly used to evaluate in advance the risks associated with specific methods and to address these three questions.

Fourth, there is an increasing emphasis on service sector and bioengineering applications of quality technology, which is not fully reflected in some of the alternative books. Therefore, this book constitutes an attempt to include more examples pertinent to service-sector jobs in accounting, education, call centers, health care, and software companies.

In this third edition, a new perspective emerges: Lean six sigma can be viewed as human-in-the-loop data-driven decision processes. To a real extent, lean six sigma training is solid preparation for machine learning and artificial intelligence work and research. Both disciplines relate to formalized decision processes and statistical concepts. Also, lean six sigma relates to more than half of procedures software in Minitab and JMP. Much of the remaining software methods in Minitab and JMP relate to human-not-in-the-loop decision-making or machine learning.

In addition, this book can be viewed as attempt to build on and refocus material in other books and research articles, including: Harry and Schroeder (1999) and Pande et al. (2000) which comprehensively cover six sigma; Montgomery (2001) and Besterfield (2001), which focus on statistical quality control; Box and Draper (1987), Dean and Voss (1999), Fedorov and Hackl (1997), Montgomery (2000), Myers and Montgomery (2001), Taguchi (1993), and Wu and Hamada (2000), which focus on design of experiments.

At least 50 books per year are written related to the "six sigma movement" which (among other things) encourage people to use SQC and DOE techniques. Most of these books are intended for a general business audience; few provide advanced readers the tools to understand modern statistical method development. Equally rare are precise descriptions of the many methods related to six sigma as well as detailed examples of applications that yielded large-scale returns to the businesses that employed them.

Unlike many popular books on "six sigma methods," this material is aimed at the college- or graduate-level student rather than at the casual reader, and includes more derivations and analysis of the related methods. As such, an important motivation of this text is to fill a need for an integrated, principled, technical description of six sigma techniques and concepts that can provide a practical guide both in making choices among available methods and applying them to real-world problems. Professionals who have earned "black belt" and "master black belt" titles may find material more complete and intensive here than in other sources.

Rather than teaching methods as "correct" and fixed, later chapters build the optimization and simulation skills needed for the advanced reader to develop new methods with sophistication, drawing on modern computing power. Design of experiments (DOE) methods provide a particularly useful area for the development of new methods. DOE is sometimes called the most powerful six sigma tool. However, the relationship between the mathematical properties of the associated matrices and bottom-line profits has been only partially explored. As a result, users of these methods too often must base their decisions and associated investments on faith. An intended unique contribution of this book is to teach DOE in a new way, as a set of fallible methods with understandable properties that can be improved, while providing new information to support decisions about using these methods.

Two recent trends assist in the development of statistical methods. First, dramatic improvements have occurred in the ability to solve hard simulation and optimization problems, largely because of advances in computing speeds. It is now far easier to "simulate" the application of a chosen method to test likely outcomes of its application to a particular problem. Second, an increased interest in six sigma methods and other formal approaches to making businesses more competitive has increased the time and resources invested in developing and applying new statistical methods.

This latter development can be credited to consultants such as Harry and Schroeder (1999), Pande et al. (2000), and Taguchi (1993), visionary business leaders such as General Electric's Jack Welch, as well as to statistical software that permits non-experts to make use of the related technologies. In addition, there is a push towards closer integration of optimization, marketing, and statistical methods into "improvement systems" that structure product-design projects from beginning to end.

Statistical methods are relevant to virtually everyone. Calculus and linear algebra are helpful, but not necessary, for their use. The approach taken here is to minimize explanations requiring knowledge of these subjects, as far as possible. This book is organized into three parts. For a single introductory course, the first few chapters in Parts One and Two could be used. More advanced courses could be built upon the remaining chapters. At The Ohio State University, I use each part for a different 11 week course.

The second edition featured expanded treatment of lean manufacturing and design for six sigma (DFSS). Specifically, many lean methods are added to Chaps. 5 and 21 is entirely new. Also, there was additional design of experiments (DOE) related introductory material in Chap. 10. The new material includes coverage of full factorials, paired t-testing, and additional coverage of analysis of variance (ANOVA). Finally, several corrections have been made particularly relating to design of experiments theory and advanced methods.

The third edition expands substantially on four major topics of increasing relevance to organizations interested in lean six sigma. First, the material is tied more directly to the Certified Quality Engineer (CQE) exam from ASQ. Second, motivation and change management are critical subjects for achieving valuable results. They are also covered on the CQE exam. Third, reliability, maintenance, and product safety are covered in order to span the CQE body of knowledge more completely. Fourth, new material provides additional to software including Minitab.

In addition, the third edition provides more introductory material on analysis of variance and multiple comparisons. Feedback from instructors suggested that more material be provided on these standard topics. The author would like to thank the National Science Foundation for Support under grant #1409214. This support helped to create awareness of the results of lean six sigma to cybersecurity.

Theodore T. Allen
Columbus, USA

References

Besterfield D (2001) Quality control. Prentice Hall, Columbus
Box GEP, Draper NR (1987) Empirical model-building and response surfaces. Wiley, New York
Breyfogle FW (2003) Implementing six sigma: smarter solutions® using statistical methods, 2nd edn. Wiley, New York
Dean A, Voss DT (1999) Design and analysis of experiments. Springer, Berlin
Fedorov V, Hackl P (1997) Model-oriented design of experiments. Springer, Berlin
Harry MJ, Schroeder R (1999) Six Sigma, the breakthrough management strategy revolutionizing the world's top corporations. Bantam Doubleday Dell, New York
Montgomery DC (2000) Design and analysis of experiments, 5th edn. Wiley, Hoboken
Montgomery DC (2001) Statistical quality control, 4th edn. Wiley, Hoboken
Myers RH, Montgomery DA (2001) Response surface methodology, 5th edn. Wiley, Hoboken
Pande PS, Neuman RP, Cavanagh R (2000) The six sigma way: how GE, motorola, and other top companies are honing their performance. McGraw-Hill, New York
Taguchi G (1993) Taguchi methods: research and development. In: Konishi S (ed) Quality engineering series, vol 1. The American Supplier Institute, Livonia
Wu CFJ, Hamada M (2000) Experiments: planning, analysis, and parameter design optimization. Wiley, New York

Acknowledgements

I thank my wife, Emily, for being wonderful. I thank my sons, Andrew and Henry, for putting good efforts into their schooling. I also thank my parents, George and Jodie, for being exceptionally good parents. Both Emily and Jodie provided important editing and conceptual help. In addition, Austin Mount-Campbell helped greatly in generating much of the new content for the second edition. Also, Sonya Humes and editors at Springer Verlag including Kate Brown and Anthony Doyle provided valuable editing and comments. Discussions with Jack Feng were helpful in planning the third edition. In particular, his knowledge and experiences relating to leadership and the ASQ Certified Quality Engineering exam were inspirational.

Gary Herrin, my advisor, provided valuable perspective and encouragement. Also, my former Ph.D. students deserve high praise for helping to develop the conceptual framework and components for this book. In particular, I thank Liyang Yu for proving by direct test that modern computers are able to optimize experiments evaluated using simulation, which is relevant to the last four chapters of this book, and for much hard work and clear thinking. Also, I thank Mikhail Bernshteyn for his many contributions, including deeply involving my research group in simulation optimization, sharing in some potentially important innovations in multiple areas, and bringing technology in Part II of this book to the marketplace through Sagata Ltd., in which we are partners. I thank Charlie Ribardo for teaching me many things about engineering and helping to develop many of the welding-related case studies in this book. Waraphorn Ittiwattana helped to develop approaches for optimization and robust engineering in Chap. 14. Navara Chantarat played an important role in the design of experiments discoveries in Chap. 18. I thank Deng Huang for playing the leading role in our exploration of variable fidelity approaches to experimentation and optimization. I am grateful to James Brady for developing many of the real case studies and for playing the leading role in our related writing and concept development associated with six sigma, relevant throughout this book.

Also, I would like to thank my former M.S. students, including Chaitanya Joshi, for helping me to research the topic of six sigma. Chetan Chivate also assisted in the development of text on advanced modeling techniques (Chap. 16). Also, Gavin Richards and many other students at The Ohio State University played key roles in providing feedback, editing, refining, and developing the examples and problems. In particular, Mike Fujka and Ryan McDorman provided the student project examples.

In addition, I would like to thank all of the individuals who have supported this research over the last several years. These have included first and foremost

Allen Miller, who has been a good boss and mentor, and also Richard Richardson and David Farson who have made the welding world accessible; it has been a pleasure to collaborate with them. Jose Castro, John Lippold, William Marras, Gary Maul, Clark Mount-Campbell, Philip Smith, David Woods, and many others contributed by believing that experimental planning is important and that I would some day manage to contribute to its study.

Also, I would like to thank Dennis Harwig, David Yapp, and Larry Brown both for contributing financially and for sharing their visions for related research. Multiple people from Visteon assisted, including John Barkley, Frank Fusco, Peter Gilliam, and David Reese. Jane Fraser, Robert Gustafson, and the Industrial and Systems Engineering students at The Ohio State University helped me to improve the book. Bruce Ankenman, Angela Dean, William Notz, Jason Hsu, and Tom Santner all contributed through discussions.

Also, editors and reviewers played an important role in the development of this book and publication of related research. First and foremost of these is Adrian Bowman of the Journal of the Royal Statistical Society Series C: Applied Statistics, who quickly recognized the value of the EIMSE optimal designs (see Chap. 13). Douglas Montgomery of Quality and Reliability Engineering International and an expert on engineering statistics provided key encouragement in multiple instances. In addition, the anonymous reviewers of this book provided much direct and constructive assistance including forcing the improvement of the examples and mitigation of the predominantly myopic, US-centered focus.

Finally, I would like to thank seven people who inspired me, perhaps unintentionally: Richard DeVeaux and Jeff Wu, both of whom taught me design of experiments according to their vision, Max Morris, who forced me to become smarter, George Hazelrigg, who wants the big picture to make sense, George Box, for his *many* contributions, and Khalil Kabiri-Bamoradian, who taught me many things. Scott Sink brings passion, ideas, and experience that are much valued by me and other members of our department. His comments have helped with developing this book.

Contents

Part III Optimization and Management

Acronyms

ANOVA Analysis of Variance is a set of methods for testing whether factors affect system output dispersion (variance) or, alternatively, for guarding against Type I errors in regression

BBD Box Behnken designs are commonly used approaches for structuring experimentation to permit fitting of second-order polynomials with prediction accuracy that is often acceptable

CCD Central composite designs are commonly used approaches to structure experimentation to permit fitting of second order polynomials with prediction accuracy that is often acceptable

DFSS Design for six sigma is a set of methods specifically designed for planning products such that they can be produced smoothly and with very high levels of quality

DOE Design of experiments methods are formal approaches for varying input settings systematically and fitting models after data have been collected

EER Experimentwise Error Rate is a probability of Type I errors relevant to achieving a high level of evidence accounting for the fact that many effects might be tested simultaneously

EIMSE The Expected Integrated Mean Squared Error is a quantitative evaluation of an input pattern or "DOE matrix" to predict the likely errors in prediction that will occur, taking into account the effects of random errors and model misspecification or bias

FMEA Failure Mode and Effects Analysis is a technique for prioritizing critical output characteristics with regard to the need for additional investments

GAs Genetic Algorithms are a set of methods for heuristically solving optimization problems that share some traits in common with natural evolution

IER Individual Error Rate is a probability of Type I errors relevant to achieving a relatively low level of evidence not accounting for the multiplicity of tests

ISO 9000: 2000 The International Standards Organization's recent approach for documenting and modeling business practices

KIV Key Input Variable is a controllable parameter or factor whose setting is likely to affect at least one key output variable

KOV Key Output Variable is a system output of interest to stakeholders

LCRSM Low Cost Response Surface Methods are alternatives to standard RSM, generally requiring fewer test runs

OEMs Original Equipment Manufacturers are the companies with well-known names that typically employ a large base of suppliers to make their products

OFAT One-Factor-at-a-Time is an experimental approach in which, at any given iteration, only a single factor or input has its settings varied with other factor settings held constant

PRESS PRESS is a cross-validation-based estimate of the sum of squares errors relevant to the evaluation of a fitted model such as a linear regression fitted polynomial

QFD Quality Function Deployment are a set of methods that involve creating a large table or "house of quality" summarizing information about competitor system and customer preferences

RDPM Robust Design Using Profit Maximization is one approach to achieve Taguchi's goals based on standard RSM experimentation, i.e., an engineered system that delivers consistent quality

RSM Response Surface Methods are the category of DOE methods related to developing relatively accurate prediction models (compared with screening methods) and using them for optimization

SOPs Standard Operating Procedures are documented approaches intended to be used by an organization for performing tasks

SPC Statistical Process Control is a collection of techniques targeted mainly at evaluating whether something unusual has occurred in recent operations

SQC Statistical Quality Control is a set of techniques intended to aid in the improvement of system quality

SSE Sum of Squared Errors is the additive sum of the squared residuals or error estimates in the context of a curve fitting method such as regression

TLS Theory of Constraints Lean Six Sigma which is the method of combining TOC, lean and six sigma

TOC Theory of Constraints is a method involving the identification and tuning of bottleneck subsystems

TPS The Toyota Production System is the way manufacturing is done at Toyota, which inspired lean production and Just In Time manufacturing

TTD Total Travel Distance is an approximate length that quantifies how far items or information pass through a facility

VIF Variance Inflation Factor is a number that evaluates whether an input pattern can support reliable fitting of a model form in question, i.e., it can help clarify whether a particular question can be answered using a given data source

VSM Value Stream Mapping is a variant of process mapping with added activities inspired by a desire to reduce waste and the Toyota Production System

Introduction

T. T. Allen, *Introduction to Engineering Statistics and Lean Six Sigma*,
https://doi.org/10.1007/978-1-4471-7420-2_1

1.1 Purpose of This Book

In this chapter, six sigma is defined as a method for problem solving. It is perhaps true that the main benefits of six sigma are: (1) the method slows people down when they solve problems, preventing them from prematurely jumping to poor recommendations that lose money; and (2) six sigma forces people to evaluate quantitatively and carefully their proposed recommendations. These evaluations can aid by encouraging adoption of project results and in the assignment of credit to participants. The main goal of this book is to encourage readers to increase their use of six sigma and its associated "sub-methods." Many of these sub-methods fall under the headings "statistical quality control" (SQC) and "design of experiments" (DOE), which, in turn, are associated with systems engineering and statistics.

"Experts" often complain that opportunities to use these methods are being missed. Former General Electric CEO Jack Welch, e.g., wrote that six sigma is relevant in any type of organization from finance to manufacturing to healthcare. When there are "routine, relatively simple, repetitive tasks," six sigma can help improve performance, or if there are "large, complex projects," six sigma can help them go right the first time (Welch and Welch 2005). In this book, later chapters describe multiple true case studies in which students and others saved millions of dollars using six sigma methods in both types of situations.

Facilitating competent and wise application of the methods is also a goal. Incompetent application of methods can result in desirable outcomes. However, it is often easy to apply methods competently, i.e., with an awareness of the intentions of methods' designers. Also, competent application generally increases the chance of achieving positive outcomes. Wisdom about how to use the methods can prevent over-use, which can occur when people apply methods that will not likely repay the associated investment. In some cases, the methods are incorrectly used as a substitute for rigorous thinking with subject-matter knowledge, or without properly consulting a subject-matter expert. These choices can cause the method applications to fail to return on the associated investments.

In ▶ Sect. 1.2, several terms are defined in relation to generic systems. These definitions emphasize the diversity of the possible application areas. People in all sectors of the world economy are applying the methods in this book and similar books. These sectors include health care, finance, education, and manufacturing. Next, in ▶ Sect. 1.3, problem-solving methods are defined. The definition of six sigma is then given in ▶ Sect. 1.4 in terms of a method, and a few specific principles and the related history are reviewed in ▶ Sect. 1.5. Finally, an overview of the entire book is presented, building on the associated definitions and concepts.

1.2 Systems and Key Input Variables

We define a "system" as an entity with "input variables" and "output variables." Also, we use "factors" synonymously with input variables and denote them $x_1, \ldots, x_m$. In our definition, all inputs must conceivably be directly controllable by some potential

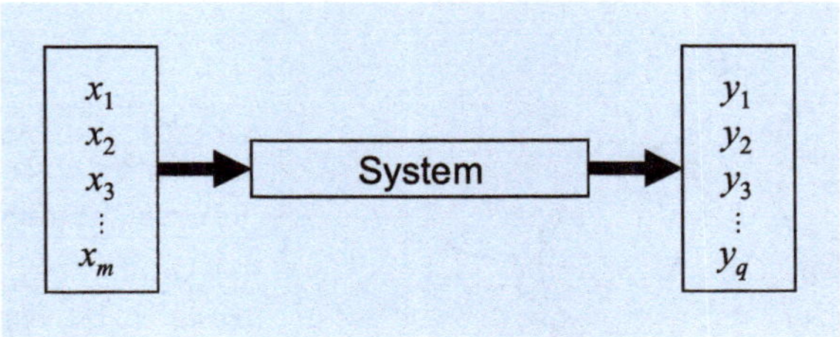

Fig. 1.1 Diagram of a generic system

participant on a project team. We use responses synonymously with output variables and denote them $y_1, \ldots, y_q$. Figure 1.1 shows a generic system.

Assume that every system of interest is associated with at least one output variable of prime interest to you or your team in relation to the effects of input variable changes. We will call this variable a "**key output variable**" (KOV). Often, this will be the monetary contribution of the system to some entity's profits. Other KOV are variables that are believed to have a reasonably strong predictive relationship with at least one other already established KOV. For example, the most important KOV could be an average of other KOVs.

"**Key input variables**" (KIVs) are directly controllable by team members, *and* when they are changed, these changes will likely affect at least one key output variable. Note that some other books use the terms "key process input variables" (KPIVs) instead of key input variables (KIVs) and "key process output variables" (KPOVs) instead of key output variables (KOVs). We omit the word "process" because sometimes the system of interest is a product design and not a process. Therefore, the term "process" can be misleading.

A main purpose of these generic-seeming definitions is to emphasize the diversity of problems that the material in this book can address. Understandably, students usually do not expect to study material applicable to all of the following: (1) reducing errors in administering medications to hospital patients, (2) improving the welds generated by a robotic welding cell, (3) reducing the number of errors in accounting statements, (4) improving the taste of food, and (5) helping to increase the effectiveness of pharmaceutical medications. Yet, the methods in this book are currently being usefully applied in all these types of situations around the world.

Another purpose of the above definitions is to clarify this book's focus on choices about the settings of factors that we can control, i.e., key input variables (KIVs). While it makes common sense to focus on controllable factors, students often have difficulty clarifying what variables they might reasonably be able to control directly in relation to a given system. Commonly, there is confusion between inputs and outputs because, in part, system inputs can be regarded as outputs. The opposite is generally not true.

The examples that follow further illustrate the diversity of relevant application systems and job descriptions. These examples also clarify the potential difficulty associated with identifying KIVs and KOVs. Figure 1.2 depicts objects associated with the examples, related to the medical, manufacturing, and accounting sectors of the economy.

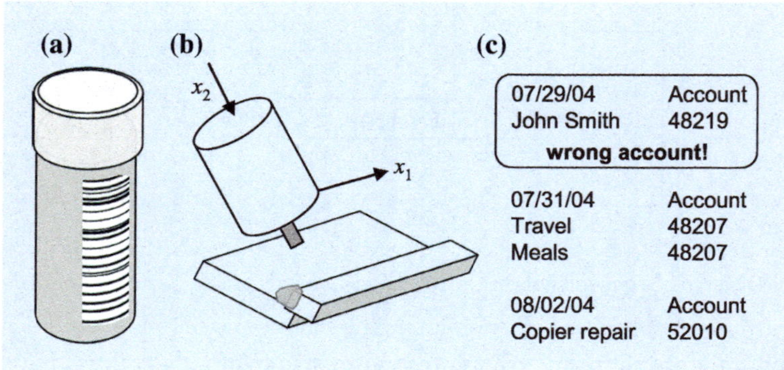

Fig. 1.2 **a** Pill box with bar code, **b** weld torch, and **c** accounting report

Example 1.2.1 Bar-Coding Hospital System

Question

A hospital is trying to increase the quality of drug administration. To do this, it is considering providing patients with bar-coded wristbands and labeling unit dose medications with barcodes to make it easier to identify errors in patient and medication names, doses, routes, and times. Your team is charged with studying the effects of bar-coding by carefully watching 250 episodes in which drugs are given to patients without bar-coding and 250 episodes with bar-coding. Every time a drug is administered, you will check the amount, if any, of discrepancy between what was supposed to be given and what was given. List KIVs and KOVs and their units.

Answer

Possible KIVs and KOVs are listed in Table 1.1. Note also that the table is written implying that there is only one type of drug being administered. If there were a need to check the administration of multiple drugs, more output variables would be measured and documented. Then, it might be reasonable to assign a KOV as a weighted sum of the mistake amounts associated with different drugs.
In the above example, there was an effort made to define KOVs specifically associated with episodes and input combinations. In this case, it would also be standard to say that there is only one output variable "mistake amount" that is potentially influenced by bar-coding, the specific patient, and administration time.
In general, it is desirable to be explicit so that it is clear what KOVs are and how to measure them. The purpose of the next example is to show that different people can see the same problem and identify essentially different systems. With more resources and more confidence with methods, people tend to consider simultaneously more inputs that can be adjusted.

Table 1.1 Key input and output variables for the first bar-code investigation

KIV	Description	KOV	Description
x_1	Bar-coding (Y or N)	y_1	Mistake amount patient #1 with x_1 = N
		y_2	Mistake amount patient #2 with x_1 = N
		⋮	⋮
		y_{501}	Average amount with bar-coding
		y_{502}	Average amount without bar-coding

Example 1.2.2 Bar-Coding System Version 2

Question

Another hospital decides to launch a relatively thorough investigation of bar-coding, including evaluation of 1000 episodes in which drugs are given to patients. In addition to considering installing bar-coding, investigators simultaneously consider (1) the use of sustained-release medications that can be administered at wider intervals, (2) assigning fewer patients to each nurse, (3) setting a limit on how much time nurses can spend chatting with patients, and (4) shortening the nurses shift hours. They plan on testing 10 combinations of these inputs multiple times each. In addition to correct dosage administration, they also want to evaluate the effects of changes on the job satisfaction of the 15 current nurses. Patient satisfaction is a possible concern, but no one on the team really believes that bar-coding affects it. Define and list KIVs and KOVs and their units.

Answer

Possible KIVs and KOVs are listed in Table 1.2. Patient satisfaction ratings are not included as KOVs. This follows despite the fact that all involved believe they are important. However, according to the definition here, key output variables must be likely to be affected by changes in the inputs being considered or believed to have a strong predictive relationship with other KOVs. Also, note that the team cannot control exactly how much time nurses spend with patients. However, the team could write a policy such that nurses could tell patients, "I cannot spend more than *X* minutes with you according to policy."

Note in the above example that average differences in output averages conditioned on changes in inputs could be included in the KOV list. Often, developing statistical evidence for the existence of these differences is the main goal of an investigation. The list of possible KOVs is rarely exhaustive, in the sense that more could almost always be added. Yet, if an output is mentioned directly or indirectly as important by the customer, subject matter expert, or team member, it should be included in the list.

1

Table 1.2 The list of inputs and outputs for the more thorough investigation

KIV	Description	KOV	Description
x_1	Bar-coding (Y or N)	y_1	Mistake amount patient-combo. #1 (cc)
x_2	Spacing on tray (mm)	y_2	Mistake amount patient-combo. #2 (cc)
x_3	Number of patients (#)	⋮	⋮
x_4	Nurse-patient time[a] (min)	y_{1000}	Mistake amount patient-combo. #1000 (cc)
x_5	Shift length (h)	y_{1002}	Nurse #1 rating for input combo. #1
		⋮	⋮
		y_{1150}	Nurse #15 rating for input combo. #20

[a]Stated policy is less than *X*

The next example illustrates a case in which an input is also an output. Generally, inputs are directly controllable, and at least one output under consideration is only indirectly controllable through adjustments of input variable setting selections. Admittedly, the distinctions between inputs and outputs in virtual or simulated world can be blurry. Yet, in this book we focus on the assumption that inputs are controllable, and outputs, with few exceptions, are not. The next example also constitutes a relatively "traditional" system, in the sense that the methods in this book have historically not been primarily associated with projects in the service sector.

Example 1.2.3 Robotic Welding System

Question
The shape of welds strongly relates to profits, in part because operators commonly spend time fixing or reworking welds with unacceptable shapes. Your team is investigating robot settings that likely affect weld shape, including weld speed, voltage, wire feed speed, time in acid bath, weld torch contact tip-to-work distance, and the current frequency. Define and list KIVs and KOVs and their units.

Answer
Possible KIVs and KOVs are listed in Table 1.3. Weld speed can be precisely controlled and likely affects bead shape and therefore profits. Yet, since the number of parts made per minute likely relates to revenues per minute (i.e., throughput), it is also a KOV.
The final example system considered here relates to annoying accounting mistakes that many of us experience on the job. Applying systems thinking to monitor and improve accounting practices is of increasing interest in industry.

Table 1.3 Key input and output variables for the welding process design problem

KIV	Description	KOV	Description
x_1	Weld speed (min/weld)	y_1	Convexity for weld #1
x_2	Wire feed speed (m/min)	y_2	Convexity for weld #2
x_3	Voltage (V)	⋮	⋮
x_4	Acid bath time (min)	y_{501}	% Acceptable for input combo. #1
x_5	Tip distance (mm)	⋮	⋮
x_6	Frequency (Hz)	y_{550}	Weld speed (min/weld)

Table 1.4 Key input and output variables for the accounting systems design problem

KIV	Description	KOV	Description
x_1	New software (Y or N)	y_1	Number mistakes report #1
x_2	Change(Y or N)	y_2	Number mistakes report #2
		⋮	⋮
		y_{501}	Average number mistakes x_1 = Y, x_2 = Y
		y_{502}	Average number mistakes x_1 = N, x_2 = Y
		y_{503}	Average number mistakes x_1 = Y, x_2 = N
		y_{504}	Average number mistakes x_1 = N, x_2 = N

Example 1.2.4 Accounting System

Question

A manager has commissioned a team to reduce the number of mistakes in the reports generated by all company accounting departments. The manager decides to experiment with both new software and a changed policy to make supervisors directly responsible for mistakes in expense reports entered into the system. It is believed that the team has sufficient resources to check carefully 500 reports generated over two weeks in one "guinea pig" divisional accounting department where the new software and policies will be tested.

Answer

Possible KIVs and KOVs are listed in Table 1.4.

1.3 Problem-Solving Methods

The definition of systems is so broad that all knowledge workers could say that a large part of their job involves choosing input variable settings for systems, e.g., in accounting, education, health care, or manufacturing. This book focuses on activities that

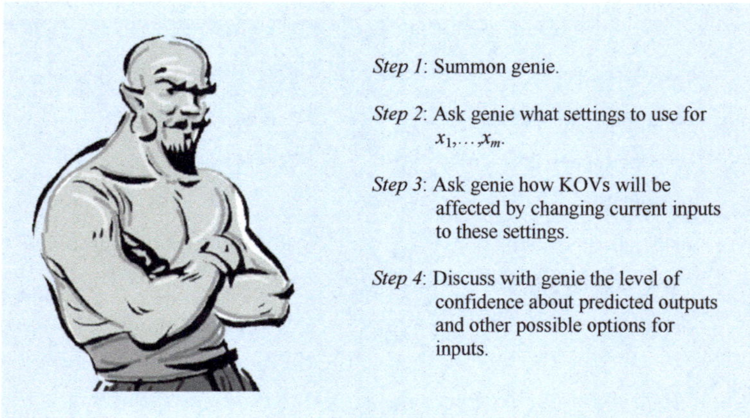

Fig. 1.3 System genie-based problem-solving method

people engage into educate themselves in order to select key input variable settings. Their goals are expressable in terms of achieving more desirable key output variable (KOV) values. It is standard to refer to activities that result in recommended inputs and other related knowledge as "**problem-solving methods**."

Imagine that you had the ability to command a "**system genie**" with specific types of powers. The system genie would appear and provide ideal input settings for any system of interest and answer all related questions. Figure 1.3 illustrates a genie based problem-solving method. Note that, even with a trustworthy genie, steps 3 and 4 probably would be of interest. This follows because people are generally interested in more than just the recommended settings. They would also desire predictions of the impacts on all KOVs as a result of changing to these settings and an educated discussion about alternatives.

In some sense, the purpose of this book is to help you and your team efficiently transform yourselves into genies for the specific systems of interest to you. Unfortunately, the transformation involves more complicated problem-solving methods than simply asking an existing system genie as implied by Fig. 1.3. The methods in this book involve potentially all of the following: collecting data, performing analysis and formal optimization tasks, and using human judgement and subject-matter knowledge.

Some readers, such as researchers in statistics and operations research, will be interested in designing new problem-solving methods. With them in mind, the term "**improvement system**" is defined as a problem-solving method. The purpose of this definition is to emphasize that methods can themselves be designed and improved. Yet methods differ from other systems, in that benefits from them are derived largely indirectly through the inputting of derived factor settings into other systems.

1.3.1 What Is "Six Sigma"?

The definition of the phrase "**six sigma**" is somewhat obscure. People and organizations that have played key roles in encouraging others to use the phrase include the authors Harry and Schroeder (1999), Pande et al. (2000), and the American Society of Quality. These groups have clarified that "six sigma" pertains to the attainment of desirable situations in which the fraction of unacceptable products produced by a system is less than 3.4 per million opportunities (PMO). In Part I of this book, the exact derivation of this number will be explained. The main point here is that a key output characteristic (KOV) is often the fraction of manufactured units that fail to perform up to expectations.

Here, the definition of six sigma is built on the one offered in Linderman et al. (2003, p. 195). Writing in the prestigious *Journal of Operations Management*, those authors emphasized the need for a common definition of six sigma and proposed a definition paraphrased below:

» Six sigma is an organized and systematic problem-solving method for strategic system improvement and new product and service development that relies on statistical methods and the scientific method to make dramatic reductions in customer defined defect rates and/or improvements in key output variables.

The authors further described that while "the name Six Sigma suggests a goal" of less than 3.4 unacceptable units PMO, they purposely did not include this principle in the definition. This followed because six sigma "advocates establishing goals based on customer requirements." It is likely true that sufficient consensus exists to warrant the following additional specificity about the six sigma method:

The six sigma method for completed projects includes as its phases either Define, Measure, Analyze, Improve, and Control (DMAIC) for system improvement or Define, Measure, Analyze, Design, and Verify (DMADV) for new system development.

Note that some authors use the term Design For Six Sigma (DFSS) to refer to the application of six sigma to design new systems and emphasize the differences compared with system improvement activities.

Further, it is also probably true that sufficient consensus exists to include in the definition of six sigma the following two principles:

» **Principle 1**: The six sigma method only fully commences a project after establishing adequate monetary justification.
Principle 2: Practitioners applying six sigma can and should benefit from applying statistical methods without the aid of statistical experts.

The above definition of six sigma is not universally accepted. However, examining it probably does lead to appropriate inferences about the nature of six sigma and of this book. First, six sigma relates to combining statistical methods and the scientific method to improve systems. Second, six sigma is fairly dogmatic in relation to the words associated with a formalized method to solve problems. Third, six sigma is very much about saving money and financial discipline. Fourth, there is an emphasis associated with six sigma on training people to use statistical tools who will never be experts and may not come into contact with experts. Finally, six sigma focuses on the relatively narrow set of issues associated with *technical* methods for improving *quantitative* measures of identified subsystems in relatively short periods of time. Many "softer" and philosophical issues about how to motivate people, inspire creativity, invoke the principles of design, or focus on the ideal endstate of systems are not addressed.

1

Example 1.3.1 Management Fad?

Question
What aspects of six sigma suggest that it might not be another passing management fad?

Answer
Admittedly, six sigma does share the characteristic of many fads in that its associated methods and principles do not derive from any clear, rigorous foundation or mathematical axioms. Properties of six sigma that suggest that it might be relevant for a long time include: (1) the method is relatively specific and therefore easy to implement, and (2) six sigma incorporates the principle of budget justification for each project. Therefore, participants appreciate its lack of ambiguity, and management appreciates the emphasis on the bottom line.

Associated with six sigma is a training and certification process. Principle 2 above implies that the goal of this process is not to create statistical experts. Other properties associated with six sigma training are:

1. Instruction is "case-based" such that all people being trained are directly applying what they are learning.
2. Multiple statistics, marketing, and optimization "component methods" are taught in the context of an improvement or "problem-solving" method involving five ordered "**activities**." These activities are either "Define" (D), "Measure" (M), "Analyze" (A), "Improve" (I), and "Control" (C) in that order (DMAIC) or "Define" (D), "Measure" (M), "Analyze" (A), "Design" (D), "Verify" (V) (DMADV).
3. An application process is employed in which people apply for training and/or projects based on the expected **profit** or return on investment from the project, and the profit is measured after the improvement system completes.
4. Training certification levels are specified as "**Green Belt**" (perhaps the majority of employees), "**Black Belt**" (project leaders and/or method experts), and "**Master Black Belt**" (training experts).

Many companies have their own certification process. In addition, the American Society of Quality (ASQ) offers the Black Belt certification. Current requirements include completed projects with affidavits and an acceptable score on a written exam.

1.3.2 What Is "Lean Manufacturing"?

The simple principle of "lean manufacturing" is to create more value with less work and zero waste. Therefore, ultimate efficiently is the goal or desired end state of the methods and principles associated with the terms lean manufacturing, lean production, or alternatively, "lean" for short. The methods focus on the modeling and optimizing the flows of materials and information through systems.

Historically, some of the lean methods are rooted from the ideas of Fredrick Winslow Taylor, the father of Industrial and Systems Engineering. Taiichi Ohno and others at Toyota developed the Toyota Production System (TPS) based on the ideas of Taylor, Henry Ford, and others. It is perhaps true that lean production is influenced by the objective of making generic systems more like the TPS.

▶ Section 5.2 describes the methods and principles of lean manufacturing in greater detail. To preview, the methods include: value-stream mapping, 5S, kanban, and poka-yoke. Value-stream mapping involves the creation of a diagram of the entire system including potentially entities, flows, production times, and waiting times. The diagram aides in identification of the process hold-ups to facilitate the elimination of the wastes associated. The 5S's are sort, straighten, sweep, standardize, and sustain. Kanban involves using an order-driven or pull-production system along with just-in-time material arrivals and work in process inventories constrained by kanban cards. Poka-yoke is error proofing the system.

1.3.3 What Is the "Theory of Constraints (ToC)"?

Eliyahu Goldratt is a consultant who wrote *The goal: a process of ongoing improvement* in 1984. This book was the first to outline his Theory of Constraints. The idea of ToC is that in a complicated system there is generally a single constraint holding up the flow of production or throughput. To increase throughput on unavoidably involves elevating or alleviating the constraining subsystem or bottleneck. Dr. Goldratt outlines five steps to do this, these steps are: Identify the constraints, decide how to exploit those constraints, subordinate all other processes, elevate the constraint, if the constraint has moved within the system go to the new location and start over. Additional details about ToC are in ▶ Sect. 3.4.1.

1.3.4 What Is "Theory of Constraints Lean Six Sigma (TLS)"?

The "lean sigma" phrase in the title of this book reflects the combination of six sigma and lean manufacturing working together. Yet, others have already applied and tested combinations of Lean Manufacturing, six sigma, and ToC. Here, the combination is termed Theory of Constraints Lean Six Sigma (TLS). According to Industry Week Magazine 2007 Survey, the implementation of the three combined has shown a typical return on investment of 10–1. Roughly speaking, lean eliminates waste and focuses on the endstate, six sigma focuses on the principled process of reducing variability, and ToC helps in prioritizing subsystems for attention. Note that for this book the term lean sigma will be used interchangeably with lean six sigma and TLS.

1.4 History of "Quality" and Six Sigma

In this section, we briefly review the broader history of management, applied statistics, and the six sigma movement. The definition of "**quality**" is as obscure as the definition of six sigma. Quality is often defined imprecisely in textbooks in terms of a subjectively assessed performance level (P) of the unit in question and the expectations (E) that customers have for that unit. A rough formula for quality (Q) is:

$$Q = \frac{P}{E} \tag{1.1}$$

Often, quality is considered in relation to thousands of manufactured parts, and a key issue is why some fail to perform up to expectation and others succeed.

It is probably more helpful to think of "quality" as a catch-word associated with management and engineering decision-making using data and methods from applied statistics. Instead of relying solely on "seat-of-the-pants" choices and the opinions of experts, people influenced by quality movements gather data and apply more disciplined methods.

1.4.1 History of Management and Quality

The following history is intended to establish a context for the current quality and six sigma movements. This explanation of the history of management and quality is influenced by Womack and Jones (1996) related to so-called "Lean Thinking" and "**value stream mapping**" and other terms in the Toyota production system.

In the renaissance era in Europe, fine objects including clocks and guns were developed using "**craft**" production. In craft production, a single skilled individual is responsible for designing, building, selling, and servicing each item. Often, a craftperson's skills are certified and maintained by organizations called "**guilds**" and professional societies.

During the 1600s and 1700s, an increasing number of goods and services were produced by machines, particularly in agriculture. Selected events and the people responsible are listed in Fig. 1.4.

It was not until the early 1900s that a coherent alternative to craft production of fine objects reached maturity. In 1913, Ford developed "Model T" cars using an "assembly line" in which many unskilled workers each provided only a small contribution to the manufacturing process. The term "**mass production**" refers to a set of management policies inspired by assembly lines. Ford used assembly lines to make large

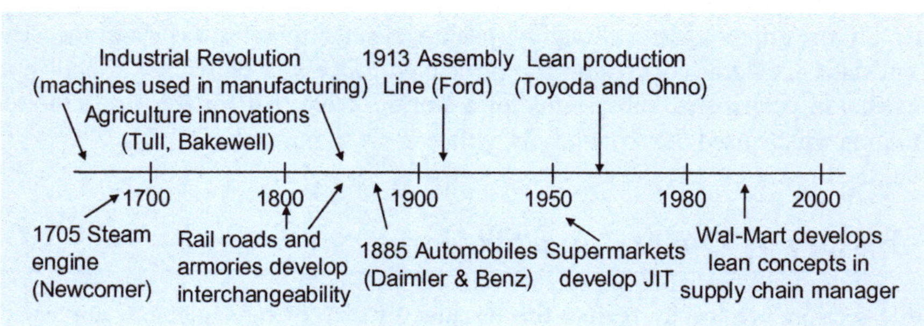

Fig. 1.4 Timeline of selected management methods (includes Toyoda at Toyota)

numbers of nearly identical cars. His company produced component parts that were "**interchangeable**" to an impressive degree. A car part could be taken from one car, put on another car, and still yield acceptable performance.

As the name would imply, another trait of mass production plants is that they turn out units in large batches. For example, one plant might make 1000 parts of one type using a press and then change or "set up" new dies to make 1000 parts of a different type. This approach has the benefit of avoiding the costs associated with large numbers of change-overs.

Significant accountability for product performance was lost in mass production compared with craft production. This follows because the people producing the product each saw only a very small part of its creation. Yet, benefits of mass production included permitting a huge diversity of untrained people to contribute in a coordinated manner to production. This in turn permitted impressive numbers of units to be produced per hour. It is also important to note that both craft and mass production continue to this day and could conceivably constitute profitable modes of production for certain products and services.

Mass production concepts contributed to intense specialization in other industrial sectors besides manufacturing and other areas of the corporation besides production. Many companies divided into departments of marketing, design engineering, process engineering, production, service, purchasing, accounting, and quality. In each of these departments people provide only a small contribution to the sale of each unit or service. The need to counteract the negative effects of specialization at an organizational level has led to a quality movement called "**concurrent engineering**" in which people from multiple disciplines share information. The interaction among production, design engineering, and marketing is considered particularly important, because design engineering often largely determines the final success of the product. Therefore, design engineers need input about customer needs from marketing and production realities from production.

The Toyota production system invented in part by Toyoda and Ohno, also called "lean production" and "just-in-time" (JIT), built in part upon innovations in U.S. supermarkets. The multiple further innovations that Toyota developed in turn influenced many areas of management and quality-related thinking including increased outsourcing in supply-chain management. In the widely read book *The Machine that Changed the World*, Womack et al. (1991) explained how Toyota, using its management approach, was able to transform quickly a failing GM plant to produce at least as many units per day with roughly one half the personnel operating expense and with greatly improved quality by almost any measure. This further fueled the thirst in the U.S. to learn from all things Japanese.

JIT creates accountability by having workers follow products through multiple operations in "U"-shaped cells (i.e., machines laid out in the shape of a "U") and by implementing several policies that greatly reduce work-in-process (WIP) inventory. To the maximum extent possible, units are made in **batches of size one**, i.e., creating a single unit of one type and then switching over to a single of another type of unit and so on. This approach requires frequent equipment set-ups. To compensate, the workers put much effort into reducing set-up costs, including the time required for set-ups. Previously, many enterprises had never put effort into reducing set-ups because they did not fully appreciate the importance.

1

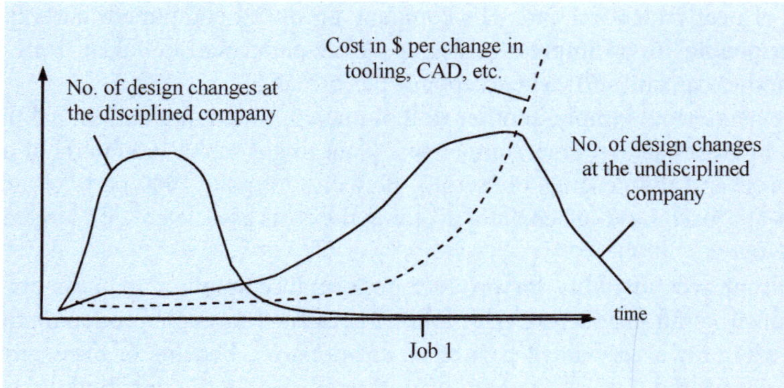

Fig. 1.5 Formal planning can reduce costs and increase agility

Also, the total inventory at each stage in the process is generally regulated using kanban cards. When the cards for a station are all used up, the process shuts down the upstream station, which can result in shutting down entire supply chains. The benefit is increased attention to the problems causing stoppage and (hopefully) permanent resolution. Finally, lean production generally includes an extensive debugging process; when a plant starts up with several stoppages, many people focus on and eliminate the problems. With small batch sizes, "U" shaped cells, and reduced WIP, process problems are quickly discovered before nonconforming units accumulate.

Example 1.4.1 Lean Production of Paper Airplanes

Question
Assume that you and another person are tasked with making a large number of paper airplanes. Each unit requires three operations: (1) marking, (2) cutting, and (3) folding. Describe the mass and lean ways to deploy your resources. Which might generate airplanes with higher quality?

Answer
A mass production method would be to have one person doing all the marking and cutting and the other person doing all the folding. The lean way would have both people doing marking, cutting, and folding to make complete airplanes. The lean way would probably produce higher quality because, during folding, people might detect issues in marking and cutting. That information would be used the next time to improve marking and cutting with no possible loss associated with communication. (Mass production might produce units more quickly, however.)
In addition to studying Toyota's lean production, observers compare many types of business practices at European, Japanese, and U.S. companies. One finding at specific companies related to the timing of design changes at automotive companies. In the automotive industry, "Job 1" is the time when the first production car roles off the line. A picture emerged, shown in Fig. 1.5.

Figure 1.5 implies that at certain automotive companies in Japan, much more effort was spent investigating possible design changes long before Job 1. At certain U.S. car companies, much more of the effort was devoted after Job 1 reacting to problems experience by customers. This occurred for a variety of reasons. Certain Japanese companies made an effort to institutionalize a forward-looking design process with "**design freezes**" that were taken seriously by all involved. Also, engineers at these specific companies in Japan were applying design of experiments (DOE) and other formalized problem-solving methods more frequently than their U.S. counterparts. These techniques permit the thorough exploration of large numbers of alternatives long before Job 1, giving people more confidence in the design decisions.

Even today, in probably all automotive companies around the world, many engineers are in "**reactive mode**," constantly responding to unforeseen problems. The term "**fire-fighting**" refers to reacting to these unexpected occurrences. The need to fire-fight is, to a large extent, unavoidable. Yet the cost per design change plot in Fig. 1.5 is meant to emphasize the importance of avoiding problems rather than fire-fighting. Costs increase because more and more tooling and other coordination efforts are committed based on the current design as time progresses. Formal techniques taught in this book can play a useful role in avoiding or reducing the number of changes needed after Job 1, and achieving benefits including reduced tooling and coordination costs and decreased need to fire-fight.

Another development in the history of quality is "**miniaturization**." Many manufactured items in the early 2000s have literally millions of critical characteristics, all of which must conform to specifications in order for the units to yield acceptable performance. The phrase "**mass customization**" refers to efforts to tailor thousands of items such as cars or hamburgers to specific customers' needs. Mass customization, like miniaturization, plays an important role in the modern work environment. Ford's motto was, "You can have any color car as long as it is black." In the era of global competition, customers more than ever demand units made to their exact specifications. Therefore, in modern production, customers introduce additional variation to the variation created by the production process.

Example 1.4.2 Freezing Designs

Question

With respect to manufacturing, how can freezing designs help quality?

Answer

Often the quality problem is associated with only a small fraction of units that are not performing as expected. Therefore, the problem must relate to something different that happened to those units, i.e., some variation in the production system. Historically, engineers "tweaking" designs has proved to be a major source of variation and thus a cause of quality problems.

1.4.2 History of Documentation and Quality

The growing role of documentation of standard operating procedures (SOP s) also relates to management history. The International Standards Organization (ISO) developed in Europe but was influenced in the second half of the twentieth century by U.S. military standards. The goals of ISO included the development of standard ways that businesses across the world could use to document their practices. ISO standards for documenting business practices, including "ISO 9000: 1994" and "ISO 9000: 2000" document series aimed to reduce variation in production.

ISO 9000: 1994 emphasized addressing 20 points and the basic guideline "Do what you say and say what you do." In other words, much emphasis was placed on whether or not the company actually used its documented policies, rather than on the content of those policies. ISO 9000: 2000 added more requirements for generating models to support and improve business subsystems. Companies being accredited pay credentialed auditors to check that they are in compliance at regular intervals. The results include operating manuals at many accredited institutions that reflect truthfully, in some detail, how the business is being run.

Perceived benefits of ISO accreditation include: (1) reducing quality problems of all types through standardization of practices, and (2) facilitating training when employees switch jobs or leave organizations. Standardization can help by forcing people to learn from each other and to agree on a single approach for performing a certain task. ISO documentation also discourages engineers from constantly tinkering with the design or process.

Another perceived benefit of ISO documentation relates to the continuing trend of companies outsourcing work formerly done in-house. This trend was also influenced by Toyota. In the 1980s researchers noticed that Toyota trusted its suppliers with much more of the design work than U.S. car makers did, and saved a great deal of money as a result. Similar apparent successes with these methods followed at Chrysler and elsewhere, which further encouraged original equipment manufacturers (OEMs) to increase outsourcing. The OEMs have now become relatively dependent on their "**supply chain**" for quality and need some way to assure intelligent business practices are being used by suppliers.

While ISO and other documentation and standardization can eliminate sources of variation, the associated "**red tape**" and other restrictive company policies can also, of course, sometimes stifle creativity and cost money. Some authors have responded by urging careful selection of employees and a "culture of discipline" (Collins 2001). Collins suggests that extensive documentation can, in some cases, be unnecessary because it is only helpful in the case of a few problem employees who might not fit into an organization. He bases his recommendations on a study of policies at exceptional and average companies based on stock performance.

1.4.3 History of Statistics and Quality

Corporations routinely apply statistical methods, partly in response to accountability issues, as well as due to the large volume of items produced, miniaturization, and mass customization. An overview of selected statistical methods is provided in ◘ Fig. 1.6.

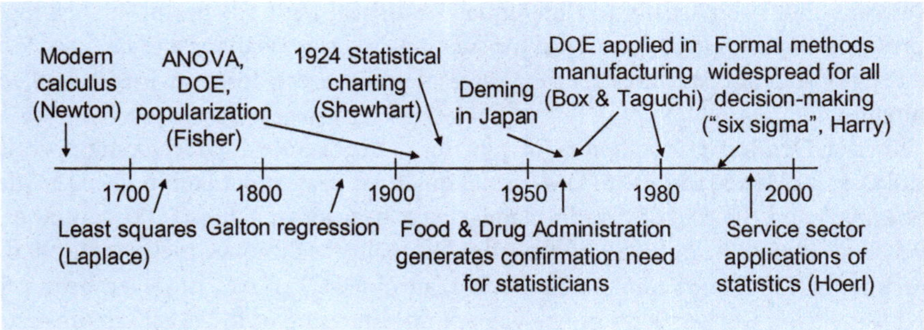

Fig. 1.6 Timeline of selected statistical methods

The origin of these methods dates back at least to the invention of calculus in the 1700s. Least squares regression estimation was one of the first optimization problems addressed in the calculus/optimization literature. In the early 1900s, statistical methods played a major role in improving agricultural production in the U.K. and the U.S. These developments also led to new methods, including fractional factorials and analysis of variance (ANOVA) developed by Sir Ronald Fisher (Fisher 1925).

The realities of mass production led W. Shewhart working in 1924 at Bell Laboratories to propose statistical process control (SPC) methods (see ▶ www.research.att.com/areas/stat/info/history.html). The specific "**X-Bar and R**" charts he developed are also called "Shewhart" charts. These methods discourage process tinkering unless statistical evidence of unusual occurrences accrues. Shewhart also clarified the common and harmful role that variation plays in manufacturing, causing a small fraction of unit characteristics to wander outside their specification limits. The implementation of Shewhart charts also exposed many unskilled workers to statistical methods.

In the 1950s, the U.S. Food and Drug Administration required companies to hire "statisticians" to verify the safety of food and drugs. Many universities developed statistics departments largely in response to this demand for statisticians. Perhaps as a result of this history, many laypeople tend to associate statistical methods with proving claims to regulatory bodies.

At the same time, there is a long history of active uses of statistical methods to influence decision-making unrelated to regulatory agencies. For example, many kinds of statistical methods were used actively in formal optimization and the science of "**operations research**" for the military during and after World War II. During the war Danzig and Wood used linear programming—developed for crop optimization—in deploying convoys. Monte Carlo simulation methods were also used for a wide variety of purposes ranging from evaluating factory flows to gaming nuclear attacks and predicting fallout spread.

George Box, Genichi Taguchi, and many others developed design of experiments (DOE) methods and new roles for statisticians in the popular consciousness besides verification. These methods were intended to be used early in the process of designing products and services. In the modern workplace, people in all departments, including marketing, design engineering, purchasing, and production, routinely use applied statistics methods. The phrases "business statistics" and "engineering statistics" have come into use partially to differentiate statistical methods useful for helping to improve profits from methods useful for such purposes as verifying the safety of foods and drugs ("standard statistics"), or the assessment of threats from environmental contaminants.

Edward Deming is credited with playing a major role in developing so-called "**Total Quality Management**" (TQM). Total quality management emphasized the ideas of Shewhart and the role of data in management decision-making. TQM continues to increase awareness in industry of the value of quality techniques including design of experiments (DOE) and statistical process control (SPC). It has, however, been criticized for leaving workers with only a vague understanding of the exact circumstances under which the methods should be applied and of the bottom line impacts.

Because Deming's ideas were probably taken more seriously in Japan for much of his career, TQM has been associated with technology transfer from the U.S. to Japan and back to the U.S. and the rest of the world. Yet in general, TQM has little to do with Toyota's lean production, which was also technology transfer from Japan to the rest of the world. Some credible evidence has been presented indicating that TQM programs around the world have resulted in increased profits and stock prices (Kaynak 2003). However, a perception developed in the 1980s and 1990s that these programs were associated with "anti-business attitudes" and "muddled thinking."

This occurred in part because some of the TQM practices such as "quality circles" have been perceived as time-consuming and slow to pay off. Furthermore, the perception persists to this day that the roles of statistical methods and their use in TQM are unclear enough to require the assistance of a statistical expert in order to gain a positive outcome. Also, Deming placed a major emphasis on his "14 points," which included #8, "Drive out fear" from the workplace. Some managers and employees honestly feel that some fear is helpful. It was against this backdrop that six sigma developed.

Example 1.4.3 Japanese Technology

Question

Drawing on information from this chapter and other sources, briefly describe three quality technologies transferred from Japan to the rest of the world.

Answer

First, lean production was developed at Toyota which has its headquarters in Japan. Lean production includes two properties, among others: inventory at each machine center is limited using kanban cards, and U-shaped cells are used in which workers follow parts for many operations which instills worker accountability. However, lean production might or might not relate to the best way to run a specific operation. Second, quality circles constitute a specific format for sharing quality-related information and ideas. Third, a Japanese consultant named Genichi Taguchi developed some specific DOE methods with some advantages that will be discussed briefly in Part II of this book. He also emphasized the idea of using formal methods to help bring awareness of production problems earlier in the design process. He argued that this can reduce the need for expensive design changes after Job 1.

1.4.4 The Six Sigma Movement

The six sigma movement began in 1979 at Motorola when an executive declared that "the real problem [is]...quality stinks." With millions of critical characteristics per integrated circuit unit, the percentage of acceptable units produced was low enough that these quality problems obviously affected the company's profits.

In the early 1980s, Motorola developed methods for problem-solving that combined formal techniques, particularly relating to measurement, to achieve measurable savings in the millions of dollars. In the mid-1980s, Motorola spun off a consulting and training company called the "Six Sigma Academy" (SSA). SSA president Mikel Harry led that company in providing innovative case-based instruction, "black belt" accreditations, and consulting. In 1992, Allied Signal based its companywide instruction on Six Sigma Academy techniques and began to create job positions in line with Six Sigma training levels. Several other companies soon adopted Six Sigma Academy training methods, including Texas Instruments and ABB.

Also during the mid-1990s, multiple formal methodologies to structure product and process improvement were published. These methodologies have included Total Quality Development (e.g., see Clausing 1994), Taguchi Methods (e.g., see Taguchi 1993), the decision analysis-based framework (e.g., Hazelrigg 1996), and the so-called "six sigma" methodology (Harry and Schroeder 1999). All these published methods developments aim to allow people involved with system improvement to use the methods to structure their activities even if they do not fully understand the motivations behind them.

In 1995, General Electric (GE) contracted with the "Six Sigma Academy" for help in improving its training program. This was of particular importance for popularizing six sigma because GE is one of the world's most admired companies. The Chief Executive Officer, Jack Welch, forced employees at all levels to participate in six sigma training and problem-solving approaches. GE's approach was to select carefully employees for Black Belt instruction, drawing from employees believed to be future leaders. One benefit of this approach was that employees at all ranks associated six sigma with "winners" and financial success. In 1999, GE began to compete with Six Sigma Academy by offering six sigma training to suppliers and others. In 2000, the American Society of Quality initiated its "black belt" accreditation, requiring a classroom exam and signed affidavits that six sigma projects had been successfully completed.

Montgomery (2001) and Hahn et al. (1999) have commented that six sigma training has become more popular than other training in part because it ties standard statistical techniques such as control charts to outcomes measured in monetary and/or physical terms. No doubt the popularity of six sigma training also derives in part from the fact that it teaches an assemblage of techniques already taught at universities in classes on applied statistics, such as gauge repeatability and reproducibility (R&R), statistical process control (SPC), design of experiments (DOE), failure modes and effects analysis (FMEA), and cause and effect matrices (C&E).

All of the component techniques such as SPC and DOE are discussed in Pande et al. (2000) and defined here. The techniques are utilized and placed in the context of a methodology with larger scope, i.e., the gathering of information from engineers and customers and the use of this information to optimize system design and make informed decisions about the inspection techniques used during system operation.

Pande et al. (2000) contributed probably the most complete and explicit version of the six sigma methods in the public domain. Yet even their version of the methodology (perhaps wisely) leaves implementers considerable latitude to tailor approaches to applications and to their own tastes. This lack of standardization of methodologies explains, at least in part, why the American Society for Quality still has only recently introduced a six sigma "black belt" certification process. An exception is a proprietary process at General Electric that "green belt" level practitioners are certified to use competently.

Example 1.4.4 Lean Sigma

Question
How do six sigma and lean production relate?

Answer
Six sigma is a generic method for improving systems or designing new products, while lean manufacturing has a greater emphasis on the best structure, in Toyota's view, of a production system. Therefore, six sigma focuses more on *how* to implement improvements or new designs using statistics and optimization methods in a structured manner. Lean manufacturing focuses on *what* form to be implemented for production systems, including specific high-level decisions relating to inventory management, purchasing, and scheduling of operations, with the goal of emulating the Toyota Production System. That being said, there are "kaizen events" and "value stream mapping" activities in lean production. Still, the overlap is small enough that many companies have combined six sigma and lean manufacturing efforts under the heading "lean sigma."

1.5 The Culture of Discipline

The purpose of this section is to summarize the practical reasons for considering using any formal SQC or DOE techniques rather than trial and error. These reasons can be helpful for motivating engineers and scientists to use these methods, and for overcoming human tendencies to avoid activities requiring intellectual discipline. This motivation might help to build something like the data-driven "culture of discipline" identified by Collins (2001).

The primary reason for formality in decision-making is the common need for extremely high quality levels. This follows from growing international competition in all sectors of the economy. Also, miniaturization and mass customization can make problems hard to comprehend. Often, for the product to have a reasonable chance of meeting customer expectations, the probability that each quality characteristic will satisfy expectations (the "yield") must be greater than 99.99%. Workers in organizations often discover that competitor companies are using formal techniques to achieve the needed quality levels with these tough demands.

Why might formal methods be more likely than trial and error to achieve these extreme quality levels? Here, we will use the phrase "One-Factor-at-a-Time" (OFAT) to refer to trial-and-error experimentation, following the discussion in Czitrom (1999). Intuitively, one performs experimentation because one is uncertain which alternatives will give desirable system outputs. Assume that each alternative tested thoroughly offers a roughly equal probability of achieving process goals. Then the method that can effectively thoroughly test more alternatives is more likely to result in better outcomes.

Formal methods (1) spread tests out inside the region of interest where good solutions are expected to be and (2) provide a thorough check of whether changes help. For example, by using interpolation models, e.g., linear regressions or neural nets, one can effectively thoroughly test all the solutions throughout the region spanned by these experimental runs.

OFAT procedures have the advantages of being relatively simple and permitting opportunistic decision-making. Yet, for a given number of experimental runs, these procedures effectively test far fewer solutions, as indicated by the regions in Fig. 1.7. Imagine the dashed lines indicate contours of yield as a function of two control factors, x_1 and x_2. The chance that the OFAT search area contains the high yield required to be competitive is far less than the formal method search area.

A good engineer can design products that work well under ideal circumstances. It is far more difficult, however, to design a product that works well for a range of conditions, i.e., noise factor settings as defined originally by Taguchi. This reason is effectively a restatement of the first reason because it is intuitively clear that it is noise factor variation that causes the yields to be less than 100.00000%. Something must be changing in the process and/or the environment. Therefore, the designers' challenge, clarified by Taguchi, is to design a product that gives performance robust to noise factor variation. To do this, the experimenter must consider an expanded list of factors including both control and noise factors. This tends to favor formal methods because

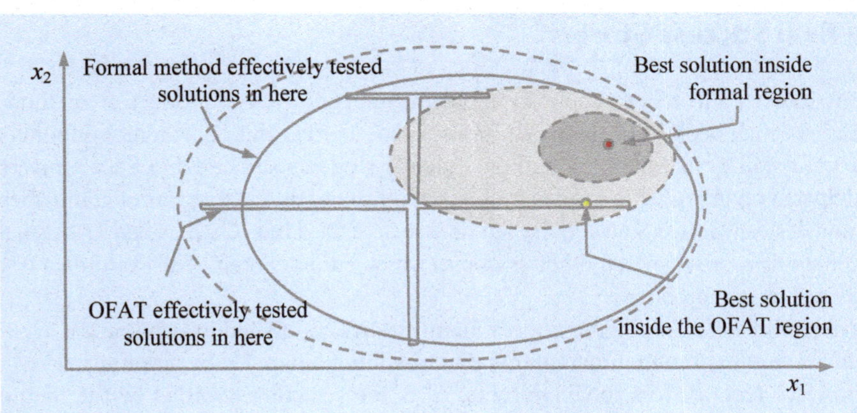

Fig. 1.7 Formal procedures search much larger spaces for comparable costs

typically the marginal cost of adding factors to the experimental plan in the context of formal methods (while achieving comparable method performance levels, e.g., probabilities of successful identification or prediction errors) is much less than for OFAT.

Often there is a financial imperative to “freeze” an engineering design early in the design process. Then it is important that this locked in design be good enough, including robust enough, such that stakeholders do not feel the need to change the design later in the process. Formal methods can help to establish a disciplined product and/or process development timeline to deliver high quality designs early.

The financial problem with the wait-and-see attitude based on tinkering and not upfront formal experimentation is that the costs of changing the design grow exponentially with time. This follows because design changes early in the process mainly cost the time of a small number of engineers. Changes later in the process cause the need for more changes, with many of these late-in-the-process changes requiring expensive retooling and coordination costs. Also, as changes cause the need for more changes, the product development time can increase dramatically, reducing the company’s “agility” in the marketplace.

Example 1.5.1 Convincing Management

Question
What types of evidence are most likely to convince management to invest in training and usage of formal SQC and DOE techniques?

Answer
Specific evidence that competitor companies are saving money is most likely to make management excited about formal techniques. Also, many people at all levels are impressed by success stories. The theory that discipline might substitute for red tape might also be compelling.

1.6 Real Success Stories

Often students and other people are most encouraged to use a product or method by stories in which people like them had positive experiences. This book contains four complete case studies in which the author or actual students at The Ohio State University participated on teams which added millions of dollars to the bottom line of companies in the midwestern United States. These studies are described in ▶ Chaps. 9 and 17. Also, this text contains more than 100 other examples which either contain real-world data or are motivated by real problems.

An analysis of all six sigma improvement studies conducted in two years at a medium-sized midwestern manufacturer is described in ▶ Chap. 21. In that study, 25 of the 34 projects generated reasonable profits. Also, the structure afforded by the methods presented in this book appeared to aid in the generation of extremely high profits in two of the cases. The profits from these projects alone could be viewed as strong justification for the entire six sigma program.

1.7 Statistics

In this section, we briefly describe the field of **statistics**, a branch of mathematics that is critical to the developing field of machine learning. In many formal definitions, statistics involves the analysis, interpretation, and presentation of historical, current, and future numerical data. Formally, "**data**" is a set of symbols which are often numbers. The term data is sometimes used synonymously with facts. Yet, the statistical applications emphasize the prediction and analysis of variations in data which are particularly important in quality improvement, a major focus of this book. The phrase "**variables data**" refers to numbers, which are effectively continuous such as lengths (e.g., 0.92 mm). "**Attributes data**" are counts of events such as manufactured items failing to conform to specifications (e.g., 22 out of 1,000 or 49 out 1,000). Attributes data are a type of "**discrete data**," which can assume only a countable number of values.

1.7.1 Inference and Probability

The field of statistics can be divided into inferential statistics and probability theory. The first of these subfields, "**inferential statistics**," is associated with historical data and estimates relating to unobserved large sets of data called **populations**. The data are collected and summarized using numbers from subsets or "**samples**" called **sample statistics**. Even though populations are generally not fully observed, they can be characterized by assumed quantities called **parameters**.

Example 1.7.1: Samples and Parameter Estimates

A population composed of the measured lengths of 1,000 screws is studied using a sample of $n=3$ data points: 2.1, 2.4, and 2.2 mm. The sample average length is $(2.1+2.4+2.2)$ mm $\div$ $3=2.2$ mm. The sample average might be used to estimate that the population mean is also 2.2 mm. After a "leap of faith," one might assume that expected distances from future populations would have a mean value of 2.2 mm.

The sample mean is used here as a "**point estimate**" or specific numerical value from data which is used to estimate the population parameter. ◘ Table 1.5 compares

◘ **Table 1.5** Sample statistics and their corresponding population parameters defined

Sample statistic of symbol	Population parameter
Sample of size n: $X_1, \ldots, X_n$	Population of size N
Sample average $X_{bar} = (X_1 + \cdots + X_n)/n$	Mean μ ("mu")
Sample standard deviation $s = \sqrt{(X_1 - X_{bar})^2 + \cdots + (X_n - X_{bar})^2/(n-1)}$	Standard deviation σ ("sigma")
Sample skew (often for evaluating normality)	Skew κ
Sample **mode** is the most common number	Mode has highest probability

statistics from inference and parameters from populations. Important methods in future chapters use statistics to attempt to estimate population parameters.

Imagined changes in the population parameters are often important for recommending actions that could be responses to environmental changes. In Part 2, a major issue relates to monitoring statistics to create evidence that population parameters may be changing in ways that local authorities may want to address. Part 3 relates to planning experiments to observe statistics to create predictive models and to use these models to help design systems and population properties.

1.7.2 Inferential Statistics

Creating statistics and visualizing data is often critical for decision-making. Many types of charting techniques are discussed in the later chapters of this book. Here, we describe a standard inferential statistics method to derive the **confidence interval** for the mean of a random variable based on data.

In our notation, the data are written $X_1, X_2, \ldots, X_n$ where n is the number of data points. This approach creates an interval having a reasonably high and regulated probability of containing the true mean under specific assumptions given by the parameter α ("alpha"). The ability to create appropriate confidence intervals is considered essential to much of the simulation theory described in this book.

Confidence Interval Construction Method
Step 1. Calculate the sample mean (Xbar) using:

$$\text{Xbar} = (1/n) \sum_{i=1,\ldots,n} X_i \tag{1.2}$$

Step 2. Calculate the sample standard deviation (s) using:

$$s = \{[\sum_{i=1,\ldots,n} (X_i - \text{Xbar})^2]/(n-1)\}^{1/2} \tag{1.3}$$

where $\{\}^{1/2}$ means take the square root of the quantity inside {}.
Step 3. Calculate the half width of the confidence interval using:

$$\text{Half width} = t_{\alpha/2,n-1} s / \left(n^{1/2}\right) \tag{1.4}$$

and declare that the interval equals Xbar ± half width, where typically $\alpha = 0.05$ and the value of $t_{\alpha/2,n-1}$ is found by consulting ◘ Table 2.1. In applying the formula in (2.7), with an overall value of $\alpha = 0.05$ and n data, we would look for the values with 0.025 and $df = n - 1$.
Step 4. (Optional). Check that it is reasonable to assume that the individual data derive independently, identically distributed (IID) from a (single) normal distribution. If not, then do not trust the interval. Batching described in ► Chap. 4 might provide a useful way to derive trustworthy intervals for some cases.

Note that the symbol "*df*" stands for degrees of freedom, which has a geometric interpretation in the context of various statistical methods such as analysis of variance. It is only an index to help us pick the right value from Table 1.6. Step 4 is often not included in descriptions of confidence intervals but we argue that this test is practically important.

In an election systems example, imagine that the input analysis phase has progressed yielding the data in Table 1.6. These would hypothetically come from $n=9$ registered voters from time trials with a stopwatch in a mock election.

If we use $X_1=7.4$, …, $X_9=5.4$ and apply the confidence interval construction method, we will have our prediction. This is based on the assumption that the future election will be similar to our mock election. Also, being good analysts, we will have error bars on our estimate. We will derive these from the confidence interval construction method applying the derived half widths (Table 1.7).

Table 1.6 Critical values of the *t* distribution ($t_{\alpha,df}$)

df	α				*df*	α			
	0.01	0.025	0.05	0.1		0.01	0.025	0.05	0.1
1	31.82	12.71	6.31	3.08	7	3	2.36	1.89	1.41
2	6.96	4.30	2.92	1.89	8	2.9	2.31	1.86	1.4
3	4.54	3.18	2.35	1.64	9	2.82	2.26	1.83	1.38
4	3.75	2.78	2.13	1.53	10	2.76	2.23	1.81	1.37
5	3.36	2.57	2.02	1.48	20	2.53	2.09	1.72	1.33
6	3.14	2.45	1.94	1.44					

Table 1.7 Registration and direct recording equipment (DRE) times in election systems example

Person	Registration time (min)	Voting using DRE machine time (min)	Total (min)
Fred	0.2	7.2	7.4
Aysha	2.1	4.5	6.6
Juan	0.4	8.1	8.5
Mary	0.8	9.2	10
Henry	1.1	4.2	5.3
Larry	0.3	12.3	12.6
Bill	0.8	15.1	15.9
Jane	0.2	6.2	6.4
Catalina	0.6	4.8	5.4

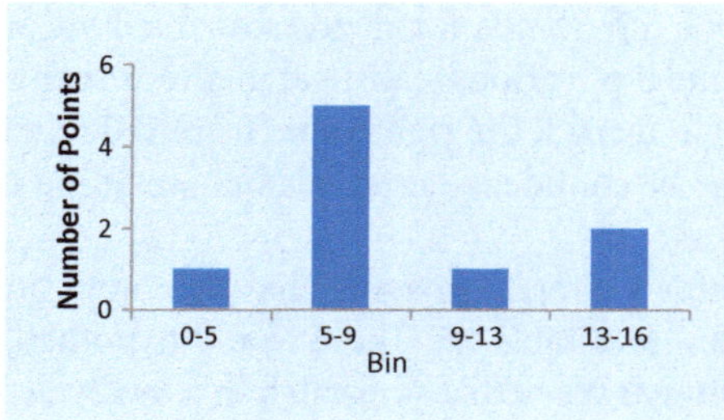

Fig. 1.8 Histogram for the registration plus voting time data showing two peaks

Step 1. We calculate the sample mean (Xbar) using:

$$\text{Xbar} = (1/n) \sum_{i=1,\ldots,n} X_i = (1/9)[7.4 + 6.6 + \cdots + 5.4] = 8.68 \text{ min.} \tag{1.5}$$

Step 2. We calculate the sample standard deviation (s) using:

$$s = \{[\sum_{i=1,\ldots,n} (X_i - \text{Xbar})^2]/(n-1)\}^{1/2} = \left\{(7.4-8.68)^2 + \cdots + (5.4-8.68)^2/8\right\}^{1/2}$$
$$= 3.58 \text{ min.} \tag{1.6}$$

Step 3. We calculate the half width of the confidence interval using:

$$\text{Half width} = t_{a/2,n-1} s/\left(n^{1/2}\right) = (2.31)(3.58)/\left(9^{1/2}\right) = 2.75 \text{ min.} \tag{1.7}$$

The interval is 8.68 ± 2.75 min or (5.92–11.43 min).

Step 4. Figure 1.8 shows a histogram of the 9 summed times. Such distribution testing is generally only trustworthy with 20 or more data points. Yet, here we merely say that the normal distribution is probably not a great fit. The observation of a second hump suggests that the true distribution might not be governed by a single bell shape. Strictly speaking, we know that the famous normal distribution is rarely (if ever) a perfect fit for a real process. However, in this case, the fit is "extra" unreasonable.

1.7.3 Probability Theory

A "**random variable**" (X) is a number or symbol whose value is not known at time of planning by the planner. While the value is unknown, generally the planner is comfortable with assuming a distribution function to summarize his or her beliefs about the random variable. If the random variable is discrete, i.e., it can assume only a countable number of values, then the distribution function is called a **probability mass function**, $\Pr\{X = x_i\}$ for $i = 1, \ldots, n$. For example, X might represent the number of fingers (not including my thumb) that I am holding up behind my back. John Doe, a student, might have beliefs corresponding to:

$$
\begin{aligned}
&\Pr\{X = 0\} = 0.1,\\
&\Pr\{X = 1\} = 0.2,\\
&\Pr\{X = 2\} = 0.3,\\
&\Pr\{X = 3\} = 0.3, \text{ and}\\
&\Pr\{X = 4\} = 0.1.
\end{aligned}
\tag{1.8}
$$

It is perhaps true that no one can tell John Doe that he is wrong in his current beliefs, although things that John learns later might change his beliefs and his distribution. The above distribution has no name other than "**discrete distribution**" in that it is not Poisson or binomial (two famous discrete distributions). It is particular to John and his current state of beliefs. Yet, if John Doe declares the first 4 probabilities and then declares a value for $P\{X=4\}$ other than 0.1, we might reasonably say that John Doe is incompetent in his ability to apply probability theory. Next, for discrete random variables, the universally acknowledged definition for the mean or expected value is given by:

$$\text{mean} \equiv \mathrm{E}[X] \equiv \mu = \sum_{i=1,\ldots,n} x_i \Pr\{X = x_i\}. \tag{1.9}$$

For example, if we apply the distribution in Eq. (2.1) above, then the expected value is:

$$\mathrm{E[X]} = (0)(0.1) + (1)(0.2) + (2)(0.3) + (3)(0.3) + (4)(0.1) = 2.1 \text{ fingers}. \tag{1.10}$$

Philosophically, the expected value is just as subjective as the distribution. It is not wrong to calculate an expected value and then to change one's mind about his or her distribution. Yet, if one has a change of mind about the distribution in (1.8), logically (1.10) must be changed accordingly.

Similarly, if a random variable is continuous, it can (hypothetically) take on any value on at least some section of the real line. Continuous random variables are characterized by density functions also called "**distribution functions**" and are written as $f(x)$. Distribution functions assign values proportional to the subjective likelihood of a random variable, X, achieving a value in the neighborhood of the value x. Just as appropriate or proper discrete density functions must have their probabilities sum to 1.0, appropriate continuous function distribution functions must have their values from $-\infty$ to ∞ integrate to 1.0.

As an example, consider that John Doe may state that his beliefs about the temperature in his home are characterized by the distribution function in ◘ Fig. 1.9. This distribution function does not resemble any frequently encountered, i.e., "famous" probability distribution and it is particular to John Doe at the time when he makes his declaration. It reflects his concern that the heater might breakdown, i.e., he believes that there is at least some chance of subzero temperatures.

The definition for the expected value of a random variable X given a continuous probability density function, $f(x)$, is defined as:

$$\mathrm{E[X]} = \mu = \int_{-\infty}^{\infty} x f(x) dx. \tag{1.11}$$

1

A custom distribution as shown in Fig. 1.9 provides special challenges for estimating the expected value. This follows because there are only a finite number of cases for which we have the anti-derivatives to directly calculate the value in Eq. (1.11). There is no well-known antiderivative for the function in Fig. 1.9. Custom distributions are important in discrete event simulation because generally the random variable, whose expected value one is estimating, does not resemble any frequently encountered continuous distribution. It might be the waiting time of a voter which is influenced by many factors such as machine breakdowns and the arrival rates of other voters.

A visual scan might yield an expected value of approximately 15 °C, which appears to lie roughly at the center of mass of the distribution. In practice, we often attempt to find the closest "famous" distribution that roughly fits our beliefs. Using these famous distributions, we gain access to preestablished formulas relating the parameters that describe our assumed distributions to the mean and other properties of these distributions. In this book, we focus on four well-studied or "famous" continuous distributions. These are the following:

- **Uniform** (equally likely to be anywhere between a and b) written $U[a, b]$,
- **Triangular** (must be greater than a, is most likely to be m, and must be less than b) written $TRIA(a, m, b)$,
- **Exponential** (could be anywhere in the ball-park of $1/\lambda$) written $EXPO(1/\lambda)$, and
- **Normal** (somewhere within 3σ of the assumed mean μ) written $N[\mu, \sigma]$.

Note that much of the statistical literature writes the normal distribution as $N[\mu, \sigma^2]$, where σ^2 is the distribution variance. We choose to follow the excel "=NORMDIST()"

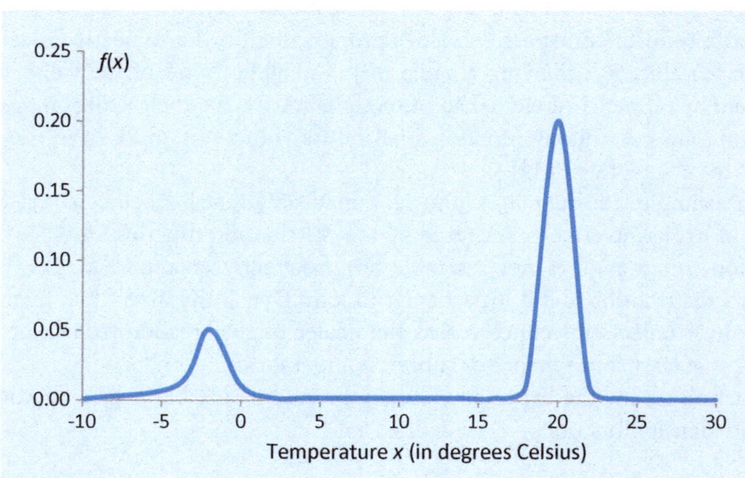

Fig. 1.9 John Doe's density function describing his beliefs on room temperature

conventions using the standard deviation instead of the mean. Also, the exponential is typically written in terms of its parameter λ and the reciprocal, 1/λ, is the exponential mean or expected value.

1.7.4 Events and Independence

An "**event**" is a set of values that a random variable can assume (outcomes from an experiment). For example, a random variable X might be the number of nonconforming items in a subgroup. An event might be the set of values corresponding to four or more nonconformities. Two events A and B are said to be "**independent**" if their intersection probability satisfies: $\Pr\{A \text{ and } B\} = \Pr\{A \cap B\} = \Pr\{A\} \times \Pr\{B\}$.

Independence often relates to the absence of a causal relationship. For example, the numbers of nonconforming items from different supplies might be reasonably assumed to be independently distributed. In Part 3 of this book, we investigate linear regression and sample correlations. Significant sample correlation can be an evidence that independence does not apply.

A Venn diagram shows all possible logical sets. When applied in probability theory, it depicts overlaps of probability events. Areas in Venn diagrams correspond to probabilities and, for a spanning set of events, the sum of the area must be 1. Logically, two events must have zero or positive overlap as shown in ◘ Fig. 1.10a. Consider that each event must either happen or its "**complement**" or opposite must happen. In common notation, the complement of the event A is A^C, see ◘ Fig. 1.10b. Logically also, the positive overlap possibility leads to the following "**Bonferroni**" equality (named after the mathematician Carlo Emilio Bonferroni):

$$\Pr\{A \cap B\} = 1 - \Pr\left\{A^c\right\} - \Pr\{B^c\}. \tag{1.12}$$

The Bonferroni inequality is helpful for making statements about events for cases in which the independence assumption is not considered to be appropriate.

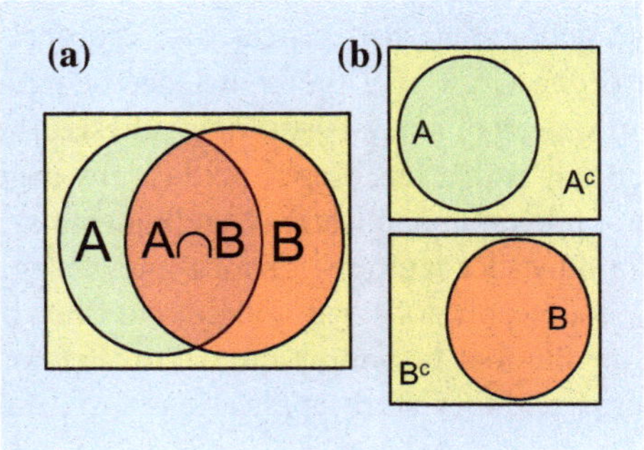

◘ **Fig. 1.10** a Venn diagrams with: overlapping events, **b** events and complements

1.8 Overview of This Book

This book is divided into three major parts. The first part describes many of the most widely used methods in the area of study called "statistical quality control" (SQC). The second part described formal techniques for data collection and analysis. These techniques are often referred to as "design of experiments" (DOE) methods. Model fitting after data are collected is an important subject by itself. For this reason, many of the most commonly used model-fitting methods are also described in this part with an emphasis on linear regression.

Part III concludes with a description of optimization methods, including their relationship to the planning of six sigma projects. Optimization methods can play an important role both for people working on a six sigma project and for the design of novel statistical methods to help future quality improvement projects.

Case studies are described near the end of each major part and are associated with exercises that ask the reader "What would you have done?" These studies were based largely on my own experiences working with students at midwestern companies during the last several years. In describing the case studies, the intent is to provide the same type of real-world contextual information encountered by students, from the engineering specifics and unnecessary information to the millions of dollars added to the bottom line.

It is important for readers to realize that only a minimal amount of "statistical theory" is needed to gain benefits from most of the methods in this book. Theory is helpful mainly for curious readers to gain a deeper appreciation of the methods and for designing new statistical and optimization methods. For this reason, statistical theory is separated to a great extent from a description of the methods. Readers wary of calculus and probability need not be deterred from using the methods.

In the 1950s, a committee of educators met and defined what is now called "Bloom's Taxonomy" of knowledge (Bloom 1956). This taxonomy is often associated with both good teaching and six sigma-related instruction. Roughly speaking, general knowledge divides into: (1) *knowledge* of the needed terminology and the typical applications sequences, (2) *comprehension* of the relevant plots and tables, (3) experience with *application* of several central approaches, (4) an ability for *analysis* of how certain data collection plans are linked to certain model-fitting and decision-making approaches, and (5) the *synthesis* needed to select an appropriate methodology for a given problem, in that order. Critiquing the knowledge being learned and its usefulness is associated with the steps of analysis and/or synthesis. The central thesis associated with Bloom's Taxonomy is that teaching should ideally begin with the knowledge and comprehension and build up to applications, ending with synthesis and critique.

Thus, Bloom's "taxonomy of cognition" divides knowledge and application from theory and synthesis, a division followed roughly in this book. Admittedly, the approach associated with Bloom's taxonomy does not cater to people who prefer to begin with general theories and then study applications and details.

1.9 Problems

In general, pick the correct answer that is most complete.

1. Consider the toy system of paper airplanes. Which of the following constitute possible design KIVs and KOVs?
   a. KIVs include time unit flies dropped from 2 m and KOVs include wing fold angle.
   b. KIVs include wing fold angle in design and KOVs include type of paper in design.
   c. KIVs include wing fold angle and KOVs include time unit flies assuming a 2 m drop.
   d. Answers in parts "a" and "b" are both correct.
   e. Answers in parts "a" and "c" are both correct.
2. Consider a system that is your personal homepage. Which of the following constitute possible design KIVs and KOVs?
   a. KIVs include background color and KOVs include time it takes to find your resume.
   b. KIVs include expert rating (1–10) of site and KOVs include amount of flash animation.
   c. KIVs include amount of flash animation and KOVs include expert rating (1–10) of site.
   d. Answers in parts "a" and "b" are both correct.
   e. Answers in parts "a" and "c" are both correct.
3. Assume that you are paid to aid with decision-making about settings for a die casting process in manufacturing. Engineers are frustrated by the amount of flash or spill-out they must clean off the finished parts and the deviations of the part dimensions from the nominal blueprint dimensions. They suggest that the preheat temperature and injection time might be changeable. They would like to improve the surface finish rating (1–10) but strongly doubt whether any factors would affect this. Which of the following constitute KIVs and KOVs?
   a. KIVs include deviation of part dimensions from nominal and KOVs include surface finish rating.
   b. KIVs include preheat temperature and KOVs include deviation of part dimensions from nominal.
   c. KIVs include surface finish rating and KOVs include deviation of part dimensions from nominal.
   d. Answers in parts "a" and "b" are both correct.
   e. Answers in parts "a" and "c" are both correct.

4. You are an industrial engineer at a hospital trying to reduce waiting times of patients in emergency rooms. You are allowed to consider the addition of one nurse during peak hours as well as subscription to a paid service that can reduce data entry times. Which of the following constitute KIVs and KOVs?
   a. KIVs include subscription to a data entry service and KOVs include waiting times.
   b. KIVs include number of nurses and KOVs include average waiting times for patients with AIDS.
   c. KIVs include average waiting times and KOVs include number of nurses.
   d. Answers in parts "a" and "b" are both correct.
   e. Answers in parts "a" and "c" are both correct.
5. Consider your friend's system relating to grade performance in school. List two possible KIVs and two possible KOVs.
6. Consider a system associated with international wire transfers in personal banking. List two possible KIVs and two possible KOVs.
7. According to this chapter, which of the following should be included in the definition of six sigma?
   a. Each project must be cost justified.
   b. For new products, project phases should be organized using DMADV.
   c. 3.4 unacceptable units per million opportunities is the generic goal.
   d. Answers in parts "a" and "b" are both correct.
   e. Answers in parts "a" and "c" are both correct.
8. According to this chapter, which of the following should be included in the definition of six sigma?
   a. Fear should be driven out of the workplace.
   b. Participants do not need to become statistics experts.
   c. Thorough SOP documentation must be completed at the end of every project.
   d. Answers in parts "a" and "b" are both correct.
   e. Answers in parts "a" and "c" are both correct.
9. How does six sigma training differ from typical university instruction? Explain in two sentences.
10. List two perceived problems associated with TQM that motivated the development of six sigma.
11. Which of the following is the lean production way to making three sandwiches?
    a. Lay out six pieces of bread, add tuna fish to each, add mustard, fold all, and cut.
    b. Lay out two pieces of bread, add tuna fish, mustard, fold, and cut. Repeat.
    c. Lay out the tuna and mustard, order out deep-fat fried bread and wait.
    d. Answers in parts "a" and "b" are both correct.
    e. Answers in parts "a" and "c" are both correct.
12. Which of the following were innovations associated with mass production?
    a. Workers did not need much training since they had simple, small tasks.
    b. Guild certification built up expertise among skilled tradesmen.
    c. Interchangeability of parts permitted many operations to be performed usefully at one time without changing over equipment.
    d. Answers in parts "a" and "b" are both correct.
    e. Answers in parts "a" and "c" are both correct.

13. In two sentences, explain the relationship between mass production and lost accountability.
14. In two sentences, explain why Shewhart invented control charts.
15. In two sentences, summarize the relationship between lean production and quality.
16. Give an example of a specific engineered system and improvement system that might be relevant in your work life.
17. Provide one modern example of craft production and one modern example of mass production. Your examples do not need to be in traditional manufacturing and could be based on a task in your home.
18. Which of the following are benefits of freezing a design long before Job 1?
    a. Your design function can react to data after Job 1.
    b. Tooling costs more because it becomes too easy to do it correctly.
    c. It prevents reactive design tinkering and therefore reduces tooling costs.
    d. Answers in parts "a" and "b" are both correct.
    e. Answers in parts "a" and "c" are both correct.
19. Which of the following are benefits of freezing a design long before Job 1?
    a. It encourages people to be systematic in attempts to avoid problems.
    b. Design changes cost little since tooling has not been committed.
    c. Fire-fighting occurs more often.
    d. Answers in parts "a" and "b" are both correct.
    e. Answers in parts "a" and "c" are both correct.
20. Which of the following are perceived benefits of being ISO certified?
    a. Employees must share information and agree on which practices are best.
    b. Inventory is reduced because there are smaller batch sizes.
    c. Training costs are reduced since the processes are well documented.
    d. Answers in parts "a" and "b" are both correct.
    e. Answers in parts "a" and "c" are both correct.
21. Which of the following are problems associated with gaining ISO accreditation?
    a. Resources must be devoted to something not on the value stream.
    b. Managers may be accused of "tree hugging" because fear can be useful.
    c. Employees rarely feel stifled because of a bureaucratic hurdles are eliminated.
    d. Answers in parts "a" and "b" are both correct.
    e. Answers in parts "a" and "c" are both correct.
22. According to Bloom's Taxonomy, which of the following is true?
    a. People almost always learn from the general to the specific.
    b. Learning of facts, application of facts, and the ability to critique, in that order, is easiest.
    c. Theory is critical to being able to apply material.
    d. Answers in parts "a" and "b" are both correct.
    e. Answers in parts "a" and "c" are both correct.
23. According to Bloom's Taxonomy which of the following would be effective?
    a. Give application experience, and then teach them theory as needed.
    b. Ask people to critique your syllabus content immediately, and then teach facts.
    c. Start with facts, then application, then some theory, and then ask for critiques.
    d. Answers in parts "a" and "b" are both correct.
    e. Answers in parts "a" and "c" are both correct.

24. Suppose one defines two basic levels of understanding of the material in this book to correspond to "green belt" (lower) and "black belt" (higher). Considering Bloom's Taxonomy, and inspecting this book's table of contents, what types of knowledge and abilities would a green belt have and what types of knowledge would a black belt have?
25. Suppose you were going to teach a fifteen year old about your specific major and its usefulness in life. Provide one example of knowledge for each level in Bloom's Taxonomy.
26. According to the chapter, which is correct and most complete?
    a. TQM has little to do with technology transfer from Europe to the U.S.
    b. The perception that TQM is anti-business developed in the last five years.
    c. One of Deming's 14 points is that fear is a necessary evil.
    d. All of the above are correct.
    e. All of the above are correct except (a) and (d).
27. Which is correct and most complete? Population characteristics are
    a. Statistics.
    b. Parameters.
    c. Variables.
    d. All of the above are correct.
    e. All of the above are correct except (a) and (d).
28. Which is correct and most complete? Population characteristics are
    a. Modes.
    b. Sample means.
    c. Often constant when monitoring parameters.
    d. All of the above are correct.
    e. All of the above are correct except (a) and (d).
29. Which is correct and most complete? Continuous variables data
    a. Is a type of attribute data.
    b. Can assume only a countable number of values.
    c. Can assume any value along a stretch of the real line.
    d. All of the above are correct.
    e. All of the above are correct except (a) and (d).
30. Which is correct and most complete? Attribute data
    a. Is a type of discrete data.
    b. Often relates to counts of negative events such as nonconforming items.
    c. Often involves whole number counts.
    d. All of the above are correct.
    e. All of the above are correct except (a) and (d).
31. Suppose one supplier supplies parts for 80% of items and another supplies parts for 90% of units. Assuming that the events are independent (unrelated), the probability that a unit contains parts from both suppliers is
    a. 70% or 0.70.
    b. 72% or 0.72.
    c. 80% or 0.80.
    d. Less than 10% by the Bonferroni inequality.
    e. All of the above are correct except (a) and (d).

32. Suppose one supplier supplies parts for 70% of items and another supplies parts for 80% of units. Assuming that the related events are independent, the probability that a unit contains parts from both suppliers is
    a. 56% or 0.56.
    b. 70% or 0.72.
    c. 80% or 0.80.
    d. Less than 20% by the Bonferroni inequality.
    e. All of the above are correct except (a) and (d).
33. Suppose that the probability that car #1 will have a massive failure in 10 years is 10% and the event that car #2 will have a massive failure in 10 years is 20%.
    a. The probability that both fail is 4% assuming independence.
    b. The probability that both fail assuming independence is 2%.
    c. The probability that neither fails is greater than 70% by the Bonferroni inequality.
    d. The probability that both fail is greater than 70%.
    e. All of the above are correct except (a) and (d).
34. Suppose that the probability that car #1 will have a massive failure in 10 years is 10% and the event that car #2 will have a massive failure in 10 years is 10%.
    a. The probability that both fail is 2% assuming independence.
    b. The probability that both fail assuming independence is 10%.
    c. The probability that neither fails is greater than 80% by the Bonferroni inequality.
    d. The probability that both fail is greater than 90%.
    e. All of the above are correct except (a) and (d).

References

Bloom BS (ed) (1956) Taxonomy of educational objectives. In: Handbook I: Cognitive domain. Longmans, Green and Co., New York

Clausing D (1994) Total quality development: a step-by-step guide to world-class concurrent engineering. ASME Press, New York

Collins J (2001) Good to great: why some companies make the leap… and others don't. Harper-Business, New York

Czitrom V (1999) One-factor-at-a-time versus designed experiments. Am Stat 53(2):126–131

Fisher RA (1925) Statistical methods for research workers. Oliver and Boyd, London

Hahn GJ, Hill WJ, Hoer RW, Zinkgraft SA (1999) The impact of six sigma improvement—a glimpse into the future of statistics. Am Stat 532:208–215

Harry MJ, Schroeder R (1999) Six sigma, the breakthrough management strategy revolutionizing the world's top corporations. Bantam Doubleday Dell, New York

Hazelrigg G (1996) System engineering: an approach to information-based design. Prentice Hall, Upper Saddle River, NJ

Kaynak H (2003) The relationship between total quality management practices and their effects on firm performance. J Oper Manag 21:405–435

Linderman K, Schroeder RG, Zaheer S, Choo AS (2003) Six sigma: a goal-theoretic perspective. J Oper Manag 21:193–203

Montgomery DC (2012) Statistical quality control, 7th edn. Wiley, New York.

Pande PS, Neuman RP, Cavanagh R (2000) The six sigma way: how GE, Motorola, and other top companies are honing their performance. McGraw-Hill, New York

Taguchi G (1993) Taguchi methods: research and development. In: Konishi S (ed) Quality engineering series, vol 1. The American Supplier Institute, Livonia, MI

Welch J, Welch S (2005) Winning. HarperBusiness, New York

Womack JP, Jones DT (1996) Lean thinking. Simon & Schuster, New York

Womack JP, Jones DT, Roos D (1991) The machine that changed the world: the story of lean production. Harper-Business, New York

Part I

Statistical Quality Control

Quality Control and Six Sigma

T. T. Allen, *Introduction to Engineering Statistics and Lean Six Sigma*,
https://doi.org/10.1007/978-1-4471-7420-2_2

2.1 Introduction

2

The phrase “**statistical quality control**” (SQC) refers to the application of statistical methods to monitor and evaluate systems and to determine whether changing key input variable (KIV) settings is appropriate. Specifically, SQC is associated with Shewhart’s statistical process charting (SPC) methods. These SPC methods include several charting procedures for visually evaluating the consistency of key process outputs (KOVs) and identifying unusual circumstances that might merit attention.

In common usage, however, SQC refers to many problem-solving methods. Some of these methods do not relate to monitoring or controlling processes and do not involve complicated statistical theory. In many places, SQC has become associated with all of the statistics and optimization methods that professionals use in quality improvement projects and in their other job functions. This includes methods for design of experiments (DOE) and optimization. In this book, DOE and optimization methods have been separated out mainly because they are the most complicated quality methods to apply and understand.

In ► Sect. 2.2, we preview some of the SQC methods described more fully later in this book. ► Section 2.3 relates these techniques to possible job descriptions and functions in a highly formalized organization. Next, ► Sect. 2.4 discusses the possible roles the different methods can play in the six sigma problem-solving method.

The discussion of organizational roles leads into the operative definition of quality, which we will define as conformance to design engineering’s specifications. ► Section 2.5 explores related issues including the potential difference between nonconforming and defective units. ► Section 2.6 concludes the chapter by describing how standard operating procedures capture the best practices derived from improvement or design projects.

2.2 Method Names as Buzzwords

The names of problem-solving methods have become “buzzwords” in the corporate world. The methods themselves are diverse; some involve calculating complicated statistics and others are simple charting methods. Some of the activities associated with performing these methods can be accomplished by a single person working alone, and others require multidisciplinary teams. The following is an abbreviated list of the methods to illustrate the breadth and purposes of these methods:

Acceptance Sampling involves collecting and analyzing a relatively small number of KIV measurements to make “accept or reject” decisions about a relatively large number of units. Statistical evidence is generated about the fraction of the units in the lot that are acceptable.

Control Planning is an activity performed by the “owners” of a process to assure that all process KOV variables are being measured in a way that assures a high degree of quality. This effort can involve application of multiple methods.

Design of Experiments (DOE) methods are structured approaches for collecting response data from varying multiple KIVs to a system. After the experimental tests

yield the response outputs, specific methods for analyzing the data are performed to establish approximate models for predicting outputs as a function of inputs.
Failure Mode and Effects Analysis (FMEA) is a method for prioritizing response measurements and subsystems addressed with highest priority.
Formal Optimization is itself a diverse set of methods for writing technical problems in a precise way and for developing recommended settings to improve a specific system or product, using input-output models as a starting point.
Gauge Repeatability and Reproducibility (R&R) involves collecting repeated measurements on an engineering system and performing complicated calculations to assess the acceptability of a specific measurement system. ("Gage" is an alternative spelling.)
Process Mapping involves creating a diagram of the steps involved with an engineering system. The exercise can be an important part of waste reduction efforts and lean engineering and can aid in identifying key input variables.
Regression is a curve-fitting method for developing approximate predictions of system KOVs (usually averages) as they depend on key input variable settings. It can also be associated with proving statistically that changes in KIVs affect changes in KOVs if used as part of a DOE method.
Statistical Process Control (SPC) **charting** includes several methods to assess visually and statistically the quality and consistency of process KOVs and to identify unusual occurrences. Therefore, SPC charting is useful for initially establishing the value and accuracy of current settings and confirming whether recommended changes will consistently improve quality.
Quality Function Deployment (QFD) involves creating several matrices that help decision-makers better understand how their system differs from competitor systems, both in the eyes of their customers and in objective features.

In the chapters that follow, these and many other techniques are described in detail, along with examples of how they have been used in real-world projects to facilitate substantial monetary savings.

Example 2.2.1 Methods and Statistical Evidence

Question

Which of the following methods involve generating statistical evidence?

a. Formal optimization and QFD generally create statistical evidence.
b. Acceptance sampling, DOE, regression, and SPC create evidence.
c. Process mapping and QFD generally create statistical evidence.
d. Answer in parts "a" and "b" are both correct.
e. Answer in parts "a" and "c" are both correct.

Answer

(b) Acceptance sampling, DOE, regression, and SPC can all easily be associated with formal statistical tests and evidence. Formal optimization, process mapping, and QFD generate numbers that can be called statistics, but they generally do not develop formal proof or statistical evidence.

2.3 Where Methods Fit into Projects

In many textbooks, statistical methods are taught as "stand alone" entities and their roles in the various stages of a system improvement or design project are not explained. It is perhaps true that one of the most valuable contributions of the six sigma movement is the association of quality methods with project phases. This association is particularly helpful to people who are learning statistics and optimization methods for the first time. These people often find it helpful to know which methods are supposed to be used at what stage.

In the six sigma literature, system improvement projects are divided into five phases or major activities (e.g., see Harry and Schroeder 1999; Pande et al. 2000):

1. **Define** terminates when specific goals for the system outputs are clarified and the main project participants are identified and committed to project success.
2. **Measure** involves establishing the capability of the technology for measuring system outputs and using the approved techniques to evaluate the state of the system before it is changed.
3. **Analyze** is associated with developing a qualitative and/or quantitative evaluation of how changes to system inputs affect system outputs.
4. **Improve** involves using the information from the analyze phase to develop recommended system design inputs.
5. **Control** is the last phase in which any savings from using the newly recommended inputs is confirmed, lessons learned are documented, and plans are made and implemented to help guarantee that any benefits are truly realized.

Often, six sigma improvement projects last three months, and each phase requires only a few weeks. Note that for new system design projects, the design and verify phases play somewhat similar roles to the improve and control phases in improvement projects. Also, the other phases adjust in intuitive ways to address the reality that in designing a new system, potential customer needs cannot be measured by any current system.

While it is true that experts might successfully use any technique in any phase, novices sometimes find it helpful to have more specific guidance about which techniques should be used in which phase. ◘ Table 2.1 is intended to summarize the associations of methods with major project phases most commonly mentioned in the six sigma literature.

Example 2.3.1 Basic Method Selection

Question

A team is trying to evaluate the current system inputs and measurement system. List three methods that might naturally be associated with this phase.

Answer

From the above definitions, the question pertains to the "measure" phase. Therefore, according to ◘ Table 2.1, relevant methods include Gauge R&R, SPC charting, and QFD.

■ **Table 2.1** Abbreviated list of methods and their role in improvement projects

Method	Phases
Acceptance sampling	Define, measure, control
Benchmarking	Define, measure, analyze
Control planning	Control, verify
Design of experiments	Analyze, design, improve
Failure mode and effects analysis (FMEA)	Analyze, control, verify
Formal optimization	Improve, design
Gauge R&R	Measure, control
Process mapping	Define, analyze
Quality function deployment (QFD)	Measure, analyze, improve
Regression	Define, analyze, design, improve
SPC charting	Measure, control

2.4 Organizational Roles and Methods

Sometimes, methods are used independently from any formal system improvement or design project. In these cases, the methods could be viewed as stand-alone projects. These applications occur in virtually all specialty departments or areas. In this section, the roles of specializations in a typical formalized company are described, together with the methods that people in each area might likely use.

■ Figure 2.1 shows one possible division of a formalized manufacturing company into specialized areas. Many formalized service companies have similar department divisions. In general, the marketing department helps the design engineering department understand customer needs. Design engineering translates input information from marketing into system designs. ► Section 2.5 will focus on this step, because design engineers often operationally define quality for other areas of company. Also, the designs generated by these engineers largely determine quality, costs of all types, and profits. Procurement sets up an internal and external supply chain to make the designed products or services. Process engineering sets up any internal processes needed for producing units, including tuning up any machines bought by procurement. Production attempts to build products to conform to the expectations of design engineering, using parts from procurement and machines from process engineering. Sales and logistics work together to sell and ship the units to customers.

■ Figure 2.1 also shows the methods that people in each area might use. Again, it is true that anyone in any area of an organization might conceivably use any method. However, ■ Fig. 2.1 does correctly imply that methods described in this book are potentially relevant throughout formalized organizations. In addition, all areas have potential impacts on quality, since anyone can conceivably influence the performance of units produced and/or the expectations of customers.

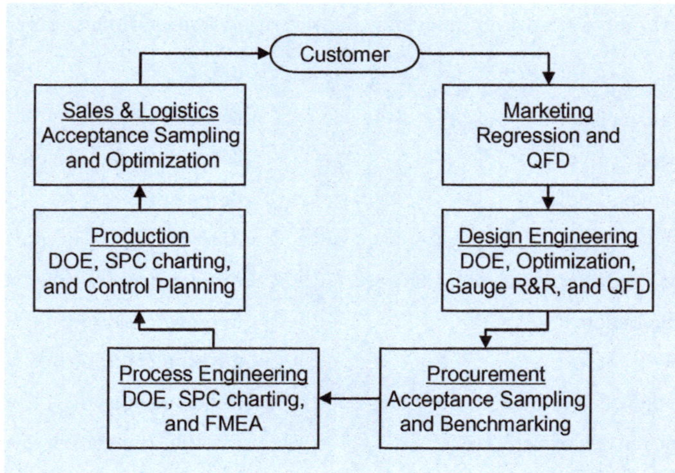

Fig. 2.1 Methods which might most likely be used by each department group

Example 2.4.1 Departmental Methods Selection

Question
In addition to the associations in Fig. 2.1, list one other department that might use acceptance sampling. Explain in one sentence.

Answer
Production might use acceptance sampling. When the raw materials or other input parts show up in lots (selected by procurement), production might use acceptance sampling to decide whether to reject these lots.

2.5 Specifications: Nonconforming Versus Defective

In manufacturing, design engineering generates a blueprint. Similar plans could be generated for the parameters of a service operation. Usually, a blueprint contains both target or "**nominal**" settings for each key input variable (KIV) and acceptable ranges. Figure 2.2 shows an example blueprint with three KIVs. The screw diameter is x_1, the percentage carbon in the steel is x_2, and x_3 is the angle associated with the third thread from the screw head.

Key input variables with acceptable ranges specified on blueprints or similar documents are called "**quality characteristics**." The minimum value allowed on a blueprint for a quality characteristic is called the lower specification limit (LSL). The maximum value allowed on a blueprint for a characteristic is called the upper specification limit (USL). For example, the LSL for x_1 is 5.00 mm for the blueprint in Fig. 2.2 and the USL for x_3 is 80.7°. For certain characteristics, there might be only an LSL or a USL but not both. For example, the characteristic x_2 in Fig. 2.2 has USL = 10.5% and no LSL.

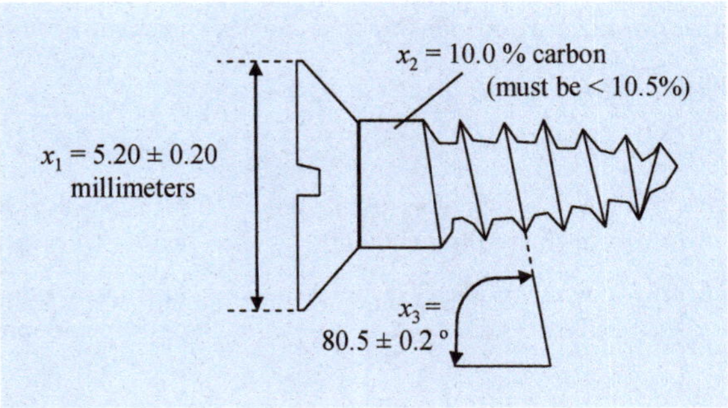

Fig. 2.2 Part of blueprint for custom designed screw with two KIVs

Note that nominal settings of quality characteristics are inputs, in the sense that the design engineer can directly control them by changing numbers, usually in an electronic file. However, in manufacturing, the actual corresponding values that can be measured are uncontrollable KOVs. Therefore, quality characteristics are associated with nominals that are KIVs (*x*s) and actual values that are KIVs (*y*s).

In many real-world situations, the LSL and USL define quality. Sometimes these values are written by procurement into contracts. A "**conforming**" part or product has all quality characteristic values, within the relevant specification limits. Other parts or products are called "**nonconforming,**" since at least one characteristic fails to conform to specifications. Manufacturers use the term "**nonconformity**" to describe each instance in which a part or product's characteristic value falls outside its associated specification limit. Therefore, a given part or unit might have many nonconformities. A "**defective**" part or product yields performance sufficiently below expectations such that its safe or effective usage is prevented. Manufacturers use the term "**defect**" to describe each instance in which a part or product's characteristic value causes substantially reduced product performance. Clearly, a defective unit is not necessarily nonconforming and vice versa. This follows because designers can make specifications without full knowledge of the associated effects on performance.

Table 2.2 shows the four possibilities for any given characteristic of a part or product. The main purpose of Table 2.2 is to call attention to the potential fallibility of specifications and the associated losses. The arguably most serious case occurs when a part or product's characteristic value causes a defect but meets specifications. In this case, a situation could conceivably occur in which the supplier is not contractually obligated to provide an effective part or product. Worse still, this case likely offers the highest chance that the defect might not be detected. The defect could then cause problems for customers.

Another kind of loss occurs when production and/or outside suppliers are forced to meet unnecessarily harsh specifications. In these cases, a product characteristic can be nonconforming, but the product is not defective. This can cause unnecessary expense because efforts to make products consistently conform to specifications can require additional tooling and personnel expenses. This type of waste, however, is to a great extent unavoidable.

2

Table 2.2 Possibilities associated with any given quality characteristic value

Conformance status	Performance related status	
	Defective	Non-defective
Nonconforming	Bad case—if not fixed, the unit could harm the customer	Medium case—unnecessary expense fixing unit might occur
Conforming	Worst case—likely to slip through and harm customer	Best case—unit fosters good performance and meets specs

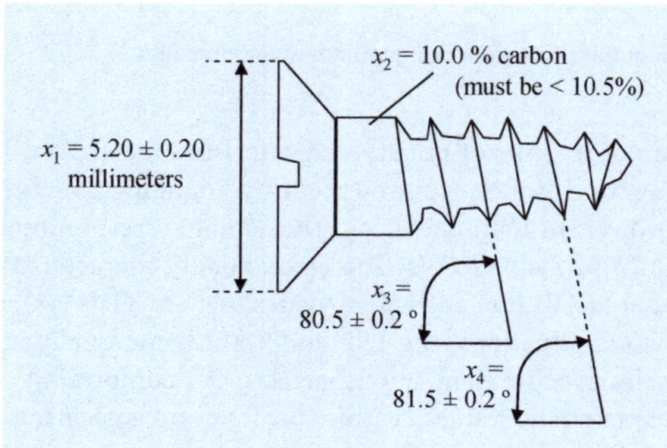

Fig. 2.3 Augmented blueprint with the additional characteristic x_4

Note that a key input variable (KIV) in the eyes of engineering design can be a key output variable (KOV) for production, because engineering design is attempting to meet customer expectations for designed products or services. To meet these expectations, design engineering directly controls the ideal nominal quality characteristic values and specifications. Production tries to manipulate process settings so that the parts produced meet the expectations of design engineering in terms of the quality characteristic values. Therefore, for production, the controllable inputs are settings on the machines, and the characteristics of units that are generated are KOVs. Therefore, we refer to "quality characteristics" instead of KIVs or KOVs.

Example 2.5.1 Screw Design Specifications

Question

Propose an additional characteristic and the associated specification limits for the screw example in Fig. 2.2. Also, give a value of that characteristic which constitutes a nonconformity and a defect.

Answer

Figure 2.3 shows the added characteristic x_4. The LSL is 81.3° and the USL is 81.7°. If x_4 equalled 95.0°, that would constitute both a nonconformity, because 95.0° > 81.7°, and a defect, because the customer would have difficulty inserting the screw.

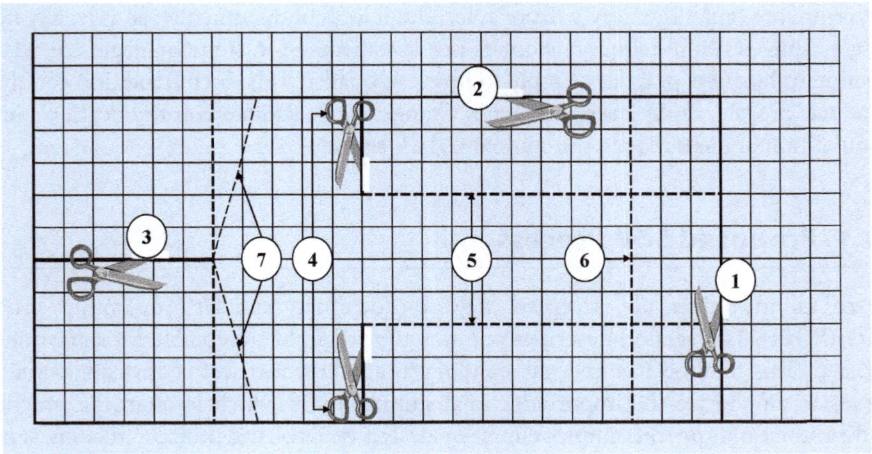

Fig. 2.4 Helicopter cut (—) and fold (--) lines (not to scale, grid spacing = 1 cm)

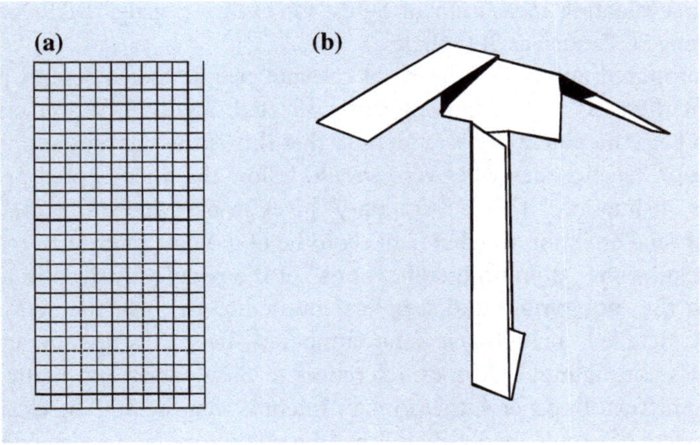

Fig. 2.5 **a** Paper with cut and fold lines (grid spacing is 1 cm); **b** desired result

2.6 Standard Operating Procedures (SOPs)

Currently, potential customers can enter many factories or service facilities and ask to view the International Standards Organization (ISO) manuals and supporting documentation. In general, this documentation is supposed to be easily available to anyone in these companies and to reflect accurately the most current practices. Creating and maintaining these documents requires significant and ongoing expense. Also, companies generally have specific procedures that govern the practices that must be documented and the requirements for that documentation.

Multiple considerations motivate these documentation efforts. First, customer companies often simply require ISO certifications of various types from all suppliers. Second, for pharmaceutical companies, hospitals, and many other companies where government regulations play a major role, a high level of documentation is legally necessary. Third, even if neither customers nor laws demand it, some managers decide to document business practices simply to improve quality. This documentation can limit product, process, and/or service design changes and facilitate communication and a competition of ideas among the company's best experts.

2.6.1 Proposed SOP Process

There is no universally accepted way to document standard operating procedures(SOPs). This section describes one way that *might* be acceptable for *some* organizations. This method has *no* legal standing in any business sector. Instead, it mainly serves to emphasize the importance of documentation, which is often the practical end-product of a process improvement or design engineering project. In some sense, the precise details in SOPs are the system inputs that project teams can actually control and evaluate. If your company has thorough and well-maintained SOPs, then the goals of SQC and DOE methods are to evaluate and improve the SOPs. There are specific methods for evaluating measurement SOPs, for example, gauge R&R for evaluating manufacturing SOPs such as SPC charts.

In the proposed approach, a team of relevant people assemble and produce the SOP so that there is "buy-in" among those affected. The SOP begins with a "title," designed to help the potential users identify that this is the relevant and needed SOP. Next, a "**scope**" section describes who should follow the documented procedures in which types of situations. Then a "**summary**" gives an overview of the methods in the SOP, with special attention to what is likely to be of greatest interest to readers. Next, the SOP includes the "**training qualifications**" of the people involved in applying the method and the "**equipment and supplies**" needed to perform the SOP. Finally, the "**method**" is detailed, including specific numbered steps. This documentation might include tables and figures. If it does, references to these tables and figures should be included in the text. In general, the primary intent is that the SOP be clear enough to insure the safety of people involved and that the operations be performed consistently enough to ensure good quality. Visual presentation and brevity are preferred when possible.

Example 2.6.1 Detailed Paper Helicopter Manufacturing SOP

Question

Provide a detailed SOP for producing paper helicopters.

Answer

Table 2.3 contains a SOP for paper helicopter manufacturing.

Note that not all information in a blueprint, including specification limits, will necessarily be included in a manufacturing SOP. Still, the goal of the SOP is, in an important sense, to make products that consistently conform to specifications.

Table 2.3 Detailed version of a paper helicopter SOP

Title: Detailed SOP for paper helicopter manufacturing
Scope: For use by college and graduate students
Summary: A detailed method to make a "base-line" paper helicopters is provided
Training Qualifications: None
Equipment and Supplies: Scissors, metric ruler, A4 paper
Method: The steps below refer to Fig. 2.4 1. Make cut ① 23 cm from lower left paper corner 2. Make cut ② 10 cm from bottom 3. Make cut ③ 5 cm down from the end of cut 2 4. Make 2 cuts, both labelled ④ in Fig. 2.4, 3 cm long each 5. Fold both sides of the base inwards along the crease lines labelled ⑤ 6. Fold the bottom up along the crease line labelled ⑥ 7. Fold wings in opposite directions along crease lines labelled ⑦

The fact that there are multiple possible SOPs for similar purposes is one of the central concepts of this book. The details of the SOPs could be input parameters for a system design problem. For example, the distances 23 and 5 cm in the above paper helicopter example could form input parameters x_1 and x_2 in a system design improvement project. It is also true that there are multiple ways to document what is essentially the same SOP. The example below is intended to offer an alternative SOP to make identical helicopters.

Example 2.6.2 Concise Paper Helicopter Manufacturing SOP

Question

Provide a more concise SOP for producing paper helicopters.

Answer

Table 2.4 contains a concise SOP for paper helicopter manufacturing.

With multiple ways to document the same operations, the question arises: what makes a good SOP? Many criteria can be proposed to evaluate SOPs, including cost of preparation, execution, and subjective level of professionalism. Perhaps the most important criteria in a manufacturing context relate to the performance that a given SOP fosters in the field. In particular, if this SOP is implemented in the company divisions, how desirable are the quality outcomes? Readability, conciseness, and level of detail may affect the outcomes in unexpected ways. The next chapters describe how statistical process control (SPC) charting methods provide thorough ways to quantitatively evaluate the quality associated with manufacturing SOPs.

Table 2.4 The concise version of a paper helicopter SOP

Title: Concise SOP for paper helicopter manufacturing
Scope: For use by college and graduate students
Summary: A concise method to make a "base-line" paper helicopters is provided
Training Qualifications: None
Equipment and Supplies: Scissors, metric ruler, A4 paper
Method: Cut on the solid lines and fold on the dotted lines as shown in Fig. 2.5a to make a helicopter that looks like Fig. 2.5b

Table 2.5 Paper helicopter measurement SOP

Title: SOP for measuring paper helicopter for student competition
Scope: For use by college and graduate students
Summary: A method is presented to measure the time in air for a student competition
Training Qualifications: None
Equipment and Supplies: Chalk, chair, stopwatch, meter stick, and two people
Method: 1. Use meter stick to measure 2.5 m up a wall and mark spot with chalk 2. Person 1 stands on chair approximately 1 m from wall 3. Person 1 orients helicopter so that base is down and wings are horizontal 4. Person 2 says "start" and Person 1 drops helicopter and Person 2 starts timer 5. Person 2 stops timer when helicopter hits the ground 6. Steps 2–5 are repeated three times, and average time in seconds is reported

2.6.2 Measurement SOPs

Quite often, SOPs are written to regulate a process for measuring a key output variable (KOV) of interest. For example, a legally relevant SOP might be used by a chemical company to measure the Ph in fluid flows to septic systems. In this book, the term "measurement SOPs" refers to SOPs where the associated output is a number or measurement. This differs from "production SOPs" where the output is a product or service. An example of a measurement SOP is given below. In the next chapters, it is described how gauge R&R methods provide quantitative ways to evaluate the quality of measurement SOPs.

Example 2.6.3 Paper Helicopter Measurement SOP

Question

Provide an SOP for measuring the quality of paper helicopters.

Answer

Table 2.5 describes a measurement SOP for timing paper helicopters.

2.7 Problems

In general, pick the correct answer that is most complete or inclusive.

1. A company is trying to design a new product and wants to systematically study its competitor's products. Which methods are obviously helpful (i.e., the method description mentions related goals)?
   a. Gauge R&R.
   b. QFD.
   c. Formal Optimization.
   d. Answers in parts "a" and "b" are both correct.
   e. Answers in parts "a" and "c" are both correct.
2. A company has implemented a new design into production. Now it is interested in prioritizing which inspection areas need more attention and in documenting a complete safety system. Which method s are obviously helpful (i.e., the method description mentions related goals)?
   a. FMEA.
   b. QFD.
   c. Control planning.
   d. Answers in parts "a" and "b" are both correct.
   e. Answers in parts "a" and "c" are both correct.
3. Which method s are obviously helpful for evaluating measurement systems (i.e., the method description mentions related goals)?
   a. Gauge R&R.
   b. DOE.
   c. Formal Optimization.
   d. Answers in parts "a" and "b" are both correct.
   e. Answers in parts "a" and "c" are both correct.
4. A company is trying to design a new product and wants to study input combinations to develop input-output predictive relationships. Which methods are obviously helpful (i.e., the method description mentions related goals)?
   a. Regression.
   b. DOE.
   c. Control planning.
   d. Answers in parts "a" and "b" are both correct.
   e. Answers in parts "a" and "c" are both correct.
5. A team is in a problem-solving phase in which the objectives and responsibilities have been established but the state of the current system has not been measured. According to this chapter, which method(s) would be obviously helpful (i.e., the method description mentions related goals)?
   a. SPC charting.
   b. Gauge R&R.
   c. DOE.
   d. Answers in parts "a" and "b" are both correct.
   e. Answers in parts "a" and "c" are both correct.

6. A team has created approximate regression models to predict input-output relationships and now wants to decide which inputs to recommend. According to this chapter, which method(s) would be obviously helpful?
   a. SPC charting.
   b. Gauge R&R.
   c. Formal optimization.
   d. Answers in parts "a" and "b" are both correct.
   e. Answers in parts "a" and "c" are both correct.
7. A team is in a problem-solving phase in which recommendations are ready but have not been fully confirmed and checked. According to this chapter, which method(s) would be obviously helpful?
   a. SPC charting.
   b. DOE.
   c. Formal optimization.
   d. Answers in parts "a" and "b" are both correct.
   e. Answers in parts "a" and "c" are both correct.
8. A large number of lots have shown up on a shipping dock, and their quality has not been ascertained. Which method(s) would be obviously helpful?
   a. Acceptance sampling.
   b. DOE.
   c. Formal optimization.
   d. Answers in parts "a" and "b" are both correct.
   e. Answers in parts "a" and "c" are both correct.
9. Based on Table 2.1, which methods are useful in the first phase of a project?
10. Based on Table 2.1, which methods are useful in the last phase of a project?
11. Which department could possibly use DOE?
    a. Design engineering.
    b. Production.
    c. Process engineering.
    d. All of the above are correct.
12. Which department(s) could possibly use SPC charting?
    a. Production.
    b. Marketing.
    c. Sales and logistics, for monitoring delivery times of truckers.
    d. All of the above are correct.
13. According to this chapter, which would most likely use acceptance sampling?
    a. Sales and logistics.
    b. Design engineering.
    c. Procurement.
    d. Answers in parts "a" and "b" are both correct.
    e. Answers in parts "a" and "c" are both correct.
14. According to the chapter, which would most likely use formal optimization?
    a. Design engineering.
    b. Production engineering.
    c. Process engineering.
    d. All of the above are correct.

15. Which of the following is true about engineering specification limits?
    a. They are associated with the "±" given on blueprints.
    b. They can fail to reflect actual performance in that nonconforming ≠ defective.
    c. They are always written in dimensionless units.
    d. Answers in parts "a" and "b" are both correct.
    e. Answers in parts "a" and "c" are both correct.
16. Which of the following is correct about engineering specifications?
    a. They are sometimes made up by engineers who do not know the implications.
    b. They are often used in contracts between procurement and suppliers.
    c. They could be so wide as to raise no production concerns.
    d. All of the above are correct.
    e. Only answers in parts "a" and "c" are correct.
17. Create a blueprint of an object you design including two quality characteristics and associated specification limits.
18. Propose an additional quality characteristic for the screw design in ◘ Fig. 2.3 and give associated specification limits.
19. Which of the following is true about manufacturing SOPs?
    a. They take the same format for all organizations and all applications.
    b. They can be evaluated using SPC charting in some cases.
    c. They are always written using dimensionless units.
    d. Answers in parts "a" and "b" are both correct.
    e. Answers in parts "a" and "c" are both correct.
20. Which of the following is true about manufacturing SOPs?
    a. They are sometimes made up by engineers who do not know the implications.
    b. According to the text, the most important criterion for SOPs is conciseness.
    c. They cannot contain quality characteristics and specification limits.
    d. Answers in parts "a" and "b" are both correct.
    e. Answers in parts "a" and "c" are both correct.
21. Which of the following is true about measurement SOPs?
    a. They are sometimes made up by engineers who do not know the implications.
    b. They describe how to make products that conform to specifications.
    c. They can be evaluated using gauge R&R.
    d. Answers in parts "a" and "b" are both correct.
    e. Answers in parts "a" and "c" are both correct.
22. Which of the following is true about measurement SOPs?
    a. They take the same format at all organizations for all applications.
    b. They are always written using dimensionless units.
    c. The same procedure can be documented in different ways.
    d. Answers in parts "a" and "b" are both correct.
    e. Answers in parts "a" and "c" are both correct.
23. Write an example of a manufacturing SOP for a problem in your life.
24. Write an example of a measurement SOP for a problem in your life.
25. In two sentences, critique the SOP in ◘ Table 2.3. What might be unclear to an operator trying to follow it?
26. In two sentences, critique the SOP in ◘ Table 2.5. What might be unclear to an operator trying to follow it?

References

Harry MJ, Schroeder R (1999) Six sigma, the breakthrough management strategy revolutionizing. The World's Top Corporations. Bantam Doubleday Dell, New York

Pande PS, Neuman RP, Cavanagh R (2000) The six sigma way: how GE, Motorola, and other top companies are honing their performance. McGraw-Hill, New York

Define Phase and Strategy

T. T. Allen, *Introduction to Engineering Statistics and Lean Six Sigma*,
https://doi.org/10.1007/978-1-4471-7420-2_3

3

3.1 Introduction

This chapter focuses on the definition of a project, including the designation of *who* is responsible for *what* progress by *when*. By definition, those applying six sigma methods must answer some or all of these questions in the first phase of their system improvement or new system design projects. Also, according to what may be regarded as a defining principle of six sigma, projects must be cost-justified or they should not be completed. Often in practice, the needed cost justification must be established by the end of the "**define**" phase.

A central theme in this chapter is that the most relevant strategies associated with answering these questions relate to identifying so-called "**subsystems**" and their associated key input variables (KIVs) and key output variables (KOVs). Therefore, the chapter begins with an explanation of the concept of systems and subsystems. Then, the format for documenting the conclusions of the define phase is discussed, and strategies are briefly defined to help in the identification of subsystems and associated goals for KOVs.

Next, specific methods are described to facilitate the development of a project charter, including benchmarking, meeting rules, and Pareto charting. Finally, one reasonably simple method for documenting significant figures is presented. Significant figures and the implied uncertainty associated with numbers can be important in the documentation of goals and for decision-making.

As a preliminary, consider that a first step in important projects involves searching the available literature. Search engines such as google and yahoo are relevant. Also, technical indexes such as the science citation index and compendex are relevant. Finally, consider using governmental resources such as the National Institute of Standards (NIST) and the United States Patent Office web sites.

3.2 Systems and Subsystems

A system is an entity with inputs and outputs. A "**subsystem**" is itself a system that is wholly contained in a more major system. The subsystem may share some inputs and outputs with its parent system. Figure 3.1 shows three subsystems inside a system. The main motivation for the "define phase" is to identify specific subsystems and to focus attention on them. The main deliverable from the define phase of a project is

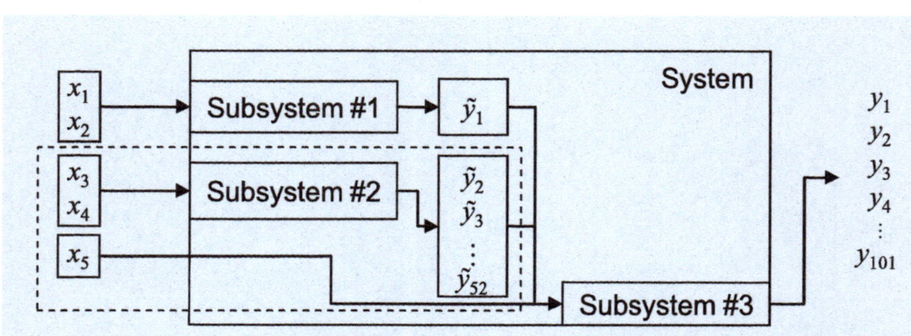

Fig. 3.1 Example of subsystems inside a system

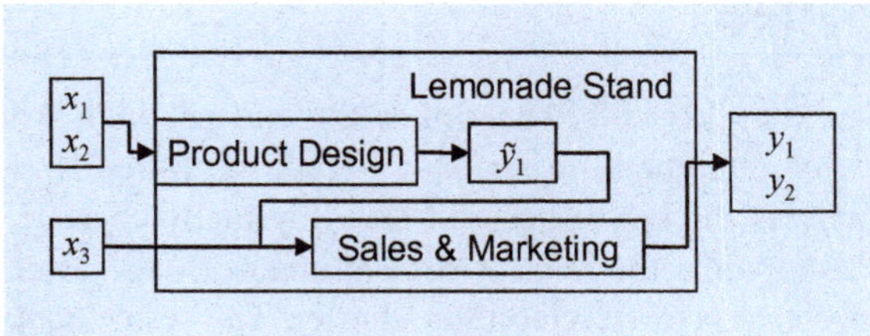

Fig. 3.2 Lemonade stand system with two subsystems

often a so-called "**charter**," defined in the next section. This charter is often expressed in terms of goals for subsystem outputs.

For example, in relation to a chess game system, one strategy to increase one's chance of success is to memorize recommended lists of responses to the first set of "opening" moves. The first opening set of moves constitutes only a fraction of the inputs needed for playing an entire game and rarely by itself guarantees victory. Yet, for a novice, focusing attention on the chess opening subsystem is often a useful strategy.

Figure 3.1 shows some output variables, $\tilde{y}_1$, …, $\tilde{y}_{52}$, from the subsystems that are not output variables for the whole system. We define "**intermediate variables**" as key output variables (KOVs) from subsystems that are inputs to another subsystem. Therefore, intermediate variables are not directly controllable by people in one subsystem but might be controllable by people in the context of a different subsystem. For example, scoring in chess is an intermediate variable which assigns points to pieces that are captured. From one phase of chess, one might have a high score but not necessarily win the game. However, points often are useful in predicting the outcome. Also, experts studying the endgame phase might assign combinations with specific point counts to different players. In general, winning or losing is generally the key output variable for the whole system.

Example 3.2.1 Lemonade Stand

Question

Consider a system in which children make and sell lemonade. Define two subsystems, each with two inputs and outputs and one intermediate variable.

Answer

Figure 3.2 shows the two subsystems: (1) Product Design and (2) Sales and Marketing. Inputs to the product design subsystem are: x_1, percentage of sugar in cup of lemonade and x_2, flavoring type (natural or artificial). An intermediate variable is the average of (1–10) taste ratings from family members, $\tilde{y}_1$. The Sales and Marketing subsystem has as inputs the taste rating and the advertising effective monetary budget, x_3. Key output variables from the Sales and Marketing subsystem include the profit, y_1, and average customer satisfaction rating (1–10), y_2.

3.3 Project Charters

3

In many cases, a written "**charter**" constitutes the end product of the first phase of a project. The charter documents *what* is to be accomplished by *whom* and *when*. ◘ Figure 3.3 summarizes the key issues addressed by many charters. Clarifying *what* can be accomplished within the project time frame with the available resources is probably the main concern in developing a charter. The term "**scope**" is commonly used in this context to formally describe *what* is to be done. The term "**deliverables**" refers to the outcomes associated with a project scope. Strictly speaking, "tangible" deliverables must be physical objects, not including documents. However, generally, deliverables could be as intangible as an equation or a key idea.

Note that the creation of a charter is often a complicated, political process. Allocating the selected team for the allotted time is essentially an expensive, risky investment by management. Management is betting that the project deliverables will be generated and will still be worthwhile when they are delivered. The main motivation for a formal design phase is to separate the complicated and management-level decision-making from the relatively technical, detailed decision-making associated with completing the remainder of the project and deliverables.

Therefore, developing a charter involves establishing a semi-formal contract between management and the team about what is "**in scope**" and what is "**out of scope**" or unnecessary for the project to be successful. As a result of this contract, team members have some protection against new, unanticipated demands, called "**scope creep**," that might be added during the remaining project phases. Protection against scope creep can foster a nurturing environment and, hopefully, increase the chances of generating the deliverables on time and under budget.

The project goals in the charter are often expressed in terms of targets for key output variables (KOVs). Commonly, the scope of the project requires that measureable KOVs must be intermediate variables associated with subsystems. The principle of cost

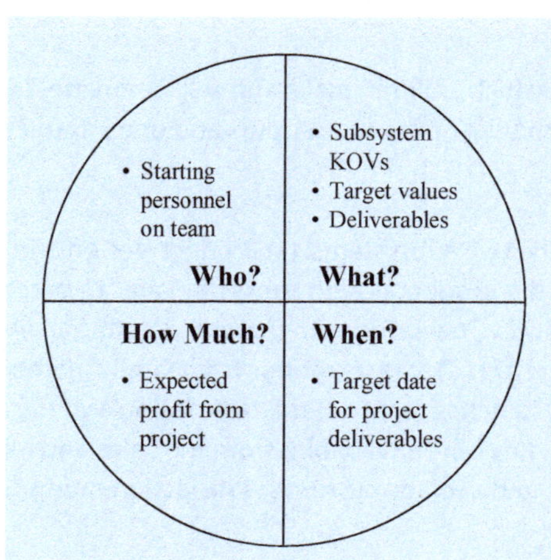

◘ **Fig. 3.3** Possible issues addressed by a project charter

justification dictates that at least one KOV for these subsystems must have a likely relationship to bottom-line profits for the major system.

Example 3.3.1 Lemonade Design Scope

Question

Because of customer complaints, an older sibling tasks a younger sibling with improving the recipe of lemonade to sell at a lemonade stand. Clarify a possible project scope including one deliverable, one target for a KOV, and one out-of-scope goal.

Answer

The younger sibling seeks to deliver a recipe specifying what percentage of sweetener to use (x_1) with a target average taste rating ($\tilde{y}_1$) increase greater than 1.5 units as measured by three family measures on a 1–10 scale. It is believed that taste ratings will drive sales, which will in turn drive profits. In the approved view of the younger sibling, it is not necessary that the older sibling will personally prefer the taste of the new lemonade recipe.

In defining *who* is on the project team, common sense dictates that the personnel included should be representative of people who might be affected by the project results. This follows in part because affected people are likely to have the most relevant knowledge, giving the project the best chance to succeed. The phrase "**not-invented-here syndrome**" (NIHS) refers to the powerful human tendency to resist recommendations by outside groups. This does not include the tendency to resist orders from superiors, which constitutes insubordination, not NIHS. NIHS implies resistance to fully plausible ideas that are resisted purely because of their external source. By including on the team people who will be affected, we can sometimes develop the "**buy-in**" need to reduce the effects of the not-invented-here syndrome. Scope creep can be avoided by including all of these people on the team.

In defining *when* a project should be completed, an important concern is to complete the project soon enough so that the deliverables are still relevant to the larger system needs. Many six sigma experts have suggested project timeframes between 2 and 6 months. For projects on the longer side of this range, charters often include a schedule for deliverables rendered before the final project completion. In general, the project timeframe limits imply that discipline is necessary when selecting achievable scopes.

There is no universally used format for writing project charters. The following example, based loosely on a funded industrial grant proposal, illustrates one possible format. One desirable feature of this format is its brevity. In many cases, a 3-month timeframe permits an effective one-page charter. The next subsection focuses on a simple model for estimating expected profits from projects.

Example 3.3.2 Snap Tab Project Charter

Question

Your team (a design engineer, a process engineer, and a quality engineer, each working 25% time) recently completed a successful 6-month project. The main deliverable was a fastener design in 3D computer aided design (CAD) format. The result achieved a 50% increase in pull-apart strength by manipulating five KIVs in the design. The new design is saving $300K/year by reducing assembly costs for two product lines (not including project expense). A similar product line uses a different material. Develop a charter to tune the five KIVs for the new material, if possible.

Answer

Scope - Develop tuned design for new material

Deliverables - One-page report clarifying whether strength increase is achievable

A 3D CAD model that includes specifications for the five KIVs

Personnel - One design engineer, one process engineer, one quality engineer

Timing - One-page report after 2 months

3D CAD model after 3 months and completion

Expected profit - $281K (see below)

3.3.1 Predicting Expected Profits

Often projects focus only on a small subsystem that is not really autonomous inside a company. Therefore, it is difficult to evaluate the financial impact of the project on the company bottom line. Yet an effort to establish this linkage is generally considered necessary. In this section, a formula is presented for predicting expected profits, specific to a certain type of production system improvement project. However, some of the associated reasoning may have applications to profit modeling in other cases.

The term "**rework**" refers to efforts to fix units not conforming to specifications. The term "**scrap**" refers to the act of throwing away nonconforming items that cannot be effectively reworked. Often, the rework and scrap costs constitute the most tangible monetary figure associated with an improvement project. However, the direct costs associated with rework and scrap may not reflect the true losses from nonconformities, for three reasons. First, parts failing inspection can cause production delays. These delays, in turn, can force sales employees to quote longer lead times, i.e., the time periods between the customer order and delivery. Longer lead times can cause lost sales. Second, reworked units may never be as good as new units, and could potentially cause failures in the field. Third, for every unit found to be nonconforming, another unit might conceivably fail to conform but go undetected. By reducing the need for rework, it is likely that the incidence of field failures will decrease. Failures in the field also can result in lost sales in the future.

Let "RC" denote the current rework and scrap costs on an annual basis. Let "f" denote the fraction of these costs that the project is targeting for reduction. Note that f equaling 1.0 (or 100%) reduction is usually considered unrealistic. Assuming that RC is known, a simple model for the expected savings is (Eq. 3.1):

$$\text{Expected Savings} = G \times f \times (2.0 \times \text{RC}) \tag{3.1}$$

where the 2.0 derives from considering savings over a 2-year horizon and G is a "**fudge factor**" designed to account for indirect savings from increasing the fraction of conforming units. Often, $G = 1.0$ which conservatively accounts only for directly measurable savings. Yet, in some companies, $G = 4.0$ is routinely used out of concern for indirect losses including production disruption and lost sales.

Note that the model in (3.1) only crudely addresses the issue of discounting future savings by cutting all revenues off after 2 years. It is also only applicable for improvement projects related primarily to rework or scrap reduction.

Often, salary expenses dominate expenses both for rework and running a project. The term "**person-years**" refers to the time in years it would take one person, working full time, to complete a task. A rule of thumb is to associate every person-year with \$100K in costs including benefits and the cost of management support. This simple rule can be used to estimate the rework costs (RC) and other project expenses. With these assumptions, a crude model for the expected profit is:

$$\text{Expected Profit} = \text{Expected Savings} - (\text{Project Person} - \text{Years}) \times \$100\text{K} \quad (3.2)$$

where "Project Person-Years" is the total number of person-years planned to be expended by all people working on a project.

Example 3.3.3 Snap Tab Expected Profits

Question

Your team (a design engineer, a process engineer, and a quality engineer, each working 25% time) recently completed a successful 6-month project. The main deliverable was a fastener design in 3D computer aided design (CAD) format. The result achieved a 50% increase in pull-apart strength by manipulating five KIVs in the design. The new design is saving \$300K/year by reducing assembly costs for two product lines (not including project expense). A similar product line uses a different material. Estimate the expected profit from this project, assuming a 2-year horizon.

Answer

As savings do not derive from rework and scrap reductions, we cannot use Eq. (3.1). However, since \$300K/year was saved on two product lines in similar circumstances, it is likely that \$150K/year in costs could be reduced through application to a single new product line. Therefore, expected savings over a 2-year horizon would be 2.0 years $\times$ \$150K/year = \$300K. With three engineers working 25% time for 0.25 year, the person-years of project expense should be $3 \times 0.25 \times 0.25 = 0.1875$. Therefore, the expected profits from the model in Eq. (3.2) would be \$300K – \$18.73K = \$281K.

3.4 Strategies for Project Definition

Identifying the subsystem to improve or design is probably the most important decision in a project. Much relevant literature on this subject is available in different disciplines, including research published in *Management Science* and the *Journal of Product Innovation Management.* Here, only a sampling of the associated ideas is presented, relating specifically to bottleneck subsystems and near-duplicate subsystems.

3.4.1 Bottleneck Subsystems

In their influential book *The Goal*, Goldratt and Cox (2016) offer ideas relevant to subset selection. It is perhaps fair to rephrase their central thesis as follows:

3

1. In a large system, there is almost always one "**bottleneck**" subsystem, having a single intermediate, truly key output variable that directly relates to total system profits.
2. Improvements to other subsystems that do not affect the bottleneck's truly key output variable have small effects (if any) on total system profits.

Therefore, the Goldratt and Cox (2016) "**Theory of Constraints**" (TOC) improvement process involves identifying the bottleneck subsystems and improving the truly key output variables. Working on the appropriate subsystem is potentially critical to the six sigma principle of affecting total system profits.

Many people do not get the opportunity to work on bottleneck subsystems. As a result, TOC implies that it is unlikely their efforts will strongly and directly affect the bottom line. Also, any bottom-line savings predicted by people not working on these subsystems should ultimately be suspect. TOC does provide some reassurance for this common occurence of improving non-bottleneck subsystems, however. After other people improve the bottleneck system or "alleviate the bottleneck," there is a chance that the subsystem under consideration will become a bottleneck.

3.4.2 Go-No-Go Decisions

The term "**categorical factor**" refers to inputs that take on only qualitatively different settings. The term "**design concept**" is often used to refer to one level setting for a categorical factor. For example, one design concept for a car could be a rear engine, which is one setting of the categorical factor of engine type. In the development of systems and subsystems, only a finite number of design concepts can be considered at any one time due to resource limitations. The phrase "**go-no-go decisions**" refers to the possible exclusion from consideration of one or more design concepts or projects. For example, an expensive way to arc weld aluminum might be abandoned in favor of cheaper methods because of a go-no-go decision. The benefits of go-no-go decisions are similar to the benefits of freezing designs described in ▶ Chap. 1.

One relevant goal of improvement or design projects is to make go-no-go decisions decisively. For example, the design concept snap tabs might be competing with the design concept screws for an automotive joining design problem. The team might explore the strength of snap tabs to decide which concept should be used.

A related issue is the possible existence of subsystems that are nearly identical. For example, many product lines could benefit potentially from changing their joining method to snap tabs. This creates a situation in which one subsystem may be tested, and multiple go-no-go decisions might result. The term "**worst-case analysis**" refers to the situation in which engineers experiment with the subsystem that is considered the most challenging. Then they make go-no-go decisions for all the other nearly duplicate systems.

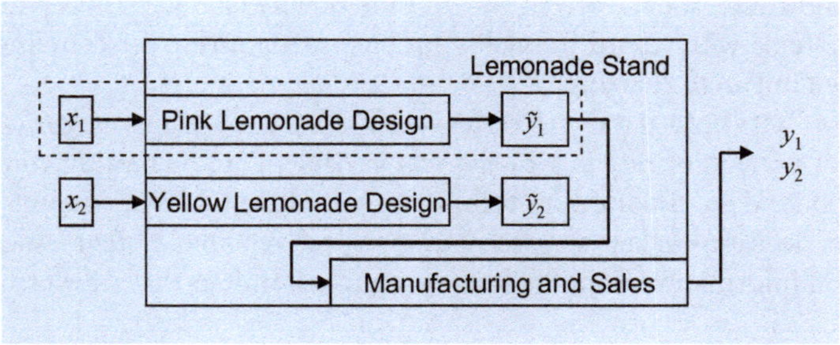

Fig. 3.4 Lemonade stand subsystems and strategy

Example 3.4.1 Lemonade Stand Improvement Strategy

Question

Children are selling pink and yellow lemonade on a busy street with many possible customers. The fraction of sugar is the same in pink and yellow lemonade, and the word-of-mouth is that the lemonades are both too sweet, particularly the pink type, which results in lost sales. Materials are available at negligible cost. Making reference to TOC and worst-case analysis, suggest a subsystem for improvement with 1 KIV and 1 KOV.

Answer

TOC suggests focusing on the apparent bottlenecks, which are the product design subsystems, as shown in Fig. 3.4. This follows because manufacturing costs are negligible and the potential customers are aware of the products. A worst-case analysis strategy suggests further focusing on the pink lemonade design subsystem. This follows because if the appropriate setting for the fraction of sugar input factor, x_1, is found for that product, the design setting would likely improve sales of both pink and yellow lemonade. A reasonable intermediate variable to focus on would be the average taste rating, $\tilde{y}_1$.

3.5 Methods for Define Phases

Many problem-solving methods are useful in the define phase. In this chapter, we include only three: Pareto charting, benchmarking, and meeting rules. However, several methods addressed in later chapters, including process mapping or value stream mapping, can aid in the development of project charters. Process mapping in ▶ Chap. 5 is particularly relevant in identifying bottleneck subsystems. Efforts to identify specific bottlenecks can also find a role in the analyze phase.

3.5.1 Pareto Charting

In general, different types of nonconformities are associated with different KOVs. Also, different KOVs or quality characteristics are associated with different subsystems. The method of Pareto charting involves a simple tabulation of the types of nonconformities generated by a production system. This is helpful in project definition because it constitutes a data-driven way to rank quality characteristics and their associated subsystems with regard to quality problems. Algorithm 3.1 contains an outline of steps taken in Pareto charting.

The term "**attribute data**" refers to values associated with categorical variables. Since type of nonconformity is a categorical variable, Pareto charting constitutes one type of attribute data visualization technique. Visualizing a large amount of attribute data permits decision-makers to gain more perspective about system issues than simply relying on information from the last few nonconformities that were created.

Algorithm 3.1. Pareto charting

Step 1. List the types of nonconformities or causes associated with failing units.
Step 2. Count the number of nonconformities of each type or cause.
Step 3. Sort the nonconformity types or causes in descending order by the counts.
Step 4. Create a category called "other," containing all counts associated with nonconformity or cause counts subjectively considered to be few in number.
Step 5. Bar-chart the counts using the type of nonconformity or causal labels.

Note that sometimes it is desirable to refer to types of nonconformities using a causal vocabulary. For example, assume a metal part length exceeded the upper specification because of temperature expansion. We could view this as a single "part length" nonconformity or as caused by temperature. Note also, the term "frequency" is often used in place of "count of nonconformities" in Pareto charts.

The phrase "**Pareto rule**" refers to the common occurrence in which 20% of the causes are associated with greater than 80% of the nonconformities. In these cases, the subsystems of greatest interest, which may be system bottlenecks, often become clearly apparent from inspection of the Pareto bar charts. Surprisingly, the people involved in a system often are shocked by the results of Pareto charting. This occurs because they have lost perspective and are focused on resolving the latest cause and not the most important cause. This explains how applying Pareto charting or "**Pareto analysis**" can be eye-opening.

Sometimes the consequences in terms of rework costs or other results can be much greater for some types of nonconformities than for others. One variant of Pareto charting uses subjectively assessed weights for the various nonconformities. For example, *Step 2* above could become "Sum the number of weighted nonconformities of each type or cause" and *Step 3* would become "Sort by weighted sum." Another variant of Pareto charting called "**cost Pareto chart**" involves to a tabulation of the costs associated with nonconformity types or causes, listed in Algorithm 3.2.

Algorithm 3.2. Cost Pareto charting

Step 1. Find list of the costs of nonconformities including types or causes.
Step 2. Sum the costs of nonconformities of each type or cause.
Step 3. Sort the nonconformity types or causes in descending order by the costs.
Step 4. Create a category called "other" containing all costs associated with nonconformity or cause counts subjectively considered to be small in costs.
Step 5. Bar-chart the costs using the associated type of nonconformity or causal labels.

The "Pareto rule" for cost Pareto charts is that often 20% of the causes are associated with greater than 80% of the costs. The implications for system design of cost Pareto charts are similar to those of ordinary Pareto charts.

Example 3.5.1 Pacemaker Nonconformities

Question

Consider the following hypothetical list of non-fatal pacemaker failures, with rework or medical costs in parentheses: battery life ($350), irregular heart beat ($350), battery life ($350), electromagnetic shielding ($110K), battery life ($350), discomfort ($350), battery life ($350), battery life ($350), battery life ($350), lethargy ($350), battery life ($350), battery life ($350), battery life ($350), battery life ($350), battery life ($150K), battery life ($350), and irregular heart beat ($350). Construct a Pareto chart and a cost Pareto chart, and comment on implications for project scope.

Answer

Table 3.1 shows the results of *Steps 1–3* for both charting procedures. Note that there are probably not enough nonconformity types to make it desirable to create "other" categories. Figure 3.5 shows the two types of Pareto charts. The ordinary chart shows that focusing the project scope on the KOV battery life and the associated subsystem will probably affect the most people. The second chart suggests that shielding issues, while rare, might also be prioritized highly for attention.

Note that the Pareto rule applies in the above example, since 80% of the nonconformities are associated with one type of nonconformity or cause, battery life. Also, this hypothetical example involves **ethical issues** since serious consequences for human patients are addressed. While quality techniques can be associated with callousness in general, they often give perspective that facilitates ethical judgements. In some cases, failing to apply methods can be regarded as ethically irresponsible.

A "**check sheet**" is a tabular compilation of the data used for Pareto charting. In addition to the total count vs nonconformity type or cause, there is also information about the time in which the nonconformities occurred. This information can aid in identifying trends and the possibility that a single cause might be generating multiple nonconformities. A check sheet for the pacemaker example is shown in Table 3.2. From the check sheet, it seems likely that battery issues from certain months might have causing all other problems except for shielding. Also, these issues might be getting worse in the summer.

3.5.2 Benchmarking

The term "**benchmarking**" means setting a standard for system outputs that is useful for evaluating design choices. Often, benchmarking standards come from evaluations of competitor offerings. For this reason, companies routinely purchase competitor pro ducts or services to study their performance. In school, studying your fellow student's

3

Table 3.1 Tabulation of the relevant nonconformity counts and costs

Nonconformity	1	2	3	4	5	6	7	8	9	10	11	12	Count	Sum ($)
Battery life	150	350	350	350	350	350	350	350	350	350	350	350	12	113,850
Irreg. heart beat	350	350	–	–	–	–	–	–	–	–	–	–	2	700
Electro. shielding	110,000	–	–	–	–	–	–	–	–	–	–	–	1	110,000
Discomfort	350	–	–	–	–	–	–	–	–	–	–	–	1	350
Lethargy	350	–	–	–	–	–	–	–	–	–	–	–	1	350

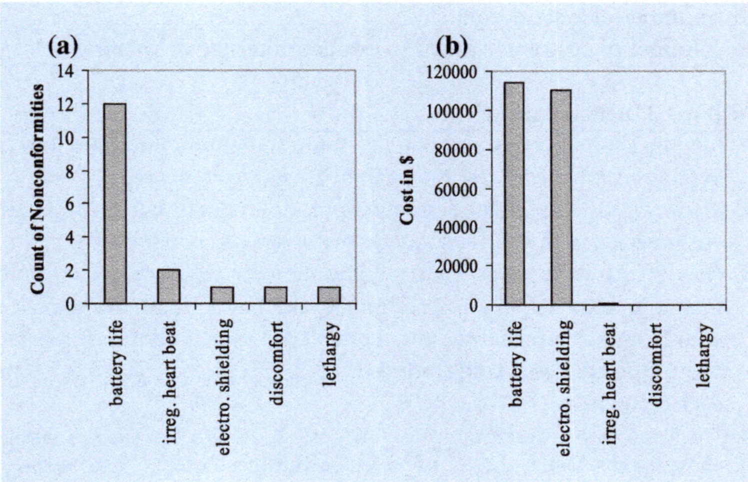

Fig. 3.5 **a** Pareto chart and **b** cost Pareto chart of hypothetical nonconformities

Table 3.2 Check sheet for pacemaker example

Nonconformity	Production date							Total
	Jan.	Feb.	May	April	May	June	July	
Battery life			3			4	5	12
Irregular heart beat						2		2
Electromagnetic shielding	1							1
Discomfort						1		1
Lethargy						1		1

homework solutions is usually considered cheating. In some cases, benchmarking in business can also constitute illegal corporate espionage. Often, however, benchmarking against competitor products is legal, ethical, and wise. Consult with lawyers if you are unclear about the rules relevant to your situation.

The version of benchmarking that we describe here, listed in Algorithm 3.3, involves creating two different matrices following Clausing (1994, p. 66). These matrices will fit into a larger "**Quality Function Deployment**" "**House of Quality**" that will be described fully in ▶ Chap. 6. The goal of the exercise is to create a visual display inform project definition decision-making. Specifically, by creating the two matrices, the user should have a better idea about which key input variables (KIVs) and key output variables (KOVs) should be focused on to stimulate sales in a competitive marketplace. Note that many people would refer to filling out either of the two tables, even only partially, as benchmarking. The symbols used are:

1. q_c is the number of customer issues.
2. q is the number of system outputs.

3. m is the number of system inputs.
4. n is the number of customers asked to evaluate alternative systems.

Algorithm 3.3. Benchmarking

Step 1. Identify alternative systems or units from competitors, including the current default system. Often, only three alternatives are compared.

Step 2. Identify q_c issues with the system outputs or products that are important to customers, described in a language customers can understand.

Step 3. Next, n customers are asked to rate the alternative systems or units through focus groups or surveys. The customers rate the products on a scale of 1–10, with 10 indicating that the system or unit completely addresses the issue being studied. The average ratings, $Y_{customer,1}, \ldots, Y_{customer,qC}$, are calculated for each competitor.

Step 4. The same competitor systems or units are studied to find the key input variable (KIV) settings, $x_1, \ldots, x_m$, and key outputs variable (KOV) settings, $Y_1, \ldots, Y_q$. Often, q and q_c are between 3 and 10.

Step 5. Display the data in two tables. The first lists the customer criteria as rows and the company ratings as columns, and the second lists the alternatives as rows and the input and outputs as columns.

Step 6. Study the information in the tables and make subjective judgements about inputs and outputs to focus on. Also, when appropriate, use information about competitor products to set benchmark targets on key output variablesOutput variables

3

Example 3.5.2 Benchmarking Welding Procedures

Question

Study the following benchmarking tables and recommend two KIVs and two intermediate variables for inclusion in project scope at ACME, Inc. Include one target for an intermediate variable (INT). Explain in three sentences.

Answer

Table 3.3 shows that Runner, Inc. is dominant with respect to addressing customer concerns. Table 3.4 suggests that Runner, Inc.'s success might be attributable to travel speed, preheat factor settings, and an impressive control of the fixture gap.

Table 3.3 Three customer issues ($q_c = 3$) and average ratings from ten customers

Customer issue	Competitor system		
	ACME, Inc.	Runner, Inc.	Coyote, Inc.
Structure is strong because of joint shape	4.7	9.0	4.0
Surface is smooth requiring little rework	5.0	8.6	5.3
Clean factory floor, little work in process	4.3	5.0	5.0

Table 3.4 Benchmark key input variables (KIV), intermediate variables (INT), and key output variables (KOVs)

Company	KIV-travel speed (ipm)	KIV-weld area	KIV-tip-to-work (mm)	KIV-wire diameter (mm)	KIV-heating pretreatment	INT-avg. offset (mm)	INT-avg. gap (mm)	INT-sheet flatness (–)	KOV-support flatness (–)	KOV-final flatness (–)
ACME, Inc.	35.0	8.0	15.0	2.0	N	1.1	0.9	1.1	3.5	1.1
Runner, Inc.	42.0	9.2	15.0	2.0	Y	0.9	0.2	1.2	4.0	1.2
Coyote, Inc.	36.0	9.5	15.0	2.5	N	0.9	0.9	1.0	1.5	1.0

These should likely be included in the study subsystem as inputs x_1 and x_2 and output $\tilde{y}_1$ respectively with output target $\tilde{y}_1 < 0.2$ mm.

3.6 Formal Meetings

People hold meetings in virtually all organizations and in all phases of projects. Meetings are perhaps particularly relevant in the define phase of projects, because information from many people is often needed to develop an effective charter.

Meetings satisfy the definition of problem-solving methods in that they can generate recommended decisions, which are inputs to systems. Also, they can involve an exchange of information. Further, there are many ways to hold meetings, each of which, in any given situation, might generate different results. The term "**formal meeting**" is used here to refer to a specific method for holding meetings. The proposed method is a hybrid of approaches in Martin et al. (1997), Robert et al. (2000), and Streibel (2002). A main purpose is to expose readers to potentially new ways of structuring meetings.

The term "**agenda**" refers to a list of activities intended to be completed in a meeting. The term "**facilitator**" refers to a person with the charge of making sure that the meeting rules and agenda are followed. The facilitator generally acts impartially and declares any biases openly and concisely as appropriate.

Algorithm 3.4. Steps for a formal meeting using rules

Step 1 The facilitator suggests, amends, and documents the meeting rules and agenda based on participant ideas and approval.
Step 2. The facilitator declares default actions or system inputs that will go into effect unless they are revised in the remainder of the meeting. If appropriate, these defaults come from the ranking management.
Step 3. The facilitator implements the agenda, which is the main body of the meeting.
Step 4. The facilitator summarizes meeting results including (1) actions to be taken and the people responsible, and (2) specific recommendations generated, which usually relate to inputs to some system.
Step 5. The facilitator solicits feedback about the meeting rules and agenda to improve future meetings.
Step 6. Participants thank each other for attending the meeting and say good-bye.

The phrase "**meeting wrap-up**" can be used to refer simultaneously to *Steps 5* and *6*. The term "**brainstorming**" refers to an activity in which participants propose creative solutions to a problem. For example, the problem could be to choose inputs and outputs for study in a project. Since creativity is desired, it can be useful to document the ideas generated in a supportive atmosphere with minimal critiquing. The term "**filtering**" refers here to a process of critiquing, tuning, and rejecting ideas generate in a brainstorming process. Since filtering is a critical, negative activity, it is often separated temporarily from brainstorming. The pair of activities, brainstorming and filtering, might appear together on an agenda in relation to a particular topic.

The phrase "**have a go-round**" is used here to refer to an activity in which many or all of the meeting participants are asked to comment on a particular issue. Having a go-round can be critical to learning information from shy people and making a large number of people feel "buy-in" or involvement in a decision. Also, having a go-round can be combined with activities such as brainstorming and filtering.

Example 3.6.1 Teleconferencing with Europe

Question

An engineer in China is teleconferencing with two shy engineers in Europe who work in the same company. European engineers have greater familiarity with related production issues. The meeting objective is to finalize the KIVs, KOVs, and targets for a project charter. The Chinese engineer has e-mailed a proposed list of these previously. Use this information to suggest defaults and a meeting agenda.

Answer

Default actions: Use the e-mailed list of KIVs, KOVs, and targets.

1. Review the e-mailed list of KIVs, KOVs, and targets.
2. Using a go-round, brainstorm possible KIVs, KOVs, and targets not included.
3. Critique results of brainstorm using one or two go-rounds.
4. Summarize results.
5. Wrap up.

Reported benefits of running formal meetings using rules include:

- Better communication, which can result in shared historical information;
- Better communication, which can result in less duplication of future efforts;
- Improved "buy-in" because everyone feels that they have been heard; and
- Increased chance of actually accomplishing meeting objectives.

These benefits often outweigh the awkwardness and effort associated with running a formal meeting.

3.7 Significant Figures

The subject of "**significant figures**" relates to defining what is mean by specific numbers. The topic can relate to specifying project goals but is relevant in perhaps all situations combining business and technical issues. This section includes one convention for the interpretation and documentation of numbers. This convention is associated with a method for deriving the uncertainty of the results of calculations. The interpretation of numbers can be important in any phase of a technical project and in many other situations. In general, there are at least three ways to document uncertainty: (1) by implication, (2) with explicit ranges written either using "$\pm$" or (low, high), or (3) using a distribution function and probability theory as described in ▸ Chap. 10. This section focuses on the former two documentation methods.

The term "**significant figures**" refers to the number of digits in a written number that can be trusted by implication. Factors that can reduce trust include the possibility of round-off errors and any explicit expression of uncertainty. Unless specified otherwise, all digits in a written number are considered significant. Also, whole numbers generally have an infinite number of significant figures unless uncertainty is expressed explicitly. The "**digit location**" of a number is the power of 10 that would generate a 1 digit in the right-most significant digit.

Example 3.7.1 Significant Figures and Digit Location

Question

Consider the two written numbers 2.38 and 50.21 $\pm$ 10.0. What are the associated significant figures and digit locations?

Answer

The significant figures of 2.38 are 3. The digit location of 2.38 is –2 since $10^{-2} = 0.01$. The number of significant figures of 50.21 $\pm$ 1.0 is 1 since the first digit in front of the decimal cannot be trusted. If it were $\pm$ 0.49, then the digit could be trusted. The digit location of 50.21 $\pm$ 10.0 is 1 because $10^1 = 10$.

In the convention here, the phrase "**implied uncertainty**" of the number x is 0.5 $\times$ $10^{\text{digit location}(x)}$. This definition was also used in Lee et al. (2000). Those authors explored other conventions not included here. The following method is proposed here to calculate the implied uncertainty of the result of calculations. In our notation, $x_1, \ldots, x_n$ are the input numbers with implied or explicit uncertainties known and the result of the calculation, y. The goal is to derive both a number for y and for its implied uncertainties.

Algorithm 3.5. Formal derivation of significant figures

Step 1. Develop ranges (low, high) for all inputs $x_1, \ldots, x_n$ using explicit uncertainty if available or implied uncertainty if not otherwise specified.

Step 2. Perform calculations using all 2^n combinations of range values.

Step 3. The ranges associated with the output number are the highest and lowest numbers derived in *Step 2*.

Step 4. Write the product either using all available digits together with the explicit range or including only significant digits.

If only significant digits are reported, then rounding should be used in the formal derivation of significant figures method. Also, it is generally reasonable to apply some degree of rounding in reporting the explicit ranges. Therefore, the most explicit, correct representation is in terms of a range such as (12.03, 12.13) or 12.07 $\pm$ 0.05. Still, 12.1 is also acceptable, with the uncertainty being implied.

Example 3.7.2 Significant Figures of Sums and Products

For each question, use the steps outlined above.

Sum Question
$y = 2.51 + (10.2 \pm 0.5)$. What is the explicit uncertainty of y?

Sum Answer
In *Step 1*, the range for x_1 is (2.505, 2.515) and for x_2 is (9.7, 10.7).
In *Step 2*, the $2^2 = 4$ sums are: $2.505 + 9.7 = 12.205$, $2.505 + 10.7 = 13.205$, $2.515 + 9.7 = 12.215$, and $2.515 + 10.7 = 13.215$. The ranges in *Step 3* are (12.205, 13.215). Therefore, the sum can be written 12.71 with range (12.2, 13.2) with rounding. This can also be written 12.71 ± 0.5.

Product Question
$y = 2.51 \times (10.2 \pm 0.5)$. What is the explicit uncertainty of y?

Product Answer
In *Step 1*, the range for x_1 is (2.505, 2.515) and for x_2 is (9.7, 10.7). In *Step 2*, the $2^2 = 4$ products are: $2.505 \times 9.7 = 24.2985$, $2.505 \times 10.7 = 26.8035$, $2.515 \times$ 9.ds7 $=$ 24.3955, and $2.515 \times 10.7 = 26.9105$. The ranges in *Step 3* are (24.2985, 26.9105). Therefore, the product can be written 25.602 with uncertainty range (24.3, 26.9) with rounding. This could be written 25.602 ± 1.3.

Whole Number Question
$y = 4$ people $\times$ 2 (jobs/person). What is the explicit uncertainty of y?

Whole Number Answer
In *Step 1*, the range for x_1 is (4, 4) and for x_2 is (2, 2) since we are dealing with whole numbers. In *Step 2*, the $2^2 = 4$ products are: $4 \times 2 = 8$, $4 \times 2 = 8$, $4 \times 2 = 8$, and $4 \times$

Table 3.5 Calculation for the formal derivation of significant figures example

x_1	x_2	x_3	$x_1 \times \exp(x_2 \times x_3)$
2.45	5.15	2.05	94238.9
2.55	5.15	2.05	98085.4
2.45	5.25	2.05	115680.6
2.55	5.25	2.05	120402.2
2.45	5.15	2.15	157721.8
2.55	5.15	2.15	164159.4
2.45	5.25	2.15	195553.3
2.55	5.25	2.15	203535.0
2.45	5.25	2.15	195553.3

2 = 8. The ranges in *Step 3* are (8, 8). Therefore, the product can be written as 8 jobs with uncertainty range (8, 8). This could be written 8 ± 0.000.

Note that in multiplication or product situations, the uncertainty range does not usually split evenly on either side of the quoted result. Then, the notation (–, +) can be used. One attractive feature of the "Formal Derivation of Significant Figures" method proposed here is that it can be used in cases in which the operations are not arithmetic in nature, which is the purpose of the next example.

Example 3.7.3 Significant Figures of "General" Cases

Sum Question

$y = 2.5 \times exp(5.2 \times 2.1)$. What is the explicit uncertainty of y?

Sum Answer

In *Step 1*, the range for x_1 is (2.45, 2.55), for x_2 is (5.15, 5.25), and for x_3 is (2.05, 2.15). In *Step 2*, the $2^3 = 8$ results are: $2.45 \times exp(5.15 \times 2.05) = 94{,}238.9, \ldots, 2.55 \times exp(5.25 \times 2.15) = 203{,}535.0$ (see Table 3.5). The ranges in *Step 3* are (94238.9, 203535.0). Therefore, the result can be written 132,649.9 with range (94,238.9, 203,535.0) with rounding. This can also be written 132,649.9 (–38,411.0, +70,885.1) or 148,886.95 ± 54,648.05.

In the general cases, it is probably most helpful and explicit to give the calculated value ignoring uncertainty followed by (+, –) to generate a range, e.g., 132,649.9 (–38,411.0, +70,885.1). Quoting the middle number in the range followed by "±" is also acceptable and is relatively concise, e.g., 148,886.95 ± 54,648.05.

In some cases, it is not necessary to calculate all 2^n products, since it is predictable which combinations will give the minimum and maximum in *Step 3*. For example, in all of the above examples, it could be deduced that the first combination would give the lowest number and the last would give the highest number. The rigorous proof of these facts is the subject of an advanced problem at the end of this chapter.

The formal derivation of significant figures method proposed here does *not* constitute a world standard. Mullis and Lee (1998) and Lee et al. (2000) propose a coherent convention for addition, subtraction, multiplication, and division operations. The desirable properties of the method in this book are: (1) it is relatively simple conceptually, (2) it is applicable to all types of calculations, and (3) it gives sensible results in some problems that certain methods in other books do not. One limitation of the method proposed here is that it might be viewed as exaggerating the uncertainty, since only the extreme lows and highs are reported. Statistical tolerancing based on Monte Carlo simulation described in Parts II and III of this book generally provides the most realistic and relevant information possible. Statistical tolerancing can also be applied to all types of numerical calculations.

Finally, many college students have ignored issues relating to the implied uncertainty of numbers in prior course experiences. Perhaps the main point to remember is that in business or research situations where thousands of dollars hang in the balance, it is generally advisable to account for uncertainties in decision-making. In the remainder of this book, the formal derivation of significant figures method is not always applied. However, there is a consistent effort to write numbers in a way that approximately indicates their implied uncertainty. For example, 4.521 will not be written when what is meant is 4.5 ± 0.5.

3.8 Chapter Summary

This chapter describes the goals of the define phase of a six sigma project. Possible goals include identifying subsystems with associated key output variables and target objectives for those variables. Also, it is suggested that project **charters** can constitute the documented conclusion of a define phase; possible contents of these charters are described.

Next, both general strategies and specific problem-solving methods are described, together with their possible roles in the development of project charters. Specifically, the theory of constraints (TOC) and worst-case analysis strategies are described and argued to be relevant in the identification of **bottleneck** subsystems and in setting targets for KOVs. Pareto charting and formal meeting rule methods are described and related to the selection of KIVs and KOVs.

Finally, a method for deriving and reporting the **significant figures** and related uncertainty associated with the results of calculations is proposed. The purpose of this method is to assure that reported quantitative results are expressed with the appropriate level of uncertainty.

Example 3.8.1 Defining Bottlenecks in Cellular Relay Towers

Question

A cellular relay tower manufacturer has a large order for model #1. The company is considering spending $2.5M to double capacity to a reworked line or, alternatively, investing in a project to reduce the fraction nonconforming of the machine line feeding into the reworked line. Currently, 30% of units are nonconforming and need to be reworked. Recommend a project scope, including the key intermediate variable(s).

Answer

The bottleneck is clearly not in sales, since a large order is in hand. The rework capacity is a bottleneck. It is implied that the only way to increase that capacity is

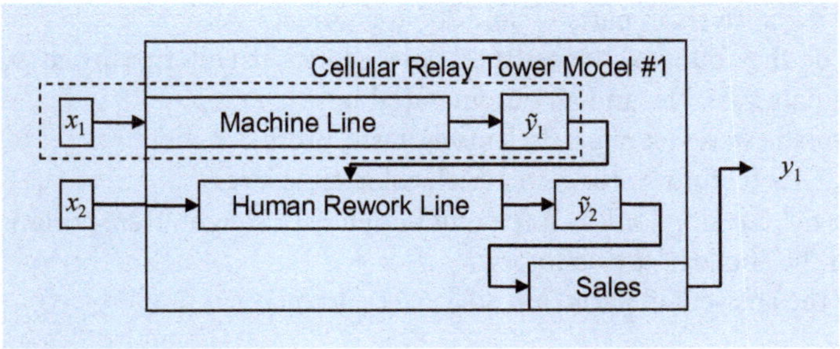

Fig. 3.6 Cellular relay tower manufacturing system

through expending \$2.5M, which the company would like to avoid. Therefore, the manufacturing line is the relevant bottleneck subsystem, with the key intermediate variable being the fraction nonconforming going into rework, $\tilde{y}_1$, in Fig. 3.6. Reducing this fraction to 15% or less should be roughly equivalent to doubling the rework capacity.

Example 3.8.2 Cellular Relay Tower Bottlenecks Continued

Question
Suppose the team would like to put more specific information about subsystem KIVs and KOVs into the project charter. Assume that much of the information about KIVs is known only by hourly workers on the factory floor. How could Pareto charts and formal meeting rules aid in collecting the desired information?

Answer
Using formal meeting rules could be useful in facilitating communication between engineers and line workers for eliciting the needed KIV information. Otherwise, communication might be difficult because of the different backgrounds and experiences of the two groups. Pareto charting could aid mainly through prioritizing the specific KOVs or causes associated with the nonconforming units.

3.9 Problems

In general, provide the correct and most complete answer.

1. According to the text, which of the following is true of six sigma projects?
   a. Projects often proceed to the measure phase with no predicted savings.
   b. Before the define phase ends, the project's participants are agreed upon.
   c. Project goals and target completion dates are generally part of project charters.
   d. All of the above are true.
   e. Only the answers in parts "b" and "c" are correct.
2. According to the text, which of the following is true of subsystems?
   a. They cannot share an input and an output with a major system.
   b. They are contained systems within a larger system.
   c. The subsystem concept is not relevant to the derivation of project charters.
   d. All of the above are true.
   e. Only the answers in parts "a" and "b" are correct.
3. Which of the following constitutes an ordered list of two input variables, an intermediate variable, and an output variable?
   a. Lemonade stand (% sugar, % lemons, taste, profit).
   b. Shoe sales (comfort rating, material, color, shoe sizes).
   c. Sandwich making for lunch (peanut butter, jelly, weight, transportation cost).
   d. All of the above fit the definition.
   e. Only the answers in parts "a" and "c" are correct.

4. Which of the following constitutes an ordered list of two input variables, two intermediate variables, and an output variable?
   a. Lemonade stand (% sugar, % lemons, taste rating, material cost, total profit).
   b. Chair manufacturing (wood type, saw type, stylistic appeal, waste, profit).
   c. Chip manufacturing (time in acid, % silicon, % dopant, % acceptable, profit).
   d. All of the above fit the definition.
   e. Only the answers in parts "a" and "b" are correct.
5. A potential scope for the sales subsystem for a lemonade stand is:
   a. Improve the taste of a different type of lemonade by adjusting the recipe.
   b. Increase profit through reducing raw optimizing over the price.
   c. Reduce "cycle time" between purchase of materials and final product delivery.
   d. All of the above fit the definition as used in the text.
   e. Only the answers in parts "b" and "c" are scope objectives.
6. Which constitute relevant tangible deliverables from a taste improvement project?
   a. A gallon of better-tasting lemonade.
   b. Documentation giving the improved recipe.
   c. An equation predicting the taste rating as a function of ingredients.
   d. All of the above are tangible deliverables.
   e. Only the answers in parts "a" and "b" are tangible deliverables.
7. Which of the following are possible deliverables from a wood process project?
   a. Ten finished chairs.
   b. Posters comparing relevant competitor chairs.
   c. Settings that minimize the amount of wasted wood.
   d. All of the above are tangible deliverables.
   e. Only the answers in parts "a" and "c" are tangible deliverables.
8. A new management demand—reducing paper consumption—is placed on an improvement team, in addition to improving report quality. This demand constitutes:
   a. An added KOV to focus on and improve quality values.
   b. Scope creep.
   c. Loss of "buy in" by the team.
   d. All of the above are relevant.
   e. Only the answers in parts "a" and "b" are relevant.
9. At a major retailer, a new accounting system is resisted even before it is tested. This would likely be caused by:
   a. The "Not-Invented-Here Syndrome".
   b. Scope creep restricting the team to work on the original charter.
   c. Information from intermediate variables supporting adoption.
   d. All of the above are possible causes.
   e. Only the answers in parts "a" and "b" are possible causes.
10. Which are symptoms of the "Not-Invented-Here Syndrome"?
   a. Acceptance of input from new coworkers.
   b. Rejection of input from new coworkers.
   c. Acceptance of recommendations developed by the people affected.
   d. Only the answers in parts "a" and "c" are correct.
   e. Only the answers in parts "b" and "c" are correct.

11. Write a charter for a project relevant to your life.
12. Why might using rework and scrap costs to evaluate the cost of nonconformities be inaccurate?
    a. Rework generally does not require expense.
    b. If there are many nonconforming units, some inevitably reach customers.
    c. Production defects increase lead times, resulting in lost sales.
    d. All of the above are possible reasons.
    e. Only the answers in parts "b" and "c" are correct.

The following paragraph is relevant for answering Questions 13–15.

Your team (two design engineers and one quality engineer, working for 4 months, each at 25% time) works to achieve $250k total savings over three different production lines (assuming a 2-year payback period). A new project requiring all three engineers is proposed for application on a fourth production line with similar issues to the ones previously addressed.

13. Assuming the same rate as for the preceding projects, the total number of person-years likely needed is approximately:
    a. 0.083.
    b. 0.075.
    c. 0.006.
    d. –0.050.
    e. 0.125.
14. According to the chapter, expected savings over 2 years is approximately:
    a. $83.3K.
    b. $166.6K.
    c. $66.6K.
    d. –$0.7K, and the project should not be undertaken.
15. According to the chapter, the expected profit over 2 years is approximately:
    a. $158.3K.
    b. $75K.
    c. $66.6K.
    d. –$16.7K, and the project should not be undertaken.
16. In three sentences or less, describe a system from your own life with a bottleneck.
17. Which statement is correct and most complete?
    a. Subsystems can be bottlenecks. KOVs can be outputs of subsystems.
    b. According to TOC, a large system usually has more than one bottleneck subsystem.
    c. Improving bottleneck systems almost always improves at least one total system KOV.
    d. Only the answers in parts "b" and "c" are correct.
    e. Only the answers in parts "a" and "c" are correct.
18. According to the text, which is true of the theory of constraints (TOC)?
    a. Workers on non-bottleneck subsystems have zero effect on the bottom line.
    b. Identifying bottleneck subsystems can help in selecting project KOVs.
    c. Intermediate variables cannot relate to total system profits.
    d. All of the above are true of the theory of constraints.
    e. Only the answers in parts "b" and "c" are correct.

19. Which is a categorical factor? (Give the correct and most complete answer.)
    a. Temperature used within an oven.
    b. The horizontal location of the logo on a web page.
    c. Type of tire used on a motorcycle.
    d. All of the above are categorical factors.
    e. All of the above are correct except (a) and (d).
20. Why are go-no-go decisions utilized?
    a. Eliminating design concepts early in a design process can save tooling costs.
    b. More design concepts exist than can be investigated, due to budget limitations.
    c. Decisive choices can be made, potentially related to multiple product lines.
    d. All of the above are possible uses.
    e. Only the answers in parts "b" and "c" are possible uses.

The following information will be used in Questions 21 and 22.

A hospital is losing business because of its reputation for long patient waits. It has similar emergency and non-emergency patient processing tracks, with most complaints coming from the emergency process. Patients in a hospital system generally spend the longest time waiting for lab test results in both tracks. Data entry, insurance, diagnosis, triage, and other activities are generally completed soon after the lab results become available.

21. According to TOC, which subsystem should in general be improved first?
    a. The data entry insurance subsystem for the nonemergency track.
    b. The lab testing subsystem.
    c. The subsystem controlling cost of the measurement systems used.
    d. Only the answers in parts "a" and "b" represent possible bottlenecks.
22. According to worst-case analysis, which subsystem should be addressed first?
    a. The slower testing subsystem for the emergency track.
    b. The insurance processing subsystem for the nonemergency track.
    c. The raw materials subsystem, because a medication's weight is the most significant factor in patient satisfaction.
    d. All of the above are possible worst-case-analysis decisions.
    e. Only the answers in parts "a" and "c" constitute a worst-case-analysis strategy.
23. An engineer might use a Pareto chart to uncover what type of information?
    a. Prioritization of nonconformity types identify the relevant subsystem.
    b. Pareto charts generally highlight the most recent problems discovered on the line.
    c. Pareto charting does not involve attribute data.
    d. All of the above are correct.
    e. Only the answers in parts "b" and "c" result from a Pareto chart.

◘ Figure 3.7 is helpful for answering Questions 24–26. It shows the hypothetical number of grades not in the "A" range by primary cause as assessed by a student.

24. Which statement or statements summarize the results of the Pareto analysis?
    a. The obvious interpretation is that laziness causes most grade problems.
    b. Avoiding courses with tough grading will likely not have much of an effect on her GPA.

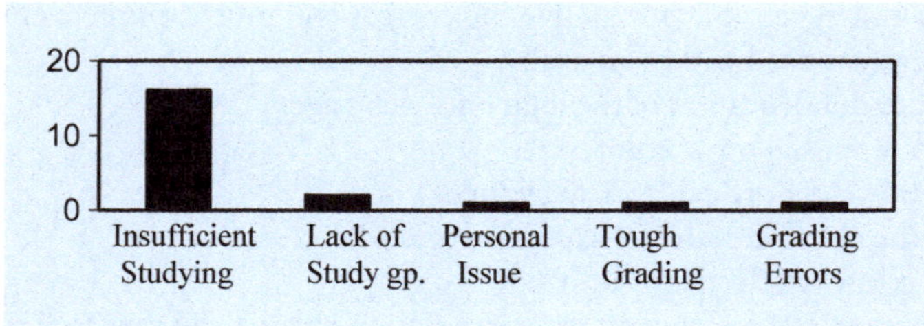

Fig. 3.7 Self-assessment of grades

   c. Personal issues with instructors' errors probably did not have much of an effect on her GPA.
   d. All of the above are supported by the analysis.

25. Which of the following are supported by the analysis?
   a. Student effort is probably not rewarded at the university.
   b. At least 80% of poor grades are explained by 20% of potential causes.
   c. The student's GPA is usually driven by tough grading.
   d. Study groups would likely never be useful in improving the student's GPA.
   e. None of the above is correct.
26. Which of the following are reasons why this analysis might be surprising?
   a. She was already sure that studying was the most important problem.
   b. She has little faith that studying hard will help.
   c. Her most vivid memory is a professor with a troubling grading policy.
   d. All of the above could be reasons.
   e. Only answers in parts (b) and (c) would explain the surprise.

The following hypothetical benchmarking data, in Tables 3.6 and 3.7, is helpful for answering Questions 27–30. Note that the tables, which are incomplete and in a non-standard format, refer to three student airplane manufacturing companies. The second table shows subjective customer ratings of products (1–10, with 10 being top quality) from the three companies.

Table 3.6 KIVs for three companies

Key input variable (KIV)	FlyRite	Hercules	Reliability
Scissor type	1	2	1
Body length (cm)	9.5	9.9	10.2
Wing length (cm)	2	4	2
Paper type (% glossy)	5.00	0.50	9.00
Arm angle at release (degrees)	15	0	10
Arm height (elbow to ground)	0.9	0.9	2
Paper thickness (mm)	2	2	2

Table 3.7 Customer issues for three companies

Customer issues	FlyRite	Hercules	Reliability
Folds have ugly rips (ripping)	3.00	7.33	2.00
Surface seems bumpy (crumpling)	5.00	5.00	3.33
Airplane flight time is short (flight time)	5.33	8.33	3.22
Airplane comes apart (flopping)	3.33	5.66	4.33
Airplane looks funny (aesthetics)	3.66	7.00	2.67

27. How many customer issues are analyzed?
    a. 3.
    b. 4.
    c. 5.
    d. 6.
28. How many controllable KIVs are considered?
    a. 4.
    b. 6.
    c. 7.
    d. 15.
29. Based on customer ratings, which company has "best in class" quality?
    a. FlyRite.
    b. Hercules.
    c. Reliability.
    d. Ripping.
    e. None dominates all others.
30. At FlyRite, which KIVs seem the most promising inputs for further study (focusing on emulation of best in class practices)?
    a. Scissor type, wing length, and arm release angle.
    b. Scissor type, wing length, and paper thickness.
    c. Paper thickness only.
    d. Aesthetics and crumpling.
    e. Scissor type and aesthetics.
31. Formal meeting rules in agreement with those from the text include:
    a. Facilitators should not enforce the agenda.
    b. Each participant receives three minutes to speak at the start of the meeting.
    c. No one shall speak without possession of the conch shell.
    d. All of the above are potential meeting rules.
    e. All of the above are correct except (a) and (d).
32. In three sentences, describe a scenario in which Pareto charting could aid in making ethical judgements.
33. In three sentences, describe a scenario in which benchmarking could aid in making ethical judgements.

34. How many significant figures are in the number 2.534?
    a. 3.
    b. 4.
    c. 5.
    d. 6.
35. What is the digit location of 2.534?
    a. 1.
    b. 2.
    c. −1.
    d. −2.
    e. −3.
36. What is the implied uncertainty of 2.534?
    a. 0.5.
    b. 5.
    c. 0.05.
    d. 0.005.
    e. 0.5×10^{-3}.
37. What is the explicit uncertainty of $4.2 + 2.534$ (permitting accurate rounding)?
    a. 6.734.
    b. 6.734 ± 0.1.
    c. 6.734 ± 0.0051.
    d. 6.734 ± 0.0505.
    e. 6.734 ± 0.0055.
38. What is the explicit uncertainty of $4.35 + 2.541$ (permitting accurate rounding)?
    a. 6.891.
    b. 6.891 ± 0.0055.
    c. 6.891 ± 0.0060.
    d. 6.891 ± 0.01.
    e. 6.890 ± 0.0060.
39. What is the explicit uncertainty of 4.2×2.534 (permitting accurate rounding)?
    a. 10.60 ± 0.129.
    b. 10.60 ± 0.10.
    c. 10.643 ± 0.129.
    d. 10.65 ± 0.10.
    e. 10.643 ± 0.100.
40. What is the explicit uncertainty of 4.35×2.541 (permitting accurate rounding)?
    a. 11.05 ± 0.01.
    b. 11.053 ± 0.01.
    c. 11.053 ± 0.10.
    d. 11.054 ± 0.015.
    e. 11.053.
41. What is the explicit uncertainty of $y = 5.4 \times \exp(4.2 - 1.3)$ (permitting accurate rounding)?
    a. 98.14 ± 0.015.
    b. 98.14 (−10.17, +10.33).
    c. 98.15 (−10.16, +10.33).

d. 98.14 (–10.16, +11.33).
e. 98.15 (–10.17, +11.33).

42. What is the explicit uncertainty of $y = 50.4 \times \exp(2.2 - 1.3)$ (permitting accurate rounding)?
    a. 123.92 ± 11.
    b. 123.96 (–11.9, +13.17).
    c. 123.92 (–11.9, +13.17).
    d. 123.96 (–9.9, +13.17).
    e. 123.92 (–9.9, +13.17).

References

Clausing D (1994) Total quality development: a step-by-step guide to world-class concurrent engineering. ASME Press, New York

Goldratt EM, Cox J (2016) The goal: a process of ongoing improvement. Routledge

Lee W, Mulliss C, Chiu HC (2000) On the Standard rounding rule for addition and subtraction. Chin J Phys 38:36–41

Martin P, Oddo F, Tate K (1997) The project management memory jogger: a pocket guide for project teams. Goal/Qpc, Salem, NH

Mullis C, Lee W (1998) On the standard rounding rule for multiplication and division. Chin J Phys 36:479–487

Robert SC, Robert HM III, Robert GMH (2000) Robert's rules of order, 10th edn. In: Robert HM III, Evans WJ, Honemann DH, Balch TJ (eds). Perseus Publishing, Cambridge, MA

Streibel BJ (2002) The manager's guide to effective meetings. McGraw-Hill, New York

Measure Phase and Statistical Charting

T. T. Allen, *Introduction to Engineering Statistics and Lean Six Sigma*,
https://doi.org/10.1007/978-1-4471-7420-2_4

4.1 Introduction

In ► Chap. 2, it was suggested that projects are useful for developing recommendations to change system key input variable (KIV) settings. The measure phase in six sigma for improvement projects quantitatively evaluates the current or default system KIVs, using thorough measurements of key output variables (KOVs) before changes are made. This information aids in evaluating effects of project-related changes and assuring that the project team is not harming the system. In general, quantitative evaluation of performance and improvement is critical for the acceptance of project recommendations. The more data, the less disagreement.

Before evaluating the system directly, it is often helpful to evaluate the equipment or methods used for measurements. The term "**measurement systems**" refers to the methods for deriving KOV numbers from a system, which could be anything from simple machines used by an untrained operator to complicated accounting approaches applied by teams of highly trained experts. The terms "**gauge**" and "**gage**," alternate spellings of the same word, referred historically to physical equipment for certain types of measurements. However, here gauge and measurement systems are used synonymously, and these concepts can be relevant for such diverse applications as measuring profits on financial statements and visually inspecting weld quality.

Measurement systems generally have several types of errors that can be evaluated and reduced. The phrase "**gauge repeatability and reproducibility**" (R&R) methods refers to a set of methods for evaluating measurement systems. This chapter describes several gauge R&R related methods with examples.

Thorough evaluation of system inputs generally begins after the acceptable measurement systems have been identified. Evaluation of systems must include sufficient measurements at each time interval to provide an accurate picture of performance in that interval. The evaluation must also involve a study of the system over sufficient time intervals to ensure that performance does not change greatly over time. The phrase "**statistical process control**" (SPC) charting refers to a set of charting methods offering thorough, visual evaluation of system performance and other benefits described later in this chapter.

The primary purpose of the measure phase for design projects is to systematically evaluate customers' needs. Therefore, it is helpful to study similar systems using gauge R&R and control charts. In addition, the measure phase can also include techniques described in the context of other phases that focus attention on customer needs, such as cause and effects matrix methods.

This chapter begins with a description of gauge R&R methods. Next, several SPC charting methods and associated concepts are described, including p charting, u charting, and Xbar & R charting.

Example 4.1.1 Gauge R&R and SPC Charting

Question

Describe the relationship between gauge and SPC charting.

Answer
Gauge R&R evaluates measurement systems. These evaluations can aid in improving the accuracy of measurement systems. SPC charting uses measurement systems to evaluate other systems. If the measurement systems improve, SPC charting will likely give a more accurate picture of the other systems quality.

4.2 Evaluating Measurement Systems

In this book, the phrase "**standard values**" refers to a set of numbers known to characterize correctly system outputs with high confidence and negligible errors. Standard values effectively constitute the true measurements associated with manufactured units and are believed to be the true values within the explicit or implied uncertainty. For example, a set of units of varying lengths are made using an alloy of steel with low thermal expansion properties. The standard values are then the believed true length values in millimeters at room temperature, found at great expense at a national institute of standards. The phrase "**measurement errors**" refers to the differences between the output variable values derived from a measurement system and the standard values. This section describes several methods for characterizing the measurement errors of measurement systems.

In some cases, measurement errors of measurement systems will be found to be acceptable, and these systems can in turn be trusted for evaluating other systems. In other cases, measurement systems will not be deemed acceptable. The improvement of measurement systems is considered to be a technical, investment-related matter beyond the scope of this book. Once improvements in the measurement systems have been made, however, the methods here can be used to evaluate the progress. It is likely that the associated measurement systems will eventually become acceptable. Fortunately, many of the other methods in this book can still give trustworthy results even if measurement systems are not perfect.

4.2.1 Types of Gauge R&R Methods

In all gauge R&R problems, the entities to be inspected do not need to be manufactured parts. They can be as diverse as accounting-related data or service stations. As a convenient example, however, the following sections describe the methods in relation to inspecting manufactured units or parts. In this chapter, two types of gauge R&R methods are described in detail: "comparison with standards" and gauge R&R (crossed). The phrase "**destructive testing**" refers to the process of measuring units such that the units cannot be measured again. The phrase "**non-destructive evaluation**" (NDE) refers to all other types of testing.

By definition gauge R&R (crossed) requires multiple measurements of each unit with different measurement systems. Therefore, this method cannot be applied in cases involving destructive sampling. Similarly, the comparison with standards method requires the availability of a standard. ◘ Table 4.1 shows the recommended methods for different possible cases.

Table 4.1 Suggested measurement evaluation methods for four cases

Standard values	Measurement type	
	Non-destructive evaluation	Destructive testing
Available	Comparison with standards	Comparison with standards
Not available	Gauge R&R (crossed)	Not available

4

It is apparent from Table 4.1 that gauge R&R (crossed) should be used only when standard values are not available. To understand this, a few definitions may be helpful. First, "**repeatability error**" ($\varepsilon_{\text{repeatability}}$) refers to the difference between a given observation and the average a measurement system would obtain through many repetitions. Second, "**reproducibility error**" ($\varepsilon_{\text{reproducibility}}$) is the difference between an average obtained by a relevant measurement system and the average obtained by all other similar systems (perhaps involving multiple people or equipment of similar type). In general, we will call a specific measurement system an "**appraiser**" although it might not be a person. Here, an appraiser could be a consulting company, or a computer program, or anything else that assigns a number to a system output.

Third, the phrase "**systematic errors**" ($\varepsilon_{\text{systematic}}$) refers in this book to the difference between the average measured by all similar systems for a unit and that unit's standard value. Note that in this book, reproducibility is not considered a systematic error, although other books may present that interpretation. Writing the measurement error as $\varepsilon_{\text{measurement}}$, the following equation follows directly from these definitions:

$$\varepsilon_{\text{measurement}} = \varepsilon_{\text{repeatability}} + \varepsilon_{\text{reproducibility}} + \varepsilon_{\text{systematic}} \quad (4.1)$$

Without using standard values, it is logically impossible to evaluate the "systematic" errors, i.e., those errors intrinsically associated with a given type of measurement system. Since gauge (crossed) does not use standard values, it can be regarded as a second-choice method. However, it is also usable in cases in which standard values are not available.

Another method called "**gauge R&R (nested)**" is omitted here for the sake of brevity. Gauge R&R (nested) is relevant for situations in which the same units cannot be tested by multiple measurement systems, e.g., parts cannot be shipped to different testers. Gauge R&R (nested) cannot evaluate either systematic errors or the separate effects of repeatability and reproducibility errors. Therefore, it can be regarded as a "third choice" method. Information about gauge R&R (nested) is available in standard software packages such as Minitab® and in other references (Montgomery and Runger 1994).

4.2.2 Gauge R&R: Comparison with Standards

The "**comparison with standards**" method, listed in Algorithm 4.1, is proposed formally here for the first time. However, similar approaches have been used for many years all over the world. The following defined constants are used in the method:

1. n is the number of units with pre-inspected standard values available.
2. m is the number of appraisers that can be assigned to perform tests.
3. r is the current total number of runs at any given time in the method.

The phrase "**standard unit**" refers to any of the n units with standard values available. The phrase "**absolute error**" means the absolute value of the measurement errors for a given test run. As usual, the "**sample average**" ($Y_{average}$) of $Y_1, Y_2, \ldots, Y_r$ is $(Y_1 + Y_2 + \cdots + Y_r) \div r$. The "**sample standard deviation**" (s) is given by:

$$s = \sqrt{\frac{(Y_1 - Y_{average})^2 + (Y_2 - Y_{average})^2 + \cdots + (Y_r - Y_{average})^2}{r - 1}} \quad (4.2)$$

Clearly, if destructive testing is used, each of the n standard units can only appear in one combination in *Step 1*. Also, it is perhaps ideal that the appraisers should not know which units they are measuring in *Step 2*. However, hiding information is usually unnecessary, either because the appraisers have no incentive to distort the values or the people involved are too ethical or professional to change readings based on past values or other knowledge.

In the context of comparison with standards, the phrase "**measurement system capability**" is defined as $6.0 \times$ EEAE. In some cases, it may be necessary to tell apart reliably system outputs that have true differences in standard values greater than a user-specified value. Here we use "D" to refer to this user-specified value. In this situation, the term "**gauge capable**" refers to the condition that the measurement system capability is less than D, i.e., $6.0 \times \text{EEAE} < D$. In general, acceptability can be determined subjectively through inspection of the EEAE, which has the simple interpretation of being the error magnitude the measurement system user can expect to encounter.

Algorithm 4.1 Gauge R&R: comparison with standards

Step 1. Write a listing of 20 randomly selected combinations of standard unit and appraiser. Attempt to create combinations that show no pattern. ◻ Table 4.2 indicates a listing with $n = 5$ and $m = 3$. Leave space for the measured values and absolute errors.

Step 2. Appraisers perform the remaining measurements in the order indicated in ◻ Table 4.2. Write the measured values and the absolute values of the errors in the table.

Step 3. Calculate the sample average and sample standard deviation of all absolute errors tested so far. The "**estimated expected absolute errors**" (EEAE) and "**estimated errors of the expected absolute** errors" (EEEAE) are:

$$\begin{aligned} \text{EEAE} &= \text{(sample average)} \\ \text{EEEAE} &= \text{(sample standard deviation)} \div \sqrt{r} \end{aligned} \quad (4.3)$$

Step 4. If the EEAE $\div$ EEEAE > 5, then write my "expected absolute errors are" EEAE $\pm$ EEEAE and stop. Otherwise, add five randomly selected combinations to the table from *Step 1*. Increase r by 5 and go to *Step 2*.

Table 4.2 Random-looking listing of standard unit and appraiser combinations

Run	Standard unit	Appraiser	Measured value	Absolute error
1	5	1		
2	5	3		
3	2	1		
4	4	2		
⋮	⋮ ⋮			
17	1	2		
18	2	3		
19	5	3		
20	4	1		

4

Example 4.2.1 Capability of Home Scale

Question

Use two 12.0-pound dumbbells to determine whether the following measurement standard operating procedure (SOP) is "gauge capable" of telling apart differences of 5.0 pounds.

1. Put "Taylor Metro" scale (dial indicator model) on a flat surface.
2. Adjust dial to set reading to zero.
3. Place items on scale and record the weight, rounding to the nearest pound.

Answer

Assuming that the dumbbell manufacturer controls its products far better than the scale manufacturer, we have $n = 3$ standard values: 12.0 pounds, 12.0 pounds, and 24.0 pounds (when both weights are on the scale). With only one scale and associated SOP, we have $m = 1$.

Algorithm 4.2 Scale capability example

Step 1. Table 4.3 lists the 20 randomly selected combinations.

Step 2. Table 4.3 also shows the measured values. Note that the scale was picked up between each measurement to permit the entire measurement SOP to be evaluated.

Step 3. $\text{EEAE} = (2 + 0 + \cdots + 2) \div 20 = 1.55$

$$\text{EEEAE} = \sqrt{\frac{(2-1.55)^2 + (0-1.55)^2 + \cdots + (2-1.55)^2}{20-1}} \div \sqrt{20} = 0.135$$

Step 4. Since EEAE ÷ EEEAE » 5, one writes the expected absolute errors as 1.55 ± 0.135 pounds, and the method stops. Since $6.0 \times 1.55 = 9.3$ pounds is greater than 5.0 pounds, we say that the SOP is not gauge capable.

Table 4.3 Real measurements for home scale capability study

Run	Standard unit	Appraiser	Measured value	Absolute error
1	3	1	22	2
2	2	1	12	0
3	2	1	11	1
4	1	1	11	1
5	3	1	22	2
6	3	1	22	2
7	1	1	10	2
8	2	1	11	1
9	3	1	22	2
10	2	1	10	2
11	1	1	11	1
12	2	1	10	2
13	2	1	11	1
14	1	1	11	1
15	3	1	23	1
16	2	1	10	2
17	1	1	10	2
18	2	1	10	2
19	1	1	10	2
20	3	1	22	2

In the preceding example, significant figures are less critical than usual because the method itself provides an estimate of its own errors. Note also that failure to establish the capability of a measurement system does not automatically signal a need for more expensive equipment. For example, in the home scale case, the measurement SOP was not gauge capable, but simply changing the procedures in the SOP would likely create a capable measurement system. With one exception, all the measurements were below the standard values. Therefore, one could change the third step in the SOP to read "Place items on the scale, note the weight to the nearest pound, and record the noted weight plus 1 pounds."

In general, vagueness in the documentation of measurement SOPs contributes substantially to capability problems. Sometimes simply making SOPs more specific establishes capability. For example, the second step in the home scale SOP might read "Carefully lean over the scale, then adjust the dial to set reading to zero."

Finally, advanced readers will notice that the EEAE is equivalent to a Monte Carlo integration estimate of the expected absolute errors. These readers might also apply pseudo random numbers in *Step 1* of the method. Related material is covered in ▶ Chap. 10 and in Part II of this book. This knowledge is not needed, however, for competent application of the comparison with standards method.

4.2.3 Gauge R&R (Crossed) with Xbar & R Analysis

In general, both gauge R&R (crossed) and gauge R&R (nested) are associated with two alternative analysis methods: Xbar & R and analysis of variance (ANOVA) analysis. The experimentation steps in both methods are the same. Many students find Xbar & R methods intuitively simpler, yet ANOVA methods can offer important advantages in accuracy. For example, if the entire procedure were to be repeated, the numerical outputs from the ANOVA process would likely be more similar than those from Xbar & R methods. Here, we focus somewhat arbitrarily on the Xbar & R analysis methods. One benefit is that the proposed method derives from the influential Automotive Industry Task Force (AIAG) Report (1994) and can therefore be regarded as the industry-standard method.

Algorithm 4.3 Gauge R&R (crossed) with Xbar & R Analysis

Step 1a. Create a listing of all combinations of n units or system measurements and m appraisers. Repeat this list r times, labeling the repeated trials 1, …, r.

Step 1b. Randomly reorder the list and leave one column blank to record the data. ◘ Table 4.4 illustrates the results from *Step 1* with $n=5$ parts or units, $m=2$ appraisers, and $r=3$ trials.

Step 2. Appraisers perform the measurements that have not already been performed in the order indicated in the table from *Step 1*. This data is referred to using the notation $Y_{i,j,k}$ where i refers to the unit, j refers to the appraiser involved, and k refers to the trial. For example, the measurement from Run 1 in ◘ Table 4.4 is referred to as $Y_{3,2,1}$.

Step 3. Calculate the following (i is for the part, j is for the appraiser, k is for the trial, n is the number of parts, m is the number of appraisers, r is the number of trials):

$$
\begin{aligned}
&Y_{average,i,j} = r^{-1} \sum_{k=1,\ldots,r} Y_{i,j,k} \text{ and} \\
&Y_{range,i,j} = \text{Max}\left[Y_{i,j,1}, \ldots, Y_{i,j,r}\right] - \text{Min}\left[Y_{i,j,1}, \ldots, Y_{i,j,r}\right] \\
&\qquad \text{for } i = 1, \ldots, n \text{ and } j = 1, \ldots, m, \\
&Y_{average\,parts,i} = (m)^{-1} \sum_{j=1,\ldots,m} Y_{average,i,j} \text{ for } i = 1, \ldots, n \\
&Y_{inspector\,average,j} = (n)^{-1} \sum_{i=1,\ldots,n} Y_{average,i,j} \text{ for } j = 1, \ldots, m \\
&Y_{average\,range} = (mn)^{-1} \sum_{i=1,\ldots,n} \sum_{j=1,\ldots,m} Y_{range,i,j} \\
&Y_{range\,parts} = \text{Max}\left[Y_{average\,parts,1}, \ldots, Y_{average\,parts,n}\right] \\
&\qquad - \text{Min}\left[Y_{average\,parts,1}, \ldots, Y_{average\,parts,n}\right] \\
&Y_{range\,inspect} = \text{Max}\left[Y_{inspector\,average,1}, \ldots, Y_{inspector\,average,m}\right] \\
&\qquad - \text{Min}\left[Y_{inspector\,average,1}, \ldots, Y_{inspector\,average,m}\right]
\end{aligned}
\tag{4.4}
$$

Repeatability $= K_1 Y_{average\ range}$
Reproducibility $= sqrt\{\mathrm{Max}[(K_2\, Y_{range\ inspect})^2 - (nr)^{-1}\,\mathrm{Repeatability}^2, 0]\}$
R&R $= sqrt[\mathrm{Repeatability}^2 + \mathrm{Reproducibility}^2]$
Part $= K_3 Y_{range\ parts}$
Total $= sqrt[\mathrm{R\&R}^2 + \mathrm{Part}^2]$
%R&R $= (100 \times \mathrm{R\&R}) \div \mathrm{Total}$
where "*sqrt*" means square root and $K_1 = 4.56$ for $r = 2$ trials and 3.05 for $r = 3$ trials, $K_2 = 3.65$ for $m = 2$ machines or inspectors and 2.70 for $m = 3$ machines or human appraisers, and $K_3 = 3.65, 2.70, 2.30, 2.08, 1.93, 1.82, 1.74, 1.67, 1.62$ for $n = 2, 3, 4, 5, 6, 7, 8, 9$, and 10 parts respectively.

Step 4. If %R&R < 10, then one declares that the measurement system is "gauge capable," and measurement error can generally be neglected. Depending upon problem needs, one may declare the process to be marginally gauge capable if $10 \le$ %R&R < 30. Otherwise, more money and time should be invested to improve the inspection quality.

The crossed method involves collecting data from all combinations of n units, m appraisers, and r trials each. The method is only defined here for the cases satisfying: $2 \le n \le 10$, $2 \le m \le 3$, and $2 \le r \le 3$. In general it is desirable that the total number of evaluations is greater than 20, i.e., $n \times m \times r \ge 20$. As for comparison with standards methods, it is perhaps ideal that appraisers do not know which units they are measuring.

Note that the gauge R&R (crossed) methods with Xbar & R analysis methods can conceivably generate undefined reproducibility values in *Step 3*. If this happens, it is often reasonable to insert a zero value for the reproducibility.

In general, the relevance of the %R&R strongly depends on the degree to which the units' unknown standard values differ. If the units are extremely similar, no inspection equipment at any cost could possibly be gauge capable. As with comparison with standards, it might only be of interest to tell apart reliably units that have true differences greater than a given number, D. This D value may be much larger than the unknown standard value differences of the parts that happened to be used in the gauge study. Therefore, an alternative criterion is proposed here. In this *nonstandard* criterion, a system is gauge capable if $6.0 \times \mathrm{R\&R} < D$.

Example 4.2.2 Standard Definition of Capability

Question

Suppose R&R = 32.0 and Part = 89.0. Calculate and interpret the %R&R.

Answer

%R&R $= (100\% \times \mathrm{R\&R}) \div sqrt[\mathrm{R\&R}^2 + \mathrm{Part}^2] = 33.8\%$. Using standard conventions, the gauge is not capable even with lenient standards. However, the measurement system might be acceptable if the parts in the study are much more similar than parts of future interest. Specifically, if the person only needs to tell the difference between parts with true differences greater than $6.0 \times 32.0 = 192$ units, the measurement system being studied is likely acceptable.

4

Table 4.4 Example gauge R&R (crossed) results for (a) Step 1a and (b) Step 1b

(a)			(b)				$Y_{i,j,k}$
Unit	Appraiser	Trial	Run	Unit (*i*)	Appraiser (*j*)	Trial (*k*)	
1	1	1	1	3	2	1	
2	1	1	2	2	1	1	
3	1	1	3	3	1	2	
4	1	1	4	5	1	2	
5	1	1	5	2	2	1	
1	2	1	6	2	2	2	
2	2	1	7	4	2	2	
3	2	1	8	5	2	2	
4	2	1	9	3	1	1	
5	2	1	10	2	1	3	
1	1	2	11	1	1	1	
2	1	2	12	5	1	1	
3	1	2	13	3	2	3	
4	1	2	14	5	2	3	
5	1	2	15	4	2	1	
1	2	2	16	1	1	2	
2	2	2	17	1	2	2	
3	2	2	18	4	1	2	
4	2	2	19	1	2	1	
5	2	2	20	4	1	1	
1	1	3	21	1	1	3	
2	1	3	22	5	1	3	
3	1	3	23	3	2	2	
4	1	3	24	4	2	3	
5	1	3	25	5	2	1	
1	2	3	26	2	1	2	
2	2	3	27	3	1	3	
3	2	3	28	4	1	3	
4	2	3	29	3	2	1	
5	2	3	30	2	1	1	

Table 4.5 Hypothetical undercut data for gauge study (superscripts show run order)

Software #1	Part				
	1	2	3	4	5
Trial 1	0.94^{2}	1.05^{11}	1.03^{13}	1.01^{5}	0.88^{15}
Trial 2	0.94^{7}	1.05^{8}	1.02^{20}	1.04^{18}	0.86^{27}
Trial 3	0.97^{10}	1.04^{19}	1.05^{22}	1.00^{24}	0.88^{30}
Software #2	Part				
	1	2	3	4	5
Trial 1	0.90^{6}	1.03^{1}	1.03^{12}	1.02^{9}	0.87^{3}
Trial 2	0.94^{7}	1.04^{14}	1.05^{23}	1.01^{17}	0.88^{16}
Trial 3	0.97^{10}	1.01^{26}	1.06^{28}	0.98^{29}	0.87^{25}

The following example shows a way to reorganize the data from *Step 2* that can make the calculations in *Step 3* easier to interpret and to perform correctly. It is probably desirable, however, to create a table as indicated in *Step 1* first before reorganizing. This follows because otherwise one might be tempted (1) to perform the tests in a nonrandom order, or (2) to change the results for the later trials based on past trials.

Example 4.2.3 Gauge R&R (Crossed) Arc Welding Example

Question

A weld engineer is analyzing the ability of two computer software programs to measure consistently the undercut weld cross sections. She performs *Steps 1* and *2* of the gauge R&R method and generates the data in Table 4.5. Complete the analysis.

Answer

Table 4.6 shows the calculations for *Step 3*. The process is marginally capable, i.e., %R&R = 25%. This might be acceptable depending upon the goals.

Table 4.6 Weld example gauge R&R data and calculations

	Part							
Inspector 1	1	2	3	4	5			
Trial 1	0.94	1.05	1.03	1.01	0.88		$Y_{average\ range}$	0.026
Trial 2	0.94	1.05	1.02	1.04	0.86		$Y_{range\ parts}$	0.167
Trial 3	0.97	1.04	1.05	1.00	0.88	$Y_{inspector\ average,1}$	$Y_{range\ inspect}$	0.012
$Y_{average,i,1}$	0.950	1.047	1.033	1.017	0.873	0.984	Repeatability	0.079
$Y_{range,i,1}$	0.030	0.010	0.030	0.040	0.020		Reproducibility	0.039
Inspector 2	1	2	3	4	5		R&R	0.088
Trial 1	0.90	1.03	1.03	1.02	0.87		Part	0.347
Trial 2	0.92	1.04	1.05	1.01	0.88		Total	0.358
Trial 3	0.91	1.01	1.06	0.98	0.87		%R&R	25%
$Y_{average,i,2}$	0.910	1.027	1.047	1.003	0.873	$Y_{inspector\ average,2}$		
$Y_{range,i,2}$	0.020	0.030	0.030	0.040	0.010	0.972		
$Y_{average\ parts,i}$	0.930	1.037	1.040	1.010	0.873			

4.3 Measuring Quality Using SPC Charting

In this section, several methods are described for using acceptable measurement systems to evaluate a system thoroughly. In addition to evaluating systems before a project changes inputs, these statistical process control (SPC) charting methods aid in the efficient monitoring of systems. If applied as intended, they increase the chance that skilled employees will only be tasked in ways that reward their efforts.

The section begins with a discussion of the concepts invented by Shewhart in the 1920s and refined years later by Deming and others. Next, four widely used charting procedures are described: *p*-charting, *u*-charting, demerit charting, and Xbar & R charting. Each method generates a measurement of the relevant process quality. These measures provide a benchmark for any later project phases to improve.

4.3.1 Concepts: Common Causes and Assignable Causes

In 1931, Shewhart formally proposed the Xbar & R charting method he invented while working at Bell Telephone Laboratories (see the re-published version in Shewhart 1980). Shewhart had been influenced by the mass production system that Henry Ford helped to create. In mass production, a small number of skilled laborers were mixed with thousands of other workers on a large number of assembly lines producing millions of products. Even with the advent of Toyota's lean production in the second half of the twentieth century and the increase of service sector jobs such as education, health care, and retail, many of the problems addressed by Shewhart's method are relevant in today's workplace.

Figure 4.1 illustrates Shewhart's view of production systems. On the left-hand side stands skilled labor such as technicians or engineers. These workers have responsibilities that blind them from the day-to-day realities of the production lines. They only see a sampling of possible output numbers generated from those lines, as indicated by the spread-out quality characteristic numbers flowing over the wall. Sometimes variation causes characteristic values go outside the specification limits, and units become nonconforming. Typically, most of the units conform to specifications. Therefore, skilled labor generally views variation as an "evil" or negative issue. Without it, one hundred percent of units would conform.

The phrase "**common cause variation**" refers to changes in the system outputs or quality characteristic values under usual circumstances. The phrase "**local authority**" refers to the people (not shown) working on the production lines and local skilled labor. Most of the variation in the characteristic numbers occurs because of the changing of factors that local authority cannot control. If the people and systems could control the factors and make all the quality characteristics constant, they would do so. Attempts to control the factors that produce common cause variation generally waste time and add variation. The term "**over-control**" refers to a foolish attempt to dampen common cause variation that actually increases it. Only a large, management-supported improvement project can reduce the magnitude of common cause variation.

On the other hand, sometimes unusual problems occur that skilled labor and local authority can fix or make less harmful. This is indicated in Fig. 4.1 by the gremlin on the right-hand side. If properly alerted, skilled labor can walk around the wall and scare away the gremlin. The phrase "**assignable cause**" refers to a change in the system inputs that can be reset or resolved by local authority. Examples of assignable causes include meddling engineers, training problems, unusually bad batches of materials

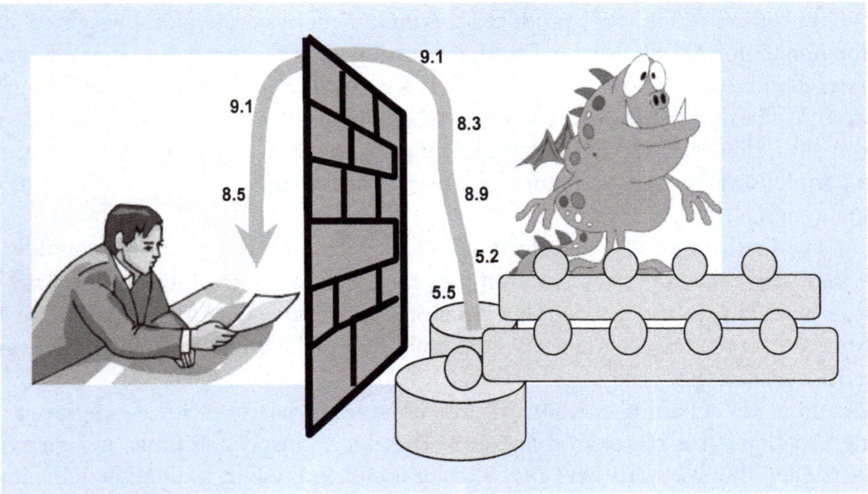

Fig. 4.1 Scarce resources, assignable causes, and data in statistical process control

from suppliers, end of financial quarters, and vacations. Using the vocabulary of common and assignable causes, it is easy to express the primary objectives of the statistical process control charts:

Evaluation of the magnitude of the common cause variation, providing a benchmark for quality improvement or design activities;

Monitoring and identification of assignable causes to alert local authority in a timely manner (something might be fixable); and

Discouraging local authority from **meddling** unless assignable causes are identified.

4

The third goal follows because local authority's efforts to reduce common cause variation are generally counterproductive.

Example 4.3.1 Theft in Retailing

Question

A retail executive is interested in benchmarking theft at five outlets prior to the implementation of new corporate anti-theft policies. List one possible source of common cause variation and assignable cause variation.

Possible Answer

Lone customers and employees stealing small items from the floor or warehouse contribute to common cause variation. A conspiracy of multiple employees systematically stealing might be terminated by local management.

4.4 Commonality: Rational Subgroups, Control Limits, and Startup

The next sections describe four charting procedures with the objectives from the last section. The charts differ in that each is based on different output variables. First, p-charting uses attribute data derived from a count of the number of nonconforming units in a subset of the units produced. Second, demerit charts plot a weighted sum of the nonconformities per item. Third, u-charting plots the number of nonconformities per item inspected; this method is described together with demerit charts. Fourth, Xbar & R charting creates two charts based on continuous quality characteristic values. A fifth, the relatively advanced "Hoteling's T" method, is described in ▶ Chap. 8 and permits simultaneous monitoring of several continuous quality characteristics on a single chart.

A "**rational subgroup**" is a selection of units from a large set, chosen carefully to represent fairly the larger set. An example of an "irrational subgroup" of the marbles in a jar would be the top five marbles. A rational subgroup would involve taking all the marbles out, spreading them evenly on a table, and picking one from the middle and one from each of the corners.

All four SPC charting methods in this book make use of rational subgroups. In some situations, it is reasonably easy and advisable to inspect all units, not merely a subset. Then, the methods here may still be useful for system evaluation and monitoring. ▶ Chapter 10 contains a theoretical discussion about how this situation called "**complete inspection**" or "100% inspection" changes the philosophical interpretation of the charts' properties.

According to the above definition, complete inspection necessarily involves rational subgroups because the complete set is representative of itself. In this chapter, the practical effects related to the hypersensitivity of charts in complete inspection situations are briefly discussed.

All charting methods in this book involve calculating an "**upper control limit**" (UCL), "**center line**" (CL), and a "**lower control limit**" (LCL). The control limits have no simple relationship to upper and lower specification limits. They relate to the goals of charting to identify assignable causes and preventing over-control of systems. It is conceivable that, on some control charts, all or none of the units involved could be nonconforming with respect to specifications.

Also, charting methods generally include a "**startup phase**" in which data is collected and the chart constants are calculated. Some authors base control charts on 30 startup or "trial" periods instead of the 25 used in this book. In general, whatever information beyond 25 that is available when the chart is being set up should be used, unless the data in question is not considered representative of the future system under usual circumstances.

In addition, all charting methods also include a "**steady state**" phase in which the limits are fixed and the chart mainly contributes through (1) identifying the occasional assignable cause and (2) discouraging people from changing the process input settings. When the charted quantities are outside the control limits, detective work begins to investigate whether something unusual and fixable is occurring. In some cases, production is shut down, awaiting detective work and resolution of any problems discovered.

In general, cases where there are many charted quantities outside the control limits often indicates that standard operating procedures (SOP) are either not in place or not being followed. Like SOPs, charts encourage consistency which only indirectly relates to producing outputs that conform to specifications. When a process is associated with charted quantities within the control limits, it is said to be "**in control**" even if it generates a continuous stream of nonconformities.

Example 4.4.1 Chart Selection for Monitoring Retail Theft

Question

Which charting procedure is most relevant for monitoring retail theft? Also, provide two examples of rational subgroups.

Possible Answer

A natural key output variable (KOV) is the amount of money or value of goods stolen. Since the amount is a single continuous variable, Xbar & R charting is the most relevant of the methods in this book. The usual inventory counts that are likely in place can be viewed as complete inspection with regard to property theft. Because inventory counts might not be gauge capable, it might make sense to institute random intense inspection of a subset of expensive items at the stores.

4.5 Attribute Data: *P*-Charting

The phrase "**go-no-go testing**" refers to evaluation of units to determine whether any of potentially several quality characteristics fail to conform to specifications. Go-no-go testing treats individual units or service applications much like go-no-go decision-making treats design concepts. If all characteristic values conform, the unit is a "go" and passes inspection. Otherwise, the unit is a "no-go" and the unit is reworked or scrapped.

The method of "***p*-charting**" involves plotting results from multiple go-no-go tests. Compared with the other charting methods that follow, *p*-charting generally requires the inspection of a much higher fraction of the total units to achieve system evaluation and monitoring goals effectively. Intuitively, this follows because go-no-go output values generally provide less information on a per-unit basis than counts of nonconformities or continuous quality characteristic values.

The quantity charted in *p*-charting is "*p*," which is the fraction of units nonconforming in a rational subgroup. Possible reasons for using *p*-charting instead of other methods in this book include:

Only go-no-go data is available because of inspection costs and preferences.

The charted quantity "*p*" is often easy to relate to rework and scrap costs.

The charted quantity "*p*" may conveniently summarize quality, taking into account multiple continuous quality characteristic inspections.

Three symbols used in documenting *p*-charting are:

1. n is the number of samples in each rational subgroup. If the sample size varies because of choice or necessity, it is written n_i, where i refers to the relevant sampling period. Then, $n_1 \neq n_2$ might occur and/or $n_1 \neq n_3$ and so on.
2. τ is the time interval between the start of inspecting one subgroup and the next.
3. p_0 is the true fraction of nonconforming units when only common cause variation is present. This is generally unknown before the procedure begins.

There are no universal standard rules for selecting n and τ. This selection is done in a pre-step before the method begins. Three considerations relevant to the selection follow. First, a rule of thumb is that n should satisfy $n \times p_0 > 5.0$ and $n \times (1 - p_0) > 5.0$. This may not be helpful, however, since p_0 is generally unknown before the method begins. Advanced readers will recognize that this is the approximate condition for p to be normally distributed. In general, this condition can be expected to improve the performance of the charts. In many relevant situations p_0 is less than 0.05 and, therefore, n should probably be greater than 100.

Second, τ should be short enough such that assignable causes can be identified and corrected before considerable financial losses occur. It is not uncommon for the charting procedure to require a period of $2 \times \tau$ before signaling that an assignable cause might be present. In general, larger n and smaller τ value shorten response times. If τ is too long, the slow discovery of problems will cause unacceptable pile-ups of nonconforming items and often trigger complaints.

Algorithm 4.4 *p*-Charting

Step 1. (Startup) Obtain the total fraction of nonconforming units or systems using 25 rational subgroups each of size n. This should require at least $25 \times \tau$ time. Tentatively, set p_0 equal to this fraction.

Step 2. (Startup) Calculate the "trial" control limits using

$$UCL_{trial} = p_0 + 3.0 \times \sqrt{\frac{p_0(1-p_0)}{n}},$$
$$CL_{trial} = p_0, \text{ and} \tag{4.5}$$
$$LCL_{trial} = \text{Maximum}\{p_0 - 3.0 \times \sqrt{\frac{p_0(1-p_0)}{n}}, 0.0\}$$

where "Maximum" means take the largest of the numbers separated by commas.

Step 3. (Startup) Identify all the periods for which p = fraction nonconforming in that period and $p < LCL_{trial}$ or $p > UCL_{trial}$. If the results from any of these periods are believed to be not representative of future system operations, e.g., because their assignable causes were fixed permanently, remove the data from the l not representative periods from consideration.

Step 4. (Startup) Calculate the total fraction nonconforming based on the remaining $25 - l$ periods and $(25 - l) \times n$ data and p_0 set equal to this number. The quantity p_0 is sometimes called the "process capability" in the context of *p*-charting. Calculate the revised limits using the same formulas as in *Step 2*:

$$UCL = p_0 + 3.0 \times \sqrt{\frac{p_0(1-p_0)}{n}},$$
$$CL = p_0, \text{ and}$$
$$LCL = \text{Maximum}\{p_0 - 3.0 \times \sqrt{\frac{p_0(1-p_0)}{n}}, 0.0\}.$$

Step 5. (Steady State) Plot the fraction nonconforming, p_j, for each period j together with the upper and lower control limits.

Third, unless otherwise specified, n is generally not large enough to represent complete inspection. One of the goals is to save sampling costs compared with complete inspection.

Note that the following method is written in terms of a constant sample size n. If n varies, then substitute n_i for n in all formulas. Then the control limits would vary subgroup to subgroup. Also, quantities next to each other in the formulas are implicitly multiplied, with the "×" omitted for brevity. The numbers 3.0 and 0.0 in the formulas are assumed to have infinite significant digits.

The resulting "*p*-chart" typically provides useful information to stakeholders (engineers, technicians, and operators) and builds intuition about the engineered system. An "**out-of-control signal**" is defined as a case in which the fraction nonconforming for a given time period, p, is either below the lower control limit ($p <$ LCL) or above the upper control limit ($p >$ UCL). From then on, technicians and engineers are

4

Table 4.7 (a) Startup period restaurant data and (b) data available after startup period

(a)				(b)	
Week	Sum not OK	Week	Sum not OK	Week	Sum not OK
1	8	14	10	1	8
2	7	15	9	2	8
3	10	16	5	3	11
4	6	17	8	4	2
5	8	18	9	5	7
6	9	19	11		
7	8	20	8		
8	11	21	9		
9	30	22	9		
10	25	23	10		
11	10	24	6		
12	9	25	9		
13	8				

discouraged from making system changes unless a signal occurs. If a signal does occur, they should investigate to see if something unusual and fixable is happening. If not, they call the signal a "**false alarm**" and again leave the system alone.

Note that applying the revised limits to the startup data could conceivably cause additional out-of-control signals to be identified. A reasonable alternative to the above method might involve investigating these new signals to see if the data is representative of future occurrences. All the control charts in this book involve the same ambiguity.

Example 4.5.1 Restaurant Customer Satisfaction

Question

An upscale restaurant chain's executive wants to start SPC charting to evaluate and monitor customer satisfaction. Every week, hosts or hostesses must record 200 answers to the question, "Is everything OK?" Table 4.7a lists the sum of hypothetical "everything is not OK" answers for 25 weeks at one location. Rare, noisy construction occurred in weeks 9 and 10. Set up the appropriate SPC chart.

Answer

It is implied that if a customer does not agree that everything is OK, then everything is not OK. Then, also, the restaurant party associated with the unsatisfied customer is effectively a nonconforming unit. Therefore, *p*-charting is relevant since the given data is effectively the count of nonconforming units. Also, the number to be inspected is constant so a fixed value of $n = 200$ is used in all calculations.

First, the trial limit calculations are

$$p_0 = \text{(total number nonconforming)} \div \text{(total number inspected)}$$
$$= 252/5000 = 0.050,$$
$$UCL_{trial} = 0.050 + 3.0 \times \text{sqrt}\,[(0.050) \times (1 - 0.050) \div 200] = 0.097,$$
$$CL_{trial} = 0.050, \text{ and}$$
$$LCL_{trial} = \text{Max}\,\{0.050 - 3.0 \times \text{sqrt}\,[(0.050) \times (1 - 0.050) \div 200], 0.0\} = 0.004.$$

Figure 4.2 shows a *p*-chart of the startup period at the restaurant location. Clearly, the *p* value in week 9 and 10 subgroups constitute out-of-control signals. These signals were likely caused by a rare assignable cause (construction) that makes the associated data not representative of future usual conditions. Therefore, the associated data are removed from consideration.

The revised limits are:

$$p_0 = \text{(total number nonconforming)} \div \text{(total number inspected)}$$
$$= 197/4600 = 0.043,$$
$$UCL = 0.043 + 3.0 \times \text{sqrt}\,[(0.043) \times (1 - 0.043) \div 200] = 0.086,$$
$$CL = 0.043, \text{ and}$$
$$LCL = \text{Max}\,\{0.043 - 3.0 \times \text{sqrt}\,[(0.043) \times (1 - 0.043) \div 200], 0.0\} = 0.000.$$

These limits should be used in future to identify assignable causes. The centerline (CL) value of 0.043 constitutes a benchmark with which to evaluate the capability improvements from new corporate policies. Local managers should feel discouraged from changing business practices unless out-of-control signals occur and assignable causes are found.

The preceding example illustrates the startup phase of control charting. The next example illustrates the steady state phase of control charting. Both phases are common to all SPC charting methods in this book.

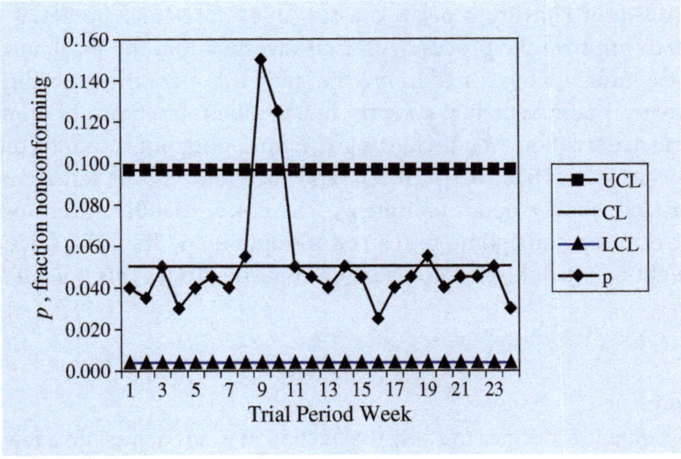

Fig. 4.2 Restaurant *p*-chart during the startup period

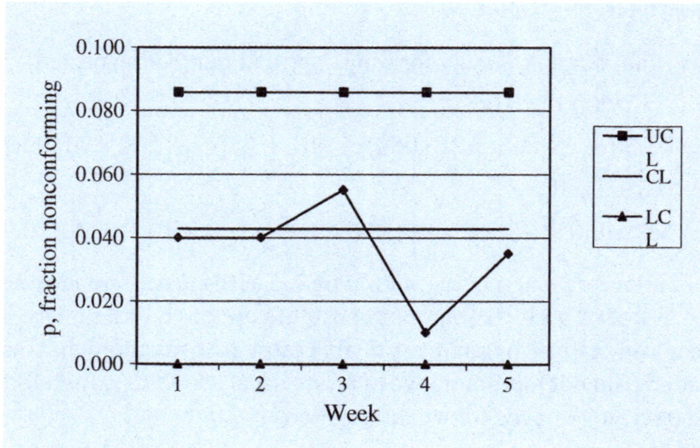

Fig. 4.3 Ongoing *p*-chart during steady state operations

Example 4.5.2 Restaurant Customer Satisfaction Continued

Question

Plot the data in Table 4.2b on the control chart derived in the previous example. Are there any out-of-control signals?

Answer

Figure 4.3 shows an ongoing *p*-charting activity in its steady state. No out-of-control signals are detected by the chart.

In the above example, the lower control limit (LCL) was zero. An often reasonable convention for this case is to consider only zero fractions nonconforming ($p = 0$) to be out-of-control signals if they occur repeatedly. In any case, values of p below the lower control limit constitute positive assignable causes and potential information with which to improve the process. After an investigation, the local authority might choose to use information gained to rewrite standard operating procedures.

Since many manufacturing systems must obtain fractions of nonconforming products much less than 1%, *p*-charting the final outgoing product often requires complete inspection. Then, $n \times p_0 \ll 5.0$, and the chart can be largely ineffective in both evaluating quality and monitoring. Therefore, manufacturers often use the charts upstream for units going into a rework operation. Then the fraction nonconforming might be much higher. The next example illustrates this type of application.

Example 4.5.3 Arc Welding Rework Charting

Question

A process engineer decides to study the fraction of welds going into a rework operation using *p*-charting. Suppose that 2500 welds are inspected over 25 days and 120 are found to require rework. Suppose one day had 42 nonconforming welds which were caused by a known corrected problem and another subgroup had 12 nonconformities but no assignable cause could be found. What are your revised limits and what is the process capability?

Answer

Assuming a constant sample size with 25 subgroups gives $n = 100$. The trial limits are $p_0 = 120/2500 = 0.048$, $UCL_{trial} = 0.110$, $CL_{trial} = 0.048$, and $LCL_{trial} = 0.000$, so there is effectively no lower control limit. We remove only the subgroup whose values are believed to be not representative of the future. The revised numbers are $p_0 = 78/2400 = 0.0325$, $UCL = 0.0857$, $CL =$ the process capability $= 0.0325$, and $LCL = 0.000$.

4.6 Attribute Data: Demerit Charting and *u*-Charting

In ► Chap. 2, the term "**nonconformity**" was defined as an instance in which a part or product's characteristic value falls outside its associated specification limit. Different types of nonconformities can have different levels of importance. For example, some possible automotive nonconformities can make life-threatening accidents more likely. Others may only cause a minor annoyance to car owners.

The term "**demerits**" here refers to a weighted sum of the nonconformities. The weights quantify the relative importance of nonconformities in the eyes of the people constructing the demerit chart. As an example, consider a single scheme in which there are two classes of nonconformities: particularly serious nonconformities with weight 5.0, and typical nonconformities with weight 1.0. The following symbols are used in the description of demerit charting:

1. n is the number of samples in each rational subgroup. If the sample size varies because of choice or necessity, it is written n_i, where i refers to the relevant sampling period. Then, $n_1 \neq n_2$ might occur and/or $n_1 \neq n_3$ etc.
2. c_{ik} is the number of nonconformities in subgroup i of severity k.

The method known as "***u*-charting**" is a special case of demerit charting, with all nonconformities having the same weight of 1.0. Traditionally, the Poisson approximation-based limits are used in u-charting instead of the more general method from Dodge (1928) in Algorithm 4.5. Therefore, the u-charting limits are $UCL_{trial} = u_0 + 3.0 \times \sqrt{u_0/n}$.

Similar considerations related to sample size selection for p-charting also apply to demerit charting. If u_0 is the average number of demerits per item, it is generally desirable that $n \times u_0 > 5$. The following method is written in terms of a constant sample size n. If n varies, then substitute n_i for n in all the formulas. Then, the control limits would vary from subgroup to subgroup. Also, quantities next to each other in the formulas are implicitly multiplied with the "×" omitted for brevity, and "/" is equivalent to "÷". The numbers 3.0 and 0.0 in the formulas are assumed to have an infinite number of significant digits.

Table 4.8 Survey results from patients leaving a hypothetical hospital wing

Day	Discharges	#Complaints		Day	Discharges	#Complaints	
		Typical	Serious			Typical	Serious
1	40	22	3	14	45	22	3
2	29	28	3	15	30	22	3
3	55	33	4	16	30	33	1
4	30	33	2	17	30	44	1
5	22	3	0	18	35	27	2
6	33	32	1	19	25	33	1
7	40	23	2	20	40	34	4
8	35	38	2	21	55	44	1
9	34	23	2	22	55	33	1
10	50	33	1	23	70	52	2
11	22	32	2	24	34	24	2
12	30	39	4	25	40	45	2
13	21	23	2				

Example 4.6.1 Monitoring Hospital Patient Satisfaction

Question

Table 4.8 summarizes the results from the satisfaction survey written by patients being discharged from a hospital wing. It is known that day 5 was a major holiday and new medical interns arrived during day 10. Construct an SPC chart appropriate for monitoring patient satisfaction.

Answer

Complaints can be regarded as nonconformities. Therefore, demerit charting fits the problem needs well since weighted counts of these are given. Using the weighting scheme suggested in this book, $C = 1.0 \times (22 + 28 + \cdots + 45) + 5.0 \times (3 + 3 + \cdots + 2) = 1030$ and $N = 40 + 29 + \cdots + 45 = 930$. Therefore, in *Step 1* $u_0 = 1030 \div 930 = 1.108$. The control limits vary because of the variable sample size.

Figure 4.4 shows the plotted u and UCLs and LCLs during the trial period. Example calculations used to make the figure include the following. The u plotted for Day 1 is derived using $(1.0 \times 22 + 5.0 \times 3) \div 40 = 0.925$. The day 1 *UCL* is given by $1.108 + 3.0 \times sqrt(1.108 \div 40) = 1.61$. The day 1 *LCL* is given by $1.108 - 3.0 \times sqrt(1.108 \div 40) = 0.61$.

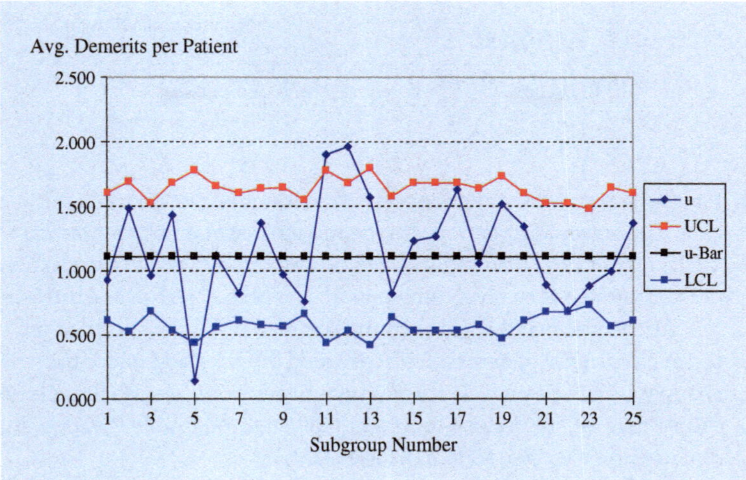

Fig. 4.4 Demerits from hypothetical patient surveys at a hospital

Algorithm 4.5 Demerit Charting

Step 1. (Startup) Obtain the weighted sum of all nonconformities, C, and count of units or systems, $N = n_1 + \cdots + n_{25}$ from 25 time periods.

Step 2. (Startup) Let d_i be the weighted total number of demerits in period i, n_i be the sample size, and c_{ik}, be the number of class k nonconformities, $k = 1, 2, \ldots, m$. If w_k is the weight of nonconformity class k, the weighted demerits d_i, and the demerit per unit D_i (which is the charted quantity and referred to in this paper as demerit per host) in period i are:

$$d_i = \sum_{k=1}^{m} w_k c_{ik} \text{ and } D_i = \frac{d_i}{n_i}. \tag{4.6}$$

Step 3. (Startup) The average number of demerits, $\overline{D}_k$, across all the p periods, for nonconformity class k is:

$$\overline{D}_k = \frac{\sum_{i=1}^{p} c_{ik}}{\sum_{i=1}^{p} n_i} \tag{4.7}$$

Then, the center line (CL) of the demerit control chart is:

$$CL = \sum_{k=1}^{m} w_k \overline{D}_k.$$

Step 4. Startup) The upper and lower control limits for period i are:

$$UCL_i = CL + M\hat{\sigma}_i \text{ and}$$
$$LCL_i = \max\left[\left(CL - M\hat{\sigma}_i\right), 0\right] \quad (4.8)$$
$$\text{Where } \hat{\sigma}_i = \sqrt{\frac{\sum_{k=1}^{m} w_k^2 \overline{D}_k}{n_i}}.$$

Step 5. (Steady State) Plot the number of nonconformities per unit, d_i, for each future period together with the revised upper and lower control limits. An out-of-control signal is defined as a case in which the fraction nonconforming for a given time period, u, is below the lower control limit ($d_i < LCL_i$) or above the upper control limit ($d_i > UCL_i$). From then on, technicians and engineers are discouraged from making minor process changes unless a signal occurs. If a signal does occur, designated people should investigate to see if something unusual and fixable is happening. If not, the signal is referred to as a false alarm

The out-of-control signals occurred on Day 5 (an unusually positive statistic) and Days 11 and 12. It might subjectively be considered reasonable to remove Day 5 since patients might have felt uncharacteristically positive due to the rare major holiday. However, it is less clear whether removing data associated with days 11 and 12 would be fair. These patients were likely affected by the new medical interns. Considering that new medical interns affect hospitals frequently and local authority might have little control over them, their effects might reasonably be considered part of common cause variation. Still, it would be wise to inform the individuals involved about the satisfaction issues and do an investigation. Pending any permanent fixes, we should keep that data for calculating the revised limits.

Removing the 3 demerits and 22 patients from Day 5 from consideration, $C = 1027$ and $N = 908$. Then, the final process capability $= CL = 1027 \div 908 = 1.131$. The revised control limits are $LCL = 1.131 - 3.0 \times \text{sqrt}[1.131 \div n_i]$ and $UCL = 1.131 + 3.0 \times \text{sqrt}[1.131 \div n_i]$ where n_i potentially varies from period to period.

4.7 Continuous Data: Xbar & R Charting

Whenever one or two continuous variable key output variables (KOVs) summarize the quality of units or a system, it is advisable to use a variables charting approach. This follows because variables charting approaches such as Xbar & R charting, described next, offer a relatively much more powerful way to characterize quality with far fewer inspections. These approaches compare favorably in many ways to the attribute charting methods such as *p*-charting and demerit charting. Some theoretical justification for these statements is described in ▶ Chap. 10. However, here it will be clear that Xbar & R charts are often based on samples sizes of $n = 5$ units inspected, compared with 50 or 100 for the attribute charting methods.

For a single continuous quality characteristic, Xbar & R charting involves the generation of two charts with two sets of control limits. Two continuous characteristics would require four charts. Generally, when monitoring any more than two continuous quality characteristics, most experts would recommend a multivariate charting method such as the "Hotelling's T^2" charting; this method along with reasons for this recommendation are described in ► Chap. 8.

There is no universal standard rule for selecting the number of samples to be included in rational subgroups for Xbar & R charting. Generally, one considers first inspecting only a small fraction of the units, such as 5 out of 200. However, Xbar & R charting could conceivably have application even if complete inspection of all units is completed. Some theoretical issues related to complete inspection are discussed in ► Chap. 10.

Two considerations for sample size selection follow. First, the larger the sample size, the closer the control limits and the more **sensitive** the chart will be to assignable causes. Before constructing the charts, however, there is usually no way to know how close the limits will be. Therefore, an iterative process could conceivably be applied. If the limits are too wide, the sample size could be increased and a new chart could be generated.

Second, n should ideally be large enough that the sample averages of the value follow a specific pattern. This will be discussed further in ► Chap. 10. The pattern in question relates to the so-called "**normal**" distribution. In many situations, this pattern happens automatically if $n \geq 4$. This common fact explains why practitioners rarely check whether n is large enough such that the sample averages are approximately normally distributed.

Algorithm 4.6 Standard Xbar & R charting

Step 1. (Startup) Measure the continuous characteristics, $X_{i,j}$, for $i = 1, \ldots, n$ units for $j = 1, \ldots, 25$ periods. Each n units is carefully chosen to be representative of all units in that period, i.e., a rational subgroup.

Step 2. (Startup) Calculate the sample averages $X_{bar,j} = (X_{1,j} + \cdots + X_{n,j})/n$ and ranges $R_j = \text{Max}[X_{1,j}, \ldots, X_{n,j}] - \text{Min}[X_{1,j}, \ldots, X_{n,j}]$ for $j = 1, \ldots, 25$. Also, calculate the average of all of the $25n$ numbers, X_{barbar}, and the average of the 25 ranges $R_{bar} = (R_1 + \cdots + R_{25})/25$.

Step 3. (Startup) Tentatively determine σ_0 using $\sigma_0 = R_{bar}/d_2$, where d_2 comes from the following table. Use linear interpolation to find d_2 if necessary. Calculate the "trial" control limits using

$$UCL_{Xbar} = X_{barbar} + 3.0 \times \frac{\sigma_0}{\sqrt{n}}$$

$$CL_{Xbar} = X_{barbar} \tag{4.9}$$

$$LCL_{Xbar} = X_{barbar} - 3.0 \times \frac{\sigma_0}{\sqrt{n}}$$

$$UCL_R = D_2 s_0$$

$$CL_R = R_{bar}$$

$$LCL_R = D_1 s_0$$

where D_1 and D_2 also come from ■ Table 4.9 and where the "trial" designation has been omitted to keep the notation readable.

Step 4. (Startup) Find all the periods for which either $X_{bar,j}$ or R_j or both are not inside their control limits, i.e., $\{X_{bar,j} < LCL_{Xbar}$ or $X_{bar,j} > UCL_{Xbar}\}$ and/or $\{R_j < LCL_R$ or $R_j > UCL_R\}$.

If the results from any of these periods are believed to be not representative of future system operations, e.g., because problems were fixed permanently, remove the data from the l not representative periods from consideration.

Step 5. (Startup) Re-calculate X_{barbar} and R_{bar} based on the remaining 25 – l periods and $(25 - l) \times n$ data. Also, calculate the revised process sigma, σ_0, using $\sigma_0 = R_{bar}/d_2$. The quantity $6\sigma_0$ is called the "process capability" in the context of Xbar & R charting. It constitutes a typical range of the quality characteristics. Calculate the revised limits using the same "trial" equations in *Step 3*.

Step 6. (Steady State, SS) Plot the sample nonconforming, $X_{bar,j}$, for each period j together with the upper and lower control limits, LCL_{Xbar} and UCL_{Xbar}. The resulting "Xbar chart" typically provides useful information to stakeholders (engineers, technicians, and operators) and builds intuition about the engineered system. Also, plot R_j for each period j together with the control limits, LCL_R and UCL_R. The resulting chart is called an "R chart".

4

The following symbols are used in the description of the method:
1. n is the number of inspected units in a rational subgroup.
2. $X_{i,j}$ refers to the i_t^h quality characteristic value in the jth time period. Note that it might be more natural to use $Y_{i,j}$ instead of $X_{i,j}$ since quality characteristics are outputs. However, $X_{i,j}$ is more standard in this context.
3. $X_{bar,j}$ is the average of the n quality characteristic values for the jth time period.
4. σ_0 is the "process sigma" or, in other words, the true standard deviation of all quality characteristics when only common cause variation is present.

Generally, n is small enough that people are not interested in variable sample sizes. In the formulas below, quantities next to each other are implicitly multiplied with the "×" omitted for brevity, and "/" is equivalent to "÷". The numbers 3.0 and 0.0 in the formulas are assumed to have an infinite number of significant digits.

An out-of-control signal is defined as a case in which the sample average, $X_{bar,j}$, or range, R_j, or both, are outside the control limits, i.e., $\{X_{bar,j} < LCL_{Xbar}$ or $X_{bar,j} > UCL_{Xbar}\}$ and/or $\{R_j < LCL_R$ or $R_j > UCL_R\}$. From then on, technicians and engineers are discouraged from making minor process changes unless a signal

Table 4.9 Constants d_2, D_1, and D_2 relevant for Xbar & R charting

Sample size (n)	d_2	D_1	D_2	Sample size (n)	d_2	D_1	D_2
2	1.128	0.000	3.686	8	2.847	0.388	5.306
3	1.693	0.000	4.358	9	2.970	0.547	5.393
4	2.059	0.000	4.698	10	3.078	0.687	5.469
5	2.326	0.000	4.918	15	3.472	1.203	5.741
6	2.534	0.000	5.078	20	3.737	1.549	5.921
7	2.704	0.204	5.204				

occurs. If a signal does occur, designated people should investigate to see if something unusual and fixable is happening. If not, the signal is referred to as a false alarm.

Note that all the charts in this chapter are designed such that, under usual circumstances, false alarms occur on average one out of 370 periods. If they occur more frequently, it is reasonable to investigate with extra vigor for assignable causes. Also, as an example of linear interpolation, consider the estimated d_2 for $n=11$. The approximate estimate for d_2 is $3.078+(1 \div 5) \times (3.472 - 3.078)=3.1568$. Sometimes the quantity "C_{pk}" (spoken "see-pee-kay") is used as a system quality summary. The formula for C_{pk} is

$$C_{pk} = \text{Min}[USL - X_{barbar}, X_{barbar} - LSL]/(3\sigma_0), \tag{4.10}$$

where σ_0 is based on the revised R_{bar} from an Xbar & R method application. Also, *USL* is the upper specification limit and *LSL* is the lower specification limit. These are calculated and used to summarize the state of the engineered system. The σ_0 used is based on *Step 4* of the above standard procedure.

Large C_{pk} and small values of $6\sigma_0$ are generally associated with high quality processes. This follows because both these quantities measure the variation in the system. We reason that *variation is responsible for the majority of quality problems* because typically only a small fraction of the units fail to conform to specifications. Therefore, some noise factor changing in the system causes those units to fail to conform to specifications. The role of variation in causing problems explains the phrase "variation is evil" and the need to eliminate source of variation.

Example 4.7.1 Fixture Gaps Between Welded Parts

Question

A Korean shipyard wants to evaluate and monitor the gaps between welded parts from manual fixturing. Workers measure 5 gaps every shift for 25 shifts over 10 days. The remaining steady state (SS) is not supposed to be available at the time this question is asked. Table 4.10 shows the resulting hypothetical data. Chart this data and establish the process capability.

Answer

From the description, $n=5$ inspected gaps between fixtured parts prior to welding, and the record of the measured gap for each in millimeters is $X_{i,j}$. The inspection interval is a production shift, so roughly $\tau=6$ h.

The calculated subgroup averages and ranges are also shown (*Step 2*) and $X_{bar\text{-}bar}=0.90$, $R_{bar}=0.34$, and $\sigma_0=0.148$. In *Step 3*, the derived values were $UCL_{X\text{-}bar}=1.103$, $LCL_{Xbar}=0.705$, $UCL_R=0.729$, and $LCL_R=0.000$. None of the first 25 periods has an out-of-control signal. In *Step 4*, the process capability is 0.889. From then until major process changes occur (rarely), the same limits are used to find out-of-control signals and alert designated personnel that process attention is needed (*Step 5*). The chart, Fig. 4.5, also prevents "over-control" of the system by discouraging changes unless out-of-control signals occur.

Table 4.10 Example gap data (in mm) to show Xbar & R charting (start-up & steady state)

Phase	j	$X_{1,j}$	$X_{2,j}$	$X_{3,j}$	$X_{4,j}$	$X_{5,j}$	$X_{bar,j}$	R_j	Phase	j	$X_{1,j}$	$X_{2,j}$	$X_{3,j}$	$X_{4,j}$	$X_{5,j}$	$X_{bar,j}$	R_j
SU	1	0.85	0.71	0.94	1.09	1.08	0.93	0.38	SU	19	0.97	0.99	0.93	0.75	1.09	0.95	0.34
SU	2	1.16	0.57	0.86	1.06	0.74	0.88	0.59	SU	20	0.85	0.77	0.78	0.84	0.83	0.81	0.08
SU	3	0.80	0.65	0.62	0.75	0.78	0.72	0.18	SU	21	0.82	1.03	0.98	0.81	1.10	0.95	0.29
SU	4	0.58	0.81	0.84	0.92	0.85	0.80	0.34	SU	22	0.64	0.98	0.88	0.91	0.80	0.84	0.34
SU	5	0.85	0.84	1.10	0.89	0.87	0.91	0.26	SU	23	0.82	1.03	1.02	0.97	1.00	0.97	0.21
SU	6	0.82	1.20	1.03	1.26	0.80	1.02	0.46	SU	24	1.14	0.95	0.99	1.18	0.85	1.02	0.33
SU	7	1.15	0.66	0.98	1.04	1.19	1.00	0.53	SU	25	1.06	0.92	1.07	0.88	0.78	0.94	0.29
SU	8	0.89	0.82	1.00	0.84	1.01	0.91	0.19	SS	26	1.06	0.81	0.98	0.98	0.85	0.936	0.25
SU	9	0.68	0.77	0.67	0.85	0.90	0.77	0.23	SS	27	0.83	0.70	0.98	0.82	0.78	0.822	0.28
SU	10	0.90	0.85	1.23	0.64	0.79	0.88	0.59	SS	28	0.86	1.33	1.09	1.03	1.10	1.082	0.47
SU	11	0.51	1.12	0.71	0.80	1.01	0.83	0.61	SS	29	1.03	1.01	1.10	0.95	1.09	1.036	0.15
SU	12	0.97	1.03	0.99	0.69	0.73	0.88	0.34	SS	30	1.02	1.05	1.01	1.02	1.20	1.060	0.19
SU	13	1.00	0.95	0.76	0.86	0.92	0.90	0.24	SS	31	1.02	0.97	1.01	1.02	1.06	1.016	0.09
SU	14	0.98	0.92	0.76	1.18	0.97	0.96	0.42	SS	32	1.20	1.02	1.20	1.05	0.91	1.076	0.29
SU	15	0.91	1.02	1.03	0.80	0.76	0.90	0.27	SS	33	1.10	1.15	1.10	1.02	1.08	1.090	0.13
SU	16	1.07	0.72	0.67	1.01	1.00	0.89	0.40	SS	34	1.20	1.05	1.04	1.05	1.06	1.080	0.16
SU	17	1.23	1.12	1.10	0.92	0.90	1.05	0.33	SS	35	1.22	1.09	1.02	1.05	1.05	1.086	0.20
SU	18	0.97	0.90	0.74	0.63	1.02	0.85	0.39									

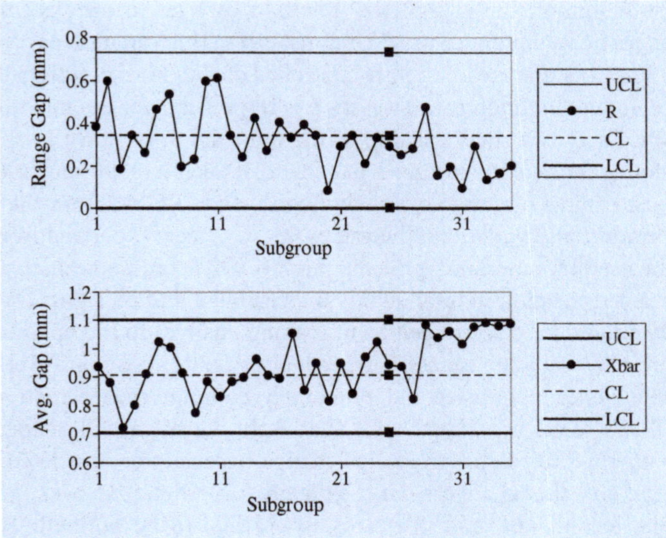

Fig. 4.5 Xbar & R charts for gap data (■ separates startup and steady state)

The phrase "sigma level" (σ_L) is an increasingly popular alternative to C_{pk}. The formula for sigma level is

$$\sigma_L = 3.0 \times C_{pk} \tag{4.11}$$

If the process is under control and certain "normal assumptions" apply, then the fraction nonconforming is less than 1.0 nonconforming per *billion* opportunities. If the mean shifts 1.5 σ_0 to the closest specification limit, the fraction nonconforming is less than 3.4 nonconforming per million opportunities. Details from the fraction nonconforming calculations are documented in ▶ Chap. 10.

The goal implied by the phrase "**six sigma**" is to change system inputs so that the σ_L derived from an Xbar & R charting evaluation is greater than 6.0.

In applying Xbar & R charting, one simultaneously creates two charts and uses both for process monitoring. Therefore, the plotting effort is greater than for *p* charting, which requires the creation of only a single chart. Also, as implied above, there can be a choice between using one or more Xbar & R charts and a single *p* chart. The *p* chart has the advantage of all nonconformity data summarized in a single, interpretable chart. The important advantage of Xbar & R charts is that generally many fewer runs are required for the chart to play a useful role in detecting process shifts than if a *p* chart is used. Popular sample sizes for Xbar & R charts are $n = 5$. Popular sample sizes for *p* charts are $n = 200$.

To review, an "assignable cause" is a change in the engineered system inputs which occur irregularly that can be affected by "local authority", e.g., operators, process engineers, or technicians. For example, an engineer dropping a wrench into the conveyor apparatus is an assignable cause.

The phrase "common cause variation" refers to changes in the system outputs or quality characteristic values under usual circumstances. This variation occurs because of the changing of factors that are not tightly controlled during normal system operation.

As implied above, common cause variation is responsible for the majority of quality problems. Typically only a small fraction of the units fail to conform to specifications, and this fraction is consistently not zero. In general, it takes a major improvement effort involving robust engineering methods including possibly RDPM from the last chapter to reduce common cause variation. The values $6\sigma_0$, C_{pk}, and σ_L derived from Xbar & R charting can be useful for measuring the magnitude of the common cause variation.

An important realization in total quality management and six sigma training is that local authority should be discouraged from making changes to the engineered system when there are no assignable causes. These changes could cause an "over-controlled" situation in which energy is wasted and, potentially, common cause variation increases.

The usefulness of both *p*-charts and Xbar & R charts partially depends upon a coincidence. When quality characteristics change because the associated engineered systems change, and this change is large enough to be detected over process noise, then engineers, technicians, and operators would like to be notified. There are, of course, some cases in which the sample size is sufficiently large (e.g., when complete inspection is used) that even small changes to the engineered system inputs can be detected. In these cases, the engineers, technicians, and operators might not want to be alerted. Then, ad hoc adjustment of the formulas for the limits and/or the selection of the sample size, n, and interval, τ, might be justified.

Example 4.7.2 Blood Pressure Monitoring Equipment

Question

Suppose a company is manufacturing blood pressure monitoring equipment and would like to use Xbar & R charting to monitor the consistency of equipment. Also, an inspector has measured the pressure compared to a reference for 100 units over 25 periods (inspecting 4 units each period). The average of the characteristic is 55.0 PSI. The average range is 2.5 PSI. Suppose that during the trial period it was discovered that one of the subgroups with average 62.0 and range 4.0 was influenced by a typographical error and the actual values for that period are unknown. Also, another subgroup with average 45.0 and range 6.0 was not associated with any assignable cause. Determine the revised limits and C_{pk}. Interpret the C_{pk}.

Answer

All units are in PSI. The trial limit calculations are:

$$X_{barbar} = 55.0, \quad R_{bar} = 2.5, \quad \sigma_0 = 2.5/2.059 = 1.21$$

$$UCL_{Xbar} = 55.0 + 3(1.21)\left(4^{-1/2}\right) = 56.8$$

$$LCL_{Xbar} = 55.0 - 3(1.21)\left(4^{-1/2}\right) = 53.2$$

$$UCL_R = (4.698)(1.21) = 5.68$$

$$LCL_R = (0.0)(1.21) = 0.00$$

The subgroup average 62.0 from the X_{barbar} calculation and 4.0 from the R_{bar} calculation are removed because the associated assignable cause was found and eliminated. The other point was left in because no permanent fix was implemented. Therefore, the revised limits and C_{pk} are derived as follows:

$$X_{barbar} = [(55.0)(25)(4) - (62.0)(4)]/[(24)(4)] = 54.7$$
$$R_{bar} = [(2.5)(25) - (4.0)]/(24) = 2.4375, \sigma_0 = 2.4375/2.059 = 1.18$$
$$UCL_{Xbar} = 54.7 + 3(1.18)\left(4^{-1/2}\right) = 56.5$$
$$LCL_{Xbar} = 54.7 - 3(1.18)\left(4^{-1/2}\right) = 53.0$$
$$UCL_R = (4.698)(1.18) = 5.56$$
$$LCL_R = (0.0)(1.21) = 0.00$$
$$C_{pk} = \text{Minimum}\{59.0–54.7, 54.7 - 46.0\}/[(3)(1.18)] = 1.21$$

Therefore, the quality is high enough that complete inspection may not be needed ($C_{pk} > 1.0$). The outputs very rarely vary by more than 7.1 PSI and are generally close to the estimated mean of 54.7 PSI, i.e., values within 1 PSI of the mean are common. However, if the mean shifts even a little, then a substantial fraction of nonconforming units will be produced. Many six sigma experts would say that the sigma level is indicative of a compawny that has not fully committed to quality improvement.

4.7.1 Alternative Continuous Data Charting Methods

The term "**run rules**" refers to specific patterns of charted quantities that may constitute an out-of-control signal. For example, some companies institute policies in which after seven charted quantities in a row are above or below the center line (CL), then the designated people should investigate to look for an assignable cause. They would do this just as if an $X_{bar,j}$ or R_j were outside the control limits. If this run rule were implemented, the second-to-last subgroup in the fixture gap example during steady state would generate an out-of-control signal. Run rules are potentially relevant for all types of charts including *p*-charts, demerit charts, Xbar charts, and R charts.

Also, many other kinds of control charts exist besides the ones described in this chapter. In general, each offers some advantages related to the need to inspect fewer units to derive comparable information, e.g., so-called "EWMA charting" in ▶ Chap. 8. Other types of charts address the problem that, because of applying a large number of control charts, multiple sets of Xbar & R charts may find many false alarms. Furthermore, the data plotted on different charts may be correlated. To address these issues, so-called "multivariate charting" techniques have been proposed, described in ▶ Chap. 8.

Finally, custom charts are possible based on the following advanced concept. Establish the distribution (see ▶ Chap. 10) of a quality characteristic with only common causes operating. This could be done by generating a histogram of values during a trial period. Then, chart any quantity associated with a hypothesis test, evaluating whether these new quality characteristic values come from the same "common causes only" distribution. Any rejection of that assumption based on a small α test (e.g., $\alpha = 0.0013$) constitutes a signal that the process is "out-of-control" and assignable causes might be present. Advanced readers will notice that all the charts in this book are based on this approach which can be extended.

4.8 Chapter Summary and Conclusions

4

This chapter describes three methods to evaluate the capability of measurement systems. Gauge R&R (comparison with standards) is argued to be advantageous when items with known values are available. Gauge R&R (crossed) is argued to be most helpful when multiple appraisers can each test the same items multiple times. Gauge R&R (nested) is relevant when items cannot be inspected by more than one appraiser.

Once measurement systems are declared "capable" or at least acceptable, these measurement systems can be used in the context of SPC charting procedures to thoroughly evaluate other systems of interest. The *p*-charting procedure was presented and argued to be most relevant when only go-no-go data are available. Demerit charting is also presented and argued to be relevant when the count of different types of nonconformities is available. The method of "*u*-charting" is argued to be relevant if all nonconformities are of roughly the same importance. Finally, Xbar & R charting is a pair of charts to be developed for cases with only one or two KOVs or quality characteristics. For a comparable or much smaller number of units tested, Xbar & R charting generally gives a more repeatable, accurate picture of the process quality than the attribute

Table 4.11 Summary of the SQC methods relevant to the measurement phase

Method	Advantages	Disadvantages
Gauge R&R: comparison with standards	Accounts for all errors including systematic errors	Requires pre-tested "standard" units
Gauge R&R (crossed)	No requirement for pre-tested "standard" units	Neglects systematic errors
Gauge R&R (nested)	Each unit only tested by one appraiser	Neglects systematic and repeatability errors
p-charting	Requires only go-no-go data, intuitive	Requires many more inspections, less sensitive
Demerit charting	Addresses differences between nonconformities	Requires more inspections, less sensitive
u-charting	Relatively simple version of demerit charts	Requires more inspections, less sensitive
Xbar & R charting	Uses fewer inspections, gives greater sensitivity	Requires 2 or more charts for single type of unit

charting methods. It also gives greater sensitivity to the presence of assignable causes. However, one unit might have several quality characteristics each requiring two charts. Table 4.11 summarizes the methods presented in this chapter and their advantages and disadvantages.

Example 4.8.1 Wire Harness Inspection Issues

Question
A wire harness manufacturer discovers voids in the epoxy that sometimes requires rework. A team trying to reduce the amount of rework find that the two inspectors working on the line are applying different standards. Team experts carefully inspect ten units and determine trustworthy void counts. What technique would you recommend for establishing which inspector has the more appropriate inspection method?

Answer
Since the issue is systematic errors, the only relevant method is comparison with standards. Also, this method is possible since standard values are available.
In the context of six sigma projects, statistical process control charts offer thorough evaluation of system performance. This is relevant both before and after system inputs are adjusted in the measure and control or verify phases respectively. In many relevant situation, the main goal of the six sigma analyze-and-improve phases is to cause the charts established in the measure phase to generate "good" out-of-control signals indicating the presence of a desirable assignable cause, i.e., the project team's implemented changes.

For example, values of p, u, or R consistently below the lower control limits after the improvement phase settings are implemented indicate success. Therefore, it is often necessary to go through two start-up phases in a six sigma project during the measure phase and during the control or verify phase. Hopefully, the established process capabilities, sigma levels, and/or C_{pk} numbers will confirm improvement and aid in evaluation of monetary benefits.

Example 4.8.2 Printed Circuit Board System Evaluation

Question
A team has a clear charter to reduce the fraction of nonconforming printed circuit boards requiring rework. A previous team had the same charter but failed because team members tampered with the settings and actually increased the fraction nonconforming. What next steps do you recommend?

Answer

Since there apparently is evidence that the process has become worse, it might be advisable to return to the system inputs documented in the manufacturing SOP prior to interventions by the previous team. Then, measurement systems should be evaluated using the appropriate gauge R&R method unless experts are confident that they should be trusted. If only the fraction nonconforming numbers are available, *p*-charting should be implemented to thoroughly evaluate the system both before and after changes to inputs.

4.9 Problems

In general, pick the correct answer that is most complete.

1. According to the text, what types of measurement errors are found in standard values?
   a. Unknown measurement errors are in all numbers, even standards.
   b. None, they are supposed to be accurate within the uncertainty implied.
   c. Measurements are entirely noise. We can't really know any values.
   d. All of the above are true.
   e. Only the answers in parts "b" and "c" are correct.
2. What properties are shared between reproducibility and repeatability errors?
   a. Both derive from mistakes made by people and/or equipment.
   b. Neither is easily related to true errors in relation to standards.
   c. Estimation of both can be made with a crossed gauge R&R.
   d. All of the above are correct.
   e. Only the answers in parts "a" and "b" are correct.
3. Which are differences between reproducibility and systematic errors?
   a. Systematic errors are between a generic process and the standard values; reproducibility errors are differences between each appraiser and the average appraiser.
   b. Evaluating reproducibility errors relies more on standard values.
   c. Systematic errors are more easily measured without standard values.
   d. All of the above are differences between reproducibility and systematic errors.
   e. Only the answers in parts "b" and "c" are true differences.

The following information and Table 4.12 are relevant for Questions 4–8. Three personal laptops are repeatedly timed for how long they require to open Microsoft Internet Explorer. Differences greater than 3.0 s should be reliably differentiated.

4. What is the EEAE (in seconds)?
   a. 1.95.
   b. 2.15.
   c. 2.20.
   d. 2.05.
   e. Answers "a" and "c" are both valid answers.

5. What is the EEEAE?
   a. 0.67.
   b. 0.15.
   c. 2.92.
   d. 0.65.
   e. 2.15.
6. What are the estimated expected absolute errors?
   a. 5.0.
   b. 14.33.
   c. 2.15 ± 0.15.
   d. 14.33 ± 0.15.
   e. 6.0.
   f. None of the above is correct.
7. What is the measurement system capability?
   a. 6.0.
   b. 9.35.
   c. 12.9.
   d. 2.39.
   e. 0.9.
   f. None of the above is correct.
8. What is needed to make the measurement system "gauge capable?"
   a. Change timing approach so that $6.0 \times \text{EEAE} < 3.0$.
   b. Make the system better so that $\text{EEAE} < 1$.
   c. The system is gauge capable with no changes.
   d. The expected absolute errors should be under 1.5.
   e. All of the above are correct except (a) and (b).

Table 4.13 on the next page is relevant to Questions 9–12.

9. Following the text formulas, solve for $Y_{range\ parts}$ (within the implied uncertainty).
   a. 0.24.
   b. 0.12.
   c. 0.36.
   d. 0.1403.
   e. 0.09.
   f. None of the above is correct.
10. Following the text formulas, solve for R&R (within the implied uncertainty).
   a. 0.140.
   b. 0.085.
   c. 0.164.
   d. 0.249.
   e. 0.200.
   f. None of the above is correct.

Table 4.12 Measurements for hypothetical laptop capability study

Run	Standard unit	Appraiser	Measured value (s)	Absolute error (s)
1	3	1	15	2
2	2	1	19	1
3	2	1	18	2
4	1	1	11	2
5	3	1	19	2
6	3	1	20	3
7	1	1	10	3
8	2	1	17	3
9	3	1	14	2
10	2	1	21	1
11	1	1	15	2
12	2	1	22	2
13	2	1	18	2
14	1	1	11	2
15	3	1	13	3
16	2	1	23	3
17	1	1	15	1
18	2	1	22	2
19	1	1	14	2
20	3	1	19	3

Table 4.13 Measurements for hypothetical gauge R&R (crossed) study

Appraiser A	Part number				
	1	2	3	4	5
Trial 1	0.24	0.35	0.29	0.31	0.24
Trial 2	0.29	0.39	0.32	0.34	0.25
Trial 3	0.27	0.34	0.28	0.27	0.26
Appraiser B					
Trial 1	0.20	0.38	0.27	0.32	0.25
Trial 2	0.22	0.34	0.29	0.31	0.23
Trial 3	0.17	0.31	0.24	0.28	0.25

11. What is the %R&R (rounding to the nearest percent)?
    a. 25%.
    b. 16%.
    c. 2.9%.
    d. 55%.
    e. 60%.
    f. None of the above is correct.
12. In three sentences or less, interpret the %R&R value obtained in Question 11.
13. Which of the following is NOT a benefit of SPC charting?
    a. Charting helps in thorough evaluation of system quality.
    b. It helps identify unusual problems that might be fixable.
    c. It encourages people to make continual adjustments to processes.
    d. It encourages a principled approach to process meddling (only after evidence).
    e. Without complete inspection, charting still gives a feel for what is happening.
14. Which of the following describes the relationship between common cause variation and local authorities?
    a. Local authority generally cannot reduce common cause variation on their own.
    b. Local authority has the power to reduce only common cause variation.
    c. Local authority shows over-control when trying to fix assignable causes.
    d. All of the above are correct.
    e. All of the above are correct except (a) and (d).
15. Which of the following is correct and most complete?
    a. False alarms are caused by assignable causes.
    b. The charts often alert local authority to assignable causes which they fix.
    c. Charts seek to judge the magnitude of average assignable cause variation.
    d. All of the above are correct.
    e. All of the above are correct except (a) and (d).

The following data set in ◘ Table 4.14 will be used to answer Questions 16–20. This data is taken from 25 shifts at a manufacturing plant where 200 ball bearings are inspected per shift.

16. Where will the center line of a *p*-chart be placed (within implied uncertainty)?
    a. 0.015.
    b. 0.089.
    c. 0.146.
    d. 431.
    e. 0.027.
    f. None of the above is correct.
17. How many trial data points are outside the control limits?
    a. 0.
    b. 1.
    c. 2.
    d. 3.
    e. 4.
    f. None of the above is correct.

Table 4.14 Hypothetical trial ball bearing numbers of nonconforming (nc) units

Subgroup #	Number nc.	Subgroup #	Number nc.	Subgroup #	Number nc.
1	15	10	25	18	14
2	26	11	12	19	16
3	18	12	14	20	18
4	16	13	17	21	20
5	19	14	19	22	22
6	21	15	15	23	24
7	24	16	17	24	12
8	10	17	9	25	10
9	30				

18. Where will the revised p-chart UCL be placed (within implied uncertainty)?
    a. 0.015.
    b. 0.089.
    c. 0.146.
    d. 0.130.
    e. 0.020.
    f. None of the above is correct.
19. Use software (e.g., Minitab® or Excel) to graph the revised p-chart, clearly showing the p_0, UCL, LCL, and percent nonconforming for each subgroup.
20. In the above example, n and τ are appropriate because:
    a. Complete inspection has been used, assuring good quality.
    b. The condition $n \times p_0 > 5.0$ is satisfied.
    c. No complaints about lack of responsiveness are reported.
    d. All of the above are correct.
    e. All of the above are correct except for (a) and (d).

The call center data in Table 4.15 will be used for Questions 21–25.

In Table 4.15, errors and their assigned weighting values are as follows: time took too long (weight is 1), value of quote given incorrectly (weight is 3), and security rules not enforced (weight is 10). Assume all the data are available when the chart is being set up.

21. What is the weighted total number of nonconformities (C)?
    a. 916.
    b. 1270.
    c. 337.
    d. 2127.
    e. 13.
    f. None of the above is correct.

Table 4.15 Types of call center errors for a charting activity

Day	Callers	Time	Value	Security	Day	Callers	Time	Value	Security
1	200	40	15	0	14	217	41	10	1
2	232	38	11	1	15	197	42	9	0
3	189	25	12	0	16	187	38	15	0
4	194	29	13	0	17	180	35	13	1
5	205	31	14	2	18	188	37	16	4
6	215	33	16	1	19	207	38	13	2
7	208	37	13	0	20	202	35	11	1
8	195	32	10	0	21	206	39	12	0
9	175	31	9	1	22	221	42	18	0
10	140	15	2	0	23	256	43	10	1
11	189	29	11	0	24	229	19	20	0
12	280	60	22	3	25	191	40	14	0
13	240	36	17	1	26	209	31	11	1

22. What is the initial centerline (μ_0)?
    a. 0.237.
    b. 0.397.
    c. 0.171.
    d. 0.178.
    e. 0.360.
    f. None of the above is correct.
23. How many out-of-control signals are found in this data set?
    a. 0.
    b. 1.
    c. 2.
    d. 3.
    e. 4.
    f. None of the above is correct.
24. Create the trial or startup demerit chart in Microsoft® Excel, clearly showing the *UCL*, *LCL*, *CL*, and process capability for each subgroup.
25. In steady state, what actions should be taken for out-of-control signals?
    a. Always immediately remove them from the data and recalculate limits.
    b. Do careful detective work to find causes before making recommendations.
    c. In some cases, it might be desirable to shut down the call center.
    d. All of the above are correct.
    e. All of the above are correct except (a) and (d).

4

Table 4.16 Hypothetical airplane flight data (S = Subgroup)

No.	X_1	X_2	X_3	X_4	X_5	No.	X_1	X_2	X_3	X_4	X_5
1	2.1	1.8	2.3	2.6	2.6	14	2.4	2.2	1.9	2.8	2.3
2	2.7	1.5	2.1	2.5	1.9	15	2.2	2.4	2.5	2.9	1.5
3	2.0	1.7	1.6	1.9	2.0	16	2.5	1.8	1.7	2.4	2.4
4	1.6	2.0	2.1	2.2	2.1	17	2.9	2.6	2.6	2.2	2.2
5	2.1	2.1	2.6	2.2	2.1	18	3.1	2.8	3.1	3.4	3.3
6	2.0	2.8	2.5	2.9	2.0	19	3.5	3.2	2.9	3.5	3.4
7	2.7	1.7	2.4	2.5	2.8	20	3.3	3.0	3.1	2.7	2.8
8	2.2	2.0	2.4	2.1	2.4	21	2.8	3.6	3.3	3.3	3.2
9	1.8	1.9	1.7	2.1	2.2	22	3.1	3.4	3.4	3.1	3.4
10	2.2	2.1	2.9	1.7	2.0	23	2.0	2.5	2.4	2.3	2.4
11	1.4	2.6	1.8	2.0	2.4	24	2.7	2.3	2.4	2.8	2.1
12	2.3	2.5	2.4	1.8	1.9	25	2.5	2.2	2.5	2.2	2.0
13	2.4	2.3	1.9	2.1	2.2						

26. Which of the following is (are) true of *u*-chart process capability?
    a. It is the usual average number of nonconformities per item.
    b. It is the fraction of nonconforming units under usual conditions.
    c. It necessarily tells less about a system than *p*-chart process capability.
    d. All of the above are true.
    e. All of the above are correct except (a) and (d).

Table 4.16 will be used in Questions 27–32. Paper airplanes are being tested, and the critical characteristic is time aloft. Every plane is measured, and each subgroup is composed of five successive planes.

27. What is the starting centerline for the Xbar chart (within the implied uncertainty)?
    a. 2.1.
    b. 2.3.
    c. 2.4.
    d. 3.3.
    e. 3.5.
    f. None of the above is correct.
28. How many subgroups generate out-of-control signals for the trial R chart?
    a. 0.
    b. 1.
    c. 2.
    d. 3.
    e. 4.
    f. None of the above is correct.

29. How many subgroups generate out-of-control signals for the trial Xbar chart?
    a. 0.
    b. 1.
    c. 2.
    d. 3.
    e. 4.
    f. None of the above is correct.
30. What is the UCL for the revised Xbar chart? (Assume that assignable causes are found and eliminated for all of the out-of-control signals in the trial chart.)
    a. 1.67.
    b. 1.77.
    c. 2.67.
    d. 2.90.
    e. 3.10.
    f. None of the above is correct.
31. What is the UCL for the revised R chart? (Assume that assignable causes are found and eliminated for all of the out-of-control signals in the trial chart.)
    a. 0.000.
    b. 0.736.
    c. 1.123.
    d. 1.556.
    e. 1.801.
    f. None of the above is correct.
32. Control chart the data and propose limits for steady state monitoring. (Assume that assignable causes are found and eliminated for all of the out-of-control signals in the trial chart.)
33. Which of the following relate run rules to six sigma goals?
    a. Run rules generate additional false alarms improving chart credibility.
    b. Sometimes $LCL = 0$ so run rules offer a way to check project success.
    c. Run rules signal assignable causes in start-up, improving capability measures.
    d. All of the above are correct.
    e. All of the above are correct except (a) and (d).
34. A medical device manufacturer is frustrated about the variability in rework coming from the results of different inspectors. No standard units are available and the test methods are nondestructive. Which methods might help reduce rework variability? Explain in four sentences or less.
35. Some psychologists believe that self-monitoring a person's happiness can reduce the peaks and valleys associated with manic depressive behavior. Briefly discuss common and assignable causes in this context and possible benefits of charting.
36. It is often true that project objectives can be expressed in terms of the limits on R charts before and after changes are implemented. The goal can be that the limits established through an entirely new start-up and steady state process should be narrower. Explain briefly why this goal might be relevant.

Table 4.17 Hypothetical data for gauge study (superscripts show run order)

Software #1	*Part*				
	1	2	3	4	5
Trial 1	0.92^{2}	1.07^{11}	1.05^{13}	1.08^{5}	0.85^{15}
Trial 2	0.92^{7}	1.07^{8}	1.07^{20}	1.05^{18}	0.86^{27}
Trial 3	0.95^{10}	1.06^{19}	1.03^{22}	1.02^{24}	0.89^{30}
Software #2	*Part*				
	1	2	3	4	5
Trial 1	0.94^{6}	1.05^{1}	1.06^{12}	1.02^{9}	0.89^{3}
Trial 2	0.94^{7}	1.08^{14}	1.03^{23}	1.05^{17}	0.80^{16}
Trial 3	0.92^{10}	1.01^{26}	1.09^{28}	0.94^{29}	0.82^{25}

Table 4.18 Hypothetical customer satisfaction data

Week	Sum not OK	Week	Sum not OK
1	5	14	9
2	4	15	8
3	8	16	4
4	3	17	3
5	40	18	8
6	42	19	6
7	6	20	6
8	6	21	7
9	10	22	5
10	9	23	4
11	8	24	2
12	2	25	4
13	1		

37. Describe a false alarm in a sporting activity in which you participate.
38. A weld engineer is analyzing the ability of two computer software programs to measure consistently a continuous characteristic called "undercut" in weld cross sections. She performs *Steps 1* and *2* of the crossed gauge R&R method and generates the data in Table 4.17. Analyze the data and state conclusions.

39. Define precisely the term "measurement error" as explained in this chapter.
40. Which charting method might you use to monitor a child's grades at school and why would you pick that chart?
41. An automotive dealership chain's executive wants to start SPC charting to evaluate and monitor customer satisfaction. Every week, salesmen must record 200 answers to the question, "Is everything OK?" Table 4.18 lists the sum of hypothetical "everything is not OK" answers for 25 weeks at one location. Unusual and corrected power outages occurred in weeks 5 and 6. Analyze the data using the appropriate SPC chart and draw conclusions about the system quality level.

References

Automotive Industry Task Force (AIAG) (1994) Measurement systems analysis reference manual. Chrysler, Ford, General Motors Supplier Quality Requirements Task Force

Dodge HF (1928) A method of rating manufactured product. Bell Syst Tech J 7(2):350–368

Montgomery DC, Runger GC (1994) Gauge capability and designed experiments (Basic Methods, Part I). Quality Engineering 6:115–135

Shewhart WA (1980) Economic quality control of manufactured product, 2nd edn. ASQ, Milwaukee, WI

Analyze Phase

T. T. Allen, *Introduction to Engineering Statistics and Lean Six Sigma*,
https://doi.org/10.1007/978-1-4471-7420-2_5

5.1 Introduction

In ▶ Chap. 3, the development and documentation of project goals was discussed. ▶ Chapter 4 described the process of evaluating relevant systems, including measurement systems, before any system changes are recommended by the project team. The analyze phase involves establishing cause-and-effect relationships between system inputs and outputs.

Analyze phase methods are based on a variety of data sources and generate many types of visual information for decision-makers. Methods that can be relevant for these analyses include the design of experiments (DOE) methods covered in Part II of this book and previewed here. Also, QFD Cause and Effects Matrices, process mapping, value-stream mapping, spaghetti diagrams and lean methods are addressed. Note that DOE methods include steps for both analysis and development of improvement recommendations. Lean manufacturing techniques areas provide potentially powerful methods and concepts for making systems more efficient. As usual, all methods could conceivably be applied usefully in any project phase or occasion.

5.2 Lean Production

An admittedly simplified history of the Japanese technology was provided in ▶ Sect. 1. The specific relationship between automotive history and lean explained in Womack et al. (1991) is briefly reviewed here. The automobile originated from horse carriages with motors and a steering wheel. The inventors behind what was later Mercedes-Benz and Maybach designed engines and attached them to horse carriages. Car ownership was generally for the rich and slowly increased in the late 1800s. Then, Henry Ford implemented effectively assembly lines and made cars for the masses. The related "mass production" methods gave rise to work function specialization and the divisions between skilled and unskilled labor that largely motivated the usage of control charts in ▶ Chap. 4. In 1914, Ford produced cars every 15 mins using eight times less manpower or work hours than was previously possible (Georano 2000).

Toyota was founded in 1937 by Kiichiro Toyoda. Kiichiro studied the ideals of Henry Ford and W. Edwards Demming when trying to advance his company. This resulted in the creation and refinement of the Toyota Production System (TPS) between 1948 and 1975 by Taiichi Ohno, Shigeo Shingo, Eiji Toyoda, and others. The Toyota Production System is known generically as lean manufacturing. Originally the system was called "Just-in-Time" which now often refers only to production scheduling (Shigeo 1989). Womack et al. (1991) reported that lean manufacturing and the TPS used approximately four times less manpower or work hours than the potentially corrupted version of mass production used in the 1980s in some manufacturing plants in the United States.

5.2.1 Process Mapping and Value Stream Mapping

The "**process mapping**" method involves creating flow diagrams of systems. Among other benefits, this method clarifies possible causal relationships between subsystems, which are represented by blocks. Inputs from some subsystems might influence outputs of downstream subsystems. This method is particularly useful in the define phase because subsystem inputs and outputs are identified during the mapping. Bottlenecks can also be clarified. Value stream mapping is also relevant in the analyze phase for similar reasons.

Also, process mapping helps in setting up discrete event simulation models, which are a type of Monte Carlo method. This topic is described thoroughly in Law and Kelton (2000) and Banks et al. (2000). Discrete event simulation plays a similar role to process mapping. Simulation also permits an investigation of bottlenecks in hypothetical situations with additional resources added such as new machines or people. To develop simulation models, data about processing times are generally required.

The "**Value Stream Mapping**" (VSM) method can be viewed as a variant of process mapping with added activities. In VSM, engineers inspect the components of the engineered system, e.g., all process steps including material handling and rework, focusing on steps which could be simplified or eliminated. The most popular references for this activity appear to be Womack and Jones (1996, 1999). Other potentially relevant references are Suzaki (1987) and Liker (1998). Therefore, value stream mapping involves analysis and also immediately suggests improvement through re-design. Other outputs of value stream mapping process described here include an expanded list of factors or system inputs to explore in experiments, as well as an evaluation of the documentation that supports the engineered system.

To paraphrase, the definition of "**value stream**" provided in Womack and Jones (1999) is the minimum amount of processing steps, from raw materials to the customer, needed to deliver the final product or system output. These necessary steps are called "**value added**" operations. All other steps are waste. For example, in making a hot dog, one might say that the minimum number of steps needed are two, heating the meat in water and packaging it in a bun. All material transport and movements related to collection of payments are not on the value stream and could conceivably be eliminated.

In this book, VSM is presented as a supplement to process mapping, with certain steps used in both methods and the later steps used in VSM only. The version here is relatively intensive. Many people would describe Step 1 by itself as process mapping.

Algorithm 5.1. Process mapping and value stream mapping

Step 1. (Both methods) Create a "□" for each predefined operation. Note if the operation does not have a standard operating procedure. Use a double-box shape for automatic processes and storage. Create a "◇" for each decision point in the overall system. Use an oval for the terminal operation (if any). Use arrows to connect the boxes with the relevant flows.

Step 2. (Both methods) Under each box, label any noise factors that may be varying uncontrollably, causing variation in the system outputs, with a "*z*" symbol. Also, label the controllable factors with an "*x*" symbol. It may also be desirable to identify any gaps in standard operating procedure (SOP) documentation.

Step 3. (VSM only) Identify which steps are "value added" or truly essential. Also, note which steps do not have documented standard operating procedures.

Step 4. (VSM only) Draw a map of an ideal future state of the process in which certain steps have been simplified, shortened, combined with others, or eliminated. This step requires significant process knowledge and practice with VSM.

A natural next step is the transformation of the process to the ideal state. This would likely be considered as part of an improvement phase.

Examination of process mapping and VSM activities both facilitate the identification of bottlenecks and the theory of constraints approach described in ▶ Chap. 2. This also provides lists of inputs and outputs for other analysis activities such as cause and effect matrices and design of experiments. However, process mapping cannot by itself establish statistical evidence to indicate that changes in certain inputs change the average output settings. Part II of this book focuses on establishing such statistical evidence. Logically, however, process mapping can preclude downstream subsystem inputs from affecting upstream subsystem outputs.

Example 5.2.1 Mapping an Arc Welding and Rework System

Question

A Midwest manufacturer making steel vaults has a robotic arc welding system which is clearly the manufacturing bottleneck. A substantial fraction of the units require intensive manual inspection and rework. VSM the system and describe the possible benefit of the results.

Answer

◘ Figure 5.1 is based on an actual walkthrough of the manufacturer. Possible benefits include clarification that several occurring subsystem processes have no documented standard operating procedures (SOPs). Given the importance of these processes as part of the bottleneck subsystem, lack of SOPs likely contributes to variation, nonconformities, and wasted capacity, directly affecting the corporate bottom line. In addition, several inputs and noise factors identified for continued study and process facilitate both goal-setting and benchmarking. Goal-setting is aided because elimination of any of the non-value added subsystems might be targeted, particularly if these constitute bottlenecks in the larger welding system. Benchmarking is facilitated because comparison of process maps with competitors might provide strong evidence that eliminating specific non-value-added tasks is possible. ◘ Figure 5.2 shows an ambitious ideal future state. Another plan might also include necessary operations like transport.

The above example illustrates that process mapping can play an important role in calling attention to non-value-added operations. Often, elimination of these operations is not practically possible. Then documenting and standardizing the non-value-added operation can be useful and is generally considered good practice. In the Toyota vocabulary, the term "necessary" can be used to described these operations. Conveyor transportation in the above operations might be called necessary.

5.2.2 5S

The 5S method relates to organizing the workplace to increase efficiency and eliminate waste. 5S seeks to increase employee morale, as well as making/keeping the employee

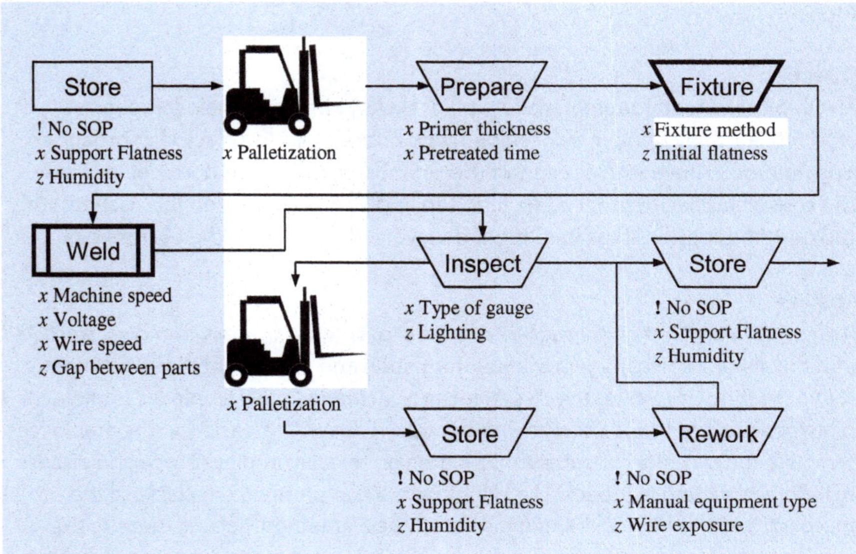

Fig. 5.1 A process map with processing times and factors indicated

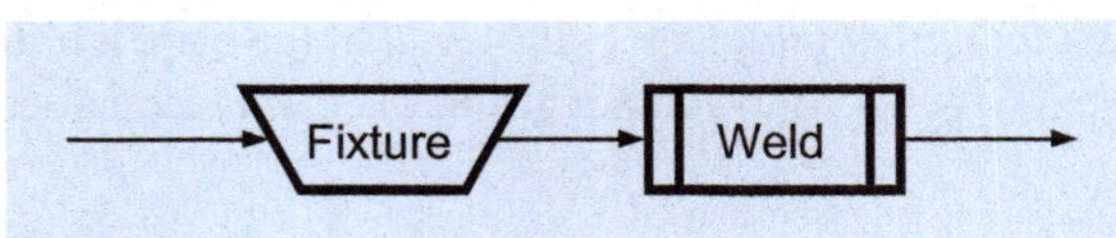

Fig. 5.2 A diagram of an ideal future state

safe and efficient. Commonly when a 5S program is instituted it is started by engineers and managers and taught to the technicians and operators to involve the entire production system. As mentioned previously, the 5S's are sort, straighten, sweep, standardize, sustain in Algorithm 5.2 derived from Imai (1997).

Algorithm 5.2. 5S steps or algorithm

Step 1. Sort—go through everything in a work area and remove all unneeded items.

Step 2. Straighten—arrange everything in the work area to promote faster work. For example tools used could be place in order by amount of usage; closest tool is used most often.

Step 3. Sweep—keep the work area clean and neat at all times. You pick up after each activity and at the end of each shift.

Step 4. Standardize—make sure everyone knows their job precisely and where everything goes. The SOPs described in ► Sect. 5.2.6 may be used here.

Step 5. Sustain—apply previous 4S's continually with care.

*Note: Each activity should be performed with efficiency and safety in mind

Example 5.2.1 Cabinate Maker

Question

An Ohio based metal cabinet making plant seeks to improve their companies' productivity. They bring in a consultant who notices that the place is unorganized, work in process inventory is everywhere, and the environment does not look safe. The consultant decides that a first task is to apply 5S. Make reasonable assumptions and describe specific steps that might be involved in applying the 5S method.

Answer

The consultant might go through all non-employee specific areas, and does steps 1–3 with the aid of local management for information and equipment. Next, he might hold a meeting describing the 5S system and encourage each employee implement it in his/her own work area using the non-employee specific areas. After the employees complete the task, the consultant might then go to each employee area and discuss including providing feedback. This might have a two pronged effects. First, the approach might reinforce the training about the 5S method. Second, getting the employee think about efficiency, safety and his/her work area might provide long term benefits and improved working environment. Following this the consultant might lead implementation of Step 4 for the entire plant. Once all this is completed, the employees would likely have a better understanding of the plant and may have ideas in the future for further improvements. Now that the plant is organized and safe, others can consider applying additional improvement techniques.

5.2.3 Kanban

"**Kanban**" cards regulate the maximum amount of inventory that production cells are allowed to have in queue before the upstream cell must shut down. This approach results in lost production capacity, but it has the benefit of forcing operators, technicians, engineers, and management to focus their attention immediately on complete resolution of problems when they occur. For example, consider a station buffer that has three kanban cards allocated to it. If the preceding station gets ahead by producing three units that are put into the buffer or inventory, the preceding station must shut down. Reducing the numbers of kanban cards throughout the system is often part of lean sigma system improvement efforts.

5.2.4 Poka-Yoke

Poka-yoke translates from Japanese to English as mistake-proofing. The term, as mentioned earlier was created my Toyota as part of the lean system. A way to understand the idea of poka-yoke better is to look at its original term, Baka-yoke meaning idiot-proofing (Dillon 1989). A non-manufacturing example of this is when Toyota stopped allowing the lock button on a car door to be locked when the door is open. This way the door needs to be closes when it locks. This makes it more difficult to lock one's keys in the car. A manufacturing example would be attaching a digital counter to a welder to insure the welder operator does the right number of welds (Shimbun 1988). Poka-yoke can be implemented anywhere errors can occur in manufacturing systems. Shigeo Shingo developed a poka-yoke method associated with three types of defects described in Algorithm 5.3.

Algorithm 5.3. Poka-Yoke

Type 1: Contact—by testing products one can identify physical defects like size, color, shape.

Type 2: Fixed value—operator is alerted if the proper number of movements were not made.

Type 3: Motion-steps—determines if steps of a procedure were followed appropriately.

Question

In an automotive factory, one person tightens four bolts with a manual tensioner. Two of the 4 bolts require a tension of 150 N m, and the other two require 50 N m. It is a common issue that undesirable tensions are put on the bolts that fail to conform to engineering specifications because of mix-ups in bolt types. Therefore, commonly some bolts are too loose while others are too tight, bending the bracket to which the bolts attach. How might poka-yoke be applied and which type of defect is relevant?

Answer

This is probably a motion-step defect because it relates to the incorrect application of operating procedures. One solution would be to split up the task, having one person tighten the two high tension bolt and have another tighten the two light tension bolts. Another option would be use different bolt head sizes for the two different tension amounts. A third option would be to color code the bolt heads according to the different tension amounts. It might be helpful to evaluate the costs of each approach systematically and look for similar issues elsewhere.

5.3 The Toyota Production System

To gain a clearer view of "ideal" systems, it is perhaps helpful to know more about Toyota. The "**Toyota Production System**" used by Toyota to make cars has evolved over the last 50-plus years, inspiring great admiration among competitors and academics. Several management philosophies and catch phrases have been derived from practices at Toyota, including "**just in time**" (JIT) manufacturing, "**lean production**", and "**re-engineering**". Toyota uses several of these novel policies in concert as described in Womack and Jones (1999). The prototypical Toyota system includes:

- "**U-shaped cells**," which has workers follow parts through many operations. This approach appears to build back some of the "craft" accountability lost by the Ford mass-production methods. Performing operations downstream, workers can gain knowledge of mistakes made in earlier operations.
- "**Mixed production**" implies that different types of products are made alternatively so that no batches greater than one are involved. This results in huge numbers of "set-ups" for making different types of units. However, the practice forces the system to speed up the set-up times and results in greatly reduced inventory and average cycle times (the time it takes to turn raw materials into finished products). Therefore, it is often more than possible to compensate for the added set-up costs by reducing costs of carrying inventory and by increasing sales revenues through offering reduced lead times (time between order and delivery).
- "**Pull system**" implies that production only occurs on orders that have already been placed. No items are produced based on projected sales. The pull system used in concert with the other elements of the Toyota Production System appears to reduce the inventory in systems and reduce the cycle times.

Womack and Jones (1999) documented the generally higher sigma levels of production systems modeled after Toyota. However, it might be difficult to attribute these gains to individual components of the Toyota production system. It is possible that the reduced inventory, increased accountability, and reduced space requirement noted are produced by a complex interaction of all of the above-mentioned components.

Many processes involve much more complicated flows than depicted in ◘ Fig. 5.1. This is particularly true in job shop environments where part variety plays a complicating role. Irani et al. (2000) describe extensions of value stream mapping to address complicating factors such as part variety.

5.4 Process Flow and Spaghetti Diagrams

Next, we describe a diagramatic method that focuses on the physical locations of flows through systems. For example, a student might convert paper into assignments at an average yearly rate of 20 pages per month. Then, the paper material would flow through the student system with most of the flows in his or her room between the filing cabinet, the desk, and the book bag. The flow diagrams of interest in this section are called a "spaghetti" diagrams because the flows look sometimes like noodles. The diagrams show these flows overlaid on the physical layout of the facility. The spaghetti diagram is associated with a process of designing revised steps to reduce the process waste by eliminating unnecessary transportation of information or materials. Such transporting is called "parts travel" in industrial engineering because the items are parts that travel on material handling equipment.

Process observation tools such as process maps, cycle time observation, spaghetti diagrams and value stream mapping (VSM) techniques are lean tools used to identify areas of concern and wasted time (Hagg et al. 2007). Additional material about VSM is described in ► Sect. 5.2.

Creating a spaghetti diagram requires the process flow in the facility. The flow is given in facility routings. The facility routings include a location of the product defined by an area in the facility and a brief description of what is being done at that area. When analysis of the spaghetti diagram is being done you use the Total Travel Distance (TTD). TTD is the distance the parts travel throughout the facility.

Algorithm 5.4 Creating Spaghetti Diagrams

Step 1. Obtain an existing or create a facility layout diagram or drawing.
Step 2. Obtain an existing routing sheet or create one using a typical product going through the facility.
Step 3. Draw a continuous curve starting at the first location on the routing sheet to the remaining locations. (Subjectively, complicated-looking flows indicate opportunities for improvement.)
Step 4. Calculate the Total Travel Distance (TTD) by adding up the distance traveled using:

$$\text{TTD} = \Sigma_i n_i d_i$$

where n_i is number of times route traveled, and d_i is distance for the *i*th route and the sum is over all routes.

Step 5. Estimate the travel time by using distance travel at a 4.4 feet/s walking rate or the speed of device used to move product or information. Speed is multiplied by distance.
Step 6. Study the spaghetti diagram and identify flows with high part's travel distances and areas in the facility that are rarely or never used.
Step 7. Move the processes closer together or rearrange to reduce the TTD.
Step 8. Repeat steps 4–7 using new layouts as needed.

Example 5.4.1 Printing Company

Question
A hypothetical company prints large sheets and banners. Its facility layout is in Fig. 5.3. The routing information for their product XJK is in Table 5.1. The company prints one color at a time. Analyze this process and make recommendation to reduce TTD and make the process more efficient, i.e., lower the cycle time.

Answer
The layout and routing in steps 1 and 2 are given in Table 5.1. The spaghetti diagram from Step 3 is shown in Fig. 5.3.

In Step 4, the distance between cut and print is 35 feet. The distance between print and dry is 90 feet and the distance between dry and ship/inspect is 100 feet. In this problem, there are four routes and the counts are:

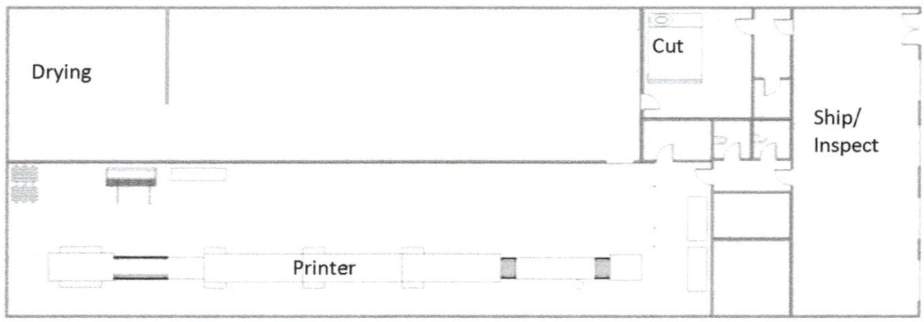

5

Fig. 5.3 A hypothetical layout for company

Table 5.1 Printer company for product XKJ

Companies routing for product XJK		
Routing location	**Location**	**Explanation**
1	Cut	37′ × 12′
2	Print	Blue for XJK
3	Dry	45 min
4	Print	Red for XJK
5	Dry	35 min
6	Print	Green for XJK
7	Dry	40 min
8	Inspect	If error send back to cutting
9	Ship	Ship to Toms Signs

$$\begin{aligned}\text{Cut to print } n_1 &= 1 \text{ time}\\ \text{Print to Dry } n_2 &= 3 \text{ times}\\ \text{Dry to print } n_3 &= 2 \text{ times}\\ \text{Dry to inspect } n_4 &= 1 \text{ time}\end{aligned}$$

Therefore, the TTD is:

$$\text{TTD} = (1)(35) + (3)(90) + (2)(90) + (1)(100) = 585 \text{ feet}$$

In Step 5, we have 585 feet divided by 4.4 feet/s is 132.95 s or 2 min 12 s.

Apparent from the diagram is the many and long flows between the printer and the drying areas. There is also a large unused area between the Cut and Drying areas. Adjusting the placement of these activities so they are closer together would be advised. In our new hypothetical layout, we suggest moving the drying area next to the cutting area as follows.

Iterating Step 4, the new layout is associated with a shortened drying-to-printer distance of 37 feet and a drying-to-inspect/ship distance of 50 feet.

The resulting TTD is:

$$(1)(35) + (3)(37) + (2)(27) + (1)(50) = 270\text{ feet}$$

Iterating Step 5, the 270 feet is divided by 4.4 feet/s giving 61 s or 1.02 min. Therefore, in this example by using a Spaghetti diagram we reduced the cycle time by 1 min and 11 s or 54%. With sufficient part volume, this could result in the ability to layoff or reallocate one or more workers to other activities and/or to produce more products with the same workforce.

Example 5.4.2 Patching Cyber Vulnerabilities

Question

Sketch the common and ideal steps relevant to patching cyber vulnerabilities?

Answer

The common procedure involves assigning the task with a ticket, then the assignee obtains access to the host. Next, the assignee searches for patch information and learned whether the patch is likely harmful. If there is a patch, it is applied and the system is tested. If it passes, the ticket is closed. A more ideal process could eliminate searching by providing the information in the assignment together with the safety of the patch (if any). If there is no patch, there would be no assigned work.

5.5 Cause and Effect Matrices

Process mapping can be viewed as a method that results in a possible list of subsystem inputs for further study. In that view, "cause and effect matrix" (C&E) methods could prioritize this list for further exploration, taking into account information from customers and current engineering insights. To apply the C&E method, it is helpful to recall the basic matrix operations of transpose and multiplication.

The transpose of a matrix **A** is written "**A**′", and it contains the same numbers moved into different positions. For every number in the transposition, the former column address is now the row, and the former row address is now the column address. For example, the following is an example of **A** and **A**′:

$$\mathbf{A} = \begin{bmatrix} 5 & 4 \\ 2 & 3 \\ -3 & 4 \end{bmatrix} \quad \text{and} \quad \mathbf{A}' = \begin{bmatrix} 5 & 2 & -3 \\ 4 & 3 & 4 \end{bmatrix}$$

When two matrices are multiplied, each entry in the result is a row in the first matrix "product into" a column in the second matrix. In this product operation, each element in the first row is multiplied by a corresponding element in the second column and the results are added together. For example:

$$\mathbf{A} = \begin{bmatrix} 5 & 4 \\ 2 & 3 \\ -3 & 4 \end{bmatrix} \quad \mathbf{B} = \begin{bmatrix} 2 & 12 \\ 6 & 9 \end{bmatrix}$$

$$\mathbf{AB} = \begin{bmatrix} 5\times2+4\times6=34 & 5\times12+4\times9=96 \\ 2\times2+3\times6=22 & 2\times12+3\times9=51 \\ -3\times2+4\times6=18 & -3\times12+4\times9=0 \end{bmatrix}$$

5

The C&E method uses input information from customers and engineers and generally requires no building of expensive prototype units. In performing this exercise, one fills out another "room" in the "House of Quality,House of Quality," described further in ► Chap. 6.

Algorithm 5.5. Constructing cause and effect matrices

Step 1. Identify and document q_c customer issues in a language customers can understand. If available, the same issues from a benchmarking chart can be used.
Step 2. Collect ratings from the customers, R_j for $j = 1, \ldots, q_C$ of the subjective importance of each of the q_c customers issues. Write either the average number or the consensus of a group of customers.
Step 3. Identify and document the system inputs, $x_1, \ldots, x_m$, and outputs $Y_1, \ldots, Y_q$ believed to be most relevant. If available, the same inputs and outputs from a benchmarking chart can be used.
Step 4. Collect and document (1–10, 10 is highest) ratings from engineers of the correlations, $C_{i,j}$ for $i = 1,\ldots, q_c, j = 1,\ldots, Cm + q$ and between the q_c customer criteria and inputs and outputs.
Step 5. Calculate the vector $\mathbf{F}' = \mathbf{R}'\mathbf{C}$ and use the values to prioritize the importance of system inputs and outputs for additional investigation.

Example 5.5.1 Software Feature Decision-Making

Question

A Midwest software company is trying to prioritize possible design changes, which are inputs for their software product design problem. They are able to develop consensus ratings from two customers and two product engineers. Construct a C&E matrix to help them prioritize with regard to nine possible features.

Answer

In discussions with software users, seven issues were identified ($q_c = 7$). Consensus ratings for the importance of each were developed. Through discussions with software engineers, the guesses were made about the correlations (were guessed) between the customer issues and the $m = 9$ inputs. No ouputs were considered ($q = 0$). ◘ Table 5.1 shows the results, including the factor ratings ($\mathbf{F}'$) in the bottom column.

The results suggest that regression formula outputs and a wizard for first-timers should receive top priority if the desires of customers are to be respected.

The above example shows how **F′** values are often displayed in line with the various system inputs and outputs. This example is based on real data from a Midwest software company. The next example illustrates how different customer needs can suggest other priorities. This could potentially cause two separate teams to take the same product in different directions to satisfy two markets.

Example 5.5.2 Targeting Software "Power Users"

Question

The software company in the previous example is considering developing a second product for "power users" who are more technically knowledgable. They ask two of these users and develop a rating vector **R′** = [5, 6, 7, 10, 7, 4, 5]. What are the highest priority features for this product?

Answer

The new factor rating vector is **F′** = [193, 195, 156, 156, 142, 183, 186, 183, 172]. This implies that the highest priorities are text delimited data analysis and logistic regression.

5.6 Design of Experiments and Regression (Preview)

Part II of this book, and much of Part III, focus on so-called design of experiment (DOE) methods. These methods are generally considered the most complicated of six sigma related methods. DOE methods all involve: (1) carefully planning sets of input combinations to test using a random run order; then, (2) tests are performed and output values are recorded; (3) an interpolation method such as "**regression**" is then used to interpolate the outputs; and (4) the resulting prediction model is then used to predict new outputs for new possible input combinations. Many regard the random run ordering in DOE as essential for the establishing "**proof**" in the statistical sense.

Regression is also relevant when the choice of input combinations has not been carefully planned. Then, the data is called "on-hand," and statistical proof is not possible in the purest sense.

DOE methods are classified into several types, which include screening using fractional factorials, response surface methods (RSM), and robust design procedures. Here, we focus on the following three types of methods (Fig. 5.4):

- **Screening using fractional factorial methods** begin with a long list of possibly influential factors; these methods output a list of factors (usually fewer in number) believed to affect the average response, and an approximate prediction model.
- **Response surface methods** (RSM) begin with factors believed to be important. These methods generate relatively accurate prediction models compared with screening methods, and also recommended engineering input settings from optimization.
- **Robust design based on process maximization** (RDPM) methods begin with the same inputs as RSM; they generate information about the control-by-noise interactions. This information can be useful for making the system outputs more desirable and consistent, even accounting for variation in uncontrollable factors.

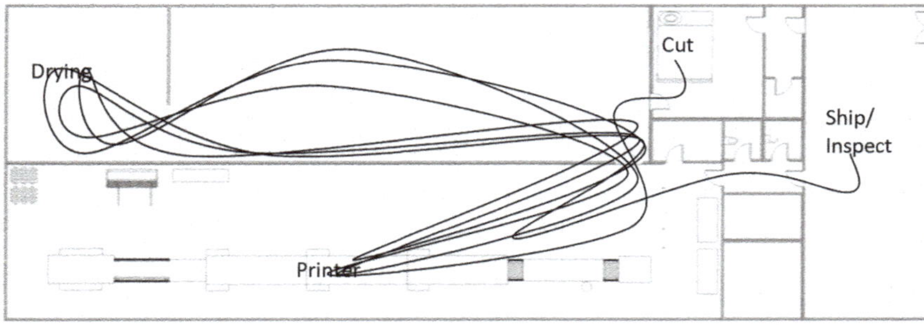

5

Fig. 5.4 Finished Spaghetti diagram for company

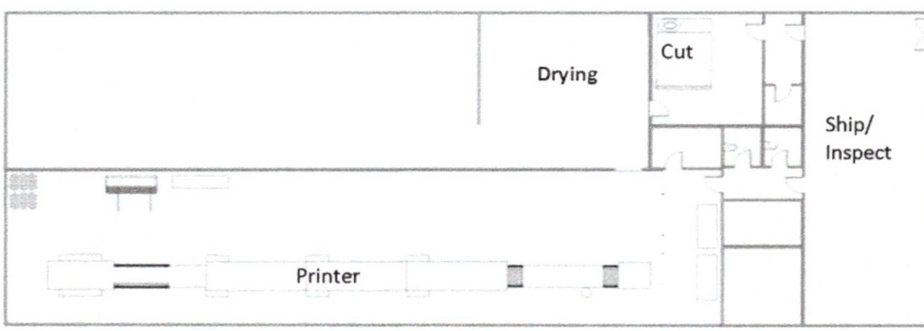

Fig. 5.5 Hypothetical new layout

Example 5.6.1 Spaghetti Meal Demand Modeling

Question

Restaurant management is interested in tuning spaghetti dinner prices because this is the highest profit item. Managers try four different prices for spaghetti dinners, \$9, \$13, \$14, and \$15, for one week each in that order. The profits from spaghetti meals were \$390, \$452, \$490, and \$402, respectively. Managers use the built-in interpolator in Excel to make the plot in Fig. 5.5. This built-in interpolator is a type of regression model. The recommended price is around \$12. Does this prove that price affects profits? Is this DOE?

Answer

This would be DOE if the runs were conducted in random order, but constantly increasing price is hardly random. Without randomness, there is no proof, only evidence. Also, several factor settings are usually varied simultaneously in DOE methods.

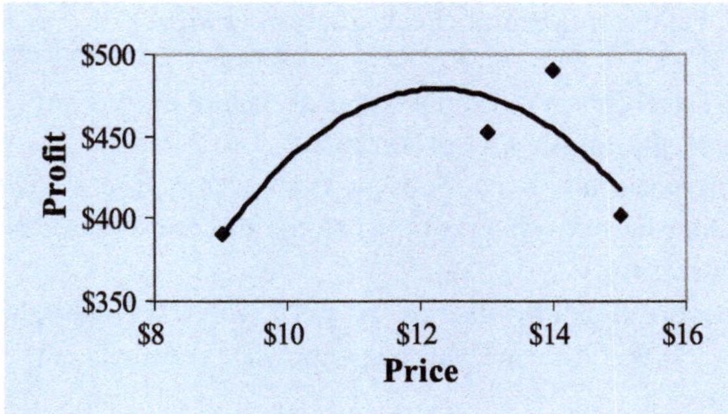

Fig. 5.6 Predicted restaurant profits as a function of spaghetti meal price

5.7 Failure Mode and Effects Analysis

The phrase "**failure mode**" refers to an inability to meet engineering specifications, expressed using a causal vocabulary. The method "**Failure Mode and Effects Analysis**" (FMEA), in Algorithm 5.3, uses ratings from process engineers, technicians, and/or operators to subjectively analyze the measurement system controls. Like cause and effect matrices, FMEA also results in a prioritized list of items for future study. In the case of FMEA, this prioritized list consists of the failure modes and their associated measured quality characteristics or key output variables.

Like gauge R&R, FMEA focuses on the measurement system s. In the case of FMEA, the focus is less on the measurement system's ability to give repeatable numbers and more on its ability to make sure nonconforming items do not reach customers. Also, FMEA can result in recommendations about system changes other than those related to measurement subsystems. Therefore, the scope of FMEA is larger than the scope of gauge R&R.

Symbols used in the definition of FMEA:

1. q is the number of customer issues and associated specifications considered.
2. S_i is the severity rating for the ith issue on a 1–10 scale, with 10 meaning serious, perhaps even life-threatening.
3. O_i is the occurrence rating of the ith issue on a 1–10 scale, with 10 meaning very common or perhaps even occurring all the time.
4. D_i is the detection rating based on current system operating procedures on a 1–10 scale, with 10 meaning almost no chance that the problem will be detected before the unit reaches a customer.

The following example illustrates the situation-dependent nature of FMEA analyses. It is relevant to the childcare situation of a certain home at a certain time and no others. Also, it shows how the users of FMEA do not necessarily need to be experts, although that is preferable. The FMEA in the next example simply represents the best guesses of one concerned parent, yet it was helpful to that parent in prioritizing actions.

Algorithm 5.6. Failure mode and effects analysis (FMEA)

Step 1. Create a list of the q customer issues or "failure modes" for which failure to meet specifications might occur.
Step 2. Document the failure modes using a causal language. Also document the potentially harmful effects of the failures and notes about the causes and current controls.
Step 3. Collect from engineers the ratings (1–10, with 10 being high) on the severity, S_i, occurrence, O_i, and detection, D_i, all for all $i = 1, \ldots, q$ failure modes.
Step 4. Calculate the risk priority numbers, RPN_i, calculated using $RPN_i = S_i O_i D_i$ for $i = 1, \ldots, q$.
Step 5. Use the risk priority numbers to prioritize the need for additional investments of time and energy to improve the inspection controls.

Example 5.7.1 Toddler Proofing One Home

Question
Perform FMEA to analyze the threats to a toddler at one time and in one home.

Answer
In this case, any harm to the toddler is considered nonconformity. Table 5.2 shows an FMEA analysis of the perceived threats to a toddler. The results suggest that the most relevant failure modes to address are TV watching, old milk, and slamming doors. The resulting analysis suggests discussion with childcare providers and efforts to limit total TV watching to no more than three episodes of the program "Blue's Clues" each day. Also, door stops should be used more often and greater care should be taken to place old sippy cups out of reach. Periodic reassessment is recommended.

The next example is more traditional in the sense that applies to a manufacturing process. Also, the example focuses more on inspection methods and controls and less on changing system inputs. Often, FMEAs result in recommendations for additional measurement equipment and/or changes to measurement standard operating procedures (SOPs).

Table 5.2 Cause and effect matrix for software feature decision-making

Customer issues	Customer rating	Include logistic regression	Import text delimited data	Include DOE	Include optimization	Include random factors	Improved stability	Improve examples	Regression formula output	Wizard for first-timers
Easy to use	5	1	1	1	1	1	5	5	7	7
Helpful user environment	6	1	6	1	1	1	10	5	6	6
Good examples	6	1	1	1	1	1	1	9	4	4
Powerful enough	5	8	9	5	5	5	6	2	2	1
Enough methods covered	8	8	4	7	7	5	1	2	1	1
Good help and support	5	1	1	1	1	1	1	6	8	9
Low price	8	7	5	7	7	7	4	2	5	4
	F′ =	182	169	159	159	143	166	181	193	185

5

Table 5.3 FMEA table for toddler home threat analysis

Failure mode	Potential effect	Severity	Potential cause	Occurrence	Current control	Detection	RPN
Electric shock from outlet	Death	10	Not watched	1	Supervision	2	20
Falling down stairs	Death	9	Escapes gate	1	Mommy wakes	3	27
Rotten milk from old cups	Tummy ache	3	Cup within reach	5	Supervision	7	105
Slipping in bath tub	Bruises	2	No matting	6	Supervision	7	84
Car accident	Death	8	Driving too far	1	Carefulness	1	8
Pinching fingers in door	Loss of finger	6	Stop not in place	2	Door stops	8	96
Watching too much TV	Obesity, less reading	4	Parents tired	7	Effort	6	168

Example 5.7.2 Arc Welding Inspection Controls

Question

Interpret the hypothetial FMEA table in ▣ Table 5.3.

Answer

Additional measurement system standard operating procedures and associated inspections are probably needed to monitor and control gaps in fixturing. It is believed that problems commonly occur that are not detected. Also, the visual and informal inspection of weld penetration is likely not sufficient. Serious consideration should be given to X-ray and/or destructive testing.

5.8 Chapter Summary

This chapter describes methods relevant mainly for studying or "analyzing" systems prior to developing recommendations for changing them. These methods include process mapping, value stream mapping, generating cause & effect matrices, design of experiments (DOE), and failure mode and effects analysis (FMEA). ▣ Table 5.4 summarizes the methods described in this chapter, along with the possible roles in an improvement or design project.

Process mapping involves careful observation and documentation of flows within a system. Possible results include the identification of subsystem bottlenecks and inputs for further study. This chapter argues that value stream mapping constitutes an augmentation of ordinary process mapping with a focus on the identification of unnecessary activities that do not add value to products or services. Additional discussion is provided about the Toyota production system and the possible ideal state of systems.

The cause and effect matrix method has the relatively specific goal of focusing attention on the system inputs or features that most directly affect the issues of importance to customers. The results are relevant to prioritizing future investigations and/or selecting features for inclusion into systems.

Design of experiments (DOE) methods are complicated enough that Part II and much of Part III of this book is devoted to them. Here, only the terms screening, response surface, and robust design are of interest. Also, the concept of using random ordering of test runs is related both to DOE methods and the concept of statistical proof.

Finally, the chapter describes how FMEA can be used to rationalize possible actions, to safeguard customers, and to prioritize measurement subsystems for improvement. FMEA is a powerful tool with larger scope than gauge R&R because it involves simultaneous evaluation of both measurement subsystems and other subsystems (▣ Table 5.5).

Table 5.4 FMEA table for arc welding process control

Controlled factors and responses	Potential failure modes	Potential failure effects	Severity	Potential causes	Occurence	Current control	Detection	RPN
Undercut	Cosmetic and stress fracture	Yield under high loads	5	Fixture gap, offset, and voltage	8	Visual and informal	2	80
Penetration	Cosmetic and yielding	Yield under high loads	5	Fixture gap, offset, and voltage	4	Visual and informal	9	180
Melt through	Cosmetic and yielding	Leakages and yielding	7	Fixture gap, offset, and voltage	4	Visual and informal	2	56
Distortion (flatness)	Out-of-spec.	Down stream $ and problems	4	Fixture gap, offset, voltage, others	10	Visual and informal	4	160
Fixture gap	Failures by undercut	All of the above	5	Operator variability and part variability	9	Visual and informal	5	225
Initial flatness	Cosmetic and yielding	All of the above plus no weld	5	Size of sheet, transportation	4	Visual and informal	5	100
Voltage variability	See undercut and penetration	See above	5	Power supply	3	Gauge and informal	2	30

Table 5.5 Summary of methods of primary relevance to the analyze phase

Method	Possible role
Process mapping	Bottleneck and input factor identification
Value stream mapping	Identifying waste for possible elimination
Cause and effect matrices	Prioritizing inputs and outputs for further study
Design of experiments	Building input-output prediction models for tuning inputs
FMEA	Prioritizing nonconformities for adding inspection controls

5.9 Problems

In general, pick the correct answer that is most complete.

1. Value Stream Mapping can be viewed as an extension of which activity?
   a. Gauge R&R.
   b. Benchmarking.
   c. Design of experiments.
   d. Process mapping.
   e. None of the above is correct.
2. What are possible benefits associated with U-shaped cells?
   a. Parts are produced before demands are placed, for readiness.
   b. Parts are produced in batches of one.
   c. Personal accountability for product quality is returned to the worker.
   d. All of the above are correct.
   e. All of the above are correct except (a) and (d).
3. Which is correct and most complete?
   a. Mixed production results in fewer setups than ordinary batch production.
   b. Kanban cards can limit the total amount of inventory in a plant at any time.
   c. U-shaped cells cause workers to perform only a single specialized task well.
   d. All of the above are correct.
   e. All of the above are correct except (a) and (d).

Table 5.6 contains hypothetical data on used motorcycles for questions 6–8.

4. What issue or issues do customers value most according to the C&E matrix?
   a. The cost is low.
   b. Performance is good.
   c. Tires seem worn down.
   d. Traction around turns is good.
   e. Customers value answers "a" and "c" with equal importance.
5. Which KIV or KOV value is probably most important to the customers?
   a. Type of durometer.
   b. Rubber width.
   c. Bead thickness.
   d. Tire diameter.
   e. All of the above KIVs or KOVs are equally important for investigation.

Table 5.6 Cause and effect matrix for used motorcycle data

Racer customer issues	Customer importance	Type of durometer	Rubber width	Bead thickness (mm)	PSI capacity	Tire height (mm)	Tire diameter (mm)
Handling feels sticky	7	2	4	7	3	6	2
Tires seem worn down	4	4	5	7	6	7	2
Handling feels stable	1	9	8	6	6	5	3
Good traction around turns	3	1	3	9	5	6	6
Cost is low	7	8	7	9	2	4	3
Installation is difficult	2	5	4	2	5	3	8
Performance is good	1	9	5	4	8	1	1
Wear-to-weight ratio is good	1	9	1	7	1	5	2
Factor rating number (F')		231	274	141	137	203	142

6. Which of the following is correct and most complete?
   a. **R′** is a 4 × 7 matrix.
   b. **R′C** is a vector with five entries.
   c. Cause and effect matrices create a prioritized list of factors.
   d. All of the above are correct.
   e. All of the above are correct except (a) and (d).

Table 5.7 will be used for Questions 9 and 10.

7. If "story is interesting" was determined to have a correlation of "3" for all factors, which would have the highest factor rating number?
   a. Money spent on script writers.
   b. Young male focus group used.
   c. Stars used.
   d. Critic approval.
   e. All of the above would have equal factor rating numbers.
8. Suppose the target audience thought the only criterion was inspiration. Which variable would be the most important to focus on?
   a. Money spent on script writers.
   b. Young male focus group used.
   c. Stars used.
   d. Critic approval.
   e. All of the above would have equal factor rating numbers.
9. List three benefits of applying cause and effect matrices.

Table 5.7 Hypothetical movie studio cause and effect matrix

Customer criteria	Customer importance	Money spent on script writers	Young male focus group used	Stars used	Critic approval
Story is interesting	8	8	5	4	7
It was funny	9	8	5	4	7
It was too long or too short	6	3	5	9	1
It made me inspired	4	6	2	4	5
It was a rush	6	6	8	6	5
It was cool	6	3	8	4	5
Factor rating number (F′)		232	219	198	205

10. Which of the following is correct and most complete?
    a. Design of experiments does not require new testing.
    b. Screening can start where C&E matrices end and further shorten the KIV list.
    c. Strictly speaking, DOE is essential for proof in the statistical sense.
    d. All of the above are correct.
    e. All of the above are correct except (a) and (d).

Table 5.8 will be used for Questions 13–14.

11. Which response or output probably merits the most attention for quality assurance?
    a. Freshness.
    b. Taste.
    c. Number of chips.
    d. Size.
    e. Burn level.
12. How many failure modes are represented in this study?
    a. 3.
    b. 4.
    c. 5.
    d. 6.
    e. 7.
13. Which of the following is correct and most complete?
    a. FMEA focuses on the manufacturing system with little regard to measurement.
    b. FMEA is based on quantitative data measured using physical equipment.
    c. FMEA helps to clarify the vulnerabilities of the current system.
    d. All of the above are correct.
    e. All of the above are correct except (a) and (d).

Table 5.9 will be used in Questions 16 and 17.

14. If the system were changed such that it would be nearly impossible for the explosion failure mode to occur (occurrence = 1) and no other failure mode was affected, the <u>highest</u> priority factor or response to focus on would be:
    a. Rubber width.
    b. Bead thickness.
    c. PSI capacity.
    d. Tire height.
15. If the system were changed such that detection of all issues was near perfect and no other issues were affected (detection = 1 for all failure modes), the <u>lowest</u> priority factor or response to focus on would be:
    a. Rubber width.
    b. Bead thickness.
    c. PSI capacity.
    d. Tire height.

Table 5.8 Hypothetical cookie-baking FMEA

Controlled factors and responses	Potential failure modes	Potential failure effects	Severity	Potential causes	Occurrence	Current control	Detection	RPN
Burn level	Cosmetic	Customer won't eat or buy	6	Over- cooked	2	Visual, informal	1	12
Texture	Too dry	Crumbs, squish easily	4	Temp too high or low	2	Touch, informal	2	16
Size	Too small	Customer might not buy more	5	End of batch	2	Visual, informal	1	10
Taste	Taste	Customer won't eat or buy	9	Wrong amount of ingredients	2	Taste	4	72
Freshness	Not stored properly	Customer won't eat or buy	8	Not stored properly	3	Taste	4	96
Number of chips	Batter not mixed	Tastes like plain cookie	2	End of batch	2	Visual, informal	4	16

Table 5.9 Hypothetical motorcycle tire FMEA

Controlled factors and responses	Potential failure modes	Potential failure effects	Severity	Potential causes	Occurrence	Current control	Detection	RPN
Rubber width	Implosion	Transportation failure, possible loss of business	6	Improper pressure requirements by motorcycle	5	Visual, experience	7	210
Bead thickness (mm)	Tire sloughing while riding	Transportation failure, possible loss of business	7	Too much rider leaning	6	Safety training	6	252
PSI capacity	Explosion	Transportation failure, possible loss of business	6	Improper pressure requirements by motorcycle	7	Visual, experience	8	336
Tire height (mm)	Compression leading to implosion	Transportation failure, possible loss of business	5	Aggressive rider	4	Engine regulation	8	160

16. Critique the toddler FMEA analysis, raising at least two issues of possible concern.
17. Which of the following is the most complete and correct?
    a. FMEA is primarily relevant for identifying nonvalue added operations.
    b. Both C&E and FMEA activities generate prioritized lists.
    c. Process mapping helps identify cause and effect relationships.
    d. All of the above are correct.
    e. All of the above are correct except (a) and (d).
18. According to the chapter, how are lean and six sigma related?
19. Apply process mapping (without the steps that are only for value stream mapping) to a system that you might improve as a student project. This system must involve at least five operations.
20. According to the chapter, what is a principle of lean manufacturing.
21. List four techniques specifically designed to identify and eliminate non-value-added steps in processes.
22. Describe the possible application of the 5S to your room where you live.
23. Perform *Steps 3* and *4* to the system identified in solving the previous problem to create a process map of an ideal future state. Assume that sufficient resources are available to eliminate all non-value added operations.
24. Describe the hypothetical application of lean sigma to improve operations in a coffee shop.
25. Discuss the five steps for theory of constraints in relation to your own life.
26. What is the primary purpose of spaghetti diagrams?
27. Describe the hypothetical application of poke-yoke to a problem in your own life.
28. How do spaghetti diagrams differ from value stream maps?
29. ◘ Table 5.10 and ◘ Fig. 5.7 describe operations and a handmade baseball bat manufacturer. The company receives blocks of wood ready for turning. Analyze the operations using a spaghetti diagram. Assume that materials arrive at the end of the dock ramp.
30. Calculate the TTD and travel time related to the previous problem. The distances are in ◘ Table 5.11. Assume that receiving starts at the dock.
31. Propose layout changes and show calculations to justify your changes related to the previous problems.
32. ◘ Table 5.12 and ◘ Fig. 5.8 describe an alternative manufacturing process for the handmade baseball bat manufacturer. Analyze the alternative process using a spaghetti diagram. Assume that receiving starts at the dock.
33. Calculate the TTD and travel time related to the previous problem. The distances are in ◘ Table 5.13. Assume that receiving starts at the dock.
34. Propose layout changes and show calculations to justify your changes for the bat company.
35. Identify a local business and describe how lean sigma tools might help them improve their operations.

5

Table 5.10 Bat Routing

Black with natural handle bat routing		
Routing location	**Location**	**Explanation**
1	Receive	Unpack from truck
2	Sort/Storage	Sort by size and store accordingly
3	Turning	Turn to specifications
4	Sort/Storage	Put in turned area by sizes
5	Painting (Seal)	Seal bat
6	Painting (Dry)	20 min
7	Painting (Paint)	Paint head of bat
8	Painting (Dry)	25 min
9	Painting (Logo)	Paint Company logo
10	Painting (Dry)	15 min
11	Sort/Storage	Put in finished bat selves by size
12	Sales floor	Display where appropriate

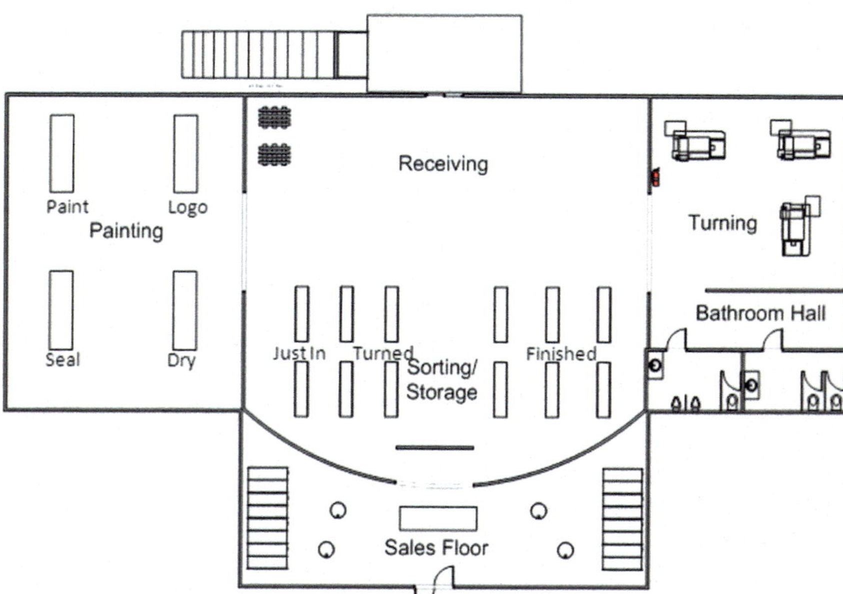

Fig. 5.7 Bat facility layout

Table 5.11 Bat facility distances

Black with natural handle bat routing		
Routing location	Location	Distance to next location (feet)
1	Receive	100
2	Sort/Storage	150
3	Turning	145
4	Sort/Storage	115
5	Painting (Seal)	50
6	Painting (Dry)	50
7	Painting (Paint)	50
8	Painting (Dry)	50
9	Painting (Logo)	50
10	Painting (Dry)	115
11	Sort/Storage	75
12	Sales floor	x

Table 5.12 Bat facility 2

Black with natural handle bat routing		
Routing location	Location	Explanation
1	Receive	Unpack from truck
2	Sort/Storage	Sort by size and store accordingly
3	Cutting	Cut to proper size
4	Sort/Storage	By size in Just in area
5	Turning	Turn to specifications
6	Sort/Storage	Put in turned area by sizes
7	Cutting	Cut ends off where lathe held bat
8	Painting (Seal)	Seal bat (waits to dry in seal area)
9	Painting (Paint)	Paint head of bat (waits to dry in paint area.
10	Cutting	Sand after first coat
11	Painting (Logo)	Paint Company logo
12	Sort/Storage	Put in finished bat selves by size
13	Sales floor	Display where appropriate

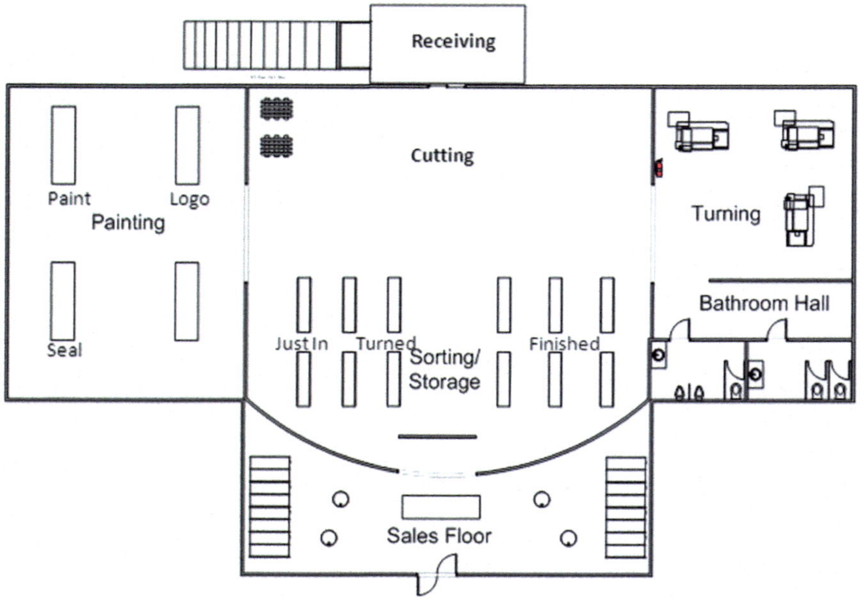

Fig. 5.8 Bat facility layout 2

Table 5.13 Bat facility 2 distances

Black with natural handle bat routing		
Routing location	**Location**	**Distance to next location (feet)**
1	Receive	100
2	Sort/Storage	75
3	Cutting	75
4	Sort/Storage	150
5	Turning	150
6	Sort/Storage	75
7	Cutting	115
8	Painting (Seal)	50
9	Painting (Paint)	115
10	Cutting	115
11	Painting (Logo)	115
12	Sort/Storage	75
13	Sales floor	x

References

Banks J, Carson JS, Nelson NB (2000) Discrete-event system simulation, 3rd edn. Pearson International, Upper Saddle River, NJ

Georgano GN (2000) Vintage cars 1886 to 1930. AB Nordbok, Sweden

Hagg H, Suskovich D, Workman-Germann J, Scachitti S, Hudson B, Swarts J, Vanni C (2007) Adaptation of lean methodologies for heathcare applications. RCHE Publications, Purdue Libraries

Irani SA, Zhang H, Zhou J, Huang H, Udai TK, Subramanian S (2000) Production flow analysis and simplification toolkit (PFAST). Int J Prod Res 38:1855–1874

Imai M (1997) Gemba kaizen: a commonsense low-cost approach to management. New York: McGraw-Hill Professional

Law A, Kelton WD (2000) Simulation modeling and analysis, 3rd edn. McGraw-Hill, New York

Liker J (ed) (1998) Becoming lean: inside stories of U.S. manufacturers. Productivity Press, Portland, OR

Nikkan KS (1988) Improving product quality by preventing defects. Productivity Press

Shigeo S (1989) A study of the Toyota production system, Productivity Press

Suzaki K (1987) The new manufacturing challenge. Simon & Schuster, New York

Womack JP, Jones DT, Roos D (1991) The machine that changed the world: the story of lean production. Harper-Business, New York

Womack JP, Jones DT (1996) Lean thinking. Simon & Schuster, New York

Womack JP, Jones DT (1999) Learning to See, Version 1.2. Lean Enterprises Instititute Incorporated

Improve or Design Phase

T. T. Allen, *Introduction to Engineering Statistics and Lean Six Sigma*,
https://doi.org/10.1007/978-1-4471-7420-2_6

6.1 Introduction

In ► Chap. 5, methods were described with goals that included clarifying the input-output relationships of systems. The purpose of this chapter is to describe methods for using the information from previous phases to tune the inputs and develop tentative recommendations. The phrase "**improvement phase**" refers to the situation in which an existing system is being improved. The phrase "**design phase**" refers to the case in which a new product is being designed.

The recommendations derived from the improve or design phases are generally considered tentative. This follows because usually the associated performance improvements must be confirmed or verified before the associated standard operating procedures (SOPs) or design guidelines are changed or written.

Here, the term "**formality**" refers to the level of emphasis placed on data and/or computer assistance in decision-making. The methods for improvement or design presented in this chapter are organized by their level of formality. In cases where a substantial amount of data is available and there are a large number of potential options, people sometimes use a high level of formality and computer assistance. In other cases, less information is available and/or a high degree of subjectivity is preferred. Then "**informal**" describes the relevant decision-making style. In general, statistical methods and six sigma are associated with relatively high levels of formality.

Note that the design of experiments (DOE) methods described in ► Chap. 5 and in Part II of this book are often considered to have scope beyond merely clarifying the input-output relationships. Therefore, other books and training materials sometimes categorize them as improvement or design activities. The level of formality associated with DOE-supported decision-making is generally considered to be relatively high.

This section begins with a discussion of informal decision-making including so-called "**seat-of-the-pants**" judgments. Next, moderately formal decision-making is presented, supported by so-called "QFD House of Quality," which combines the results of benchmarking and C&E matrix method applications. Finally, relatively formal "**optimization**" and "**operations research**" approaches are briefly described. Part III of this book describes these topics in greater detail.

6.2 Informal Optimization

It is perhaps true that the majority of human decisions are made informally. It is also true that formality in decision processes generally costs money and time and may not result in improved decisions. The main goals of this book are to make available to users relatively formal methods to support decision-making and to encourage people to use these methods.

With continual increases in competitive pressures in the business world, throrough investigation of options and consideration of various issues can be necessary to achieve profits and/or avoid bankruptcy. Part III of this book reviews results presented in Brady (2005) including an investigation of 39 six sigma projects at a Midwest company over two years. Every project decision process involved a high degree of formality. Also, approximately 60% of projects showed profits, average per project profits exceeded $140,000, and some showed extremely high profits.

The degree of formality varies among informal methods. The phrase "**anecdotal information**" refers to ideas that seem to be supported by a small number of stories, some of which might be factual. "**Seat-of-the-pants**" decision-making uses subjective judgments, potentially supported by anecdotal information, to propose new system inputs or designs. Recommendations from seat-of-the-pants approaches are rarely accompanied by any objective empirical conformation of improvement.

Relatively formal approaches involve subjective decision-making supported by the generation of tables or plots. For example, a cause and effects matrix might be generated and encourage the addition of new features to a software product design. Since no computers were used to systematically evaluate a large number of alternatives and no data collection process was used to confirm the benefits, this process can still be regarded as informal. The following example illustrates informal decision-making supported by a formal data collection and display method.

Example 6.2.1 High Vacuum Aluminum Welds

Question

At one energy company, dangerous substances are stored in high vacuum aluminum tubes. The process to weld aluminum produces a non-negligible fraction of nonconforming and potentially defective welds. Develop tentative recommendations for process and measurement system design changes supported by the hypothetical FMEA shown in Table 6.1. Assume that engineers rate the detection of complete X-ray inspection as a "2".

Answer

An engineer at the energy company might look at Table 6.1 and decide to implement complete inspection. This tentative choice can be written x_1 = complete inspection. With this choice, fractures caused by porosity might still cause problems but they would no longer constitute the highest priority for improvement.

The next example illustrates a deliberate choice to use informal methods even when a computer has assisted in identifying possible input settings. In this case, the best price for maximizing expected profits is apparently known under certain assumptions. Yet, the decision-maker subjectively recommends different inputs.

Example 6.2.2 Spaghetti Meal Revisited

Question

Use the DOE and regression process described in the example in ▶ Chap. 5 to support an informal decision process.

Answer

Rather than selecting a $12 menu price for a spaghetti dinner because this choice formally maximizes profit for this item, the manager might select $13. This could occur because the price change could fit into a general strategy of becoming a higher "class" establishment. The predicted profits indicate that little would be lost from this choice.

Table 6.1 Hypothetical FMEA for aluminum welding process in energy application

Failure mode	Potential effect	Severity	Potential cause	Occurrence	Current control	Detection	RPN
Porosity caused fracture	Hazardous leakage	10	Dirty metal	3	Partial X-ray and on-line	5	150
Bead shape caused frac.	Hazardous leakage	10	Fixturing	1	Visual inspection	1	10
Contamination of vacuum	Significant expense	4	Spatter	4	Visual and mirror	4	64
Joint distortion	Production delays	3	Fixturing	2	Visual	1	6

6.3 Quality Function Deployment (QFD)

The method "**Quality Function Deployment**" (QFD) is a popular formal approach to support what might be called moderately formal decision-making. QFD involves creating a full "**House of Quality**" (HOQ), which is a large matrix that contains much information relevant to decision-making. This HOQ matrix constitutes an assemblage of the results of the benchmarking and cause and effect matrices methods together with additional information. Therefore, information is included both on what makes customers happy and on measurable quantities relevant to engineering and profit maximization. The version of QFD presented here is inspired by "enhanced QFD" in Clausing (1994, pp. 64–71).

Algorithm 6.1. Quality function deployment (QFD)

Step 1. Perform the QFD benchmarking method described in ► Chap. 4 in the measurement phase.
Step 2. Perform the QFD cause and aeffect matrix method described in ► Chap. 5 in the analysis phase.
Step 3. Consult engineers to set default targets for the relevant engineering outputs $Y_1, \ldots, Y_q$ and certain inputs, $x_1, \ldots, x_m$. In some cases, inputs will not have targets. In other cases, inputs such as production rate of a machine are also outputs with targets, e.g., machine speed in welding. These target numbers are entered below entries from the benchmarking table in the columns corresponding to quantitative output variables.
Step 4. (Optional) Poll engineers informally estimate the correlations between the process inputs and outputs. These numbers form the "roof" of the House of Quality, but may not be needed if DOE methods have already established these relationships using real data from prototype systems.
Step 5. Inspect the diagram and revise the default input settings subjectively. If one of the companies dominates the ratings, consider changing the KIV factor settings, particularly those with highest factor rating numbers, to emulate that company.

Example 6.3.1 Arc Welding Process QFD

Question
Interpret the information in Table 6.2 and make recommendations.

Answer
A reasonable set of choices in this case might be to implement all of the known settings for Company 3. This would seem to meet the targets set by the engineers. Then, tentative recommendations might be: $x_1=40, x_2=9, x_3=5, x_4=15, x_5=2.5, x_6=8, x_7=19, x_8=0.8, x_9=9.7, x_{10}=1.0, x_{11}=3.0, x_{12}=23, x_{13}=$Yes, $x_{14}=$Yes, and $x_{15}=1.5$, where the input vector is given in the order of the inputs and outputs in Table 6.2. Admittedly, it is not clear that these choices maximize the profits, even though these choice seem most promising in relation to the targets set by company engineers and management.

Table 6.2 An example of the House of Quality

Customer criterion	Customer importance (R)	Travel speed (ipm)	Weld area (WFS/TS)	Trip-to-work dist. (mm)	Wire diameter (mm)	Gas type (%CO2)	Travel angle (°)	Fixture average offset (mm)	Fixture average gap (mm)	Initial flatness (mm)
Incidence crevices (undercut)	5	2	10	9	2	10	2	6	3	1
Quality of the surface (porosity)	8	9	3	2	7	1	6	9	8	4
Incidence of joint leakage (fusion)	10	5	10	10	4	2	2	9	8	10
Average strength of joints (penetration)	9	10	10	9	2	2	2	6	7	5
Excessive fusion/ holes (melt through)	3	9	5	4	1	1	1	1	2	10
Incidence of cracks (cracks)	5	3	2	1	2	8	1	5	6	5
Incidence of HAZ cracks (HAZ)	4	2	4	2	1	1	1	5	5	9
Incidence of buckling (buckling)	7	10	9	2	3	1	10	9	5	3
Incidence of unsighty spatter (spatter)	5	8	10	8	3	1	5	4	7	2

(continued)

Table 6.2 (continued)

Customer criterion	Customer importance (R)	Travel speed (ipm)	Weld area (WFS/TS)	Trip-to-work dist. (mm)	Wire diameter (mm)	Gas type (%CO2)	Travel angle (°)	Fixture average offset (mm)	Fixture average gap (mm)	Initial flatness (mm)
Out-of-plane distortion (distortion)	10	7	10	10	10	2	8	9	10	8
Factor rating number (**F′**)		452	518	421	277	175	283	467	448	384
Company 1	35.0		8.0	15.0	2.0	0.1	15.0	1.1	0.9	1.1
Company 2	42.0		9.2	15.0	2.0	0.0	0.0	0.9	1.0	1.2
Company 3	40.0		9.5	15.0	2.5	0.1	10.0	NA	NA	1.0
Targets	55		–	–	2.5	–	–	0.5	0.5	0.8

Customer criterion	Plate thickness (mm)	Arc length (voltage)	Palletized or not	Heating pretreatment	Flatness of support (cm)	% Meeting AVVS specs	Company 1	Company 2	Company 3
Incidence crevices (undercut)	1	7	1	2	2	10	3.3	4.0	8.0
Quality of the surface (porosity)	10	6	4	1	7	10	5.0	5.0	5.3
Incidence of joint leakage (fusion)	2	1	1	5	4	10	6.0	9.3	8.3

(continued)

Table 6.2 (continued)

Customer criterion	Plate thickness (mm)	Arc length (voltage)	Palletized or not	Heating pretreatment	Flatness of support (cm)	% Meeting AVVS specs	Company 1	Company 2	Company 3
Average strength of joints (pen-etration)	5	6	3	1	5	8	9.0	9.0	9.0
Excessive fusion/ holes (melt through)	9	8	2	2	2	10	6.0	8.0	9.0
Incidence of cracks (cracks)	1	8	1	2	2	10	9.0	9.7	10.0
Incidence of HAZ cracks (HAZ)	7	8	3	2	2	10	8.0	9.0	9.0
Incidence of buckling (buckling)	3	5	1	1	9	3	4.0	4.7	7.0
Incidence of unsighty spatter (spatter)	2	1	1	6	3	7	8.3	8.0	7.7

(continued)

Table 6.2 (continued)

Customer criterion	Plate thickness (mm)	Arc length (voltage)	Palletized or not	Heating pretreatment	Flatness of support (cm)	% Meeting AVVS specs	Company 1	Company 2	Company 3
Out-of-plane distortion (distortion)	8	10	3	9	8	3	3.0	4.0	7.0
Factor rating number (F')	321	383	139	228	333	508			
Company 1	2.0	20.0	No	No	3.5	15.0			
Company 2	2.0	19.0	No	No	4.0	10.0			
Company 3	2.0	23.0	Yes	Yes	1.5	Na			
Targets	–	–	–	No	1.5	5			

In the preceding example, the choice was made to copy another company to the extent possible. Some researchers have provided quantitative ways to select settings from QFD activities; see Shin et al. (2002) for a recent reference. These approaches could result in recommendations having settings that differ from all of the benchmarked alternatives. Also, decision-makers can choose to use QFD simply to aid in factor selection in order to perform followup design of experiments method applications or to make sure that a formal decision process is considering all relevant issues. One of the important benefits of applying QFD to support decision-making is increasing confidence that the project team has thoroughly accounted for many types of considerations. Even when applying QFD, information can still be incomplete, making it unadvisable to copy best in-class competitors without testing and/or added data collection.

6.4 Formal Optimization

In some situations, people are comfortable with making sufficient assumptions such that their decision problems can be written out very precisely and solved with a computer. For example, suppose that a person desires to maximize the profit, $g(x_1)$, from running a system as a function of the initial investment, x_1. Further, suppose one believes that the following relationship between profit and x_1 holds $g(x_1) = -11 + 12x_{1-2}x_1^2$ and that one can only control x_1 over the range from $x_1 = 2$ to $x_1 = 5$. Functional relationships like $g(x_1) = -11 + 12x_{1-2}x_1^2$ can be produced from regression modeling potentially derived from applying design of experiments.

The phrase "**optimization solver**" refers to the approach the computer uses to derive recommended settings. In the investment example, it is possible to apply the Excel spreadsheet solver with default settings to derive the recommended initial investment. This problem can be written formally as the following "**optimization program**" which constitutes a precise way to record the relevant problem:

$$\begin{aligned} &\text{Maximize: } g(x_1) = -11 + 12x_1 - 2x_1^2 \\ &\text{Subject to: } x_1 \in [2, 5] \end{aligned} \tag{6.1}$$

where $x_1 \in [2, 5]$ is called a "**constraint**" because it limits the possibilities for x_1. It can also be written $2 \leq x_1 \leq 5$.

The term "**optimization formulation**" is synonymous with optimization program. The term "**formulating**" refers to the process of transforming a word problem into a specific optimization program that a computer could solve. The study of "**operations research**" focuses on the formulation and solutions of optimization programs to tune systems for more desirable results. This is the study of perhaps the most formalized decision processes possible.

Figure 6.1 shows the application of the Excel solver to derive the solution of this problem, which is $x_1 = 3.0$. The number 3.0 appears in cell "A1" upon pressing the "Solve" button. To access the solver, one may need to select "Tools", then "Add-Ins...", then check the "Solver Add-In" and click OK. After the Solver is added in, the "Solver..." option should appear on the "Tools" menu.

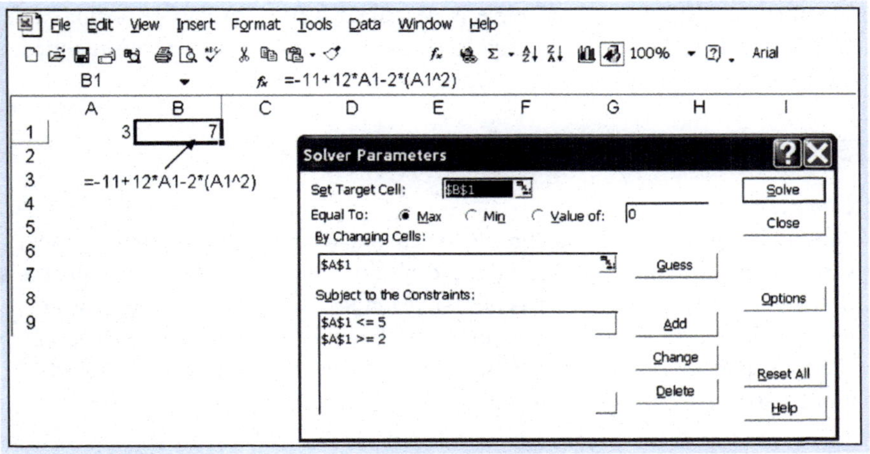

■ **Fig. 6.1** Screen shot showing the application of the excel solver

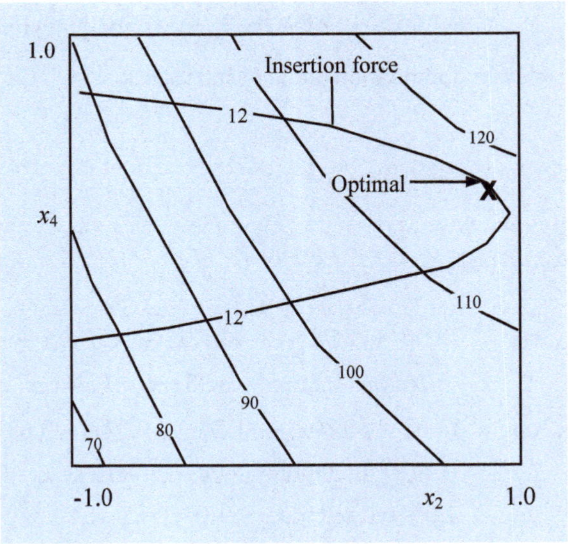

■ **Fig. 6.2** The insertion force constraint on pull force contours with $x_1 = 1$ and $x_3 = 1$

The term "**optimal solution**" refers to the settings generated by solvers when there is high confidence that the best imaginable settings have been found. In the problem shown in ■ Fig. 6.2, it is clear that $x_1 = 3.0$ is the optimal solution since the objective is a parabola reaching its highest value at 3.0.

Parts II and III of this book contain many examples of optimization formulations. In addition, Part III contains computer code for solving a wide variety of optimization problems. The next example illustrates a real-world decision problem in which the prediction models come from an application of design of experiments (DOE) response surface methods (RSM). This example is described further in Part II of this book. It illustrates a case in which the settings derived by the solver were recommended and put into production with largely positive results.

Example 6.4.1 Snap Tab Formal Optimization

Question

Suppose a design team is charged with evaluating whether plastic snap tabs can withstand high enough pull-apart force to replace screws. Designers can manipulate design inputs x_1, x_2, x_3, and x_4 over allowable ranges −1.0 to 1.0. These inputs are dimensions of the design in scaled units. Also, a requirement is that each snap tab should require less than 12 lb (386 N). From RSM, the following models are available for pull-apart force ($y_{est,1}$) and insertion force ($y_{est,2}$) in pounds:

6

$$\begin{aligned} y_{est,1}(x_1,x_2,x_3,x_4) &= 72.06 + 8.98x_1 + 14.12x_2 + 13.41x_3 + 11.85x_4 + 8.52x_1^2 \\ &\quad - 6.16x_2^2 + 0.86x_3^2 + 3.93x_1x_2 - 0.44x_1x_3 - 0.76x_2x_3 \\ y_{est,2}(x_1,x_2,x_3,x_4) &= 14.62 + 0.80x_1 + 1.50x_2 - 0.32x_3 - 3.68x_4 - 0.45x_1^2 \\ &\quad - 1.66x_3^2 + 7.89x_4^2 - 2.24x_1x_3 - 0.33x_1x_4 + 1.35x_3x_4. \end{aligned}$$

Formulate the relevant optimization problem and solve it.

Answer

The optimization formulation is:

$$\begin{aligned} &\text{Maximize:} \quad y_{est,1}(x_1,x_2,x_3,x_4) \\ &\text{Subject to:} \quad y_{est,2}(x_1,x_2,x_3,x_4) \le 12.0\,\text{lb}. \end{aligned}$$

$$\begin{aligned} y_{est,1}(x_1,x_2,x_3,x_4) &= 72.06 + 8.98x_1 + 14.12x_2 + 13.41x_3 + 11.85x_4 + 8.52x_1^2 \\ &\quad - 6.16x_2^2 + 0.86x_3^2 + 3.93x_1x_2 - 0.44x_1x_3 - 0.76x_2x_3 \\ y_{est,2}(x_1,x_2,x_3,x_4) &= 14.62 + 0.80x_1 + 1.50x_2 - 0.32x_3 - 3.68x_4 - 0.45x_1^2 \\ &\quad - 1.66x_3^2 + 7.89x_4^2 - 2.24x_1x_3 - 0.33x_1x_4 + 1.35x_3x_4 \\ &\quad -1.0 \le x_1,x_2,x_3,x_4 \le 1.0 \end{aligned}$$

The solution derived using a standard spreadsheet solver was $x_1 = 1.0$, $x_2 = 0.85$, $x_3 = 1.0$, and $x_4 = 0.33$.

Note that plots of the objectives and constraints can aid in building human appreciation of the solver results. People generally want more than just a recommended solution or set of system inputs. They also want some appreciation of how sensitive the objectives and constraints are to small changes in the recommendations. In some cases, plots can spot mistakes in the logic of the problem formulation or the way in which data was entered into the solver.

Figure 6.2 shows a plot of the objective contours and the insertion force constraint for the snap tab optimization example. Note that dependence of objectives and constraints can only be plotted as a function of two input factors in this way. The plot shows that a conflict exists between the goals of increasing pull-apart forces and decreasing insertion forces.

An important concern with applying formal optimization is that information requirements are often such that all relevant considerations cannot be included in the formulation. For example, in the restaurant problem there was no obvious way to include into the formulation information about the overall strategy to raise prices. On the other hand, the information requirements of applying formal optimization can be an advantage. They can force people from different areas in the organization to agree on the relevant assumptions and problem data. This exercise can encourage communication that may be extremely valuable.

One way to account for additional considerations is to add constraints to formulations. These added constraints can force the optimization solver to avoid solutions that are undesirable because of considerations not included in the formulation. In general, some degree of informality is needed in translating solver results into recommended design inputs.

6.5 Chapter Summary

This chapter describes methods and decision-processes for generating tentative recommended settings for system inputs. The methods range from informal seat-of-the-pants decision-making based on anecdotal evidence to computer-assisted formal optimization.

The method of quality function deployment (QFD) is introduced. QFD constitutes an assemblage of benchmarking tables (from ▶ Chap. 4), cause and effects matrices (from ▶ Chap. 5), and additional information from engineers. The term "House of Quality" (HOQ) was introduced and used to describe the full QFD matrices. Inspection of the HOQ matrices can help decision-makers subjectively account for a variety of considerations that could potentially influence design input selection.

Examples of decision processes in this chapter include subjective assessments informed by inspecting failure mode and effects analysis (FMEA), quality function deployment (QFD) tables, and regression model predictions. Also, more formal approached based on "optimization solver" results are also described together with possible limitations.

While decision processes range in the degree of formality involved, the end product is the same. The results are tentative recommendations for system inputs pending validation from the "confirm" or "verify" project phase. In general, this validation is important to examine before the standard operating procedures (SOPs) and/or design guidelines are changed.

6.6 Problems

In general, pick the correct answer that is most complete.

1. According to the chapter, recommendations from the improve phase are:
   a. Necessarily derived from formal optimization.
   b. Tentative pending empirical confirmation or verification.
   c. Derived exclusively from seat-of-the-pants decision-making.
   d. All of the above are correct.
   e. All of the above are correct except (a) and (d).
2. According to this chapter, the study of the most formal decision processes is called:
   a. Quality Function Deployment (QFD).
   b. Optimization solvers.
   c. Operations Research (OR).
   d. Theory of Constraints (TOC).
   e. Design of Experiments (DOE).
3. Management of a trendy leather goods shop decides upon a 250% markup on handbags using no data and judgment only. This represents:
   a. Formal decision-making.
   b. A House of Quality application.
   c. Anecdotal information about the retail industry.
   d. None of the above is correct.
4. Which of the following is correct and most complete?
   a. Seat-of-the-pants decision-making is rarely (if ever) supported by anecdotal information.
   b. Inspecting a HOQ while making decisions is moderately formal.
   c. Performing DOE and using a solver to generate recommendations is informal.
   d. All of the above are correct.
   e. All of the above are correct except (a) and (d).

Questions 5–7 are based on Table 6.3.

5. Which company seems to dominate in the ratings?
   a. Company 1.
   b. Company 2.
   c. Company 3.
   d. None.
6. Which of the setting changes for Company 1 seems most supported by the HOQ?
   a. Change the area used per page to 9.5.
   b. Change the arm height from 2 to 1 m.
   c. Change to batched production (batched or not set to yes).
   d. Change paper thickness to 1 mm.
7. Which of the following is correct and most complete?
   a. Emulating the best-in-class competitor in the HOQ might not work.
   b. The customer ratings might not be representative of the target population.
   c. The HOQ can help in picking KIVs and KOVs for a DOE application.
   d. All of the above are correct.
   e. All of the above are correct except (a) and (d).

Table 6.3 Hypothetical HOQ for paper airplane design

Customer criterion	Customer importance	Scissor speed (in. p.m.)	Area used per page per plane (in. square)	Weight (g)	Scissor diameter (mm)	Paper type (% glossy)	Arm angle at release (degrees)	Arm height (elbow to ground in meters)	Paper thickness (mm)	Batched or not	Company 1	Company 2	Company 3
Paper failure at fold (ripping)	4	2	4	8	3	1	2	1	4	1	3.33	4	8
Surface roughness (crumpling)	2	4	1	6	2	1	2	1	5	2	5	5	5.33
Immediate flight failure (falls)	8	5	3	4	2	2	1	2	1	1	8	7.66	8.33
Holes in wings (design flaw)	6	7	8	6	2	2	1	1	1	1	6	8	9
Wings unfold (flopping)	5	1	7	6	5	1	1	2	1	2	9	9.66	10
Ugly appearance (aesthetics)	9	2	9	8	2	1	1	1	1	1	3	4	7
Factor rating number F′	–	121	206	214	87	48	40	47	54	41			
Company 1	–	35	8	15	2	2.00%	15	2	2	No			
Company 2	–	42	9.2	12	2	0.10%	0	1	2	No			
Company 3	–	42	9.5	18	3	8.00%	10	2	2	Yes			

8. List two benefits of applying QFD compared with using only formal optimization.
9. In two sentences, explain how changing the targets could affect supplier selection.
10. Create an HOQ with at least four customer criterion, two companies, and three engineering inputs. Identify the reasonable recommended inputs.

Question 11 refers to the following problem formulation:

$$\text{Maximize: } g(x_1) = -5 + 6x_1 - 4x_1^2 + 0.5x_1^3$$
$$\text{Subject to: } x_1[-1, 6].$$

11. The optimal solution for x_1 is (within the implied uncertainty)
    a. 3.0.
    b. −1.0.
    c. 0.9.
    d. 6.0.
    e. None of the above is correct.

Questions 12 and 13 refer to the following problem formulation:

12. The optimal solution for x_1 is (within the implied uncertainty)

$$\text{Maximize: } g(x_1) = 0.25 + 2x_1 - 3x_1^2 + 0.5x_1^3$$
$$\text{Subject to: } x_1[-1, 1].$$

    a. −1.0.
    b. 0.2.
    c. 0.4.
    d. 0.6.
    e. 1.0.
13. The optimal objective is (within the implied uncertainty)
    a. −2.5.
    b. 0.2.
    c. 0.5.
    d. 0.6.
    e. 1.2.
14. Formulate and solve an optimization problem from your own life. State all your assumptions in reasonable detail.
15. Formal optimization often requires:
    a. Subjectively factoring considerations not included in the formulation.
    b. Clarifying as a group the assumptions and data for making decisions.
    c. Plotting the objective function in the vicinity of the solutions for insight.
    d. All of the above are correct.
    e. All of the above are correct except (a) and (d).

References

Brady JE (2005) Six sigma and the university: research, teaching, and meso-analysis. PhD dissertation, Industrial & Systems Engineering, The Ohio State University, Columbus

Clausing D (1994) Total quality development: a step-by-step guide to world-class concurrent engineering. ASME Press, New York

Shin JS, Kim KJ, Chandra MJ (2002) Consistency check of a house of quality Chart. Inte J Qual Reliab Manag 19:471–484

Control or Verify Phase

T. T. Allen, *Introduction to Engineering Statistics and Lean Six Sigma*,
https://doi.org/10.1007/978-1-4471-7420-2_7

7.1 Introduction

If the project involves an improvement to existing systems, the term "**control**" is used to refer to the final six sigma project phase in which tentative recommendations are confirmed and institutionalized. This follows because inspection controls are being put in place to confirm that the changes do initially increase quality and that they continue to do so. If the associated project involves new product or service design, this phase also involves confirmation. Since there is less emphasis on evaluating a process on an on-going basis, the term "**verify**" refers evaluation on a one-time, off-line basis.

Clearly, there is a chance that the recommended changes will not be found to be an improvement. In that case, it might make sense to return to the analyze and/or improvement phases to generate new recommendations. Alternatively, it might be time to terminate the project and ensure that no harm has been done. In general, casual reversal of the DMAIC or DMADV ordering of activities might conflict with the dogma of six sigma. Still, this can constitute the most advisable course of action.

► Chapter 6 presented methods and decision processes for developing recommended settings. Those settings were called tentative because in general, sufficient evidence was not available to assure acceptability. This chapter describes two methods for thoroughly evaluating the acceptability of the recommended system input settings.

The method of "**control planning**" refers to a coordinated effort to guarantee that steady state charting activities will be sufficient to monitor processes and provide some degree of safeguard on the quality of system outputs. Control planning could itself involve the construction of gauge R&R method applications and statistical process control charting procedures described in ► Chap. 4.

The method of "**acceptance sampling**" provides an economical way to evaluate the acceptability of characteristics that might otherwise go uninspected. Both acceptance sampling and control planning could therefore be a part of a control or verification process.

Overall, the primary goal of the control or verify phase is to provide strong evidence that the project targets from the charter have been achieved. Therefore, the settings should be thoroughly tested through weeks of running in production, if appropriate. Control planning and acceptance sampling can be useful in this process. Ultimately, any type of strong evidence confirming the positive effects of the project recommendations will likely be acceptable. With the considerable expense associated with many six sigma projects, the achievement of measurable benefits of new system inputs is likely. However, a conceivable, useful role of the control or verify phases is to determine that no recommended changes are beneficial and the associated system inputs should not be changed.

Finally, the documentation of any confirmed input setting changes in the corporate standard operating procedures (SOPs) is generally required for successful project completion. This chapter begins with descriptions of control planning and acceptance sampling methods. It concludes with brief comments about appropriate documentation of project results.

7.2 Control Planning

The method of control planning could conceivably involve many of the methods presented previously: check sheets (▶ Chap. 3), gauge R&R to evaluate measurement systems (▶ Chap. 4), statistical process control (SPC) charting (▶ Chap. 4), and failure mode and effects analysis (FMEA, ▶ Chap. 5).

The phrase "**critical characteristics**" refers to key output variables (KOVs) that are deemed important enough to system output quality that statistical process control charting should be used to monitor them. Because significant cost can be associated with a proper, active implementation of control charting, some care is generally given before declaring characteristics to be critical. An FMEA application can aid in determining which characteristics are associated with the highest risks and therefore might be declared critical and require intense monitoring and inspection efforts.

With respect to *Step* 6, a subjective evaluation of each chart is made as to whether it has the desired sensitivity and response times desired. Sensitivity relates to the proximity of the limits to each other; this determines how large the effects of assignable causes need to be for detection. If the limits are too wide, n should be increased. If the limits are needlessly close together, n might be reduced to save inspection costs.

Also, if charts would likely be too slow to usefully signal assignable cause, the inspection interval, τ, should be decreased. Depending on the effects of assignable causes, it could easily take two or three periods before the chart generates an "out-of-control" signal. For restaurant customer satisfaction issues, several weeks before alerting the local authority may be acceptable. Therefore, $\tau = 1$ week might be acceptable. For manufacturing problems in which scrap and rework are very costly, it may be desirable to know about assignable causes within minutes of occurrence. Then, $\tau = 30$ min might be acceptable.

Algorithm 7.1. Control planning

Step 1. The engineering team selects a subset of the q process outputs to be "critical characteristics" or important quality issues. Again, these are system outputs judged to be necessary to inspect and monitor using control charting.

Step 2. The gauge capability is established for each of these critical characteristics, unless there are no doubts. The cycle of gauge R&R evaluation followed by improvements followed by repeated gauge R&R evaluation is iterated until all measurement systems associated with critical characteristics are considered acceptable.

Step 3. Specific responsibilities for investigating out-of-control signals are assigned to people for each critical characteristic (a "**reaction plan**") and the chart types are selected. A check sheet might be added associated with multiple characteristics.

Step 4. The sample sizes (ns) and periods between inspections (τ) for all characteristics are tentatively determined.

Step 5. The charts are set up for each of the q critical characteristics.

Step 6. The sample sizes are increased or decreased and the periods are adjusted for any charts found to be unacceptable. If a change is made, the start up period for the appropriate characteristic is repeated as needed.

Step 7. Evaluate and record the C_{pk}, process capability, and or sigma level (σ_L) of the process with the recommended settings.

Note that after the control plan is created, it might make sense to consider declaring characteristics with exceedingly high C_{pk} values not to be critical. Nonconformities for these characteristics may be so rare that monitoring them could be a waste of money.

The following example illustrates a situation involving a complicated control plan with many quality characteristics. The example illustrates how the results in control planning can be displayed in tabular format. In the example, the word "quarantine" means to separate the affected units so that they are not used in downstream processes until after they are reworked.

Example 7.2.1 Controlling the Welding of Thin Ship Structures

Question 1

If the data in Table 7.1 were real, what is the smallest number of applications of gauge R&R (crossed) and statistical process control charting (SPC) that must have been performed?

Answer 1

At least five applications of gauge R&R (crossed) and four applications of Xbar & R charting must have been done. Additionally, it is possible that a *p*-chart was set up to evaluate and monitor the fraction of nonconforming units requiring rework, but the capability from that *p*-chart information is not included in Table 7.1.

Table 7.1 A hypothetical control plan for ship structure arc welding

Critical quality or issue	Measurement technique	Control method	% R&R	C_{pk}	Period (τ)	Sample size (n)	Reaction plan
Fixture maximum gaps	Caliper	Xbar & R charting	12.5%	1.0	1 shift	6	Adjust and check
Initial flatness (mm)	Photogrammetry	Xbar & R charting	7.5%	0.8	1 shift	4	Adjust and check
Spatter	Visual	100% insp.	9.3%	NA	NA	100%	Adjust
Distortion (rms flatness)	Photogrammetry	Xbar & R charting	7.5%	0.7	1 shift	4	Quarantine and rework
Appearance	Visual go-no-go	*p*-charting and check sheet	Seen as not needed	NA	100%	100%	Notify shift supervisor
Penetration depth (mm)	X-ray inspection	Xbar & R charting	9.2%	2.1	1 shift	4	Notify shift supervisor

Question 2

Assuming no safety issues were involved, might it be advisable to remove of the characteristics from the "critical" list and save inspection costs?

Answer 2

Gauge R&R results confirm that penetration was well measured by the associated X-ray inspection. However, considering $C_{pk} > 2.0$ and $\sigma_L > 6.0$, inspection of that characteristic might no longer be necessary. Yes, removing it from the critical list might be warranted.

Question 3

Suppose the *p*-chart shown in ◘ Table 7.1 was set up to evaluate and monitor rework. How could this chart be used to evaluate a six sigma project?

Answer 3

The centerline of the *p*-chart is approximately 14%. This "process capability" number might be below the number established in the "measure phase" before the changes were implemented. An increase in process capability can be viewed as a tangible deliverable from a project. This is particularly true in the context of *p*-charting because rework costs are easily quantifiable (► Chap. 2).

Clearly, the exercise of control planning often involves balancing the desire to guarantee a high degree of safety against inspection costs and efforts. If the control plan is too burdensome, it conceivably might not be followed. The effort implied by the control plan in the above would be appropriate to a process involving valuable raw

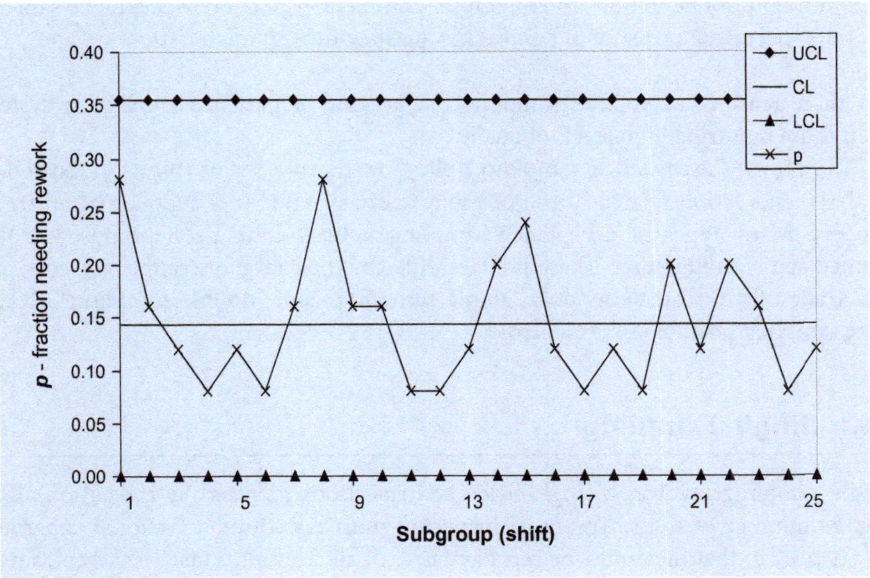

◘ **Fig. 7.1** Stage II SPC chart on total fraction nonconforming

materials and what might be regarded as "high" demands on quality. Yet, in some truly safety critical applications in industries like aerospace, inspection plans commonly are even more burdensome. In some cases, complete or 100% inspection is performed multiple times (Fig. 7.1).

7.3 Acceptance Sampling

The method of "**acceptance sampling**" involves the inspection of a small number of units to make decisions about the acceptability of a larger number of units. As for charting methods, the inspected entity might be a service rather than a manufactured unit. For simplicity, the methods will be explained in terms of units. Romig, working at Bell Laboratories in the 1920s, is credited with proposing the first acceptance sampling methods. Dodge and Romig (1959) documents much of the authors' related contributions.

7

Since not all units are inspected in acceptance sampling, acceptance sampling unavoidably involves risks. The method of "**complete inspection**" involves using one measurement to evaluate all units relevant to a given situation. Complete inspection might naturally be expected to be associated with reduced or zero risks. Yet often this is a false comparison. Reasons why acceptance sampling might be useful include:

1. The only trustworthy inspection method is "destructive" testing (▶ Chap. 4). Then complete inspection with nondestructive evaluation is not associated with zero risks and the benefits of inspection are diminishing. Also, complete inspection using destructive testing would result in zero units for sale.
2. The alternative might be no inspection of the related quality characteristic. The term "**quasi-critical characteristics**" here refers to KOVs that might be important but might not be important enough for complete inspection. Acceptance sampling permits a measure of control for quasi-critical characteristics.

For these reasons, acceptance sampling can be used as part of virtually any system, even those requiring high levels of quality.

The phrase "**acceptance sampling policy**" refers to a set of rules for inspection, analysis, and action related to the possible return of units to a supplier or upstream process. Many types of acceptance sampling policies have been proposed in the applied statistics literature. These policies differ by their level of complexity, cost, and risk trade-offs. In this book, only "**single sampling**" and "**double sampling**" acceptance sampling policies are presented.

7.3.1 Single Sampling

Single sampling involves a single batch of inspections followed by a decision about a large number of units. The units inspected must constitute a "rational subgroup" (▶ Chap. 4) in that they must be representative of all relevant units. The symbols used to describe single sampling are:

1. N is the number of units in the full "**lot**" of all units about which acceptance decisions are being made.
2. n is the number of units inspected in the rational subgroup.
3. c is the maximum number of units that can be found to be nonconforming for the lot to be declared acceptable.
4. d is the number of nonconforming found from inspection of the rational subgroup.

As for control charting processes, there is no universally accepted method for selecting the sample size, n, of the radical subgroup. In single sampling, there is an additional parameter c, which must be chosen by the method user.

The primary risk in acceptance sampling can be regarded as accepting lots with large numbers of nonconformities. In general, larger samples sizes, $n\Uparrow$, and tighter limits on the numbers nonconforming, $c\Downarrow$, decrease this risk. In ▶ Chap. 10, theory is used to provide additional information about the risks to facilitate the selection of these constants.

Algorithm 7.2. Single sampling

Step 1. Carefully select n units for inspection such that you are reasonably confident that the quality of these units is representative of the quality of the N units in the lot, i.e., they constitute a rational subgroup.

Step 2. Inspect the n units and determine the number d that do not conform to specifications.

Step 3. If $d > c$, then the lot is rejected. Otherwise the lot is "accepted" and the d units found nonconforming are reworked or scrapped.

Rejection of a lot generally means returning all units to the supplier or upstream sub-system. This return of units often comes with a demand that the responsible people should completely inspect all units and replace nonconforming units with new or reworked units. Note that the same inspections for an acceptance sampling policy might naturally fit into a control plan in the control phase of a six sigma process. One might also chart the resulting data on a p-chart or demerit chart.

Example 7.3.1 Destructive Testing of Screws

Question 1

Suppose our company is destructively sampling 40 welds from lots of 1200 welds sent from a supplier. If any of the maximum sustainable pull forces are less than 150 N, the entire lot is shipped back to the supplier and a contractually agreed penalty is assessed. What is the technical description of this policy?

Answer 1

This is single sampling with $n = 40$ and $c = 0$.

Question 2

Is there a risk that a lot with ten nonconforming units would pass through this acceptance sampling control?

Answer 2

Yes, there is a chance. In ▶ Chap. 10, we show how to calculate the probability under standard assumptions, which is approximately 0.7. An OC curve, also described in ▶ Chap. 10, could be used to understand the risks better.

7.3.2 Double Sampling

The "double sampling" method involves an optional second set of inspections if the first sample does not result in a definitive decision to accept or reject. This approach is necessarily more complicated than single sampling. Yet the risk verses inspection cost tradeoffs are generally more favorable.

7

The symbols used to describe single sampling are:

1. N is the number of units in the full "**lot**" of all units about which acceptance decisions are being made.
2. n_1 is the number of units inspected in an initial rational subgroup.
3. c_1 is the maximum number of units that can be found to be nonconforming for the lot to be declared acceptable after the first batch of inspections.
4. r is the cut-off limit on the count nonconforming after the first batch of inspections.
5. n_2 is the number of units inspected in an optional second rational subgroup.
6. c_2 is the maximum number of units that can be found to be nonconforming for the lot to be declared acceptable after the optional second batch of inspections.

As for single sampling, there is no universally accepted method for selecting the sample sizes, n_1 and n_2, of the radical subgroups. Nor is there any universal standard for selecting the parameters c_1, r_1, and c_2. In general, larger samples sizes, $n_1 \Uparrow$ and $n_2 \Uparrow$, and tighter limits on the numbers nonconforming, $c_1 \Downarrow$, $r_1 \Downarrow$, and $c_2 \Downarrow$, decrease the primary risks. In ▶ Chap. 10, theory is used to provide additional information about the risks to facilitate the selection of these constants.

Algorithm 7.3. Double sampling

Step 1. Carefully select n_1 units for inspection such that you are reasonably confident that the quality of these units is representative of the quality of the N units in the lot, i.e., inspect a rational subgroup.

Step 2. Inspect the n_1 units and determine the number d_1 that do not conform to specifications.

Step 3. If $d_1 > r$, then the lot is rejected and process is stopped. If $d_1 \leq c_1$, the lot is said to be accepted and process is stopped. Otherwise, go to *Step 4*.

Step 4. Carefully select an additional n_2 units for inspection such that you are reasonably confident that the quality of these units is representative of the quality of the remaining $N - n_1$ units in the lot, i.e., inspect another rational subgroup.

Step 5. Inspect the additional n_2 units and determine the number d_2 that do not conform to specifications.

Step 6. If $d_1 + d_2 \leq c_2$, the lot is said to be "accepted". Otherwise, the lot is "rejected".

As in single sampling, rejection of a lot generally means returning all units to the supplier or upstream sub-system. This return of units often comes with a demand that the responsible people should completely inspect all units and replace nonconforming units with units that have been reworked or are new. The double sampling method is shown in Fig. 7.2. Note that if $c_1 + 1 = r$, then there can be at most one batch of inspection, i.e., double sampling reduces to single sampling.

In general, if lots are accepted, then all of the items found to be nonconforming must be reworked or scrapped. It is common at that point to treat all the remaining units in a similar way as if they had been inspected and passed.

The selection of the parameters c_1, c_2, n_1, n_2, r, and d_2 may be subjective or based on military or corporate policies. Their values have implications for the chance that the units delivered to the customer do not conform to specifications. Also, their values have implications for bottom-line profits. These implications are studied more thoroughly in ▶ Chap. 10 to inform the selection of specific acceptance sampling methods.

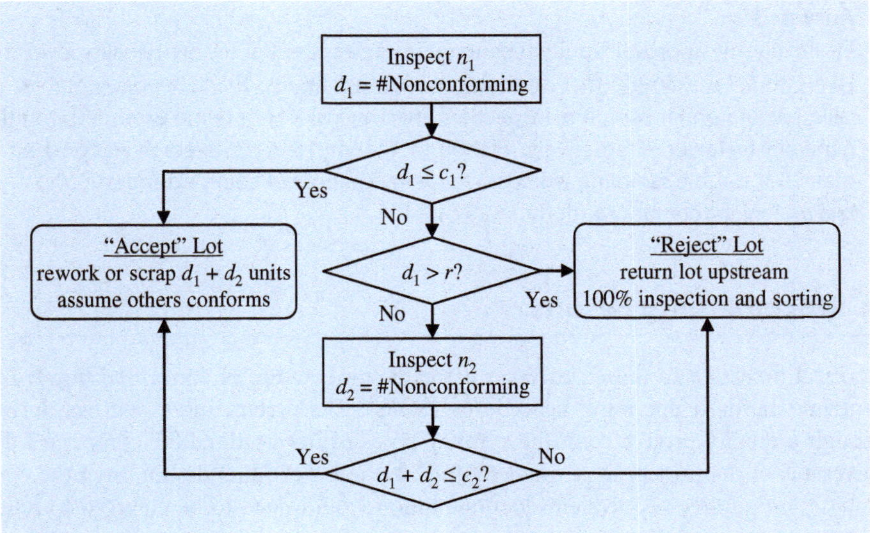

Fig. 7.2 Flow-chart of double sampling method

Example 7.3.2 Evaluating Possible New Hires at Call Centers

Question 1
Suppose a manager at a call center is trying to determine whether new hires deserve permanent status. She listens in on 20 calls, and if all except one are excellent, the employee converts to permanent status immediately. If more than four are unacceptable, the employee contract is not extended. Otherwise, the manager evaluates an additional 40 calls and requires that at most three calls be unacceptable. What method is the manager using?

Answer 1
This is double sampling with $n_1 = 20$, $c_1 = 1$, $r = 4$, $n_2 = 40$, $c_2 = 3$.

Question 2
How should the calls be selected for monitoring?

Answer 2
To be representative, it would likely help to choose randomly which calls to monitor, for example, one on Tuesday morning, two on Wednesday afternoon, etc. Also, it would be desirable that the operator would not know the call is being evaluated. In this context, this approach generates a rational subgroup.

Question 3
If you were an operator, would you prefer this approach to complete inspection? Explain.

Answer 3
Intuitively, my approval would depend on my assessment of my own quality level. If I were sure, for example, that my long-run average was less than 5% unacceptable calls, I would prefer complete inspection. Then my risk of not being extended would generally be lower. Alternatively, if I thought that my long-run average was greater than 20%, double sampling would increase my chances of being extended. (Who knows, I might get lucky with those 20 calls.)

7.4 Documenting Results

A critical final step in these processes is the documentation of confirmed inputs into company standard operating procedures (SOPs). The system input settings derived through a quality project could have many types of implications for companies. If a deliverable of the project were input settings for a new product design, then the company design guide or equivalent documentation would need to be changed to reflect the newly confirmed results.

If the project outputs are simply new settings for an existing process, the associated changes should be reflected in process SOPs. Still, improvement project recommendations might include purchasing recommendations related to the selection of suppliers or to design engineering related to changes to the product design. These changes could require significant time and effort to be correctly implemented and have an effect.

The phrase "**document control**" policies refers to efforts to guarantee that only a single set of standard operating procedures is active at any given time. When document control policies are operating, changing the SOPs requires "checking out" the active copy from a source safe and then simultaneously updating all active copies at one time.

Finally, in an organization with some type of ISO certification, it is likely that auditors would require the updating of ISO documents and careful document control for certification renewal. ▶ Chapter 2 provides a discussion of SOPs and their roles in organizations. Ideally, the efforts to document changes in company SOPs would be sufficient to satisfy the auditors. However, special attention to ISO specific documentation might be needed.

Example 7.4.1 Design Project Completion

Question
A design team led by manufacturing engineers has developed a new type of fastener with promising results in the prototyping results. What needs to happen for adequate project verification?

Answer
Since communication between production and design functions can be difficult, extra effort should be made to make sure design recommendations are entered correctly into the design guide. Also, a control planning strategy should provide confirmation of the quality and monetary benefits from actual production runs. Finally, the new fastener specifications must be documented in the active design guide, visible by all relevant divisions around the world.

7.5 Chapter Summary

This chapter describes two methods for assuring product quality: control planning and acceptance sampling. The control planning method might itself require several applications of the gauge R&R and statistical process control charting from ▶ Chap. 4. In control planning, the declaration of key output variables as being "critical quality characteristic" is generally associated with a need for both the evaluation of the associated measurement systems and statistical process control charting.

Acceptance sampling constitutes a method to provide some measure of control on "quasi-critical" characteristics that might otherwise go uninspected. This chapter contains a description of two types of acceptance sampling methods: single sampling and double sampling. Double sampling is more complicated but offers generally more desirable risk-vs-inspection cost tradeoffs.

Finally, the chapter describes how the goals of the control or verify phases are not accomplished until: (1) strong evidence shows the monetary savings or other benefits of the project; and (2) the appropriate comporate documentation is altered to reflect the confirmed recommended settings.

7.6 Problems

In general, pick the correct answer that is most complete.

1. According to the six sigma literature, a project for improving an existing system ends with which phase?
   a. Define.
   b. Analyze.
   c. Improve.
   d. Verify.
   e. Control.
2. The technique most directly associated with guaranteeing that all measurement equipment are capable and critical characteristics are being monitored is:
   a. Process mapping.
   b. Benchmarking.
   c. Design of Experiments (DOE).
   d. Control Planning.
   e. Acceptance Sampling.
3. According to this chapter, successful project completion generally requires changing or updating:
   a. Standard operating procedures.
   b. The portion of the control plan relating to gauge R&R.
   c. The portion of the control plan relating to sample size selection.
   d. All of the above are correct.
   e. All of the above are correct except (a) and (d).
4. Which of the following is most relevant to cost-effective evaluation of many units?
   a. Benchmarking.
   b. Control planning.
   c. Acceptance sampling.
   d. Design of Experiments (DOE).
   e. A reaction plan.
5. Which of the following is correct and most complete?
   a. Filling out each row of a control plan could require performing a gauge R&R.
   b. According to the text, reaction plans are an optional stage in control planning.
   c. All characteristics on blueprints are critical characteristics.
   d. All of the above are correct.
   e. All of the above are correct except (a) and (d).
6. The text implies that FMEAs and control plans are related in which way?
   a. FMEAs can help clarify whether characteristics should be declared critical.
   b. FMEAs determines the capability values to be included in the control plan.
   c. FMEAs determine the optimal reaction plans to be included in control plans.
   d. All of the above are correct.
   e. All of the above are correct except (a) and (d).

Table 7.2 A hypothetical control plan for manufacturing paper airplanes

Critical characteristic or issue	Measurement technique	Control method	%R&R	C_{pk}	Period (τ)	Sample size (*n*)	Reaction plan
Surface roughness (crumpling)	Laser	X-bar & R	7.8	2.5	1 shift	10	Adjust and re-check
Unsightly appearance (aesthetics)	Visual	Check sheet, *p*-chart	20.4	0.4	2 shifts	100%	Quarantine and rework
Unfolding (flopping)	Caliper stress test	X-bar & R	10.6	1.4	0.5 shifts	5	Notify supervisor

Questions 7–9 derive from the paper airplane control plan in Table 7.2.

7. Assume budgetary considerations required that one characteristic should not be monitored. According to the text, which one should be declared not critical?
   a. Crumpling.
   b. Aesthetics.
   c. Flopping.
   d. Calipers.
8. The above control plan implies that how many applications of gauge R&R have been applied?
   a. 0.
   b. 1.
   c. 2.
   d. 3.
   e. None of the above.
9. Which part of implementing a control plan requires the most on-going expense during steady state?
10. Complete inspection is (roughly speaking) a single sampling plan with:
   a. $n = d$.
   b. $N = c$.
   c. $N = n$.
   d. c = d.
   e. None of these describe complete inspection even roughly speaking.
11. When considering sampling policies, the risks associated with accepting an undesirable lot grows with:
   a. Larger rational subgroup size.
   b. Decreased tolerance of nonconformities in the rational subgroup (e.g., lower *c*).
   c. Increased tolerance of nonconformities in the overall lot (*e.g.*, higher *c*).
   d. Decreased overall lot size.
   e. None of the above.

For questions 12–13, consider the following scenario:

Each day, 1000 screws are produced and shipped in two truckloads to a car manufacturing plant. The screws are not sorted by production time. To determine lot quality, 150 are inspected by hand. If 15 or more are defective, the screws are returned.

12. Why is this single sampling rather than double sampling?
    a. The lost size is fixed.
    b. There is at most one decision resulting in possible acceptance.
    c. There are two occasions during which the lot might be rejected.
    d. defective is not enough for an accurate count of nonconformities.
    e. Double sampling is preferred for large lot sizes.
13. List two advantages of acceptance sampling compared with complete inspection.

For Questions 14–16, consider the following scenario:

Each shift, 1000 2′ × 2′ sheets of steel enter your factory. Your boss wants to be confident that approximately 5% of the accepted incoming steel is nonconforming.

14. Which of the following is correct and most complete for single sampling?
    a. Acceptance sampling is too risky for such a tight quality constraint.
    b. Assuming inspection is perfect, $n=950$ and $c=0$ could ensure success.
    c. Assuming inspection is perfect, $n=100$ and $c=2$ might seem reasonable.
    d. All of the above are correct.
    e. All of the above are correct except (a) and (d).
15. Which of the following is correct and most complete?
    a. Gauge R&R might indicate that destructive sampling is necessary.
    b. It is not possible to create a p-chart using single sampling data.
    c. Double sampling necessarily results in fewer inspections than single sampling.
    d. All of the above are correct.
    e. All of the above are correct except (a) and (d).
16. Design a double sampling plan that could be applied to this problem.

The following double-sampling plan parameters will be examined in questions 17–18: $N=7500$, $n_1=100$, $n_2=350$, $c_1=3$, $c_2=7$, and $r=6$.

17. What is the maximum number of units inspected, assuming the lot is accepted?
18. What is the minimum number of units inspected?
19. Why do recorded voices on customer service voicemail systems say, "This call may be monitored for quality purposes?"
20. Which of the following is correct and most complete?
    a. Correct document control requires the implementation of control plans.
    b. Often, projects complete with revised SOPs are implemented corporation-wide.
    c. Documenting findings can help capture all lessons learned in the project.
    d. All of the above are correct.
    e. All of the above are correct except (a) and (d).
21. List and explain in order each of the five steps in an alternative to the DMAIC process.

22. Suppose a manager at a call center is trying to determine whether new hires deserve permanent status. She listens in on 35 calls, and if all except one are excellent, the employee converts to permanent status immediately. If more than four are unacceptable, the employee contract is not extended. Otherwise, the manager evaluates an additional 45 calls and requires that at most three calls be unacceptable. What method is the manager using? And what are any n and c values that may apply?
23. Suppose our company is destructively sampling 15 welds from lots of 800 welds sent from a supplier. If any of the maximum sustainable pull forces are less than 130 N, the entire lot is shipped back to the supplier and a contractually agreed penalty is assessed. What is the technical description of this policy? And what are any n and c values that may apply?
24. Compare and contrast the double sampling, single sampling, and full inspection approaches. List at least a single criterion that makes double sampling seem preferable to single sample and one that makes single sampling seem preferable.

Reference

Dodge HF, Romig HG (1959) Sampling inspection tables, single and double sampling, 2nd edn. Wiley, New York

Advanced SQC Methods

T. T. Allen, *Introduction to Engineering Statistics and Lean Six Sigma*,
https://doi.org/10.1007/978-1-4471-7420-2_8

8.1 Introduction

In the previous chapters several methods are described for achieving various objectives. Each of these methods can be viewed as representative of many other similar methods developed by researchers. Many of these methods are published in such respected journals as the *Journal of Quality Technology*, *Technometrics*, and *The Bell System Technical Journal*. In general, the other methods offer additional features and advantages.

For example, the exponentially weighted moving average (EWMA) charting methods described in this chapter provide a potentially important advantage compared with Shewhart Xbar & R charts. This advantage is that there is generally a higher chance that the user will detect assignable causes associated with only a small shift in the continuous quality characteristic values that persists over time.

Also, the "multivariate charting" methods described here offer an ability to monitor simultaneously multiple continuous quality characteristics. Compared with multiple applications of Xbar & R charts, the multivariate methods (Hoteling's T^2 chart) generally cause many fewer false alarms. Therefore, there are potential savings in the investigative efforts of skilled personnel.

Yet the more basic methods described in previous chapters have "stood the test of time" in the sense that no methods exist that completely dominate them in every aspect. For example, both EWMA and Hotelling's T^2 charting are more complicated to implement than Xbar & R charting. Also, neither provide direct information about the range of values within a subgroup.

Many alternative versions of methods have been proposed to process mapping, gauge R&R, SPC charting, design of experiments, failure mode and effects analysis (FMEA), formal optimization, Quality Function Deployment (QFD), acceptance sampling, control planning. In this chapter, only two alternatives to Xbar & R charting are selected for inclusion, somewhat arbitrarily: EWMA and multivariate charting or Hoteling's T^2 chart.

8.2 EWMA Charting for Continuous Data

Roberts (1959) proposed "**EWMA charting**" to allow people to identify a certain type of assignable cause more quickly. The assignable cause in question forces the quality characteristic values to shift in one direction an amount that is small in comparison with the limit spacing on an Xbar chart. EWMA charting is relevant when the quality characteristics are continuous. Therefore, EWMA charting can be used in the same situations in which Xbar & R charting is used and can be applied based on exactly the same data as Xbar & R charts.

Generally speaking, if assignable causes only create a small shift, Xbar & R charts might require several subgroups to be inspected before an out-of-control signal. Each of these subgroups might require a large number of inspections over a long time τ. In the same situation, an EWMA chart would likely identify the assignable cause in fewer subgroups, even if each subgroup involved fewer inspections.

The symbols used in describing EWMA charting, (used in Algorithm 8.1) are:

1. n is the number of units in a subgroup. Here, n could be as lows as 1.
2. τ is the period of time between the inspection of successive subgroups.
3. $X_{i,j}$ refers to the ith quality characteristic value in the the jth time period.
4. $X_{bar,j}$ is the average of the n quality characteristic values for the jth time period.
5. λ is an adjustable "**smoothing parameter**" relevant during startup. Higher values of λ make the chart rougher and decrease the influence of past observations on the current charted quantity. Here, $\lambda = 0.20$ is suggested as a default.
6. Z_i is the quantity plotted which is an exponentially weighted moving average. In period $i - 1$, it can be regarded as a forecast for period i.

Generally, n is small enough that people are not interested in variable sample sizes. In the formulas below, quantities next to each other are implicitly multiplied with the "×" omitted for brevity. Also, "/" is equivalent to "÷". The numbers in the formulas 3.0 and 0.0 are assumed to have an infinite number of significant digits.

The phrase "EWMA chart" refers to the associated resulting chart. An out-of-control signal is defined as a case in which Z_j is outside the control limits. From then on, technicians and engineers are discouraged from making minor process changes unless a signal occurs. If a signal does occur, they should investigate to see if something unusual and fixable is happening. If not, they should refer to the signal as a false alarm.

Note that a reasonable alternative approach to the one above is to obtain X_{barbar} and σ_0 from Xbar & R charting. Then, Z_j and the control limits can be calculated using Eqs. (8.3) and (8.4) in Algorithm 8.1.

Algorithm 8.1. EWMA charting

Step 1. (Startup) Measure the continuous characteristics, $X_{i,j}$, for $i = 1, \ldots, n$ units for $j = 1, \ldots, 25$ periods.

Step 2. (Startup) Calculate the sample averages $X_{bar,j} = (X_{1,j} + \ldots + X_{n,j})/n$. Also, calculate the average of all of the 25n numbers, X_{barbar}, and the sample standard deviation of the 25n numbers, s. The usual formula is

$$s^{g} = g\sqrt{\frac{\left(X_{1,1} - X_{barbar}\right)^2 + \left(Y_{1,2} - X_{barbar}\right)^2 + \cdots + \left(X_{25,n} - X_{barbar}\right)^2}{25n - 1}}. \quad (8.1)$$

Step 3. (Startup) Set $\sigma_0 = s$ tentatively and calculate the "trial" control limits using

$$UCL_{trial,j} = X_{barbar} + 3.0\,\sigma_0\sqrt{\frac{\lambda}{(2-\lambda)}\left[1-(1-\lambda)^{2j}\right]}$$
$$CL_{trial} = X_{barbar}, \text{ and} \quad (8.2)$$
$$LCL_{trial,j} = X_{barbar} - 3.0\,\sigma_0\sqrt{\frac{\lambda}{(2-\lambda)}\left[1-(1-\lambda)^{2j}\right]},$$

Step 4. (Startup) Calculate the following:

$$Z_0 = X_{barbar}$$
$$Z_j = \lambda X_{bar,j} + (1-\lambda)Z_{(i-1)} \quad \text{for } i = 1, \ldots, 25. \tag{8.3}$$

Step 5. Investigate all periods for which $Z_j < LCL_{trial,j}$ or $Z_j > UCL_{trial,j}$. If the results from any of these periods are believed to be not representative of future system operations, e.g., because problems were fixed permanently, remove the data from the *l* not representative periods from consideration.

Step 6. (Startup) Re-calculate X_{barbar} and *s* based on the remaining 25 – *l* periods and (25 – *l*) × *n* data. Also, set $\sigma_0 = s$ and the process capability is 6.0 × σ_0. Calculate the revised limits using

$$\begin{aligned} UCL &= X_{barbar} + 3.0\sigma_0\sqrt{\frac{\lambda}{(2-\lambda)}} \\ CL &= X_{barbar}, \text{ and} \\ LCL &= X_{barbar} - 3.0\sigma_0\sqrt{\frac{\lambda}{(2-\lambda)}} \end{aligned} \tag{8.4}$$

Step 7. (Steady State, SS) Plot Z_j, for each period $j = 25, 26, \ldots$ together with the upper and lower control limits, *LCL* and *UCL*, and the center line, *CL*.

8

Example 8.2.1 Fixture Gaps Between Welded Example Revisited

Question

The same Korean shipyard mentioned in ► Chap. 4 wants to evaluate and monitor the gaps between welded parts from manual fixturing. Workers measure 5 gaps every shift for 25 shifts over 10 days. Table 8.1 shows the resulting hypothetical data including 10 data not available during the set-up process. This time, assume the process engineers believe that even small gaps cause serious problems and would like to know about any systematic shifts, even small ones, as soon as possible. Apply EWMA charting to this data and establish the process capability.

Answer

As in ► Chap. 4, $n = 5$ inspected gaps between fixtured parts prior to welding, and $\tau = 6$ h. If the inspection budget were increased, it might be advisable to inspect more units more frequently. The calculated subgroup averages are also shown (*Step 2*) and $X_{barbar} = 0.90$ and $\sigma_0 = s = 0.157$ is tentatively set. In *Step 3*, the derived values for the control limits are shown in Fig. 8.1. In *Step 4*, the Z_j are calculated and shown in Table 8.1. In *Step 5*, none of the first 25 periods yields an out-of-control signal. The *Step 6* process capability is 0.942 and control limits are shown in Fig. 8.1. From then until major process changes occur (rarely), the same limits are used to find out-of-control signals (*Step 7*). Note that nine periods into the steady state phase, the chart would signal startup suggesting that looking for a cause that has shifted the average gap higher.

Table 8.1 Example gap data in millimeters (SU = start up, SS = steady state)

Phase	j	$X_{1,j}$	$X_{2,j}$	$X_{3,j}$	$X_{4,j}$	$X_{5,j}$	$X_{\text{bar},j}$	Z_j
SU	1	0.85	0.71	0.94	1.09	1.08	0.93	0.91
SU	2	1.16	0.57	0.86	1.06	0.74	0.88	0.90
SU	3	0.80	0.65	0.62	0.75	0.78	0.72	0.87
SU	4	0.58	0.81	0.84	0.92	0.85	0.80	0.85
SU	5	0.85	0.84	1.10	0.89	0.87	0.91	0.86
SU	6	0.82	1.20	1.03	1.26	0.80	1.02	0.90
SU	7	1.15	0.66	0.98	1.04	1.19	1.00	0.92
SU	8	0.89	0.82	1.00	0.84	1.01	0.91	0.92
SU	9	0.68	0.77	0.67	0.85	0.90	0.77	0.89
SU	10	0.90	0.85	1.23	0.64	0.79	0.88	0.89
SU	11	0.51	1.12	0.71	0.80	1.01	0.83	0.88
SU	12	0.97	1.03	0.99	0.69	0.73	0.88	0.88
SU	13	1.00	0.95	0.76	0.86	0.92	0.90	0.88
SU	14	0.98	0.92	0.76	1.18	0.97	0.96	0.90
SU	15	0.91	1.02	1.03	0.80	0.76	0.90	0.90
SU	16	1.07	0.72	0.67	1.01	1.00	0.89	0.90
SU	17	1.23	1.12	1.10	0.92	0.90	1.05	0.93
SU	18	0.97	0.90	0.74	0.63	1.02	0.85	0.91
SU	19	0.97	0.99	0.93	0.75	1.09	0.95	0.92
SU	20	0.85	0.77	0.78	0.84	0.83	0.81	0.90
SU	21	0.82	1.03	0.98	0.81	1.10	0.95	0.91
SU	22	0.64	0.98	0.88	0.91	0.80	0.84	0.90
SU	23	0.82	1.03	1.02	0.97	1.00	0.97	0.91
SU	24	1.14	0.95	0.99	1.18	0.85	1.02	0.93
SU	25	1.06	0.92	1.07	0.88	0.78	0.94	0.93
SS	26	1.06	0.81	0.98	0.98	0.85	0.936	0.93
SS	27	0.83	0.70	0.98	0.82	0.78	0.822	0.91
SS	28	0.86	1.33	1.09	1.03	1.10	1.082	0.95
SS	29	1.03	1.01	1.10	0.95	1.09	1.036	0.96
SS	30	1.02	1.05	1.01	1.02	1.20	1.060	0.98
SS	31	1.02	0.97	1.01	1.02	1.06	1.016	0.99
SS	32	1.20	1.02	1.20	1.05	0.91	1.076	1.01

(continued)

Table 8.1 (continued)

Phase	j	$X_{1,j}$	$X_{2,j}$	$X_{3,j}$	$X_{4,j}$	$X_{5,j}$	$X_{bar,j}$	Z_j
SS	33	1.10	1.15	1.10	1.02	1.08	1.090	1.02
SS	34	1.20	1.05	1.04	1.05	1.06	1.080	1.03
SS	35	1.22	1.09	1.02	1.05	1.05	1.086	1.05

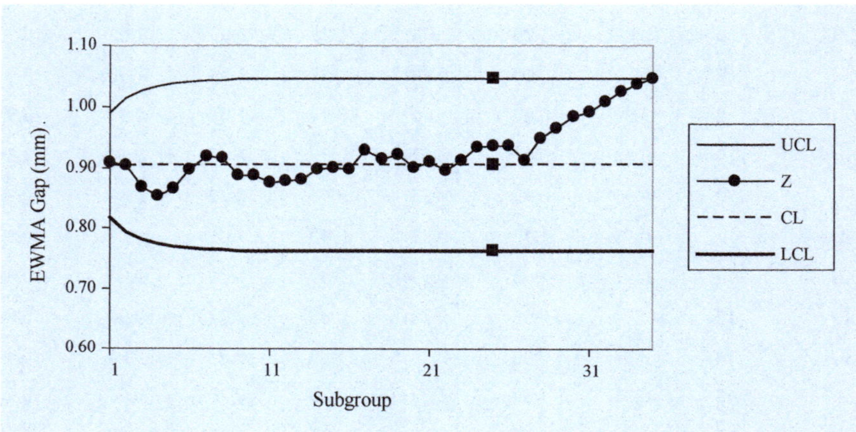

Fig. 8.1 EWMA chart for the gap data (▪ separates startup and steady state)

8.3 Multivariate Charting Concepts

Often, one person or team may have monitoring responsibilities for a large number of continuous characteristics. For example, in chemical plants a team can easily be studying thousands of characteristics simultaneously. Monitoring a large number of charts likely generates at least two problems.

False alarms may overburden personnel and/or demoralize chart users. If a single chart has a false alarm roughly once every 370 periods, one thousand charts could generate many false alarms each period. With high rates of false alarms, the people in charge of monitoring could easily abandon the charts and turn to anecdotal information.

Maintenance of a large number of charts could by itself constitute a substantial, unnecessary administrative burden. This follows because usually a small number of causes could be affecting a large number of characteristics. Logically, it could be possible to have a small number of charts, one for each potential cause.

Figure 8.2 shows a fairly simple set of assembly operations on two production lines. A single process engineer could easily be in charge of monitoring all 14 quality characteristics involved. Also, misplacement of second cylinder on the first could easily affect all quality characteristics on a given production line.

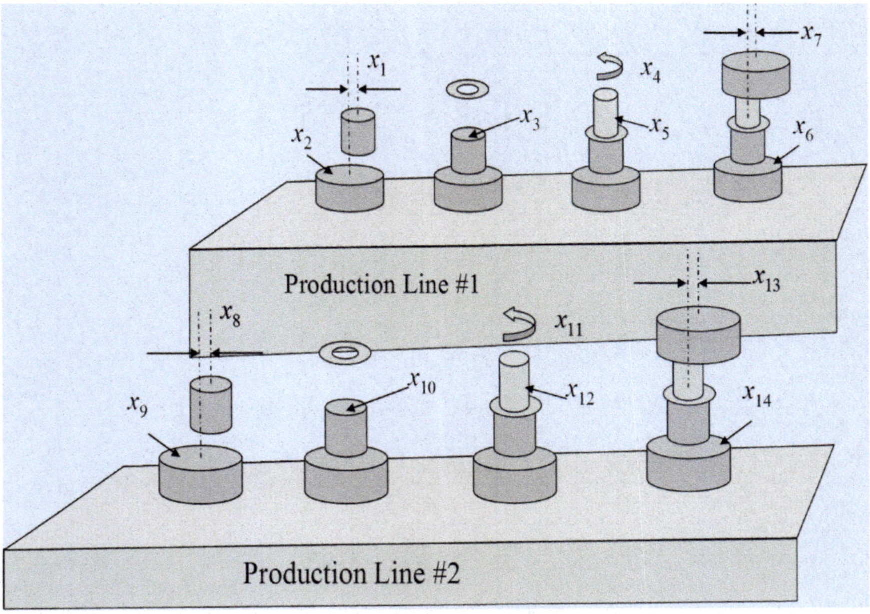

Fig. 8.2 Production sub-system involving $q = 14$ quality characteristics

The phrases "**Hotelling's T² charting**" or, equivalently, "**multivariate charting**" refer to a method proposed in Hotelling (1947). This method permits a single chart to permit simultaneous monitoring of a large number of continuous quality characteristics. This allows the user to regulate the false alarm rate directly and to reduce the burden of maintaining a large number of charts.

For single characteristic charts, the usual situation is characterized by plotted points inside the interval established by the control limits. With multiple characteristics, averages of these characteristics are plotted in a higher dimensional space than an interval on a line. The term "**ellipsoid**" refers to a multidimentional object which in two dimensions is an ellipse or a circle. Under usual circumstances, the averages of quality characteristics lie in a multidimensional ellipsoid. The following example shows a multidimensional ellipsoid with out-of-control signals outside the ellipsoid.

Example 8.3.1 Personal Blood Pressure and Weight

Question

Collect simultaneous measurements of a friend's weight (in pounds), systolic blood pressure (in mm Hg), and diastolic blood pressure (in mm Hg) three times each week for 50 weeks. Plot the weekly average weight versus the weekly average diastolic blood pressure to identify usual weeks from unusual weeks.

Answer

Table 8.2 shows real data collected over 50 weeks. The plot in Fig. 8.3 shows the ellipse that characterizes usual behavior and two out-of-control signals.

8

Table 8.2 Systolic (x_{i1k}) and diastolic (x_{i2k}) blood pressure and weight (x_{i3k}) data

k	X_{1k}	X_{21k}	X_{31k}	X_{12k}	X_{22k}	X_{32k}	X_{13k}	X_{23k}	X_{33k}	k	X_{11k}	X_{21k}	X_{31k}	X_{12k}	X_{22k}	X_{32k}	X_{13k}	X_{23k}	X_{33k}
1	127	130	143	76	99	89	172	171	170	26	159	124	147	101	93	107	172	173	172
2	127	149	131	100	95	85	170	175	172	27	147	132	146	91	94	91	172	173	173
3	146	142	138	87	93	87	172	173	172	28	135	148	152	89	96	85	171	172	173
4	156	128	126	94	89	95	171	173	170	29	154	144	136	98	95	96	172	175	173
5	155	142	129	92	100	104	170	171	170	30	139	131	133	85	91	85	172	172	172
6	125	150	125	96	96	97	170	169	171	31	140	120	142	100	88	89	173	172	172
7	133	143	123	92	113	99	169	170	171	32	131	122	138	94	88	81	174	172	171
8	147	140	121	93	102	97	170	170	171	33	136	139	130	89	91	87	171	172	172
9	137	120	135	88	100	113	170	171	170	34	130	135	135	90	89	91	173	173	172
10	138	139	148	112	104	90	170	172	172	35	137	142	149	86	98	91	175	175	174
11	146	150	129	99	105	96	172	170	170	36	127	120	140	93	93	96	171	174	172
12	129	122	150	96	90	110	170	170	172	37	144	147	141	95	104	80	172	173	174
13	146	150	129	99	105	96	172	170	170	38	126	119	122	83	94	87	173	172	173
14	128	150	151	95	110	92	170	172	172	39	144	142	133	83	102	91	171	171	172

(continued)

Table 8.2 (continued)

k	X_{1k}	X_{21k}	X_{31k}	X_{12k}	X_{22k}	X_{32k}	X_{13k}	X_{23k}	X_{33k}	k	X_{11k}	X_{21k}	X_{31k}	X_{12k}	X_{22k}	X_{32k}	X_{13k}	X_{23k}	X_{33k}
15	125	142	141	95	90	93	172	169	170	40	140	154	141	92	90	97	174	173	173
16	120	136	142	82	75	87	169	169	171	41	141	126	145	103	96	91	173	171	171
17	144	140	135	97	97	97	172	167	167	42	134	144	144	81	91	89	172	172	171
18	130	136	142	91	89	96	170	172	169	43	136	132	122	95	98	96	172	173	171
19	121	126	143	92	85	93	171	170	170	44	119	127	133	90	91	86	174	172	174
20	146	131	135	101	97	88	171	167	169	45	130	133	137	84	91	87	172	175	175
21	130	145	135	101	93	91	169	169	170	46	138	150	148	91	91	89	175	174	173
22	132	127	151	95	86	91	169	170	170	47	135	132	148	96	88	95	177	176	174
23	138	129	153	92	89	93	171	171	170	48	146	129	135	91	87	96	174	174	175
24	123	135	144	89	85	91	171	171	172	49	129	103	120	90	81	94	175	173	173
25	152	160	148	94	94	99	172	169	172	50	125	139	142	91	95	92	172	172	172

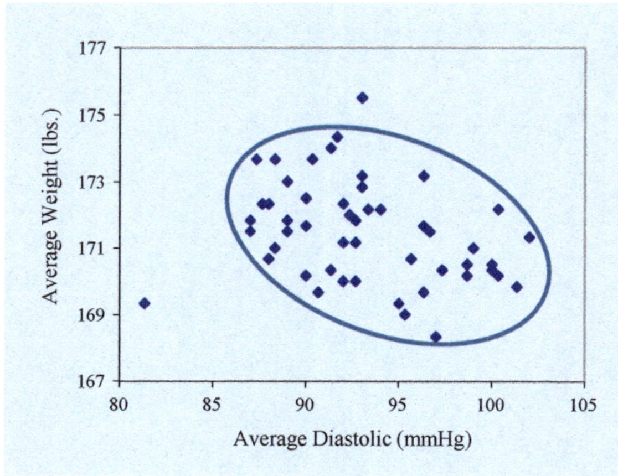

Fig. 8.3 Plot of average weight versus average diastolic blood pressure

8.4 Multivariate Charting (Hotelling's T² Charts)

In this section, the method proposed in Hotelling (1947) is described in Algorithm 8.2. This chart has been proven useful for applications that range from college admissions to processing plant control to financial performance. The method has two potentially adjustable parameters.

The symbols used are the following:

q is the number of quality characteristics being monitored.

r is the number of subgroups in the start-up period. This number could be as low as 20, which might be considered acceptable by many. The default value suggested here is $r=50$, but even higher numbers might be advisable because a large number of parameters need to be estimated accurately for desirable method performance.

α is the overall false alarm rate, is adjustable. Often, $\alpha=0.001$ is used so that false alarms occur typically once every thousand samples.

x_{ijk} is the value of the ith observation, of the jth characteristic in the kth period.

T^2 the quantity being plotted which is interpretable as a weighted distance from the center of the relevant ellipsoid.

Example 8.4.1. Personal Health Monitoring Continued

Question

Apply Hotelling's T^2 analysis to the data in Table 8.2. Describe any insights gained.

Answer

The following steps were informed by the data and consultation with the friend involved. The method offered evidence that extra support should be given to the friend during challenging situations including holiday travel and finding suitable childcare, as shown in Fig. 8.4.

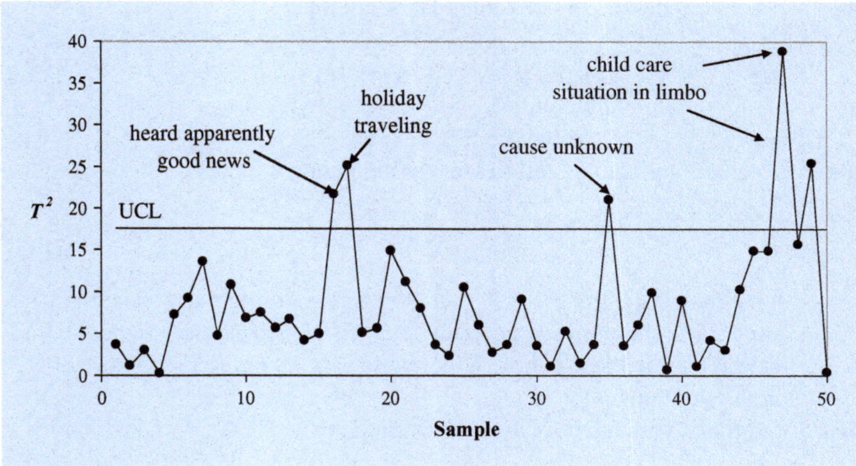

Fig. 8.4 Trial period in the blood pressure and weight example

Algorithm 8.2. Hotelling T² charting

Step 1 *(Startup)*: Measure or collect n measurements for each of the q characteristics from each of r periods, x_{ijk} for $i = 1, \ldots, n, j = 1, \ldots, q$, and $k = 1, \ldots, r$.

Step 2 *(Startup)*: Calculate all of the following:

$$\bar{x}_{jk} = \frac{1}{n}\sum_{i=1}^{n} x_{ijk} \text{for} j = 1, \ldots, q \quad \text{and} \quad k = 1, \ldots, q, \tag{8.5}$$

$$\bar{\bar{x}}_j = \frac{1}{q}\sum_{k=1}^{q} \bar{x}_{jk} \quad \text{for} \quad j = 1, \ldots, q, \tag{8.6}$$

$$S_{jhk} = \frac{1}{n-1}\sum_{i=1}^{n} (x_{ijk} - \bar{x}_{jk})(x_{ihk} - \bar{x}_{hk})$$
$$\text{for} j = 1, \ldots, q, h = 1, \ldots, q, \text{and} k = 1, \ldots, r \tag{8.7}$$

$$\bar{\mathbf{S}}_{jh} = \frac{1}{r}\sum_{k=1}^{r} S_{jhk} \quad \text{for} \quad j = 1, \ldots, q \quad \text{and} \quad h = 1, \ldots, q, \quad \text{and} \tag{8.8}$$

$$\mathbf{S} = \begin{pmatrix} \bar{\mathbf{S}}_{1,1} & \cdots & \bar{\mathbf{S}}_{1,q} \\ \vdots & \ddots & \vdots \\ \bar{\mathbf{S}}_{1,q} & \cdots & \bar{\mathbf{S}}_{q,q} \end{pmatrix}. \tag{8.9}$$

Step 3 *(Startup)*: Calculate the trial control limits using

$$UCL = \frac{q(r-1)(n-1)}{rn-r-q+1} F_{\alpha,q,rn-r-q+1} \quad \text{and} \quad LCL = 0. \tag{8.10}$$

where $F_{\alpha,q,rn-r-q+1}$ comes from ◘ Table 8.2.

Step 4 *(Startup)*: Calculate T^2 statistics for charting using

$$T^2 = n(\bar{\mathbf{x}} - \bar{\bar{\mathbf{x}}})' \mathbf{S}^{-1} (\bar{\mathbf{x}} - \bar{\bar{\mathbf{x}}}) \tag{8.11}$$

and plot. If $T^2 < LCL$ or $T^2 > UCL$, then investigate. Consider removing the associated subgroups from consideration if assignable causes are found that make it reasonable to conclude that these data are not representative of usual conditions.

Step 5 *(Startup)*: Calculate the revised limits using the remaining r^* units using

$$UCL = \frac{q(r^*+1)(n-1)}{r^*n-r^*-q+1} F_{a,q,r^*n-r^*-q+1} \quad \text{and} \quad LCL = 0, \tag{8.12}$$

where F comes from ◘ Table 8.3. Also, calculate the revised **S** matrix.

Step 6 *(Steady state)*: Plot the T^2 for new observations and have a designated person or persons investigate out-of-control signals. If and only if assignable causes are found, the designated local authority should take corrective action. Otherwise, the process should be left alone

Algorithm 8.3. Personal health monitoring continued

Step 1 *(Startup)*: The data are shown in ◘ Table 8.2 for $n = 3$ samples (roughly over the period being one week), $q = 3$ characteristics (systolic and diastolic blood pressure and weight), and $r = 50$ periods.

Step 2 *(Startup)*: The trial calculations resulted in

$$\mathbf{S} = \begin{bmatrix} 93.0 & 10.6 & 0.36 \\ 10.6 & 35.6 & 0.21 \\ 0.36 & 0.21 & 1.3 \end{bmatrix}$$

Step 3 *(Startup)*: The limits were
$UCL = 17.6$ and $LCL = 0$.

Step 4 *(Startup)*: The T^2 statistics were calculated and charted in the below using

$$T^2 = 3(\bar{\mathrm{x}} - \bar{\bar{\mathrm{x}}})' \begin{pmatrix} 0.011 & -0.003 & -0.003 \\ -0.003 & 0.029 & -0.004 \\ -0.003 & -0.004 & 0.793 \end{pmatrix} (\bar{\mathrm{x}} - \bar{\bar{\mathrm{x}}})$$

Four assignable causes were identified as described in the figure below. The process brought into clearer focus the occurrences that were outside the norm and unusually troubling or heartening.

8

Table 8.3 Critical values of the F distribution with $\alpha = 0.01$, i.e., $F_{\alpha=0.01,\nu1,\nu2}$

ν_2 \ ν_1	1	2	3	4	5	6	7	8	9	10
1	405284.1	499999.5	540379.2	562499.6	576404.6	585937.1	592873.3	598144.2	602284.0	605621.0
2	998.5	999.0	999.2	999.2	999.3	999.3	999.4	999.4	999.4	999.4
3	167.0	148.5	141.1	137.1	134.6	132.8	131.6	130.6	129.9	129.2
4	74.1	61.2	56.2	53.4	51.7	50.5	49.7	49.0	48.5	48.1
5	47.2	37.1	33.2	31.1	29.8	28.8	28.2	27.6	27.2	26.9
6	35.5	27.0	23.7	21.9	20.8	20.0	19.5	19.0	18.7	18.4
7	29.2	21.7	18.8	17.2	16.2	15.5	15.0	14.6	14.3	14.1
8	25.4	18.5	15.8	14.4	13.5	12.9	12.4	12.0	11.8	11.5
9	22.9	16.4	13.9	12.6	11.7	11.1	10.7	10.4	10.1	9.9
10	21.0	14.9	12.6	11.3	10.5	9.9	9.5	9.2	9.0	8.8
11	19.7	13.8	11.6	10.3	9.6	9.0	8.7	8.4	8.1	7.9
12	18.6	13.0	10.8	9.6	8.9	8.4	8.0	7.7	7.5	7.3
13	17.8	12.3	10.2	9.1	8.4	7.9	7.5	7.2	7.0	6.8
14	17.1	11.8	9.7	8.6	7.9	7.4	7.1	6.8	6.6	6.4
15	16.6	11.3	9.3	8.3	7.6	7.1	6.7	6.5	6.3	6.1

Step 5 *(Startup)*: The revised limits from $r^* = 46$ samples resulted in $UCL = 18.5$, $LCL = 0$, and

$$\mathbf{S} = \begin{bmatrix} 92.5 & 9.3 & -0.14 \\ 9.3 & 36.6 & 0.02 \\ -0.14 & 0.02 & 1.1 \end{bmatrix}$$

Step 6 *(Steady state)*: Monitoring continued using Eq. (12) and the revised **S**. Later data showed that new major life news caused a need to begin medication about one year after the trial period finished.

Note that identifying the assignable cause associated with an out-of-control signal on a T^2 chart is not always easy. Sometimes, software permits the viewing of multiple Xbar & R charts or other charts to hasten the process of fault diagnosis. Also, it is likely that certain causes are associated with certain values taken by linear combinations of responses. This is the motivation for many methods based on so-called "**principle components**" which involve plotting and studying multiple linear combinations of responses to support rapid problem resolution.

8

Also, note that multivariate charting has been applied to such diverse problems as college admissions and medical diagnosis. For example, admissions officers might identify several "successful seniors" to establish what the system should like under usual situations. Specifically, the officers might collect the data pertinent to the successful seniors that was available before those students were admitted. Plugging that data into the formulas in *Steps 2–5* generates rules for admissions. If a new student applies with characteristics yielding an out-of-control signal as calculated using Eq. (8.11), admission might not be granted. That student might be expected to perform in an unusual manner and/or perform poorly if admitted.

8.5 Summary

This chapter has described two advanced statistical process control (SPC) charting methods. First, exponential average moving average (EWMA) charting methods are relevant when detecting even small shifts in a single quality characteristic. They also provide a visual summary of the mean smoothed. Second, Hotelling's T^2 charts (also called multivariate control charts) permit the user to monitor a large number of quality characteristics using a single chart. In addition to reducing the burden of plotting multiple charts, the user can regulate the overall rate of false alarms.

8.6 Problems

In general, pick the correct answer that is most complete.

1. Which of the following is correct and most complete?
   a. EWMA control charts typically plot attribute data such as the number nonconforming.

   b. Hotelling T^2 charts are sometimes called multivariate control charts.
   c. Multivariate control charts offer an alternative to several applications of Xbar & R charts.
   d. All of the above are correct.
   e. All of the above are correct except (a) and (d).
2. Which of the following is correct and most complete?
   a. Multivariate charting typically involves the calculation of a large number of parameters during the startup phase.
   b. Critical characteristics can vary together because they share a common cause.
   c. EWMA charting often but not always discovers problems more quickly than Xbar & R charting.
   d. All of the above are correct.
   e. All of the above are correct except (a) and (d).
3. Which of the following is correct and most complete?
   a. The λ in EWMA charting can be adjusted based on a desire to detect small shifts slowly or big shifts more quickly.
   b. EWMA generally detects large shifts faster than Xbar & R charting.
   c. EWMA is particularly relevant when critical characteristics correlate.
   d. All of the above are correct.
   e. All of the above are correct except (a) and (d).
4. Which of the following is correct and most complete?
   a. Multiple Xbar & R charts do not help in assignable cause identification.
   b. Specific assignable causes might be associated with large values of certain linear combinations of quality characteristic values.
   c. Multivariate charting could be applied to college admissions.
   d. All of the above are correct.
   e. All of the above are correct except (a) and (d).
5. Provide two examples of cases in which multivar iate charting might apply.

References

Hotelling H (1947) Multivariate quality control, techniques of statistical analysis. Eisenhard H., Wallis (eds). McGraw-Hill, New York

Roberts SW (1959) Control chart tests based on geometric moving averages. Technometrics 1:236–250

SQC Case Studies

T. T. Allen, *Introduction to Engineering Statistics and Lean Six Sigma*,
https://doi.org/10.1007/978-1-4471-7420-2_9

9.1 Introduction

This chapter contains two descriptions of real projects in which a student played a major role in saving millions of dollars: the printed circuit board study and the wire harness voids study. The objectives of this chapter include: (1) providing direct evidence that the methods are widely used and associated with monetary savings and (2) challenging the reader to identify situations in which specific methods could help.

In both case studies, savings were achieved through the application of many methods described in previous chapters. Even while both case studies achieved considerable savings, the intent is not to suggest that the methods used were the only appropriate ones. Method selection is still largely an art. Conceivably, through more judicious selection of methods and additional engineering insights, greater savings could have been achieved. It is also likely that luck played a role in the successes.

The chapter also describes an exercise that readers can perform to develop practical experience with the methods and concepts. The intent is to familiarize participants with a disciplined approach to documenting, evaluating, and improving product and manufacturing approaches.

9.2 Case Study: Printed Circuit Boards

Printed circuit board (PCB) assemblies are used for sophisticated electronic equipment from computers to everyday appliances. Manufacturing printed circuit boards involves placing a large number of small components into precise positions and soldering them into place. Due to the emphasis on miniaturization, the tendency is to reduce the size of the components and the spacing between the components as much as electrical characteristics will allow. Therefore, both the multiplicity of possible failures and also the number of locations in the circuit boards where failures could occur continue to increase. Also, identifying the source of a quality problem is becoming increasingly difficult.

As noted in ▶ Chap. 2, one says that a unit is "nonconforming" if at least one of its associated "quality characteristics" is outside the "specification limits". These specification limits are numbers specified by design engineers. For example, if voltage outputs of a specific circuit are greater than 12.5 V or less than 11.2 V we might say that the unit is nonconforming. As usual, the company did not typically use the terms "defective" or "defect" because the engineering specifications may or may not correspond to what the customer actually needs. Also, somewhat arbitrarily, the particular company in question preferred to discuss the "**yield**" instead of the fraction nonconforming. If the "process capability" or standard fraction nonconforming is p_0, then $1 - p_0$ is called the standard yield.

Typical circuit board component process capabilities are in the region of 50 parts per million defective (ppm) for solder and component nonconformities. However, since the average board contains over 2000 solder joints and 300 components, even 50 ppm defective generates far too many boards requiring rework and a low overall capability.

In early 1998, an electronics manufacturing company with plants in the Midwest introduced to the field a new advanced product that quickly captured 83% of the market in North America, as described in Brady and Allen (2002). During the initial production period, yields (the % of product requiring no touchup or repair) had stabilized in the 70% range with production volume at 6000 units per month. In early 1999, the product was selected for a major equipment expansion in Asia. In order to meet the increased production demand, the company either needed to purchase additional test and repair equipment at a cost of $2.5 million, or the first test yield had to increase to above 90%. This follows because the rework needed to fix the failing units involved substantial labor content and production resources reducing throughput. The improvement to the yields was the preferred situation due to the substantial savings in capital and production labor cost, and, thus, the problem was how to increase the yield in a cost-effective manner.

Example 9.2.1 PCB Project Planning

Question

According to this book, which of the following is most recommended?

a. Convene experts and perform one-factor-at-a-time (OFAT) experiments because the project is not important enough for a 3–6 month scope.
b. $2.5M is a substantial enough potential payback to apply six sigma using a define, measure, analyze, design, and verify (DMADV) process.
c. $2.5M is a substantial enough potential payback to apply six sigma using a define, measure, analyze, improve, and control (DMAIC) process.

Answer

Convening experts is often useful and could conceivably result in quick resolution of problems without need for formalism. However, (a) is probably not the best choice because: (1) if OFAT were all that was needed, the yield would likely have already been improved by process engineers; and (2) a potential $2.5M payoff could pay off as many as 25 person years. Therefore, the formalism of a six sigma project could be cost justified. The answer (c) is more appropriate than (b) from the definition of six sigma in ► Chap. 1. The problem involves improving an existing system, not designing a new one.

9.2.1 Experience of the First Team

This project was of major importance to the financial performance of the company. Therefore a team of highly regarded engineers from electrical and mechanical engineering disciplines was assembled from various design and manufacturing areas throughout the company. Their task was to recommend ways to improve the production yield based on their prior knowledge and experience with similar products. *None of these engineers from top universities knew much about, nor intended to use, any*

Table 9.1 Yields achieved for 16 weeks prior to the initial team's activities

Week	Yield (%)	Week (%)	Yield (%)	Week (%)	Yield (%)	Week (%)	Yield (%)
1	71	5	87	9	66	13	63
2	58	6	68	10	70	14	68
3	69	7	71	11	76	15	76
4	77	8	59	12	82	16	67

Table 9.2 The initial team's predicted yield improvements by adjusting each factor

Factor	Yield improvement (%)
Replace vendor of main oscillator	5.3
Add capacitance to base transistor	4.7
Add RF absorption material to isolation shield	4.4
New board layout on power feed	4.3
Increase size of ground plane	3.8
Lower residue flux	3.6
Change bonding of board to heat sink	3.2
Solder reflow in air versus N_2	2.3
Raise temperature of solder tips	1.7

9

formal experimental planning and analysis technologies. Table 9.1 gives the weekly first test yield results for the 16 weeks prior to the team's activities based on a production volume of 1500 units per week.

Based on their technical knowledge of electrical circuit designs and their manufacturing experience, the assembled improvement team members critically reviewed the design and production process. They concluded that it was "poor engineering" of the circuit and manufacturing process that was at the heart of the low first test yield, thus creating the need for rework and retest. They came up with a list of 15 potential process and design changes for improvement based on their engineering judgment and anecdotal evidence. With this list in hand, they proceeded to run various **one-factor-at-a-time** (OFAT) experiments to prove the validity of their plan. Therefore, by not applying a six sigma project framework, the approach taken by the first team is arguably inappropriate and likely to lead to poor outcomes.

Due to perceived time and cost constraints, only one batch for each factor was completed with approximately 200 units in each run. Therefore, the inputs were not varied in batches not randomly ordered. Factors that showed a yield decrease below the 16-week average were discarded along with the experimental results. Table 9.2

Table 9.3 Performance after the implementation of the initial recommendations

Week	Yield (%)	Week	Yield (%)
17	62	21	40
18	49	22	41
19	41	23	45
20	42		

shows the results of the experiments with yield improvements predicted by the engineers based on their one-factor-at-a-time experiments.

Example 9.2.2 PCB First Team Experimental Strategy

Question

Which of the following could the first team most safely be accused of?

a. Stifling creativity by adopting an overly formal decision-making approach.
b. Forfeiting the ability to achieve statistical proof by using a nonrandom run order.
c. Not applying engineering principles, over-reliance on statistical methods.
d. Failing to evaluate the system prior to implementing changes.

Answer

Compared with many of the methods described in this book, team one has adopted a fairly "organic" or creative decision style. Also, while it is usually possible to gain additional insights through recourse to engineering principles, it is likely that these principles were consulted in selecting factors for OFAT experimentation to a reasonable extent. In addition, the first team did provide enough data to determine the usual yields prior to implementing recommendations. Therefore, the criticisms in (a), (c), and (d) are probably not fair. According to ▸ Chap. 5, random run ordering is essential to establishing statistical proof. Therefore, (b) is correct.

Based on their analysis of the circuit, the above experimental results and past experience, the improvement team predicted that a yield improvement of 16.7% would result from their proposed changes. All of their recommendations were implemented at the end of Week 17. Table 9.3 gives the weekly first test yields results for the 6 weeks of production after the revision.

It can be determined from the data in the tables that, instead of a yield improvement, the yield actually dropped 29%. On Week 22 it was apparent to the company that the proposed process changes were not achieving the desired outcome. Management assigned to this project two additional engineers who had been successful in the past with yield improvement activities. These two engineers both had mechanical engineering backgrounds and had been exposed to "design of experiments" and "statistical process control" tools through continued education at local universities, including Ohio State University, and company-sponsored seminars.

Example 9.2.3 PCB Second Team First Logical Steps

Question

Which is the most appropriate first action for the second team?

a. Perform design of experiments using a random run ordering.
b. Apply quality function deployment to relate customer needs to engineering inputs.
c. Return the process inputs to their values in the company SOPs.
d. Perform formal optimization to determine the optimal solutions.

Answer

Design of experiments, quality function deployment, and formal optimization all require more system knowledge than what is probably immediately available. Generally speaking, returning the system inputs to those documented in SOPs is a safe move unless there are objections from process experts. Therefore, (c) is probably the most appropriate initial step.

9.2.2 Second Team Actions and Results

The second team's first step was to construct a yield attribute control chart (a yield chart or 1-defective chart "$1 - p$") with the knowledge of the process change date (Table 9.4). From the chart, the two engineers were able to see that most of the fluctuations in yield observed before the team implemented their changes during Week 17 were, as Deming calls it, common cause variation or random noise. From this, they concluded that since around 1000 rows of data were used in each point on the chart, a significant number of samples would need to resolve yield shifts of less than 5% during a one-factor-at-a-time experiment. Control limits with $p_0 = 0.37$ and $n = 200$ units have UCL – LCL $= 0.2$ or 20% such that the sample sizes in the OFAT experiments were likely too small to spot significant differences.

The two engineers' first decision was to revert back to the original, *documented* process settings. This differed from the initial settings used by the first team because tinkering had occurred previously. The old evidence that had supported this anonymous tinkering was probably due to random noise within the process (factors changing about which the people are not aware). Table 9.4 gives the weekly test yields for the 5 weeks after this occurrence.

Table 9.4 Yields for the 5 weeks subsequent to the initial intervention

Week	Yield (%)	Week	Yield (%)	Week	Yield (%)
24	62	26	77	28	77
25	78	27	75		

Example 9.2.4 PCB Project Tools and Techniques

Question

If you were hired as a consultant to the first team, what specific recommendations would you make?

Answer

The evidence of improvement related to the team's recommended inputs is weak. It would therefore likely be beneficial to return the process to the settings documented in the standard operating procedures. Initiate a six sigma project. This project could make use of Pareto charting to prioritize which nonconformities and associated subsystems to focus on. Control charting can be useful for establishing a benchmark for the process quality and a way to evaluate possible progress. Design of experiments involving random run ordering might be helpful in providing proof that suggested changes really will help.

This approach restored the process to its previous "in control" state with yields around 75%. The increase in yield shown on the control chart (Fig. 9.1) during this time frame was discounted as a "**Hawthorne effect**" since no known improvement was implemented. The phrase "Hawthorne effect" refers to a difference caused simply by our attention to and study of the process. Next the team tabulated the percent of failed products by relevant defect code shown in Fig. 9.2. It is generally more correct to say "nonconformity" instead of "defect" but in this problem the engineers called these failures to meet specifications "defects". The upper control limit (UCL) and lower control limit (LCL) are shown calculated in a manner similar to "*p*-charts" in standard textbooks on statistical process control, e.g., Besterfield (2001), based on data before any of the teams' interventions.

The procedure of Pareto charting was then applied to help visualize the problem shown in the figure below. The total fraction of units that were nonconforming was

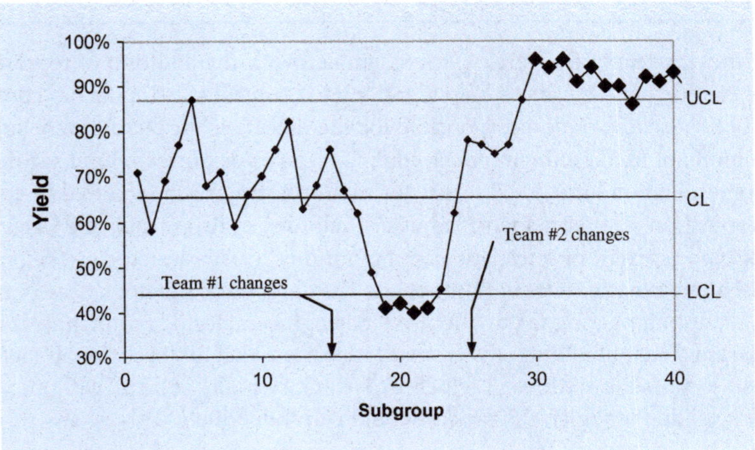

Fig. 9.1 Control chart for the entire study period

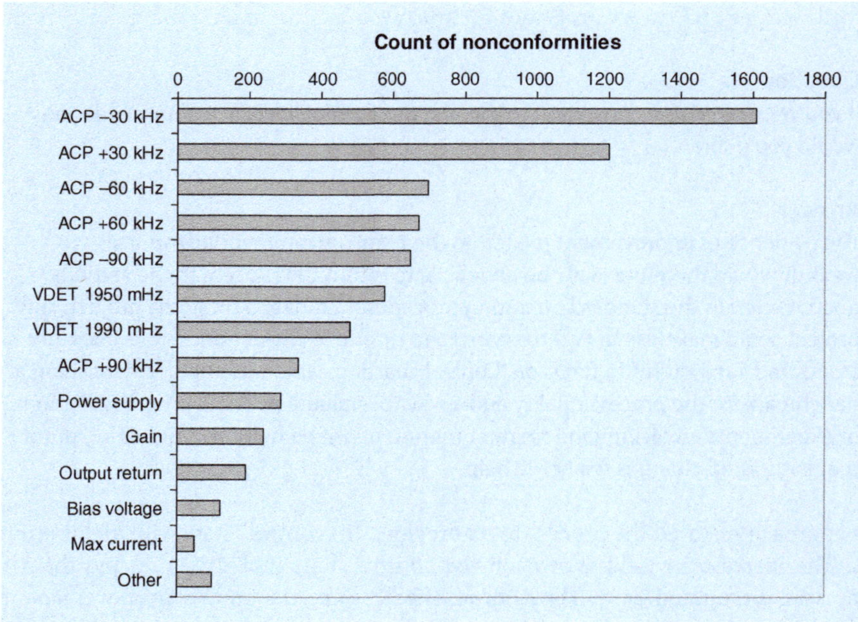

9

Fig. 9.2 Pareto chart of the nonconforming units from 15 weeks of data

30%. The total fraction of unit that were nonconforming associated with the ACP subsystem was 21.5%. Therefore, 70% of the total yield loss (fraction nonconforming) was associated with the "ACP" defect code or subsystem. The engineers then concentrated their efforts on this dominant defect code. This information, coupled with process knowledge, educated their selection of factors for the following study.

9.3 Printed Circuitboard: Analyze, Improve, and Control Phases

At this time, the team of experts was reassembled with the addition of representation from the production workers to help identify what controllable inputs or "**control factors**" might cause a variation responsible for the defects. Four factors were suggested, and two levels of each factor were selected: (1) transistor power output (at the upper or lower specification limits), (2) transistor mounting approach (screwed or soldered), (3) screw position on the frequency adjuster (half turn or two turns), and (4) transistor heat sink type (current or alternative configuration). This last factor was added at the request of a representative from production. This factor was not considered important by most of the engineering team. The two lead engineers decided to include this factor as the marginal cost of adding it was small. Also, note that all levels for all factors corresponded to possible settings at which experiments could be run (without violating any contracts) and at which the process could run indefinitely without any prohibitive effect.

Table 9.5 Data from the screening experiment for the PCB case study

Run	A	B	C	D	y_1 – Yield
1	−1	1	−1	1	92.7
2	1	1	−1	−1	71.2
3	1	−1	−1	1	95.4
4	1	−1	1	−1	69.0
5	−1	1	1	−1	72.3
6	−1	−1	1	1	91.3
7	1	1	1	1	91.5
8	−1	−1	−1	−1	79.8

Table 9.6 Analysis results for PCB screening experiment

Factor	Estimated coefficients (β_{est})	t_{Lenth}	Estimated probability of being "important"
A	−1.125	0.98	0.13170
B	−0.975	0.85	0.02081
C	−1.875	1.64	0.03732
D	9.825	8.59	0.96173

The team selected the experimental plan shown below with reference to the decision support information provided by statistical software. In this eight run "screening" experiment on the transistor circuit shown in Table 9.5, each run involved making and testing 350 units with the controllable factors adjusted according to the relevant row of the matrix. For example, in the first run, the selected transistor power output was at the lower end of the specification range (−1), the transistor mounting approach was soldered (+1), the screwed position of the frequency adjustor was two turns (+1), and the current transistor heat sink type was used (−1). The ordering of the test runs was also decided using a random number generator. The output yields or "**response values**" resulting from making and testing the units are shown in the right-hand column. We use the letter "y" to denote experimental outputs or responses. In this case, there is only a single output that is denoted y_1.

As is often the case, substantial time was required to assemble the resources and approvals needed to perform the first test run. In fact, this time was comparable to the time needed for the remaining runs after the first run was completed.

An analysis of this data based on first order linear regression and so-called **Lenth's method** (Lenth 1989) generated the statistics in the second column of Table 9.6. Note that t_{Lenth} for factor D, 8.59, is larger than the "critical value" $t_{EER,0.05,8} = 4.87$. Since the experimental runs were performed in an order determined by a so-called "**pseudo-random** number generator" (see ▸ Chaps. 3 and 5), we can say that "we have

proven with $\alpha = 0.05$ that factor D significantly affects average yield". For the other factors, we say that "we failed to find significance" because t_{Lenth} is less than 12.89. The level of "**proof**" is somewhat complicated by the particular choice of experimental plan. In ▶ Chap. 3, experimental plans yielding higher levels of evidence will be described. Intuitively, varying multiple factors simultaneously does make statements about causality dependent upon assumptions about the joint effects of factors on the response. However, the Lenth (1989) method is designed to give reliable "proof" based on often realistic assumptions.

An alternative analysis is based on the calculation of Bayesian posterior probabilities for each factor being important yields, the values shown in the last column of ◘ Table 9.6. This analysis similarly indicates that the probability that the heat sink type affects the yield is extremely high (96%). Further, it suggests that the alternative heat sink is better than the current one (the heat sink factor estimated coefficient is positive).

Based on this data (and a similar analysis), the two engineers recommended that the process should be changed permanently to incorporate the new heat sink. In the terminology needed in subsequent chapters, this corresponded to a recommended setting $x_4 = \mathrm{D} =$ the new heat sink. This was implemented during Week 29. ◘ Table 9.7 gives the weekly yield results for the period of time after the recommended change was implemented. Using the yield charting procedure, the engineers were able to confirm that the newly designed process produced a stable first test yield (no touch-up) in excess of 90%, thus avoiding the equipment purchase and saving the company \$2.5 million.

This case illustrates the benefits of our DOE technology. First, the screening experiment technology used permitted the fourth factor to be varied with only eight experimental runs. The importance of this factor was controversial because the operators had suggested it and not the engineers. If the formal screening method had not been used, then the additional costs associated with one-factor-at-a-time (OFAT) experimentation and adding this factor would likely have caused the team not to vary that factor. Then the important subsequent discovery of its importance would not have occurred. Second, in the experimental plan, which in this case is the same as the standard "design of experiments" (DOE), multiple runs are associated with the high level of each factor and multiple runs are associated with the low level of each factor.

◘ **Table 9.7** Confirmation runs establishing the process shift/improvement

Week	Yield (%)	Week	Yield (%)
29	87	36	90
30	96	37	86
31	94	38	92
32	96	39	91
33	91	40	93
34	94	41	89
35	90	42	96

For example, 1400 units were run with the current heat sink and 1400 units were run with the new heat sink. The same is true for the other factors. The reader should consider that this would not be possible using an OFAT strategy to allocate the 2800 units in the test. Finally, the investigators varied only factors that they could make decisions about. Therefore, when the analysis indicated that the new heat sink was better, they could "dial it up", i.e., implement the change.

Note that the purpose of describing this study is not necessarily to advocate the particular experimental plan used by the second team. The purpose is to point out that the above "screening design" represents an important component of one formal experimentation and analysis strategy. The reader would likely benefit by having these methods in his or her set of alternatives when he or she is selecting a methodology. (For certain objectives and under certain assumptions, this experimental plan might be optimal.) The reader already has OFAT as an option.

Example 9.3.1 PCB Improvement Project Critique

Question

While evidence showed that the project resulting system inputs helped save money, which of the following is the safest criticism of the approach used?

a. The team could have applied design of experiments methods.
b. A cause and effect matrix could have clarified what was important to customers.
c. Having a charter approved by management could have shorted the DOE time.
d. Computer assisted optimization would improve decision-making in this case.

Answer

The team did employ design of experiments methods, so answer (a) is clearly wrong. It was already clear that all stakeholders wanted was a higher yield. Therefore, no further clarification of customer needs (b) would likely help. With only a single system output or response (yield) and a first order model from the DOE activity, optimization can be done in one's head. Set factors at the high level (low level) if the coefficient is positive (negative) and significant. Answer (c) is correct, since much of the cost and time involved with the DOE related to obtaining needed approvals and no formal charter had been cleared with management in the define phase.

9.4 Wire Harness Voids Study

A Midwest manufacturer designs and builds wire harnesses for the aerospace and marine markets. Quality and continuous improvement are key drivers for new sales. Some examples of systems supported are U.S. nuclear submarines, manned space flight vehicles, satellites, launch vehicles, tactical and strategic missiles and jet engine control cables. The case study involved a Guidance Communication Cable used on a ballistic missile system.

Part of this communication cable is molded with a polyurethane compound to provide mechanical strength and stress relief to the individual wires as they enter the connector shell. This is a two-part polyurethane which is procured premixed and frozen to prevent curing. The compound cures at room temperature or can be accelerated with elevated heat. Any void or bubble larger than 0.04 in. that appears in the polyurethane after curing constitutes a single nonconformity to specifications. Whenever the void nonconformities are found, local rework on the part is needed, requiring roughly 17 min of time per void. Complete inspection is implemented, in part because the number of voids per part typically exceeds ten. Also, units take long enough to process that a reasonably inspection interval includes only one or two units.

Example 9.4.1 Charting Voids

Question

Which charting method is most relevant for measuring performance?

a. Xbar and R charting is the most relevant since there is a high quality level.
b. *p*-charting is the most important since we are given the fraction of nonconforming units only.
c. *u*-charting is the most important since we are given count of nonconformities and no weighting data.

Answer

Since each void is a nonconformity and no voids are obviously more important than others, the relevant chart from ▶ Chap. 4 is a *u*-chart. Further, a *p*-chart would not be effective because the number of runs per time period inspected was small and almost all of them had at least one void, i.e., $n \times (1 - p_0) < 5$.

9.4.1 Define Phase

A six sigma project was implemented in the Special Assembly Molded Manufacturing Cell to reduce the rework costs associated with polyurethane molding voids. A cross functional team of seven members was identified for this exercise. During the Define Phase a series of meetings were held to agree on the charter. The resulting project "kick off" or charter somewhat nonstandard form is shown in Table 9.8.

9.4.2 Measure Phase

The team counted the number of voids or nonconformities in 20 5-part runs and set up a *u*-chart as shown in Fig. 9.3. The *u*-charting start up period actually ran into January so that the recommended changes from the improvement phase went into effect immediately after the start up period finished. A *u*-chart was selected instead of, e.g., a *p*-chart. As noted above, the number of units inspected per period was small and almost all units had at least one void. Therefore, a *p*-chart would not be informative since p_0 would be nearly 1.0.

Table 9.8 Void defects in molded products—project team charter

(1) Process	Special Assembly Polyurethane Molding
(2) Predicted savings	$50,000/year
(3) Team members	2 Product engineers, 2 Process engineers, 2 Manufacturing engineers, 1 Quality assurance (representative from all areas)
(4) Quantifiable project objectives	Reduce rework due to voids by 80% on molded products Provide a "model" for other products with similar processes
(5) Intangible possible benefits	This project is chartered to increase the productivity of the special assembly molded cell. Furthermore, the project will improve the supply chain. Production planning will be improved through reducing variation in manufacturing time
(6) Benefits	Improved supply chain, Just-in-time delivery, cost savings
(7) Schedule	Define phase, Oct. 16–Oct. 29; Measure phase, Oct. 29–Nov. 19; Analyze phase, Nov. 19–Dec. 12; Improve phase, Dec. 12–Jan. 28; Control phase, Jan. 28–Feb. 18
(8) Support required	Ongoing operator and production manager support
(9) Potential barriers	Time commitment of team members
(10) Communication	Weekly team meeting minutes to champion, production manager and quality manager

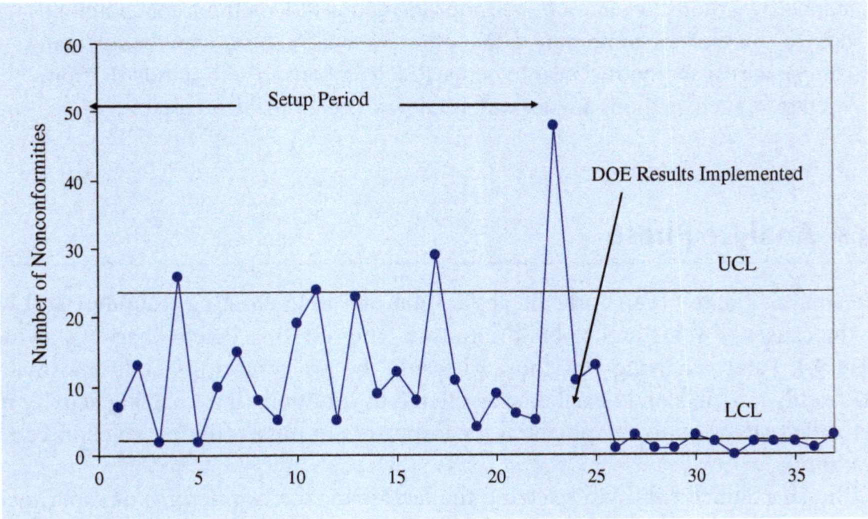

Fig. 9.3 *u*-Chart of the voids per unit used in measurement and control phases

At the end of the chart start-up period, an informal gauge R&R activity investigated the results from two inspectors. The approach used was based on units that had been inspected by the relevant subject matter expert so "standard values" were available. The results showed that one operator identified an average of 17 voids per run while the second operator identified an average of 9 voids per run based on the same parts. The specifications from the customer defined a void to be any defect 0.040″ in diameter or 0.040″ in depth. An optical measuring device and depth gage were provided to the inspectors to aid in their determination of voids. Subsequent comparisons indicated both operators to average nine voids per run.

Example 9.4.2 Wire Harness Voids Gauge R&R

Question

Which likely explains why formal gauge R&R was not used?

a. The count of voids is attribute data, and it was not clear whether standard methods were applicable.
b. The engineers were not aware of comparison with standards methods since it is a relatively obscure method.
c. The project was not important enough for formal gauge R&R to be used.
d. Answers in parts (a) and (b) are both reasonable explanations.

Answer

Gauge R&R is generally far less expensive than DOE. Usually if managers feel that DOE is cost justified, they will likely also approve gauge R&R. The attribute data nature of count data often makes engineers wonder whether they can apply standard methods. Yet, if $n \times u_0 > 5$, applying gauge R&R methods for continuous data to count of nonconformity data is often reasonable. Also, even though many companies use methods similar to gauge R&R (comparison with standards) from ► Chap. 4, such methods are not widely known. Therefore, (d) is correct.

9.4.3 Analyze Phase

The analysis phase began with the application of Pareto charting to understand better the causes of voids and to build intuition. The resulting Pareto chart is shown in ◘ Fig. 9.4. Pareto charting was chosen because the nonconformity code information was readily available and visual display often aids intuition. This charting activity further called attention to the potential for inspectors to miss counting voids in certain locations.

The chart aided subtly in selecting the factors for the two designs of experiments (DOE) applications described below (with results omitted for brevity). A first DOE was designed in a nonstandard way involving two factors one of which was qualitative at four levels. The response was not void count but something easier to measure. The DOE was designed in a nonstandard way in part because not all combinations of the

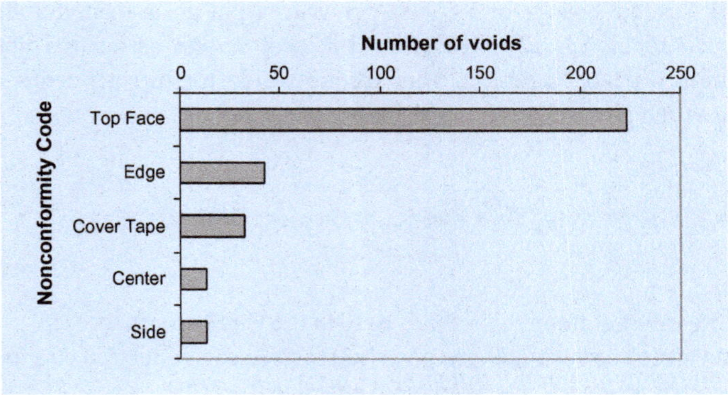

Fig. 9.4 Pareto chart of the void counts by location or nonconformity type

two factors were economically feasible. The results suggested a starting point for the next DOE.

A second application of DOE was performed to investigate systematically the effect of seven factors on the void count using an eight run fractional factorial. The results suggested that the following four factors had significant effects on the number of voids: thaw time, thaw temperature, pressure, pot life.

9.4.4 Improve Phase

The team recommended adjusting the process settings in the following manner. For all factors that had significant effects in the second DOE, the settings were selected that appeared to reduce the void count. Other factors were adjusted to settings believed to be desirable, taking into account considerations other than void count. Also, the improved measurement procedures were simultaneously implemented as suggested by the informal application of gauge R&R in the measurement phase.

9.4.5 Control Phase

The team spent 10 weeks confirming that the recommended settings did in fact reduce the void counts as indicated in Fig. 9.3. Charting was terminated at that point because it was felt that the information from charting would not be helpful with such small counts, and the distribution of void nonconformities had certainly changed. In other words, there was a long string of out-of-control signals indicating that the adoption of the new settings had a positive and sustained assignable cause effect on the system.

Four weeks were also spent documenting the new parameters into the production work instructions for the production operators and into the mold design rules for the tool engineers. At the same time, training and seminars were provided on the results of the project. The plan was for a 17-week project, with actual duration of 22 weeks. At the same time the project was projected to save $50,000 per year with actual calculated direct rework-related savings of $97,800 per year. Total costs in materials and labor were calculated to be $31,125. Therefore, the project was associated with approximately a 4-month payback period. This accounts only for direct rework-related savings, and the actual payback period was likely much sooner.

Example 9.4.3 Wire Harness Void Reduction Project Critique

Question

Which is the safest critique of methods used in the wire harness study?

a. A cost Pareto chart would have been better since cost reduction was the goal.
b. QFD probably would have been more effective than DOE.
c. Charting of void count should not have been dropped since it always helps.
d. An FMEA could have called attention to certain void locations being missed.
e. Pareto charting must always be applied in the define phase.

Answer

It is likely that a cost Pareto chart would not have shown any different information than an ordinary Pareto chart. This follows since all voids appeared to be associated with the same cost of rework and there were no relevant performance or failure issues mentioned. QFD is mainly relevant for clarifying, in one method, customer needs and competitor strengths. The realities at this defense contractor suggested that customer needs focused almost solely on cost reduction, and no relevant competitors were mentioned. DOE was probably more relevant because the relevant system inputs and outputs had been identified and a main goal was to clarify the relevant relationships. With such low void counts, u-charting would likely not have been effective since $n \times u_0 < 5$. In general, all methods can be applied in all phases, if Table 2.1 in Chap. 2 is taken seriously. This is particularly true for Pareto charting, which generally requires little expense. FMEA would likely have cost little and might have focused inspector attention on failures associated with specific void locations. Therefore, (d) is probably most correct.

9

9.5 Case Study Exercise

This section describes an exercise that readers can perform to obtain what might be called a "green belt" level of experience. This exercise involves starting with an initial standard operating procedure (SOP) and concluding with a revised and confirmed SOP. Both SOPs must be evaluated using at least one control chart. For training purposes, the sample size can be only $n = 2$, and only 12 subgroups are needed for the startup periods for each chart.

At least eight "methods" or "activities" listed in Table 2.1 must be applied. The creation of SOPs of various types can also be considered as activities counted in the total of eight. Document all results in four pages including tables and figures. This requirement on the number of methods reflects actual requirements that some original equipment manufacturers place on their own teams and supplier teams. Since design of experiments (DOE) may be unknown to the reader at this point, it might make sense not to use these methods.

Tables and figures must have captions, and these should be referred to inside the body of your report. The following example illustrates the application of a disciplined approach to the toy problem on the fabrication and flight of paper air wings. This project is based on results from an actual student project with some details changed to make the example more illustrative.

9.5.1 Project to Improve a Paper Air Wings System

Define: The primary goal of the this project is to improve the experience of making, using, and disposing of paper air wings. The initial standard operating procedure for making paper air wings is shown in ◘ Table 9.9. Performing the process mapping method generated the flowchart of the manufacturing and usage map in ◘ Fig. 9.5. The process elicited the key input variables, *x*'s, and key output variables, *y*'s, for study. Further, it was decided that the subsystem of interest would not include the initial storage, so that initial flatness and humidity were out of scope.

Measure: The measurement SOP in ◘ Table 9.10 was developed to evaluate the current manufacturing SOP and design. The initial manufacturing SOP was then evaluated using the measurement SOP and an Xbar and R charting procedure. In all, 24 air wings were built and tested in batches of two. This generated the data in ◘ Table 9.11 and the Xbar and R chart in ◘ Fig. 9.6.

As there were no out-of-control signals, the initial and revised charts were the same. The initial process capability was 0.8 s ($6\sigma_0$) and the initial average flight time was 1.7 s (Xbarbar).

Analyze: The main analysis method investigated applied was benchmarking with a friend's air wing material selection and manufacturing method. The friend was asked to make air wings, and the process was observed and evaluated. This process generated the benchmarking matrices shown in ◘ Table 9.12. The friend also serves as the

◘ **Table 9.9** Initial standard operating procedure for making paper air wings

Title:	Initial SOP for paper air wing manufacturing
Scope:	For people who like to make simple paper airplanes
Summary:	A method for making a simple air wing
Training qualifications:	None
Equipment and supplies:	One piece of notebook paper
Method: 	Tear a 2″ by 2″ square from the paper Fold the paper in half along the long diagonal and unfold partially Drop the air wing gently into the storage area

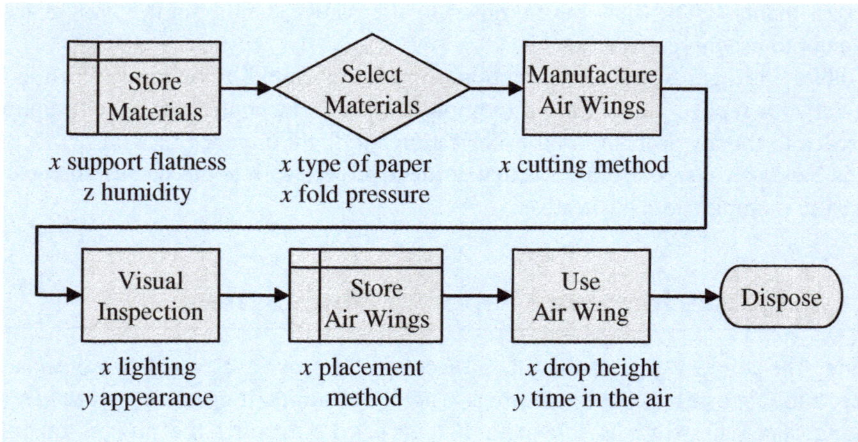

Fig. 9.5 Process map flowchart of air wing manufacturing and usage

Table 9.10 Measurement SOP for the air wing study

Title:	SOP for paper air wing time in air testing
Scope:	Focuses on the simulating usage of a single air wing
Summary:	Describes the drop height and time evaluation approach
Training:	None
Qualifications:	None
Equipment and supplies:	A paper air wing and a digital stopwatch
Method:	Hold the air wing with your right hand Hold stopwatch in left hand Bend right arm and raise it so that the hand is 60″ high Release paper airplane and simultaneously start stop-watch When the paper air wing lands, stop stopwatch Record the time

customer, generating the ratings in the tables. It was observed that the air times were roughly similar, but the appearance of the friends air wings was judged superior.

Improve: From the process mapping experience, the placement method was identified as a key input variable. Common sense suggested that careful placement might improve the appearance. Even though the air time was a little lower based on a small amount of evidence, other benchmarking results suggested that the friend's approach likely represented best practices to be emulated. This resulted in the revised standard operating procedure (SOP) in Table 9.13.

Control: To verify that the air time was not made worse by the revised SOP, Xbar and R charting based on an additional 24 air wings were constructed and tested (see Table 9.14 and Fig. 9.7).

Table 9.11 Air wing times studying the initial system for short run Xbar and R chart

Subgroup	Paper	Cutting	X_1	X_2	Average	Range
1	Notebook	Tear	1.72	1.65	1.69	0.07
2	Notebook	Tear	1.89	1.53	1.71	0.36
3	Notebook	Tear	1.73	1.79	1.76	0.06
4	Notebook	Tear	1.95	1.78	1.87	0.17
5	Notebook	Tear	1.58	1.86	1.72	0.28
6	Notebook	Tear	1.73	1.65	1.69	0.08
7	Notebook	Tear	1.46	1.68	1.57	0.22
8	Notebook	Tear	1.71	1.52	1.62	0.19
9	Notebook	Tear	1.79	1.85	1.82	0.06
10	Notebook	Tear	1.62	1.69	1.66	0.07
11	Notebook	Tear	1.85	1.74	1.80	0.11
12	Notebook	Tear	1.65	1.77	1.71	0.12

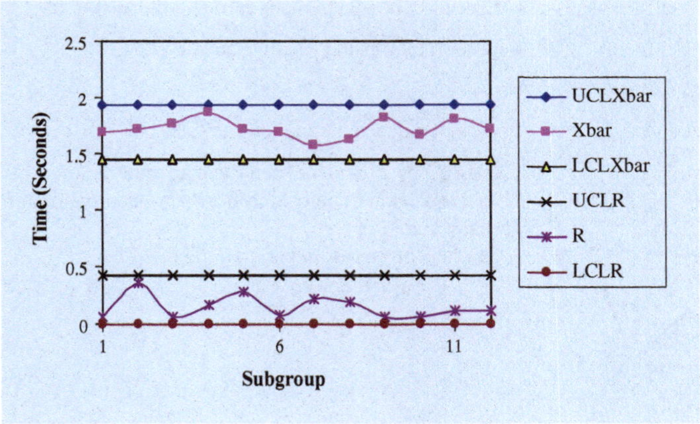

Fig. 9.6 Combined Xbar and R chart for initial system evaluation

The revised SOP was followed for the manufacturing, and the testing SOP was applied to emulate usage. The improvement in appearance was subjectively confimed. The revised average was not improved or made worse (the new Xbarbar equaled 1.7 s). At the same time the consistency improved as measured by the process capability ($6\sigma_0$ equaled 0.3 s) and the width of control limits.

Hypothetically, assume the following:

1. There was a lower specification limit on the air time equal to 1.60 s.
2. The number of units produced per year was 10,000.
3. Rework costs per item were \$1.
4. The initial SOP resulted in a $3/24 = 0.13$ fraction nonconforming.
5. The relevant fudge factor (G) is 4.0 to account for loss of good will and sales.

Table 9.12 Benchmarking matrices for the air wing study

Customer issue	Project leaders's planes	Friends's planes
Appearance seems crinkled	4	9
Appearance is consistent	6	8
Flight air time is long	8	7

Competitor	KIV-paper type	KIV-cutting method	KIV-fold method	KIV-placement method	KOV-time in the air (average of two in seconds)
Project leader	Notebook	Tearing	Regular	Dropping	1.7
Friend	Magazine	Scissors	Press firmly	Dropping	1.6

Table 9.13 Revised standard operating procedure for making paper air wings

Title:	Revised SOP for paper air wing manufacturing
Scope:	For people who like to make simple paper air wings
Summary:	A method for making a simple air wing
Training qualifications:	None
Equipment and supplies:	1 piece of magazine paper and scissors
Method: 	1. Cut a 2″ by 2″ square from the magazine 2. Fold the paper in half along the diagonal pressing firmly and partially unfold 3. Carefully place the air wing on the pile

Then, by effectively eliminating nonconformities, the project would have saved \$10,000, considering a 2-year payback period.

9.6 Chapter Summary

In this chapter, two case studies were described together with questions asking the reader to synthesize and critique. The intent was to establish the business context of the material and encourage the reader to synthesize material from previous chapters. Both case studies apparently had positive conclusions. Yet, there is little evidence that the manner in which each was conducted was the best possible. It seems likely that other methods could have been used and combined with engineering insights to achieve even better results. The chapter closed with an exercise intended to permit readers to gain experience in a low consequence environment.

Table 9.14 Air wing times studying the initial system for short run Xbar and R chart

Subgroup	Paper	Cutting	X_1	X_2	Average	Range
1	Magazine	Scissors	1.66	1.68	1.67	0.02
2	Magazine	Scissors	1.63	1.63	1.63	0.00
3	Magazine	Scissors	1.67	1.72	1.70	0.05
4	Magazine	Scissors	1.73	1.71	1.72	0.02
5	Magazine	Scissors	1.77	1.72	1.75	0.05
6	Magazine	Scissors	1.68	1.72	1.70	0.04
7	Magazine	Scissors	1.71	1.78	1.75	0.07
8	Magazine	Scissors	1.64	1.74	1.69	0.10
9	Magazine	Scissors	1.62	1.73	1.68	0.11
10	Magazine	Scissors	1.73	1.71	1.72	0.02
11	Magazine	Scissors	1.76	1.66	1.71	0.10
12	Magazine	Scissors	1.70	1.73	1.72	0.03

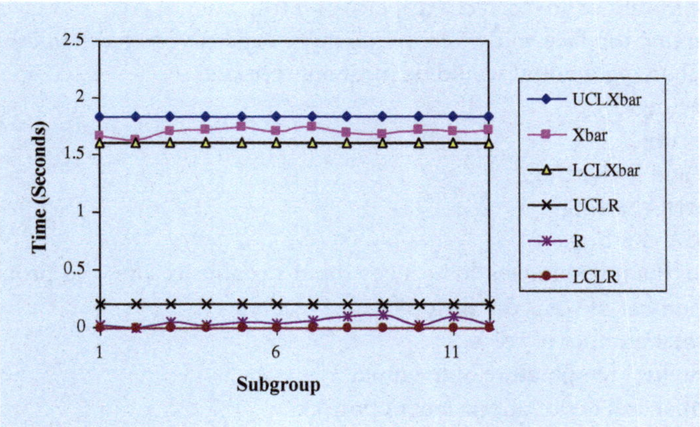

Fig. 9.7 Combined Xbar and R chart for initial system evaluation

9.7 Problems

In general, pick the correct answer that is most complete.

1. According to this book, which is (are) specifically discouraged?
   a. Planning a PCB project scope requiring 10 months or more.
   b. Starting the PCB project by performing quality function deployment.
   c. Performing a verify instead of control phase for the PCB project.
   d. All of the above are correct.
   e. All of the above are correct except (a) or (d).

2. Which of the following was true about the PCB project?
   a. Miniaturization was not relevant.
   b. Quality issues limited throughput causing a bottleneck.
   c. Design of experiments screening methods were not applied.
   d. All of the parts failing to conform to specifications were defective.
   e. All of the above are true.
3. According to this book, which is the most appropriate first project action?
   a. Quality function deployment.
   b. Design of experiments.
   c. Creating a project charter.
   d. Control planning.
   e. All of the following are equally relevant for the define phase.
4. If you were hired as a consultant to the first team, what specific recommendations would you make besides the ones given in the example?
5. In the voids project, which phase and method combinations occurred? Give the answer that is correct and the most complete.
   a. Analyze—Quality Function Deployment.
   b. Measure—gauge R&R (informal version).
   c. Define—creating a charter.
   d. All of the above are correct.
   e. All of the above are correct except (a) and (d).

6. Suppose that top face voids were much more expensive to fix than other voids. Which charting method would be most appropriate?
   a. p-charting.
   b. u-charting.
   c. Xbar and R charting.
   d. Demerit charting.
   e. EWMA charting.
7. Which of the following could be a key input variable for the void project system?
   a. The number of voids on the top face.
   b. The total number of voids.
   c. The preheat temperature of the mold.
   d. The final cost on an improvement project.
   e. None of the above.
8. In the voids case study, what assistance would FMEA most likely provide?
   a. It could have helped to identify the techniques used by competitors.
   b. It could have helped to develop quantitative input-output relationships.
   c. It could have helped select specific inspections systems for improvement.
   d. It could help guarantee that design settings were optimal.
   e. It could have helped achieve effective document control.
9. In the voids project, what change is most likely to invite scope creep?
   a. Increasing the predicted savings to $55,000.
   b. Removing two engineers from the team.
   c. Changing the quantifiable project objective to read, "to be decided".
   d. Shortening the schedule to complete in January.
   e. None of the above is relevant to scope creep.

10. Which rows of the void project charter address the not-invented-here syndrome?
    a. The predicted savings section or row.
    b. The team members' project objectives section or row.
    c. The quantifiable project objectives section or row.
    d. The intangible possible benefits section or row.
    e. None of the entries
11. Which of the following is the most correct and complete?
    a. According to the definition of six sigma, monetary benefits must be measured.
    b. Charting is often helpful for measuring benefits and translating to dollars.
    c. Gauge R&R might or might not be needed in a control plan.
    d. All of the above are correct.
    e. All of the above are correct except (a) and (d).
12. According to this text, which of the following is the most correct and complete?
    a. Pareto charting could never be used in the analyze phase.
    b. Formal optimization must be applied.
    c. Meeting, a charter, two control charting activities, a C&E matrix, informal optimization, and SOPs might constitute a complete six sigma project.
    d. All of the above are correct.
    e. All of the above are correct except (a) and (d).
13. Write a paragraph about a case study that includes at least one "safe criticism" based on statements in this book. Find the case study in a refereed journal such as *Applied Statistics*, *Quality Engineering*, or *The American Statistician*.
14. Identify at least one KOV and a target value for a system that you want to improve in a student project. We are looking for a KOV associated with a project that is measurable, of potential interest, and potentially improvable without more than ten total hours of effort by one person.
15. This exercise involves starting with an initial standard operating procedure (SOP) and concluding with a revised and confirmed SOP. Both SOPs must be evaluated using at least one control chart. For training purposes, the sample size can be only $n = 2$ and only 12 subgroups are needed for the startup periods for each chart.
    a. At least six "methods" or "activities" listed in ▫ Table 2.1 must be applied. The creation of SOPs of various types can also be considered as activities counted in the total of six. Document all results in four pages including tables and figures. Since design of experiments (DOE) may be unknown to the reader at this point, it might make sense not to use these methods.
    b. Perform the exercise described in part (a) with eight instead of six methods or activities. Again, documenting SOPs can be counted in the total of eight.

References

Besterfield D (2001) Quality control. Prentice Hall, Columbus, OH
Brady J, Allen T (2002) Case study. Am Stat 56(4):1–4
Lenth RV (1989) Quick and easy analysis of unreplicated factorials. Technometrics 31:469–473

SQC Theory

T. T. Allen, *Introduction to Engineering Statistics and Lean Six Sigma*,
https://doi.org/10.1007/978-1-4471-7420-2_10

10.1 Introduction

Some people view statistical material as a way to push students to sharpen their minds, but as having little vocational or practical value. Furthermore, practitioners of six sigma have demonstrated that it is possible to derive value from statistical methods while having little or no knowledge of statistical theory. However, understanding the implications of probability theory (assumptions to predictions) and inference theory (data to informed assumptions) can be intellectually satisfying and enhance the chances of successful implementations in at least some cases.

This chapter focuses attention on two of the most practically valuable roles that theory can play in enhancing six sigma projects. First, there are many parameters to be selected in applying acceptance sampling. In general, larger sample sizes and lower acceptable limits reduce the chances of accepting bad lots. However, it can be helpful to quantify these risks, particularly considering the need to balance the risks versus costs of inspection.

Second, control charts also pose risks, even if they are applied correctly as described in ▶ Chap. 4. These risks include the possibility that out-of-control signals will occur even when only assignable causes are operating. Then, investigators would waste their time and either conclude that a signal was a false alarm or, worse, would attempt to overcontrol the system and introduce variation. Also, there is a chance that charting will fail to identify an assignable cause. Then, large numbers of nonconforming items could be shipped to the customer. Evaluating formally these risks using probability can help
10
in making decisions about whether to apply Xbar & R charting (▶ Chap. 4) and EWMA charting or multivariate charting (▶ Chap. 8). Also, some of the risks are a function of the sample size. Therefore, quantifying dependencies can help in selecting sample sizes.

In ▶ Sect. 10.2, the fundamental concepts of probability theory are defined, including random variables, both discrete and continuous, and probability distributions. ▶ Section 10.3 focuses in particular on continuous random variables and normally distributed random variables. ▶ Section 10.4 describes discrete random variables, including negative binomial and hypergeometric random variables. Then, ▶ Sect. 10.5 builds on probability theory to aid in the assessment of control charting risks and in the definition of the average run length. ▶ Section 10.6 uses probability theory to evaluate acceptance sampling risks including graphic depictions of this risk in operating characteristic (OC) curves. ▶ Section 10.7 summarizes the chapter.

Figure 10.1 shows the relationship of the topic covered. Clearly, theory can play many roles in the application and development of statistical methods. The intent of Fig. 10.1 is to show that theory in this book is developed for specific purposes.

10.2 More on Probability Theory

The probability of an event is the subjective chance that it will happen. The purpose of this section is to define formally the words in this statement. As noted in ▶ Chap. 1, an event is a set of possible outcomes from an experiment. The phrase "**random variable**" and symbol, X, refer to a number, the value of which cannot be known precisely at the time of planning by the planner. Often, events are expressed in terms of random variables.

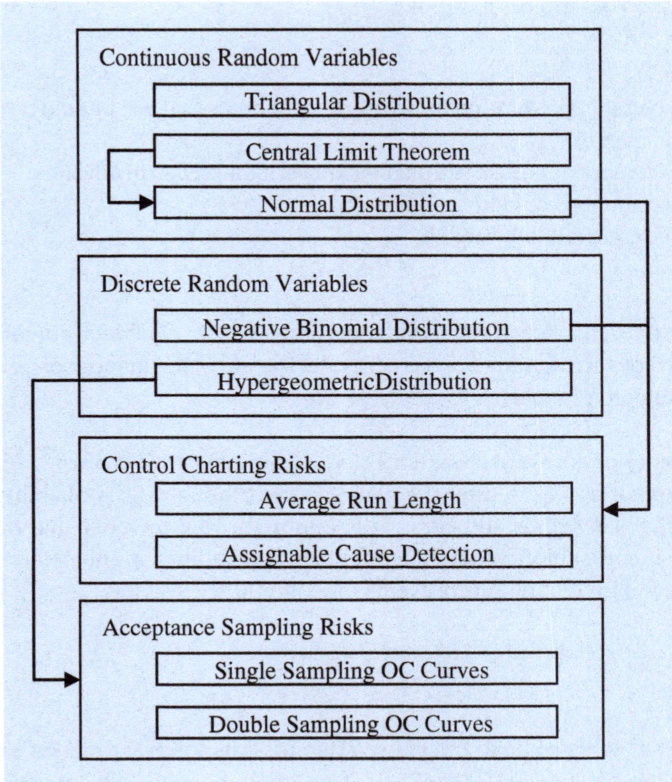

Fig. 10.1 The relationship of the topics in this chapter

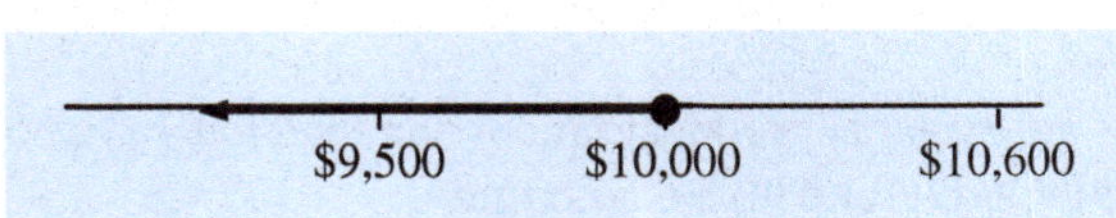

Fig. 10.2 An example of an event

Therefore, an event (A) is a set of possible values that X might assume. If X assumes the value x that is in this set, then we say that the "event occurs". For example, X could be the event that the price for boats on a certain market is below $10,000. Figure 10.2 shows this event.

The phrase "**continuous random variables**" refers to random variables that can assume an uncountably infinite (or effectively infinite) number of values. This can happen if the values can have infinity digits like real numbers, e.g., $X = 3.9027\ldots$ The boat price next month on a certain market can be treated as a continuous random number even though the price will likely be rounded to the nearest penny. The phrase "**discrete random variable**" refers to random variables that can assume only a countable number of values. Often, this countable number is medium sized, such as 30 or 40, and sometimes it is small such as 2 or 3.

Example 10.2.1 Boats Sales Random Variable

Question

What can be said about the unknown number of boats that will be sold next month at a certain market?

a. It is not random because the planner knows it for certain in advance.
b. It is a continuous random variable.
c. It is a discrete random variable.

Answer

It is a random variable, assuming that the planner cannot confidently predict the number in advance. Count of units is discrete. Therefore, the number of boats is a discrete random variable (c).

The "**probability of an event**," written $\Pr(A)$, is the subjective chance from 0 to 1 the event will happen as assessed by the planner. Even though the probability is written below in terms of integrals and sums, it is important to remember that the inputs to these formulas are subjective and therefore the probability is subjective. This holds when it comes to predicting future events for complicated systems.

Example 10.2.2 Probability of Selling Two Boats

Question

A planner has sold two boats out of two attempts last month at a market and has been told those sales were lucky. What is true about the probability of selling two more next month?

a. The probability is 1.0 since $2 \div 2 = 1$ based on last month's data.
b. The planner might reasonably feel that the probability is high, for example 0.7 or 70%.
c. Probabilities are essentially rationalizations and therefore have no value.
d. The answers in part (a) and (b) are both true.

Answer

Last month you sold two similar boats which might suggest that the probability is high, near 1. However, past data can rarely if ever be used to declare that a probability is 1.0. While probabilities are rationalizations, they can have value. For example, they can communicate feelings and cause participants in a decision to share information. The planner can judge the probability is any number between 0 and 1, and 0.7 might seem particularly reasonable. Therefore, (b) is the only true answer.

The previous example illustrates the ambiguities about assigning probabilities and their indirect relationship to data. The next example is designed to show that probability calculations can provide more valuable information to decision-makers than physical data or experience in some cases.

Example 10.2.3 Selecting An Acceptance Sampling Plan

Question

The planner has enjoyed a positive experience with a single sampling plan. A bad lot of 1200 parts was identified and no complaints were made about expected lots. A quality engineer states some reasonable-seeming assumptions and declares the following: there is a 0.6 probability of cutting the inspection costs by half and a 0.05 higher chance of detecting a bad lot using a double sampling policy. Which answer is most complete and correct?

a. In business, never trust subjective theory. Single sampling was proven to work consistently.
b. The evidence to switch may be considered trustworthy.
c. Single sampling is easier. Simplicity could compensate for other benefits.
d. Double sampling practically guarantees bad lots will not be accepted.
e. The answers in parts (b) and (c) are both correct.

Answer

Currently, many top managers feel the need to base the most important business decisions on calculated probabilities. It can be technically correct not to trust subjective theory. However, here the position is adopted that proof can only come from an experiment using randomization (see ▶ Chap. 11) or form physics or mathematical theory. In general, all acceptance sampling methods involve a risk of accepting "bad" lots. Probabilistic information may be regarded as trustworthy evidence if it is based on reasonable assumptions. Also, trading off intangibles against probabilistic benefits is often reasonable. Therefore, (e) is the most complete and correct answer.

The rigorous equations and mathematics in the next few sections should not obscure the fact that probability theory is essentially subjective in nature and is the servant of decision-makers. The main point is that even though probabilities are subjective, probability theory can take reasonable assumptions and yield surprisingly thorough comparisons of alternative methods or decision options. These calculations can be viewed as mental experiments or simulations. While not unreal in an important sense, these calculations can often offer more convincing verifications than empirical tests. Similar views were articulated by Keynes (1937) and Savage (1972).

10.3 Continuous Random Variables

Considering that continuous numbers can assume an uncountably infinite number of values, the probability that they are any particular value is zero. Beliefs about them must be expressed in terms of the chance that they are "near" any given value. The phrase "**probability density function**" or the symbol, $f(x)$, quantifies beliefs about the likelihood of specific values of the continuous random variable X. The phrase "**distribution function**" is often used to refer to probability density functions. Considering the infinities involved, integral calculus is needed to derive probabilities from probability density functions as follows:

$$\Pr(A) = \int_{x \in A} f(x)dx \tag{10.1}$$

where $x \in$ A means, in words, the value assumed by x is in the set A.

Because of the complications associated with calculus, people often only approximately express their beliefs using density functions. These approximations make the probability calculations relatively easy. The next example illustrates the use of the "triangular distribution" that is reasonably easy to work with and fairly flexible.

Example 10.3.1 Boat Prices Continued

Question

Assume an engineer believes that the price of a boat will be between $a = \$9{,}500$ and $b = \$10{,}600$, with $c = \$10{,}000$ being the most likely price of a boat he might buy next month. Develop and plot a probability density function that is both reasonably consistent with these beliefs and easy to work with.

Answer

A reasonable choice is the so-call "triangular" distribution function, in Eq. (10.2) and Fig. 10.3:

$$f(x) = \begin{cases} 0 & \text{if } x \le a \text{ or } x \ge b \\ \frac{2(x-a)}{(b-a)(c-a)} & \text{if } a < x \le c \\ \frac{2(b-x)}{(b-a)(b-c)} & \text{if } c < x < b \end{cases} . \tag{10.2}$$

Note that the total area underneath all probability density functions is 1.0. Therefore, if X is any continuous random variable and a is any number, $\Pr\{X < a\} = 1 - \Pr\{X \ge a\}$.

The next example shows that a probability can be calculated from a distribution function. It is important to remember that the distribution functions are subjectively chosen just like the probabilities. The calculus just shows how one set of subjective assumptions implies other subjective assumptions.

Example 10.3.2 Boat Price Probabilities

Question

A planner is comfortable assuming that a boat price has a triangular distribution with parameters $a = \$9{,}500$, $b = \$10{,}600$, and $c = \$10{,}000$ as in the previous example. Use calculus to derive what this assumption implies about the probability that the price will be less than \$10,000. Also, what is the probability it will be greater than \$10,000?

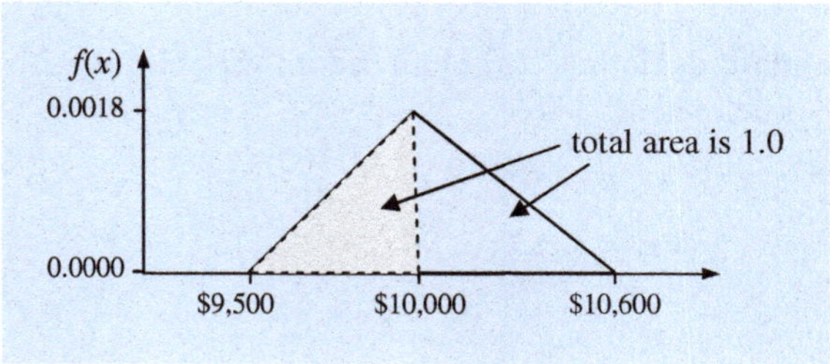

Fig. 10.3 Distribution for boat price (shaded refers to Example 10.3.2)

Answer

A is the event {*X* < \$10,000}. Based on the subjectively assumed distribution function, described in Fig. 10.3:

$$\Pr(A) = \int_{-\infty}^{10,000} f(x)dx = 0 + \int_{\$9,500}^{\$10,000} \frac{2(x - \$9,500)}{(\$10,600 - \$9,500)(\$10,000 - \$9,500)} dx \quad (10.3)$$

This integral corresponds to the shaded area in Fig. 10.3. From our introductory calculus course, we might remember that the anti-derivative of x^n is $(n+1)^{-1}x^{n+1} + K$, where K is a constant. (With computers, integrals can be done even without antiderivatives, but in this case, one is available.) Applying the anti-derivative, our expression becomes

$$\Pr(A) = \Big|_{\$9,500}^{\$10,000} \frac{2(0.5x^2 - \$9,500x)}{(\$10,600 - \$9,500)(\$10,000 - \$9,500)} \quad (10.4)$$
$$= -163.63 - (-164.09) = 0.45.$$

Therefore, the planner's subjective assumption of the triangular distribution function implies a subjective probability of 45% that the market price will be below \$10,000. The probability of being greater than \$10,000 is 1.0 − 0.45 = 0.55.

As implied above, distributions with widely known names like the triangular distribution rarely if ever exactly correspond to the beliefs of the planner. Choosing a named distribution is often done simply to make the calculations easy. Yet, computers are making calculus manipulations easier all the time so that custom distributions might grow in importance. In the future, planners will increasingly turn to oddly shaped distribution functions, *f*(*x*), that still have an area equal to 1.0 underneath them but which more closely correspond to their personal beliefs.

Using calculus, the "**mean**" (μ) or "expected value" ($E[X]$) of a random variable with probability density function, *f*(*x*), is defined as

$$E[X] = \int_{-\infty}^{\infty} x f(x)dx = \mu \quad (10.5)$$

Similarly, the "**standard deviation**" (σ) of a random variable, X, with probability density function, $f(x)$, is defined as

$$\mathrm{E}[X] = \sqrt{\int_{-\infty}^{\infty} (x-\mu)^2 f(x)dx} = \sigma \tag{10.6}$$

The next example illustrates the calculation of a mean from a distribution function.

Example 10.3.3 The Mean Boat Price

Question

A planner is comfortable assuming that a boat price has a triangular distribution with parameters $a = \$9{,}500$, $b = \$10{,}600$, and $c = \$10{,}000$, as in the previous example. Use calculus to derive what this assumption implies about the mean boat price.

Answer

$$E[X] = 0 + x\int_{\$9{,}500}^{\$10{,}000} \frac{2(x-\$9{,}500)}{(\$10{,}600-\$9{,}500)(\$10{,}000-\$9{,}500)}dx$$
$$+ x\int_{\$10{,}000}^{\$10{,}600} \frac{2(\$10{,}600-x)}{(\$10{,}600-\$9{,}500)(\$10{,}600-\$10{,}000)}dx + 0 = \$10{,}033.33 \tag{10.7}$$

10

Looking at a plot of a distribution function, it is often easy to guess approximately the mean. It is analogous to the center of gravity in physics. Looking at Fig. 10.3, it seems reasonable that the mean is slightly to the right of \$10,000.

The "uniform" probability distribution function has $f(x) = 1 \div (b-a)$ for $a \le x \le b$ and $f(x) = 0$ otherwise. In words, X is uniformly distributed if it is equally likely to be anywhere between the numbers a and b with no chance of being outside the $[a, b]$ interval. The distribution function is plotted in Fig. 10.4.

The uniform distribution is probably the easiest to work with but also among the least likely to exactly correspond to a planner's subjective beliefs. Probabilities and mean values of uniformly distributed random variables can be calculated using plots and geometry since areas correspond to probabilities.

Example 10.3.4 The Uniform Distribution

Question

Suppose a planner is comfortable with assuming that her performance rating next year, X, had a distribution $f(x) = 0.1$ for $85 \le X \le 95$. What does this imply about her believed chances of receiving an evaluation between 92 and 95?

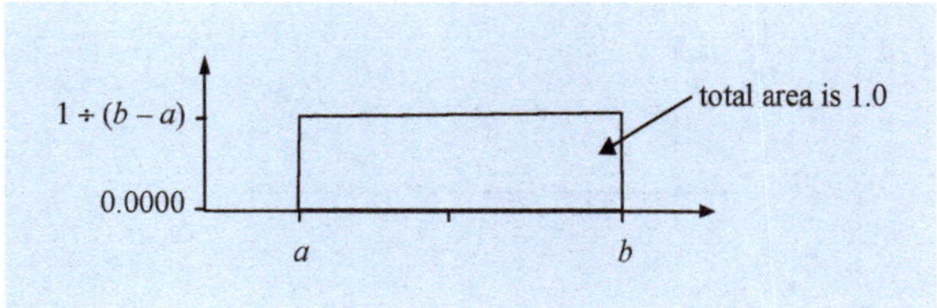

Fig. 10.4 The generic uniform distribution function

Answer
P(92 $\le X \le$ 95) is given by the area under the distribution function over the range [92, 95], which equals 0.1 $\times$ 3 = 0.3 or 30%.

As noted earlier, there are only a small number of distribution shapes with "**famous distribution functions**" names such as the triangular distribution, the uniform distribution, and the normal distribution, which will be discussed next. One can, of course, propose "**custom distribution functions**" specifically designed to express beliefs in specific situations.

10.3.1 The Normal Probability Density Function

The "**normal**" probability density function, *f*(*x*), has a special role in statistics in general and statistical quality control in particular. This follows because it is relevant for describing the behavior of plotted quantities in control charts. The reason for this relates to the central limit theory (CLT). The goals of this section are to clarify how to calculate probabilities associated with normally distributed random variables and the practical importance of the central limit theorem.

The normal probability density function is

$$f(x) = \frac{0.398942}{\sqrt{\sigma}} e^{-\frac{(x-\mu)^2}{2\sigma^2}} \tag{10.8}$$

where the parameters μ and σ also happen to be the mean and standard deviation of the relevant normally distributed random variable (Fig. 10.5).

The normal distribution is important enough that many quality engineers have memorized the probabilities shown in Fig. 10.6. The phrase "**standard normal distribution**" refers to the case $\mu = 0$ and $\sigma = 1$, which refers to both plots in Fig. 10.6.

In general, for any random variable *X* and constants μ and σ with $\sigma > 0$:

$$\Pr\{X < a\} = \Pr\{\frac{(X-\mu)}{\sigma} < \frac{(a-\mu)}{\sigma}\}. \tag{10.9}$$

This follows because the events on both sides of the equation are equivalent. When one occurs, so does the other. When one does not occur, neither does the other.

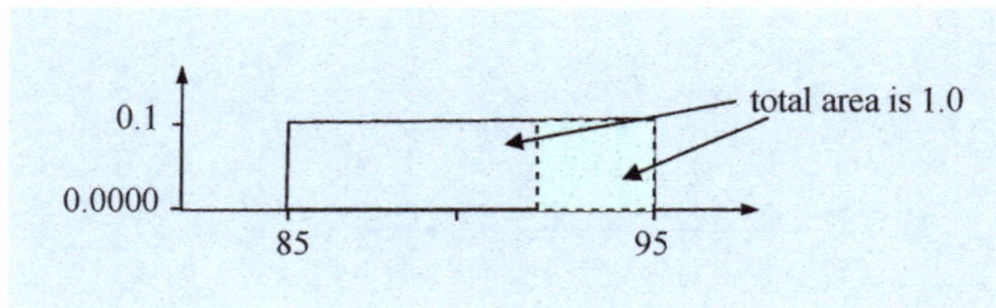

Fig. 10.5 The uniform distribution function example probability calculation

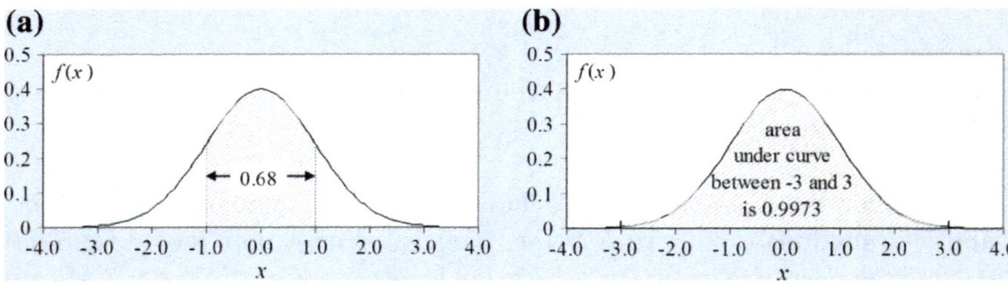

Fig. 10.6 Shows the fraction within **a** 1.0 × σ of the μ and **b** 3.0 × σ of μ

The normal distribution has three special properties that aid in hand calculation of relevant probabilities. First, the "**location scale**" property of normal probability density function guarantees that, if X is normally distributed, then

10

$$Z \sim \frac{(X-\mu)}{\sigma} \tag{10.10}$$

is also normally distributed, for any constants μ and σ. Note that, for many distributions, shifting and scaling results in a random variable from a distribution with a different name.

Second, if μ and σ are the mean and standard deviation of X respectively, then Z derived from Eq. (10.10) has mean 0.000 and standard deviation 1.000. Then, we say that Z is distributed according to the "**standard normal**" distribution.

Third, the "**symmetry property**" of the normal distribution guarantees that $\Pr\{Z<a\}=\Pr\{Z>-a\}$. One practical benefit of these properties is that probabilities of events associated with normal probability density functions can be calculated using Table 10.1. The table gives $\Pr\{Z<a\}$ where the first digit of a is on the left-hand-side column and the last digit is on the top row. For example:

$\Pr\{Z<-4.81\}=1.24\text{E}-06$ or 0.00000124.

Note that, taken together, the above imply that:

$$\begin{aligned}\Pr\{X>a\} &= \Pr\{Z>(a-\mu)\div\sigma\}\\ &= \Pr\{Z<-(a-\mu)\div\sigma\}\\ &= \Pr\{(X-\mu)\div\sigma<-(a-\mu)\div\sigma\}\\ &= \Pr\{X<2\mu-a\}.\end{aligned}$$

The examples that follow illustrate probability calculations that can be done with paper and pencil and access to Table 10.1. They show the procedure of using the

Table 10.1 If $Z \sim N[0, 1]$, then the table gives $P(Z < z)$. The first column gives the first three digits of z, the top row gives the last digit

	0.00	0.01	0.02	0.03	0.04
−6.0	9.87E−10	9.28E−10	8.72E−10	8.20E−10	7.71E−10
−4.4	5.41E−06	5.17E−06	4.94E−06	4.71E−06	4.50E−06
−3.5	0.00023	0.00022	0.00022	0.00021	0.00020
−3.4	0.00034	0.00032	0.00031	0.00030	0.00029
−3.3	0.00048	0.00047	0.00045	0.00043	0.00042
−3.2	0.00069	0.00066	0.00064	0.00062	0.00060
−3.1	0.00097	0.00094	0.00090	0.00087	0.00084
−3.0	0.00135	0.00131	0.00126	0.00122	0.00118
−2.9	0.00187	0.00181	0.00175	0.00169	0.00164
−2.8	0.00256	0.00248	0.00240	0.00233	0.00226
−2.7	0.00347	0.00336	0.00326	0.00317	0.00307
−2.6	0.00466	0.00453	0.00440	0.00427	0.00415
−2.5	0.00621	0.00604	0.00587	0.00570	0.00554
−2.4	0.00820	0.00798	0.00776	0.00755	0.00734
−2.3	0.01072	0.01044	0.01017	0.00990	0.00964
−2.2	0.01390	0.01355	0.01321	0.01287	0.01255
−2.1	0.01786	0.01743	0.01700	0.01659	0.01618
−2.0	0.02275	0.02222	0.02169	0.02118	0.02068
−1.9	0.02872	0.02807	0.02743	0.02680	0.02619
−1.8	0.03593	0.03515	0.03438	0.03362	0.03288
−1.7	0.04457	0.04363	0.04272	0.04182	0.04093
−1.6	0.05480	0.05370	0.05262	0.05155	0.05050
−1.5	0.06681	0.06552	0.06426	0.06301	0.06178
−1.4	0.08076	0.07927	0.07780	0.07636	0.07493
−1.3	0.09680	0.09510	0.09342	0.09176	0.09012
−1.2	0.11507	0.11314	0.11123	0.10935	0.10749
−1.1	0.13567	0.13350	0.13136	0.12924	0.12714
−1.0	0.15866	0.15625	0.15386	0.15151	0.14917
−0.9	0.18406	0.18141	0.17879	0.17619	0.17361
−0.8	0.21186	0.20897	0.20611	0.20327	0.20045
−0.7	0.24196	0.23885	0.23576	0.23270	0.22965
−0.6	0.27425	0.27093	0.26763	0.26435	0.26109
−0.5	0.30854	0.30503	0.30153	0.29806	0.29460

(continued)

10

Table 10.1 (continued)

	0.00	0.01	0.02	0.03	0.04
−0.4	0.34458	0.34090	0.33724	0.33360	0.32997
−0.3	0.38209	0.37828	0.37448	0.37070	0.36693
−0.2	0.42074	0.41683	0.41294	0.40905	0.40517
−0.1	0.46017	0.45620	0.45224	0.44828	0.44433
0.0	0.50000	0.49601	0.49202	0.48803	0.48405
	0.05	**0.06**	**0.07**	**0.08**	**0.09**
−6.0	7.24E−10	6.81E−10	6.40E−10	6.01E−10	5.65E−10
−4.4	4.29E−06	4.10E−06	3.91E−06	3.73E−06	3.56E−06
−3.5	0.00019	0.00019	0.00018	0.00017	0.00017
−3.4	0.00028	0.00027	0.00026	0.00025	0.00024
−3.3	0.00040	0.00039	0.00038	0.00036	0.00035
−3.2	0.00058	0.00056	0.00054	0.00052	0.00050
−3.1	0.00082	0.00079	0.00076	0.00074	0.00071
−3.0	0.00114	0.00111	0.00107	0.00104	0.00100
−2.9	0.00159	0.00154	0.00149	0.00144	0.00139
−2.8	0.00219	0.00212	0.00205	0.00199	0.00193
−2.7	0.00298	0.00289	0.00280	0.00272	0.00264
−2.6	0.00402	0.00391	0.00379	0.00368	0.00357
−2.5	0.00539	0.00523	0.00508	0.00494	0.00480
−2.4	0.00714	0.00695	0.00676	0.00657	0.00639
−2.3	0.00939	0.00914	0.00889	0.00866	0.00842
−2.2	0.01222	0.01191	0.01160	0.01130	0.01101
−2.1	0.01578	0.01539	0.01500	0.01463	0.01426
−2.0	0.02018	0.01970	0.01923	0.01876	0.01831
−1.9	0.02559	0.02500	0.02442	0.02385	0.02330
−1.8	0.03216	0.03144	0.03074	0.03005	0.02938
−1.7	0.04006	0.03920	0.03836	0.03754	0.03673
−1.6	0.04947	0.04846	0.04746	0.04648	0.04551
−1.5	0.06057	0.05938	0.05821	0.05705	0.05592
−1.4	0.07353	0.07215	0.07078	0.06944	0.06811
−1.3	0.08851	0.08691	0.08534	0.08379	0.08226
−1.2	0.10565	0.10383	0.10204	0.10027	0.09853
−1.1	0.12507	0.12302	0.12100	0.11900	0.11702

(continued)

Table 10.1 (continued)

	0.05	0.06	0.07	0.08	0.09
−1.0	0.14686	0.14457	0.14231	0.14007	0.13786
−0.9	0.17106	0.16853	0.16602	0.16354	0.16109
−0.8	0.19766	0.19489	0.19215	0.18943	0.18673
−0.7	0.22663	0.22363	0.22065	0.21770	0.21476
−0.6	0.25785	0.25463	0.25143	0.24825	0.24510
−0.5	0.29116	0.28774	0.28434	0.28096	0.27760
−0.4	0.32636	0.32276	0.31918	0.31561	0.31207
−0.3	0.36317	0.35942	0.35569	0.35197	0.34827
−0.2	0.40129	0.39743	0.39358	0.38974	0.38591
−0.1	0.44038	0.43644	0.43251	0.42858	0.42465
0.0	0.48006	0.47608	0.47210	0.46812	0.46414

equivalence of events to transform normal probability calculations to a form where answers can be looked up using the table. In situations where Excel spreadsheets can be accessed, similar results can be derived using built-in functions. For example, "=NORMDIST(5, 9, 2, TRUE)" gives the value 0.02275, where the TRUE refers to the cumulative probability that X is less than a. A false would give the probability density function value at the point $X = a$.

Example 10.3.5 Normal Probability Calculations

Question 1

Assume $X \sim N(\mu = 9, \sigma = 2)$. What is the $\Pr\{X < 5\}$?

Answer 1

$Z = (X - 9)/2$ has a standard normal distribution. Therefore,
$\Pr\{X < 5\} = \Pr\{Z < (5\text{–}9)/2\}$. Therefore,
$\Pr\{X < 5\} = \Pr\{Z < -2.00\} = 0.02275$, from Table 10.1.

Question 2

Assume $X \sim N(\mu = 20, \sigma = 5)$. What is the $\Pr\{X > 22\}$?

Answer 2

$\Pr\{X > 22\} = \Pr\{Z > (22\text{–}20)/5\} = \Pr\{Z > 0.4\}$, using the location scale property.
Also, $\Pr\{Z > 0.4\} = \Pr\{Z < -0.4\}$, because of the symmetry property of the normal distribution. $\Pr\{X > 22\} = \Pr\{Z < -0.40\} = 0.344578$, from the table.

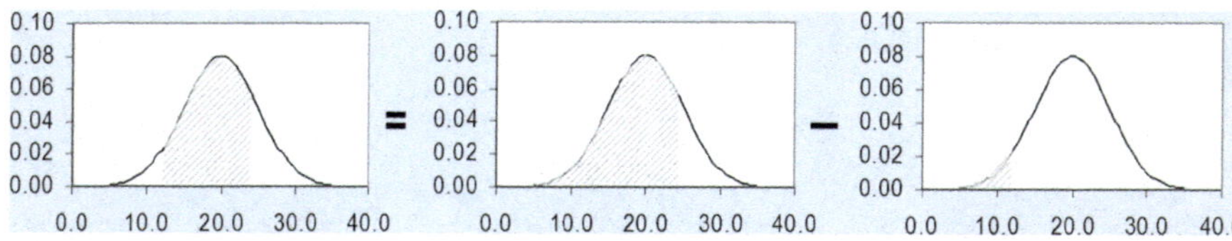

Fig. 10.7 Proof by picture of the equality of probabilities corresponding to areas

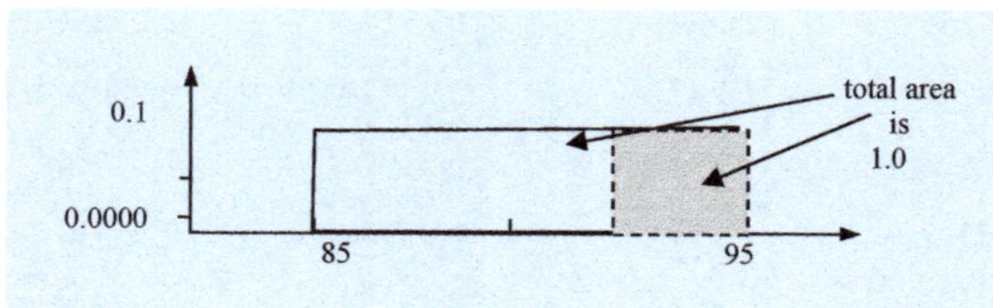

Fig. 10.8 The uniform distribution function example probability calculation

Question 3

Assume $X \sim N(\mu = 20, \sigma = 5)$. What is the $\Pr\{12 < X < 23\}$?

Answer 3

$\Pr\{12 < X < 23\} = \Pr\{X < 23\} - \Pr\{X < 12\}$, which follows directly from the definition of probability as an integral in Fig. 10.7. Next,

$$\begin{aligned}\Pr\{X < 12\} &= \Pr\{Z < (12 - 20)/5\} \\ &= \Pr\{Z < -1.60\} = 0.054799 \text{ and} \\ \Pr\{X < 23\} &= \Pr\{Z < (23 - 20)/5\} \\ &= \Pr\{Z < 0.60\} = 1 - \Pr\{Z < -0.60\} \\ &= 1 - 0.274253,\end{aligned}$$

where the location scale and symmetry properties have been used. Therefore, the answer is $1 - 0.274253 - 0.054799 = 0.671$. (The implied uncertainty of the original numbers is unclear, but quoting more than three digits for probabilities is often not helpful because of their subjective nature.)

Question 4

Suppose a planner is comfortable with assuming that her performance rating next year, X, will have a distribution $f(x) = 0.1$ for $85 \le X \le 95$. What does this imply about her believed chances of receiving an evaluation of 92–95?

Answer 4

$P(92 \le X \le 95)$ is given by the area under the distribution function over the range [92, 95] in Fig. 10.8, which equals $0.1 \times 3 = 0.3$ or 30%.

The phrase “**test for normality**” refers to an evaluation of the extent to which a decision-making can feel comfortable believing that responses or averages are normally distributed. In some sense, numbers of interest from the real world never come from normal distributions. However, if the numbers are averages of many other numbers, or historical data suggests approximate normality, then it can be of interest to assume that future similar numbers come from normal distributions. There are many formal approaches for evaluating the extent to which assuming normality is reasonable, including evaluation of skew and kurtosis and normal probability plotting the numbers as described in ▶ Chap. 15.

10.3.2 Defects Per Million Opportunities

Assume that a unit produced by an engineered system has only one critical quality characteristic, $Y_1(\mathbf{x}_c)$. For example, the critical characteristic of a bolt might be inner diameter. If the value of this characteristic falls within values called the “**specification limits**,” then the unit in question is generally considered acceptable, otherwise not. Often critical characteristics have both “**upper specification limits**” (USL) and “**lower specification limits**” (LSL) that define acceptability. For example, the bolt diameter must be between LSL = 20.5 mm and USL = 22.0 mm for the associated nuts to fit the bolt.

Suppose further that the characteristic values of items produced vary uncontrollably around an average or “mean” value, written “μ,” with typical differences between repeated values equal to the “standard deviation,” written “σ”. For example, the bolt inner diameter average might be 21.3 mm with standard deviation, 0.2 mm, i.e., $\mu = 21.3$ mm and $\sigma = 0.2$ mm.

With these definitions, one says that the “**sigma level**,” σ_L, of the process is

$$\sigma_L = \text{Minimum}[\text{USL}-\mu, \mu - \text{LSL}]/\sigma \,. \qquad \textbf{(10.11)}$$

Note that $\sigma_L = 3 \times C_{pk}$ (from ▶ Chap. 4). If $\sigma_L > 6$, then one says that the process has “**six sigma quality**.” For instance, the bolt process sigma level in the example given is 3.5. This quality level is often considered “mediocre”.

With six sigma quality and assuming normally distributed quality characteristic values under usual circumstances, the fraction of units produced with characteristic values outside the specification limits is less than 1 part per billion (PPB). If the process mean shifts 1.5σ toward the closest limit, then the fraction of “**nonconforming**” units (with characteristic values that do not conform to specifications) is less than 3.4 parts per million (PPM).

Figure 10.9 shows the probability density function associated with a process having slightly better than six sigma quality. This figure implies assumptions including that the upper specification limit is much closer to the mean than the lower specification limit.

One practical benefit of this definition is that it emphasizes the importance of achieving what might be considered high levels of process quality. This emphasis can be useful since the costs of poor quality are often hard to evaluate and much greater than the cost of fixing or “**reworking**” nonconforming units. Typically, correct

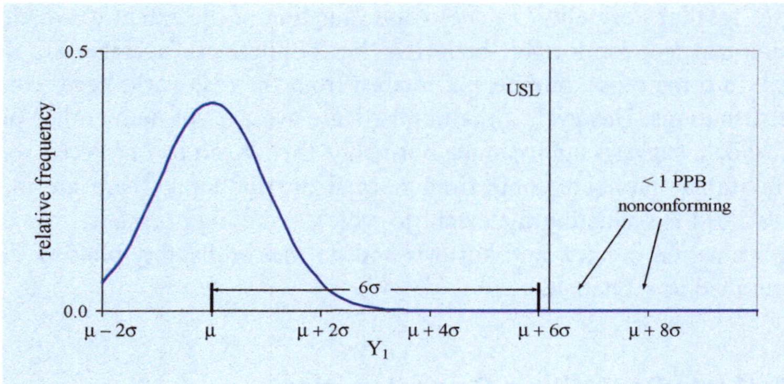

Fig. 10.9 Shows the relative frequency of parts produced with six sigma quality

accounting of the costs includes higher inventory maintenance and delayed shipment dates as well as down-the-line costs incurred when quality issues disrupt production by creating unpredictable rework and processing times.

Also, if there are "**nonconforming**" units, then customers may be upset or even injured. The losses to the company from such incidents might include lawsuit costs, lost revenue because demand may be reduced, and turnover and absenteeism costs arising from a demotivated workforce.

10.3.3 Independent, Identically Distributed and Charting

The central limit theorem (CLT) plays an important role in statistical quality control largely because it helps to predict the performance of control charts. As described in ▶ Chaps. 4 and 8, control charts are used to avoid intervention when no assignable causes are present and to encourage intervention when they are present. The CLT helps to calculate the probabilities that charts will succeed in these goals with surprising generality and accuracy. The CLT aids in probability calculations regardless of the charting method (with exceptions including R charting) or the system in question, e.g., from restaurant or hospital emergency room to traditional manufacturing lines.

To understand how to benefit from the central limit theorem and to comprehend the limits of its usefulness, it is helpful to define two concepts. First, the term "**independent**" refers to the condition in which a second random variable's probability density function is the same regardless of the values taken by a set of other random variables. For example, consider the two random variables: X_1 is the number of boats that will be sold next month and X_2 is their sales prices as determined by unknown boat sellers. Both are random variables because they are unknown to the planner in question. The planner in question assumes that they are independent if and only if the planner believes that potential buyers make purchasing decisions with no regard to price within the likely ranges. Formally, if $f(x_1, x_2)$ is the "joint" probability density function, then independence implies that it can be written $f(x_1, x_2) = f(x_1)f(x_2)$.

Second, "**identically distributed**" means that all of the relevant variables are assumed to come from exactly the same distribution. Clearly, the number of boats and

Table 10.2 Independent and identically distributed assumptions for control charting

	Reasons system outputs	
	Might be	Might not be
Independent	Units inspected are made with enough time for the system to reset	The same common causes influence successive observations the same way
Identically distributed	Only common cause variation is operating	Assignable causes changed the output pattern necessitating new assumptions

the price of boats cannot be identically distributed since one is discrete (the number) and one is continuous (the price). However, the numbers of boats sold in successive months could be identically distributed if (1) buyers were not influenced by seasonal issues and (2) there was a large enough pool of potential buyers. Then, higher or lower number of sales one month likely would not influence prospects much in the next month.

In the context of control charts, making the combined assumption that system outputs being charted are independent and identically distributed (IID) is relevant. Departures of outputs from these assumptions are also relevant. Therefore, it is important to interpret the meaning of IID in this context. System outputs could include the count of demerits on individual hospital surveys or the gaps on individual parts measured before welding.

To review: under usual circumstances, common causes force the system outputs to vary with a typical pattern (randomly with the identical, same density function). Rarely, however, assignable causes enter and change the system, thereby changing the usual pattern of values (effectively shifting the probability density function). Therefore, even under typical circumstances the units inspected will not be associated with constant measurement values of system outputs. The common cause factors affecting them will force the observations to vary up and down. If measurements are made on only a small fraction of units produced at different times by the system, then it can be reasonably assumed that the common causes will effectively reset themselves. Then, the outputs will be IID to a good approximation. However, even with only common causes operating, units made immediately after one another might not be associated with independently distributed system outputs. Time is often needed for the common causes to reset enough that independence is reasonable. Table 10.2 summarizes reasons why IID might or might not be a reasonable assumption in the context of control charting.

Example 10.3.6 Moods, Patient Satisfaction, and Assuming Independence

Question

Assume the mood of the emergency room nurse, working at a small hospital with typically ten patients per shift, affects patient satisfaction. The key output variable is the sum of demerits. Consider the following statement: "The nurse's mood is a source of common cause variation, making it unreasonable to assume that subsequent patients' assigned demerits are independently distributed." Which answer is most complete and correct?

a. The statement is entirely reasonable.
b. Moods are always assignable causes because local people can always fix them.
c. Moods fluctuate so quickly that successive demerit independence is reasonable.
d. Satisfaction ratings are always independent since patients never talk together.
e. The answers in parts (b) and (c) are both reasonable.

Answer

In most organizations, moods are uncontrollable factors. Since they are often not fixable by local authority, they are not generally regarded as assignable causes. Moods typically change at a time scale of one shift or one half shift. Therefore, multiple patients would likely be affected by the same mood. Therefore, assuming successive demerit independence is reasonable. Satisfaction ratings might not be independently distributed because the same common cause factor fluctuation might affect multiple observations. Therefore, the answer is (a).

The previous example focused on the appropriateness of the independence assumption in a case in which sequential observations might reasonably be affected by the same common cause variation. When that happens, it can become unreasonable to assume that the associate system outputs will be independently distributed.

The term "**autocorrelation**" refers to departures of charted quantities from being independently distributed. These departures are fairly common in practice and do not always substantially degrade the effectiveness of the charts. If there were no autocorrelation, the charted quantities would show no pattern at all. Each would be equally likely to be above or below the center line. If there is autocorrelation, the next observation often is relatively close to the last making a relatively smooth pattern.

Example 10.3.7 Identically Distributed Fixture Gaps

Question

An untrained welder is put on second shift and does not follow the standard operating procedure for fixturing parts, dramatically increasing gaps. Consider the following statement: "The operator's lack of training constitutes an assignable cause and could make it difficult to believe the same, identical distribution applies to gaps before and after the untrained welder starts." Which answer is most complete and correct?

a. The statement is entirely reasonable.
b. Training issues are assignable causes because local authority can fix them.
c. It is usual for assignable causes to effectively shift the output density function.
d. With only common causes operating, it is often reasonable to assume outputs continually derive from the identical distribution function.
e. All of the above answers are correct.

Answer

Training issues are often easy for local authority to fix. Generally speaking, common causes are associated with a constant or identical probability density function for system outputs. Assignable causes are associated with changes to the distribution function that make extreme values more likely. Therefore, all of the above answers are correct.

10.3.4 The Central Limit Theorem

In the context of SQC, the central limit theorem (CLT) can be viewed as an important fact that increases the generality of certain kinds of control charts. Also, it can be helpful for calculating the small adjustment factors d_2, D_1, and D_2 that are commonly used in Xbar charting. Here, the CLT is presented with no proof using the following symbols:

$X_1, X_2, \ldots, X_n$ are random variables assumed to be independent identically distributed (IID). These could be quality characteristic values outputted from a process with only common causes operating. They could also be a series of outputs from some type of numerical simulation.

$f(x)$ is the common density function of the identically distributed $X_1, X_2, \ldots, X_n$.

Xbar_n is the sample average of $X_1, X_2, \ldots, X_n$. Xbar_n is effectively the same as Xbar from Xbar charts with the "n" added to call attention to the sample size.

σ is the standard deviation of the $X_1, X_2, \ldots, X_n$, which do not need to be normally distributed.

The CLT focuses on the properties of the sample averages, Xbar_n.

If $X_1, X_2, \ldots, X_n$ are independent, identically distributed (IID) random variables from a distribution function with any density function $f(x)$ with finite mean and standard deviation, then the following can be said about the average, Xbar_n, of the random variables. Defining

$$\text{Xbar}_n = \frac{(X_1 + X_2 + \cdots + X_n)}{n} \text{ and } Z_n = \frac{\text{Xbar}_n - \int_{-\infty}^{\infty} u f(u) du}{(\sigma/\sqrt{n})}, \tag{10.12}$$

it follows that

$$\lim_{n\to\infty} \Pr(Z_n \le x) = \int_{-\infty}^{x} \frac{1}{\sqrt{2\pi}} e^{-\frac{1}{2}u^2} du. \tag{10.13}$$

In words, averages of n random variables, Xbar_n, are approximately characterized by a normal probability density function. The approximation improves as the number of quantities in the average increases. A reasonably understandable proof of this theorem, i.e., the above assumptions are equivalent to the latter assumption, is given in Grimmet and Stirzaker ((2001, Chap. 5).

To review, the expected value of a random variable is:

$$E[X] = \int_{-\infty}^{\infty} uf(u)du \quad (10.14)$$

Then, the CLT implies that the sample average converges, Xbar_n, converges to the true mean $E[X]$ as the number of random variables averaged goes to infinity. Therefore, the CLT can be effectively rewritten as

$$E[X] = \text{Xbar}_n + e_{\text{MC}}, \quad (10.15)$$

where e_{MC} is normally distributed with mean 0.000 and standard deviation σ ÷ sqrt[n] for "large enough" n. We call Xbar_n the "Monte Carlo estimate" of the mean, $E[X]$. There, with only common causes operating, the Xbar chart user is charting Monte Carlo estimates of the mean. Since σ is often not known, it is sometimes of interest to use the sample standard deviation, s:

$$s = \sqrt{\frac{\sum_{i=1}^{n} (X_i - \text{Xbar}_n)^2}{n-1}} \quad (10.16)$$

Then, it is common to use:

$$\sigma_{estimate} = s \div c_4 \quad (10.17)$$

where c_4 comes from Table 10.3. As noted in ▶ Chap. 6, the standard deviation can also be estimated using the average range, Rbar, using:

Table 10.3 Constants c_4 and d_2 relevant to Monte Carlo estimation and charting

Sample size (n)	c_4	d_2
2	0.7979	1.128
3	0.8864	1.693
4	0.9213	2.059
5	0.9400	2.326
6	0.9515	2.534
7	0.9594	2.704
8	0.9560	2.847
9	0.9693	2.970
10	0.9727	3.078
15	0.9823	3.472
20	0.9869	3.737

$$\sigma_{estimate} = \text{Rbar} \div d_2 \quad (10.18)$$

However σ is estimated, $\sigma_{estimate} \div \text{sqrt}[n]$ is called the "estimated error of the Monte Carlo estimate" or a typical difference between Xbar_n and $E[X]$.

Example 10.3.8 Identically Distributed Fixture Gaps

Question

A forging process is generating parts whose maximum distortion from nominal is the critical quality characteristic, *X*. From experience, one believes *X* has average 5.2 mm and standard deviation 2.1. Let Xbar_5 denote the average characteristic value of five parts selected and measured each hour. Which is correct and most complete?

a. Xbar_5 is normally distributed with mean 5.2 and standard deviation 0.94.
b. Xbar_5 is likely approximately normally distributed with mean 5.2 and standard deviation 0.94.
c. The credibility of assuming a normal distribution for Xbar_5 can be evaluated by studying the properties of several Xbar_5 numbers.
d. Training issues are assignable causes because local authority can fix them.
e. All of the above are correct except a.
f. All of the above answers are correct except d.

Answer

The central limit theorem only guarantees approximate normality in the limit that $n \to \infty$. Therefore, since there is no reason to believe that *X* is normally distributed, e.g., it cannot be negative, there is no reason to assume that Xbar_5 is exactly normally distributed. Yet, often with $n = 5$ approximate normality of averages holds with standard deviation approximately equal to $\sigma_0 \div \text{sqrt}[n] = 2.1 \div \text{sqrt}[5] = 0.94$. Also, the credibility of this distribution assumption can be evaluated by studying many values of Xbar_5, e.g., using a normal probability plot (see ► Chap. 15). Therefore, the most complete of the correct answers is (d).

If the distribution of the random variable stays the same (*X* is identically distributed), then the Xbar_n will be approximately normally distributed according to the same normal distribution. If the Xbar_n distribution changes, then the distribution of *X* likely changed and an assignable cause is present. Spotting assignable causes in this manner constitutes and important motivation for Xbar charting and many other kinds of charts. The following example illustrates the application of the central limit theorem for spotting unusual occurrences and the breadth of possible applications.

Example 10.3.9 Monitoring Hospital Waiting Times

Question

The time between the arrival of patients in an emergency room (ER) and when they meet with doctors, *X*, can be a critical characteristic. Assume that times are typically 20 min with standard deviation 10 min. Suppose that the average of seven consecutive patient times was 35 min. Which is correct and most complete?

a. A rough estimate for the probability that this would happen without assignable causes is 0.000004.
b. This data constitutes a signal that something unusual is happening.
c. It might be reasonable to assign additional resources to the ER.
d. It is possible that no assignable causes are present.
e. All of the above are correct.

Answer

It has not been established that the averages of seven consecutive times, Xbar7, are normally distributed to a good approximation under usual circumstances. Still, it is reasonable to assume this for rough predictions. Then, the central limit theorem gives that Xbar7, under usual circumstances, has mean 20 min and standard deviation 10 ÷ sqrt[7] = 3.8 min. The chance that Xbar7 would be greater than 35 min is estimated to be Pr{Z > (35 − 20) ÷ 3.8} = Pr{Z < −4.49} = 0.000004 from Table 10.1.

This average could theoretically happen under usual circumstances with no assignable causes but it would be very unlikely. Therefore, it might constitute a good reason to send in additional medical resources if they are available. The answer is (e), all of the above are correct.

10

10.3.5 Advanced Topic: Deriving d_2 and c_4

In this section, the derivation of selected constants used in control charting (▶ Chap. 4) is presented. The purposes are (1) to clarify the approximate nature of these constants in usual situations and (2) to illustrate so-called "Monte Carlo integration" as an application of the central limit theorem.

The constants d_2 and c_4 are used in estimating the true, usually unknown, standard deviation of individual observations, σ_0. For normally distributed $X_1, \ldots, X_n$, d_2 and c_4 are unbiased estimates in the sense that:

$$E[(\text{Max}\{X_1,\ldots,X_n\} - \text{Min}\{X_1,\ldots,X_n\}) \div d_2] = \sigma_0 \tag{10.19}$$

$$E[(\text{Sample standard deviation}\{X_1,\ldots,X_n\}) \div c_4] = \sigma_0 \tag{10.20}$$

These equations are equivalent to the following definitions:

$$d_2 \equiv E[(\text{Max}\{X_1,\ldots,X_n\} - \text{Min}\{X_1,\ldots,X_n\}) \div \sigma_0] \tag{10.21}$$

$$c_4 \equiv E[(\text{Sample standard deviation}\{X_1,\ldots,X_n\}) \div \sigma_0] \tag{10.22}$$

Many computer software such as Microsoft® Excel permit the generation of pseudo-random numbers that one can safely pretend are normally distributed with known standard deviation, σ_0. Using these pseudo-random numbers and the central limit theorem values for d_2 and c_4 can be estimated as illustrated in the following examples.

Example 10.3.10 Estimating d_2 (n = 5)

Question

Use 5000 pseudo-random normally distributed numbers to estimate d_2 for the $n=5$ sample size case. Also, give the standard error of your estimated value.

Answer

The pseudo-random numbers shown in Table 10.4 were generated using Excel (Tools Menu ⇒ Data Analysis ⇒ Random Number Generation). The distribution selected was normal with mean 0 and standard deviation $\sigma_0=1$ with random seed equal to 1 (without loss of generality). Defining $R=\text{Max}\{X_1, \ldots, X_n\} - \text{Min}\{X_1, \ldots, X_n\}$, one has 1000 effectively random variables whose expected value is d_2 according to Eqs. (10.15) and (10.21). Averaging, we obtain 2.3338 as our estimated for d_2 with Monte Carlo estimated standard error 0.8767 ÷ sqrt[1000] = 0.0278. This estimate is within one standard deviation of the true value from Table 10.3 of 2.326. Note that Table 10.4 also permits an estimate for c_4.

10.4 Discrete Random Variables

Discrete random variables are unknown numbers that can assume only a countable number of random variables. The phrase "**probability mass function**" or the symbol, $\Pr(X=x)$, quantifies beliefs about the likelihood of specific values of the continuous random variable X. The phrase "**distribution function**" can also be used to refer to probability mass functions as well as density functions.

For discrete random variables, an event is a set of values that the variables can assume. For example, a discrete random variable might be the number of nonconforming items produced in 3 h of production. The event, A, might be the event that less than or equal to two nonconforming items were produced.

Let $x_1, x_2, \ldots, x_N$ refer to possible values that X could take. Also, assume that the decision-maker is comfortable assuming certain probabilities for each of these values, written $\Pr\{X=x_1\}$, $\Pr\{X=x_2\}$,...$\Pr\{X=x_N\}$. The sum of these probabilities must be 1.0000 to guarantee interpretability. Then, the probability of an event can be written as a sum over the values of x_i in the set A:

$$\Pr(A) = \sum_{x_i \in A} \Pr\{X = x_i\}. \tag{10.23}$$

In this book, we focus on cases in which the set of possible values that X can assume are nonnegative integers 0, 1, 2, ... $(N-1)$, e.g., the number of nonconforming units in a lot of parts. An event of particular interest is the chance that X is less than or equal to a constant, c. Then, the probability of this event can be written:

$$\Pr\{X \leq c\} = \Pr\{X = 0\} + \Pr\{X = 1\} + \cdots + \Pr\{X = c\}. \tag{10.24}$$

Table 10.4 1000 simulated subgroups each with five pseudo-random numbers

Sub-group	X_1	X_2	X_3	X_4	X_5		R	S
1	−3.0230	0.1601	−0.8658	0.8733	0.2147	→	3.8963	1.5271
2	−0.0505	−0.3845	1.2589	0.9262	0.6638	→	1.6434	0.6834
3	−0.9381	1.0756	0.5549	0.0339	−0.5129	→	2.0136	0.8062
4	−2.1705	−1.3322	−0.3466	−1.0480	−0.9705	→	1.8239	0.6633
5	2.2743	−0.1366	−1.1796	−2.5994	−2.3693	→	4.8737	1.9834
6	−0.3111	0.0795	0.1794	0.2579	0.2719	→	0.5830	0.2398
7	−0.9692	0.4208	−0.1237	−0.3796	−1.5801	→	2.0009	0.7725
8	0.2733	0.7835	0.8510	0.0499	−0.5188	→	1.3698	0.5635
9	1.1551	0.6028	1.7050	1.4446	0.0988	→	1.6062	0.6497
...	...	...	...	...	...		...	...
1000	−1.3413	2.1058	−0.6665	−1.4371	0.7682		3.5429	1.5222
						Average	2.3338	0.9405
						Standard deviation	0.8767	0.3488

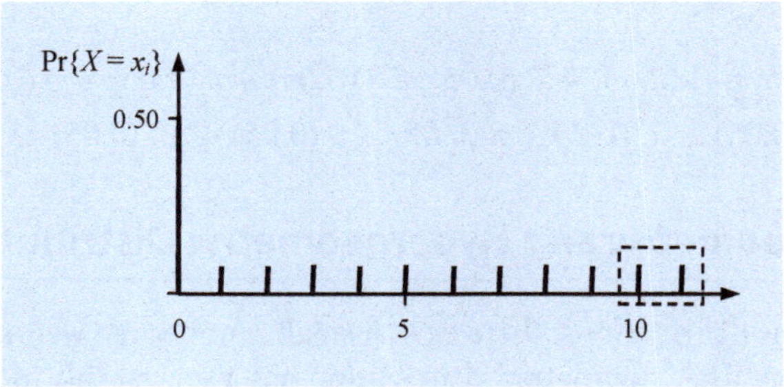

Fig. 10.10 Distribution for number of victims with selected event (*dotted lines*)

The following example illustrates the elicitation of a discrete distribution function from a verbal description. It also shows that many "no-name" distribution functions can be relevant in real world situations.

Example 10.4.1 Number of Accident Cases

Question

An emergency room nurse tells you that there is about a 50% chance that no accident victims will come any given hour. If there is at least one victim, it is equally likely that any number up to 11 (the most ever observed) will come. Plot a probability mass function consistent with these beliefs and estimate the probability that greater than or equal to 10 will come.

Answer

Figure 10.10 plots a custom distribution for this problem. The relevant sum is $\Pr\{X=10\}+\Pr\{X=11\}=0.10$, giving 10% as the estimated probability.

The expected value of a continuous random variable is expressible as a sum. This sum can be written:

$$E(X) = \sum_{x_i \in A} x_i \Pr\{X = x_i\}. \tag{10.24}$$

The following example illustrates the practical calculation of an expected value. Note that since the individual probabilities are subjective, when applied to real decision problems, the resulting expected value is also subjective.

Example 10.4.2 Expected Number of Accident Cases

Question

Using the distribution function from the previous example, calculate the expected number of accident cases in any given hour.

Answer

The expected value is

$$E[X] = 0\,(0.5) + 1\,(0.05) + 2\,(0.05) + 3\,(0.05) + 4\,(0.05) + 5\,(0.05) + 6\,(0.05) + 7\,(0.05) + 8\,(0.05) + 9\,(0.05) + 10\,(0.05) + 11\,(0.05) = 3.3. \quad (10.25)$$

10.4.1 The Geometric and Hypergeometric Distributions

As for continuous distributions, there exist a small number of "well-known" probability mass functions. The "**geometric**" distribution has a special role in statistical quality control because it can aid in analysis of the times between false alarms in applications of control charting. The geometric probability mass function is

$$\Pr\{X = x\} = \begin{cases} p_0^{(x-1)}(1-p_0) \text{ for } x = 1, \ldots, \infty \\ 0 \text{ for all other } x \text{ including } x = 0 \end{cases} \quad (10.26)$$

where p_0 is parameter. The following example shows a case in which one entertains assumptions such that the geometric distribution is perfect.

Example 10.4.3 Perfect Geometric Distribution Case

10

Question 1

Assume that one is considering a set of independent trials or tests. Each test is either a failure or a success. Assume further that the chance of success is p_0. What is the probability that the first failure will occur on trial x?

Answer 1

This is the perfect case for the geometric distribution. The distribution function is, therefore

$$\Pr\{X = x\} = p_0^{(x-1)}(1-p_0) \text{ for } x = 1, \ldots, \infty$$

Advanced readers will realize that the definition of independence of events permits the formula to be generated through the multiplication of $x - 1$ consecutive successes followed by 1 failure.

Question 2

Consider four independent trials, each with a success probability of 0.8. What is the probability that the first failure will occur on trial number four?

Answer 2

Applying the formula, $\Pr\{X=4\}=0.8^3 \times 0.2=0.1024$.

An important message of the above example is that the geometric probability mass function, while appearing to derive from elementary assumptions, is still approximate and subjective when applied to real problems. For example, in a real situation one might have several trials but yet not be entirely comfortable assuming that results are independent and are associated with the same, constant success probability, p_0. Then, the geometric probability mass function might be applied for convenience only, to gain approximate understanding.

The general formula for the expected value of a geometric random variable is:

$$E[X] = (1)p_0^{(1-1)}(1-p_0) + (2)p_0^{(2-1)}(1-p_0) + \cdots = \frac{1}{1-p_0} \tag{10.27}$$

The "**hypergeometric**" distribution also has a special role in SQC theory because it helps in understanding the risks associated with acceptance sampling methods. The hypergeometric probability mass function is

$$\Pr\{X = x\} = \begin{cases} \dfrac{\binom{M}{x}\binom{N-M}{n-x}}{\binom{N}{n}} & \text{for } x = 0, 1, \ldots, \infty \\ 0 \text{ for all other } x \end{cases} \tag{10.28}$$

where M, N, and n are parameters that must be nonnegative integers. The symbol "()" refers to the so-called "choose" operation given by

$$\begin{aligned} \binom{M}{x} &= \text{"}M \text{choose } x\text{"} \\ &= \frac{M!}{x!(M-x)!} \\ &= \frac{[M \times (M-1) \times \cdots \times 1]}{[x \times (x-1) \times \cdots \times 1] \times [(M-x) \times (M-x-1) \times \cdots \times 1]}. \end{aligned} \tag{10.29}$$

The following example shows the assumptions that motivate many applications of the hypergeometric distribution.

Example 10.4.4 Perfect Hypergeometric Distribution Case

Question

Assume that one is considering a situation with n units selected from N units where the total number of nonconforming units is M. Assume the selection is random such that each of the N units has an equal chance of being selected because a "rational subgroup" is used (see ▶ Chap. 4). Diagram this sampling situation and provide a formula for the chance that exactly x nonconforming units will be selected.

Answer

This is the perfect case for the hypergeometric distribution. The distribution function is, therefore, given by Eq. (10.28). Advanced readers can calculate this formula from the assumptions by counting all cases in which x units are selected, divided by the total number of possible selections. ◘ Figure 10.11 illustrates the selection situation.

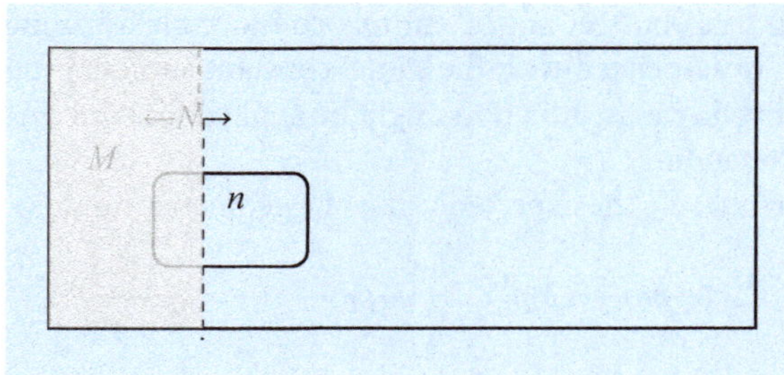

Fig. 10.11 The selection of n units from N with M total nonconforming

Calculating the probabilities from the hypergeometric distribution can be practically difficult. Factorials of a large numbers such as 100 is a very large number such as 9.3326×10^{157} that can exceed the capacity of calculators.

In cases in which $n \leq 0.1\,N$, $n \geq 20$, $M \leq 0.1\,N$, and $n \times M \leq 5 \times N$, the following "Poisson approximation" formula is often used for calculations:

$$\frac{\binom{M}{x}\binom{N-M}{n-x}}{\binom{N}{n}} \approx \frac{2.7182^{\frac{nM}{N}}\left(\frac{nM}{N}\right)^{x}}{x!}. \tag{10.30}$$

10

In Microsoft® Excel, the function "HYPGEOMDIST" generates probabilities, as illustrated in the next example.

Example 10.4.5 Chance of Finding the Nonconforming Units

Question

Assume that one is considering a situation with $n = 15$ units selected from $N = 150$ units where the total number of nonconforming units is $M = 10$. Assume the selection is random such that each of the N units has equal chance of being selected. What is the chance that exactly $x = 2$ units will be selected that are nonconforming?

Answer

The assumed beliefs are consistent with the hypergeometric mass function,

$$\begin{aligned}\frac{\binom{10}{2}\binom{150-10}{15-2}}{\binom{150}{15}} &= \frac{\frac{10!}{2!\times 8!} \times \frac{140!}{13!\times 27!}}{\frac{150!}{15!\times 35!}} \\ &= \frac{\frac{(10\times 9)}{(2\times 1)} \times \frac{1}{(1\times 1)}}{\frac{(150\times 149\times \cdots \times 141)}{(15\times 14)\times(135\times 134\times \cdots \times 128)}} = 0.19\end{aligned} \tag{10.31}$$

Note that the Poisson approximation is generally not considered accurate with $n = 15$. However, for reference the Poisson approximation gives 0.184 for the probability, which might be acceptable depending on the needs.

10.5 Xbar Charts and Average Run Length

An important role of Xbar charting and other charting procedures is to signal to local authority resources that something unusual is happening that might be fixable. In this regard, there are two kinds of errors that can occur:

1. Nothing unusual or assignable might be occurring, and local authority might be called in. This wastes time, diminishes support for charting efforts, and can increase variation. This is analogous to Type I error in hypothesis testing.
2. Something unusual and assignable is occurring, and local authority is not alerted. This is analogous to Type II error in hypothesis testing.

An analysis of these risks can provide insight to facilitate the selection of the chart sample size, n, and period between samples, τ. In this section, risks of both types are explored with reference to the normal and geometric distributions. The normal distribution is helpful in estimating the chance an individual charted point will generate an out-of-control signal.

10.5.1 The Chance of a Signal

Analysis of Xbar charting methods starts with the assumption that, with only common causes operating, individual observations are independent, identically distributed (IID) from some unknown distribution. Then, the central limit theorem guarantees that, for large enough sample size n, Xbar will be approximately normally distributed. Denoting the mean of individual observations μ and the standard deviation σ_0, the central limit theorem further guarantees that Xbar will have mean equal to μ and standard deviation approximately equal to $\sigma_0 \div \text{sqrt}[n]$.

Figure 10.12 shows the approximate distributions of the charted Xbar values for two cases. First, if only common causes are operating (the unknown distribution of the quality characteristic stays fixed), the Xbar mean remains μ and standard deviation approximately equals $\sigma_0 \div \text{sqrt}[n]$. The event of a false alarm is {Xbar > *UCL* or Xbar < *LCL*}. The probability of this event is approximately

$$\begin{aligned}\Pr\{\text{false alarm}\} &= \Pr\{\text{Xbar} > \mu + \frac{3\sigma}{\sqrt{n}}\} + \Pr\{\text{Xbar} < \mu - \frac{3\sigma}{\sqrt{n}}\} \\ &= \Pr\{Z > 3\} + \Pr\{Z < -3\} = 2 \times \Pr\{Z < -3\} = 0.0026\end{aligned} \tag{10.32}$$

where the symmetry property of the normal distribution and Table 10.1 were applied. The phrase "false alarm rate when the process is in-control" is often used to refer to the above probability.

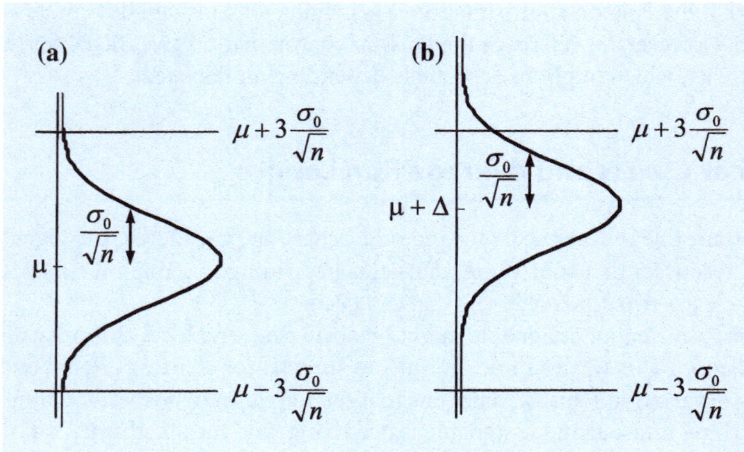

Fig. 10.12 Approximation distribution of Xbar for **a** no shift and **b** shift = + Δ

The second case considered here involved a shift of "Δ" in the mean of the distribution of the individual observations because of an assignable cause. This in turn causes a shift of Δ in the mean of Xbar as indicated by Fig. 10.12b.

The chance of false alarm is

$$\begin{aligned}\Pr\{\text{chart signal}\} &= \Pr\{\text{Xbar} > \mu + \frac{3\sigma_0}{\sqrt{n}}\} + \Pr\{\text{Xbar} < \mu - \frac{3\sigma_0}{\sqrt{n}}\} \\ &= \Pr\{Z > 3 - \frac{\Delta}{\sigma_0/\sqrt{n}}\} + \Pr\{Z < -3 - \frac{\Delta}{\sigma_0/\sqrt{n}}\} \\ &\approx \Pr\{Z < -3 + \frac{\Delta}{\sigma_0/\sqrt{n}}\}\end{aligned} \tag{10.33}$$

where the symmetry property of the normal distribution has been applied and the chance of an out-of-control signal from the limit away from the shift is neglected.

Detecting a nonzero shift ($\Delta \neq 0$) is generally considered desirable. The above formula offers one way to quantify the benefit of inspecting more units (larger n) in terms of increasing the chance that the chart will detect the shift.

Example 10.5.1 Detecting Injection Molding Weight Shifts

Question

Assume that an injection molding process is generating parts with average mass 87.5 g with standard deviation 1.2 g. Suppose an assignable cause shifts the mean to 88.5 g. Compare estimates for probabilities of detecting the shift in one subgroup with sample sizes of 5 and 10. Could that difference be subjectively considered important?

Answer

For this problem, we have $\Delta = 1.0$ g, $\sigma_0 = 1.2$ g, and $n = 5$ or $n = 10$. Applying the formula the detection "rate" or probability is

$$\Pr\{\text{chart signal}\} \approx \Pr\{Z < -3 + \frac{1.0}{1.2/\sqrt{n}}\}, \tag{10.34}$$

which gives 0.128 and 0.358 for $n = 5$ or $n = 10$ respectively. Going from roughly one-tenth chance to one-third chance of detection could be important depending on material, inspection, and other costs. With either inspection effort, there is a good chance that the next charted quantity will fail to signal the assignable cause. It will likely require several subgroups for the shift to be noticed.

10.5.2 Average Run Length

The chances of false alarms and detecting shifts associated with assignable causes are helpful for decision-making about sample sizes in charting. Next, we investigate the timing of false alarms and shift detections. Figure 10.13 shows one possible Xbar chart and the occurrence of a false alarm on subgroup 372.

"Run length" (RL) is the number of subgroups inspected before an out-of-control signal occurs. Therefore, run length is a discrete random variable because when the chart is being set up, the actual run length is unknown but must be a whole number. The expected value of the run length or "**average run length**" (ARL) is often used for subjective evaluation of alternative sample sizes (n) and different charting approaches (e.g., Xbar charting and EWMA charting from ▶ Chap. 9).

If one is comfortable in assuming that the individual quality characteristics are independent, identically distributed (IID), then these assumptions logically imply

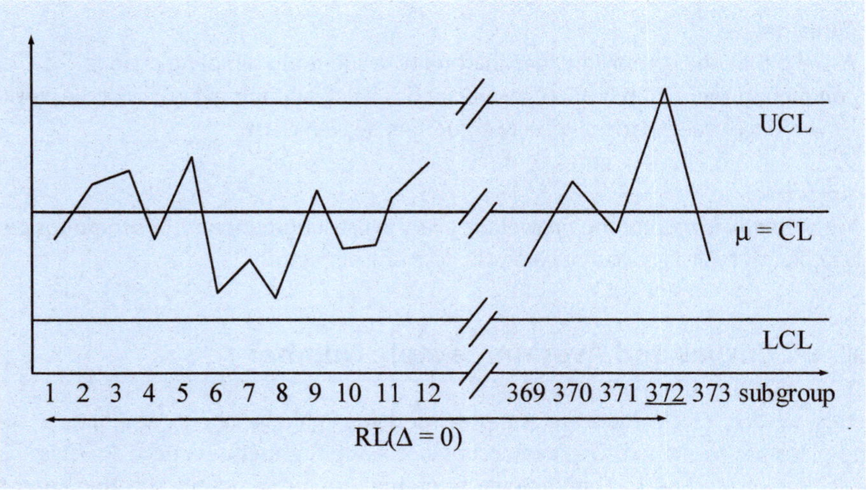

Fig. 10.13 Random run length (RL) with only common causes operating

Table 10.5 The average run lengths (ARL) in numbers of periods

n	ARL($\Delta=0$)	ARL($\Delta=1\sigma_0$)
4	370.4	3.2
7	370.4	2.2
10	370.4	1.9

a comfort with assuming that the run length is distributed according to a geometric probability mass function. Under these assumptions, the expected value of a geometric random variable is relevant, and the ARL is given as a function of the shift Δ, the quality characteristic distribution, σ_0, and the sample size n:

$$E[RL] = ARL = \frac{1}{\Pr\{Z < -3 + \frac{\Delta}{\sigma_0/\sqrt{n}}\}}. \tag{10.35}$$

Table 10.5 shows the ARLs for Xbar charts with different sample sizes given in units of τ. For example, if the period between sampling is every 2.0 h ($\tau = 2.0$ h), then the ARL($\Delta = 0$) = 370 (periods) × 2.0 (h/period) = 740 h. Therefore, false alarms will occur every 740 h. In fact, a property of all Xbar charts regardless of sample size is that the in-control run ARLs are 370.4. This in-control ARL is typical of many kinds of charts.

10

Note that chart "designer" or user could use the ARL formula to decide which sample size to use. For example, if it is important to detect $1\sigma_0$ process mean shifts within two periods with high probability such that ARL($\Delta = 1\sigma_0$) < 2.0, then sample sizes equal to or greater than 10 should be used. Note also that ARL does not depend on the true mean, μ.

Example 10.5.2 Alarms at Full Capacity Plants Operating

Question

Assume that one is applying Xbar charting with subgroup sampling periods completing every 2.0 h ($\tau = 2$ h) for a plant operating all shifts every day of the week. How often do false alarms typically occur from a given chart?

Answer

With false alarms occurring on average every 370.4 subgroups and 12 subgroups per day, alarms typically occur once per 30 days or 1 per month.

10.6 OC Curves and Average Sample Number

In this section, techniques to support decision-making about acceptance sampling plans are presented. Users of acceptance sampling methods need to select both which type of method to apply (single sampling, double sampling, or other) and the parameters of the selected method. The phase "design of sampling plan" refers to these selections. The methods presented here aid in informed decision-making by clarifying the associated risks and costs.

In applying acceptance sampling, two possible outcomes are: (1) acceptance and (2) rejection of the lot of parts being inspected. Since neither outcome is known at the time of designing the plan, a discrete random variable can be associated with these outcomes with a probability of acceptance written, p_A. The true fraction of nonconforming items in the lot, p_0, is generally not known but is of interest.

An "operating characteristic curve" or "OC curve" is a plot of the predicted percent of the lots that will be accepted ($100 \times p_A$) as a function of the assumed percentage probability of nonconforming units ($100 \times p_0$). Since the true number nonconforming is not known, the plot can be interpreted as follows. Hypothetically, if the true fraction was p_0 and the true number nonconforming was $M = p_0 \times N$, the acceptance probability would be a given amount.

For double sampling and many other types of sampling plans, the number of units that will be inspected is a discrete random variable during the time when the plan is designed. The "**average sample number**" (ASN) is the expected number of samples that will be required. For a single sampling plan with parameters n and c, the ASN $= n$ because any user of that plan always inspects n units.

10.6.1 Single Sampling OC Curves

For single sampling, the event that the lot is accepted is simply defined in terms of the number of units found nonconforming, X. If $\{X \leq c\}$, the lot is accepted, otherwise it is not accepted. Therefore, the probability of acceptance is

$$p_A = \Pr\{X = 0\} + \Pr\{X = 1\} + \cdots + \Pr\{X = c\} \quad (10.36)$$

For known lot size N, sample size n, and true number nonconforming, M, it is often reasonable to assume that $\Pr\{X = x\}$ is given by the hypergeometric distribution. Then, the probability $\Pr\{X \leq c\}$ is given by the so-called "cumulative" hypergeometric distribution (◘ Fig. 10.14).

◘ Figure 10.15 shows the calculation of the entire OC Curve for single sampling with $N = 1000$, $c = 2$, and $n = 100$. The plotting proceeds, starting with values of M, then deriving $100 \times p_0$ and $100 \times p_A$ by calculation. Because of the careful use of dollar signs, the formulas in cells B6 and C6 can be copied down to fill in the rest of the curve. Looking at the chart, the decision-maker might decide that the 0.22 probability of accepting a lot with 4% nonconforming is unacceptable. Then, increasing n and/or decreasing c might produce a more desirable set of risks.

Example 10.6.1 Transportation Safety Inspections

Question 1

An airport operator is considering using video surveillance to evaluate a team of trainees with respect to courteous and effective safety checks of passengers. (This is

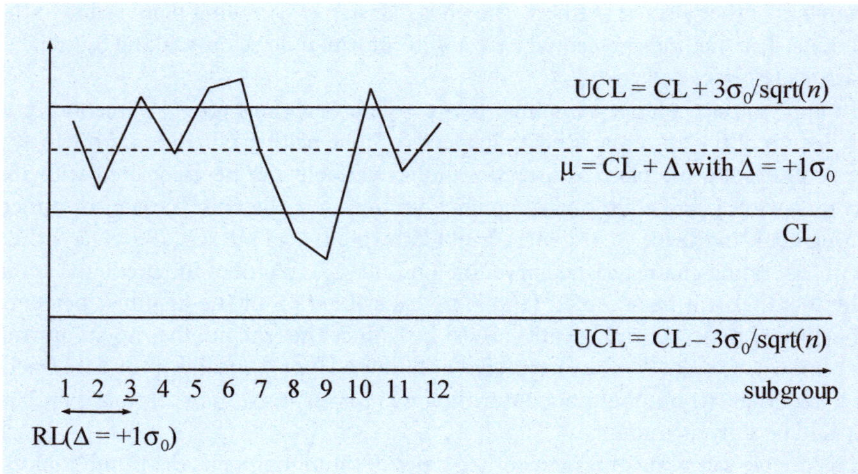

■ **Fig. 10.14** Random run length (RL) after a mean shift upwards of $\Delta = +1\sigma_0$

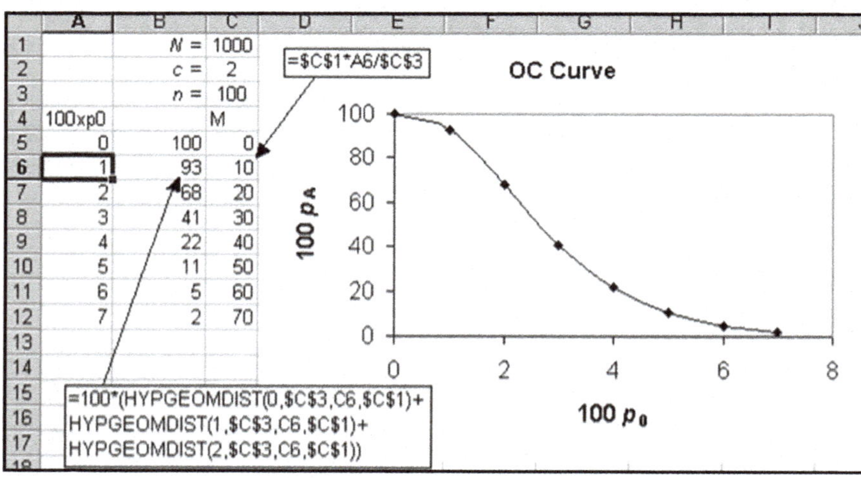

■ **Fig. 10.15** An example calculation of an entire OC curve

a case of inspecting inspectors.) The airport has only enough resources to examining surveillance tape for 150 passengers out of the 2000 inspections that occur each day. If greater than three inspections are unacceptably discourteous and/or ineffective, the entire team is flagged for re-training. Plot the OC curve for this policy and briefly describe the implications.

Answer 1

This question is based on a single sampling plan for $N = 2000$ units in a lot. It has $n = 150$ units in a sample and the rejection limit is $c = 3$.

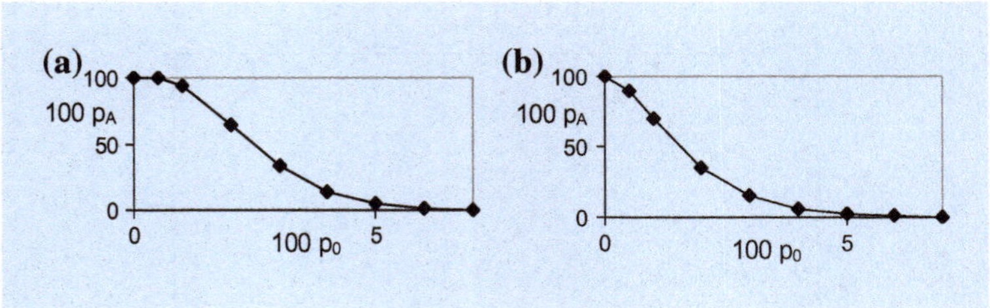

Fig. 10.16 OC curves for plans with **a** $n = 150$ units, $c = 3$ and **b** $n = 110$, $c = 1$

$$\begin{aligned}
p_0 &= 0.01 \text{ then } M = 20 \\
p_A &= \mathrm{P}(X = 0, N = 2000, n = 150, M = 20) \\
&\quad + \mathrm{P}(X = 1, N = 2000, n = 150, M = 20) \\
&\quad + \mathrm{P}(X = 2, N = 2000, n = 150, M = 20) \\
&\quad + \mathrm{P}(X = 3, N = 2000, n = 150, M = 20) \\
&= 0.94 \\
p_0 &= 0.02 \text{ then } M = 40 \Rightarrow p_A = 0.65.
\end{aligned} \tag{10.37}$$

The resulting OC curve is given in Fig. 10.16. The plot shows that the single sampling approach will effectively identify trainees yielding unacceptable inspections greater than 5% of the time, and if the fraction nonconforming is kept to less than 1%, there is almost zero chance of being found to need re-training.

Question 2

Consider an alternative single sampling plan with $n = 110$ and $c = 1$. As the customer of those parts, which plan do you feel is less or more risky? Explain.

Answer 2

The new policy is less risky in the sense that the probability of acceptance is always smaller (within two decimal places). However, relatively good teams are much more likely to be flagged for re-training, which might be considered unnecessary.

10.6.2 Double Sampling

The event that a double sampling procedure results in acceptance is relatively complex. Denote the number of units found nonconforming in the first set of inspections as X_1 and the number found nonconforming in the optional second set of inspections as X_2. Then, the acceptance occurs if $\{X_1 \le c_1\}$ or if $\{c_1 < X_1 \le r$ and $X_1 + X_2 \le c_2\}$. Therefore, the double sampling probability of acceptance is:

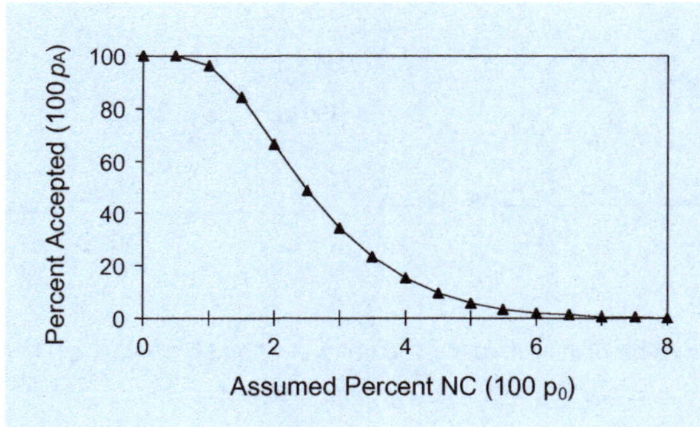

Fig. 10.17 Double sampling OC curve

$$
\begin{aligned}
p_A = {} & \Pr\{X_1 \le c_1\} + \Pr\{X_1 = c_1 + 1\} \times \Pr\{X_2 \le c_2 - (c_1 + 1)\} \\
& + \Pr\{X_1 = c_1 + 2\} \times \Pr\{X_2 \le c_2 - (c_1 + 2)\} + \cdots \\
& + \Pr\{X_1 = r\} \times \Pr\{X_2 \le c_2 - r\}
\end{aligned} \tag{10.38}
$$

which is expressed in terms of cumulative hypergeometric probabilities for assumptions that might be considered reasonable. Advanced readers will notice that the assumption of independence of X_1 and X_2 is implied by the above equation. Because of the computational challenge, it is common to apply the binomial approximation and the binomial cumulative when it is appropriate.

Example 10.6.2 Student Evaluation at a Teaching Hospital

Question

Consider a teaching hospital in which the $N = 7500$ patients are passed through a training class of medical students in a probationary period. The attending physician inspects patient interactions with $n_1 = 150$ patients. If less than or equal to $c_1 = 3$ student-patient interactions are unacceptable, the class is passed. If the number unacceptable is greater than $r = 6$, then the class must enter an intensive program (lot is rejected). Otherwise, an additional $n_2 = 350$ interactions are inspected. If the total number unacceptable is less than or equal to $c_2 = 7$, the class is passed. Otherwise, intensive training is required. Develop an OC curve and comment on how the benefit of double sampling is apparent.

Answer

Figure 10.17 shows the OC curve calculated using an Excel spreadsheet. Generally speaking, a desirable OC curve is associated with a relatively steep drop in the acceptance probability as a function of the true fraction nonconforming (compared with single sampling with the same average sample number). In this way, high quality classes of students (lots) are accepted with high probability and low quality lots are rejected with high probability.

10.6.3 Double Sampling Average Sample Number

OC curves can help quantify some of the benefits associated with double sampling and other sampling methods compared with single sampling. Yet, it can be difficult to evaluate the importance of costs associated with these benefits because the number of inspections in double sampling is random. The average sample number (ASN) is the expected number of units inspected.

Assume that the application of the sampling methods is done carefully to make the units inspected be representative of the whole lot of units. Then, the application of hypergeometric probability mass functions and the assumption of independence give the following formula:

$$\text{ASN (double sampling)} = n_1 + \Pr\{X \le c_1\} \times n_2 \tag{10.39}$$

where X is distributed according to the hypergeometric probability mass function. Note that the average ample number is a function of the true number nonconforming $M = p_0 \times N$, which must be assumed. Of course, ASN (single sampling) $= n$, independent of any assumptions. Generally speaking, comparable OC curves can be achieved by single and double sampling with ASN (single sampling) considerably higher than ASN (single sampling).

Example 10.6.3 Single Versus Double Sampling

Question

Consider a lot with $N = 2000$, single sampling with $n = 150$, and $c = 3$, and double sampling with $n_1 = 70$, $c_1 = 1$, $r = 4$, $n_1 = 190$, and $c_2 = 4$. These single and double sampling plans have comparable OC curves. Compare the average sample numbers (ASN) under the assumption that the true fraction nonconforming is 3%.

Answer

Under the standard assumption that all units in the lot have an equal chance of being selected, the hypergeometric mass function is reasonable for predicting ASN. For single sampling, the ASN is 150. Assuming 3% are nonconforming, $M = 0.03 \times 2000 = 60$. For double sampling, the ASN $= 70 + (0.1141 + 0.2562) \times 190 = 140.3$.

10.7 Chapter Summary

The purpose of this chapter is to show how probability theory can aid in the comparison of alternative methods, including the selection of specific method parameter values such as sample sizes. Applied statistics methods such as Xbar charting and acceptance sampling involve uncertainties and risks that are evaluated using theory.

The central limit theorem and the normal distribution are introduced. The primary purposes are (1) to clarify why Xbar charting is so universally applicable and (2) to permit the calculation of false alarm and the chance of correctly identifying an assignable cause. Also, application of the central limit theorem and Monte Carlo integration for deriving the charting constants d_2 and c_4 is presented.

Next, discrete random variables are introduced, with geometric and hypergeometric random variables being key examples. The concept of the average run length (ARL) is defined to quantify the typical time between false alarms or before assignable causes are correctly identified in control charting. A formula for the ARL is developed, combining normal distribution and geometric distribution calculations.

The conditions are described under which the hypergeometric probability mass function is a reasonable choice for estimating the chances of selecting certain numbers of nonconforming units in sampling. These are related to rational subgroup application such that inspected units are representative of larger lots.

Finally, the hypergeometric distribution and Poisson approximation are applied both (1) to develop operating characteristic curves (OC curves) and (2) to estimate the average sample number (ASN) for double sampling. OC curves give decision-makers information about how a given acceptance sampling plan will react to different hypothetical (imagined true) quality levels.

10.8 Problems

In general, choose the correct answer that is most complete.

1. Which is correct and most complete?
   a. Random variables are unknown by the planner at time of planning.
   b. Applying any quality technology is generally associated with some risk that can be estimated using probability theory and/or judgment.
   c. Even though method evaluation involves subjectivity, the same methods can be compared thoroughly using the same assumptions.
   d. All of the above are correct.
   e. All of the above are correct except (a) and (d).
2. Which is correct and most complete?
   a. In usual situations, probabilities of events are assigned without subjectivity.
   b. An application of probability theory is to evaluate alternatives.
   c. In making up a distribution function, the area under the curve must be 1.00.
   d. All of the above are correct.
   e. All of the above are correct except (c) and (d).
3. Which is correct and most complete?
   a. If X follows a triangular distribution, X is a continuous random variable.
   b. If X follows a binomial distribution, X is a continuous random variable.
   c. If X follows a normal distribution, X is a discrete random variable.
   d. All of the above are correct.
   e. All of the above are correct except (a) and (d).
4. Assume $X \sim N(\mu = 10, \sigma = 2)$. What is the $\Pr\{X > 16\}$?
5. Suppose someone tells you that she believes that revenues for her product line will be between \$2.2M and \$3.0M next year with the most likely value equal to \$2.7M. She says that \$2.8M is much more likely than \$2.3M. Define a distribution function consistent with her beliefs.

6. If X is uniformly distributed between 10 and 20, what is $\Pr\{X < 14\}$?
7. Which is correct and most complete?
   a. If X is N(10, 3), then $Z = (X - 10) \div 3$ is N(0, 1).
   b. If X is triangular with parameter $a = 10$, $b = 11$, and $c = 14$, then $Z = (X - 10) \div 4$ is triangular with parameters $a = 0$, $b = 0.2$, and $c = 2$.
   c. If X is N(10, 3), then $Pr\{X > 13\} = 0.259$ (within implied uncertainty).
   d. All of the above are correct.
   e. All of the above are correct except parts (a) and (d).
8. Assume that individual quality characteristics are normally distributed with mean equal to the average of the specifications limits. Also, assume that Z is distributed according to the standard normal distribution. Is it true that 2 $\Pr(Z < -3.0\ C_{pk})$?
9. Which is correct and most complete? (Assume C_{pk} is known.)
   a. Changing 3s to 2s in UCL formulas would cause more false alarms.
   b. Autocorrelation often increases the chance of false alarms in Xbar charting because the standard deviation is underestimated during the trial period.
   c. Absence of assignable causes alone guarantees Xbar is normally distributed.
   d. All of the above are correct.
   e. All of the above are correct except parts (a) and (d).
10. If X is hypergeometrically distributed with parameters $N = 10$, $n = 2$, and $M = 1$, what is $\Pr\{X = 0\}$?
11. Which is correct and most complete?
    a. Cumulative normal distribution probabilities are difficult to compute by hand, i.e., without using software or a table.
    b. The central limit guarantees that all random variables are normally distributed.
    c. The central limit theorem does not apply to discrete random variables.
    d. All of the above are correct.
    e. All of the above are correct except (c) and (d).
12. Which is correct and most complete?
    a. Except for labels on axes, probability density functions for normal random variables all look pretty much the same.
    b. The normal distribution is not the only one with the location-scale property.
    c. The central limit theorem implies the Xbar variance decreases as n increases.
    d. All of the above are correct.
    e. All of the above are true except parts (b) and (d).
13. Which is correct and most complete? (Assume n is large enough.)
    a. False alarm chance for the next subgroup on an Xbar chart is $\Pr\{Z < -3\}$.
    b. Chance of 2 false alarms in the next 3 subgroups is 0.66 $\Pr\{Z < -3\}$.
    c. False alarm chance for the next subgroup on an Xbar chart is $2\Pr\{Z < -3\}$.
    d. All of the above are correct.
    e. All of the above are correct except (a) and (d).
14. Which of following is the most correct and complete?
    a. The number of nc. units in a subgroup is often viewed as a discrete random variable.
    b. If you do not know exactly how many nonconforming units in a lot, the hypergeometric distribution can be used to calculate the chance of finding X many nonconforming units in a subgroup of size n.

c. If $n \times p_0 < 5.0$, then it can be useful to apply the normal distribution cumulate to estimate probabilities of events involving discrete random variables.
d. All of the above are correct.
e. All of the above are correct except parts (c) and (d).

15. Assume that $N = 2000$, $n = 150$, and $M = 20$.
    a. Estimate $\Pr\{X \leq 3\}$ using the hypergeometric distribution.
    b. Estimate $\Pr\{X \leq 3\}$ using the Poisson approximation.
16. Plot a single sampling OC curve for $N = 3000$, $n = 200$, and $c = 2$.
17. Plot a single sampling OC curve for $N = 4000$, $n = 150$, and $c = 3$.
18. Which of the above two policies is more likely to do the following:
    a. Accept lots with large fractions of nonconforming units.
    b. Accept lots with small fractions of nonconforming units.
19. What is the shape of an ideal acceptance sampling curve? Explain briefly.

References

Grimmet GR, Stirzaker DR (2001) Probability and random processes, 3rd edn. Oxford University Press, Oxford
Keynes JM (1937) General theory of employment. Quart J Econ
Savage LJ (1972) The foundations of statistics, 2nd edn. Dover Publications Inc, New York

Part II

Design of Experiments (DOE) and Regression

DOE: The Jewel of Quality Engineering

T. T. Allen, *Introduction to Engineering Statistics and Lean Six Sigma*,
https://doi.org/10.1007/978-1-4471-7420-2_11

11.1 Introduction

Design of experiments (DOE) methods are among the most complicated and useful of statistical quality control techniques. DOE methods can be an important part of a thorough system optimization, yielding definitive system design or redesign recommendations. These methods all involve the activities of experimental planning, conducting experiments, and fitting models to the outputs. An essential ingredient in applying DOE methods is the use of procedure called "randomization" which is defined at the end of this chapter. To preview, randomization involves making many experimental planning decisions using a random or unpatterned approach.

The primary purpose of this chapter is to preview the various DOE methods described in Part II of this book. All of these DOE methods involve changing key input variable (KIV) settings which are directly controllable (called factors) using carefully planned patterns, and then observing outputs (called responses). This chapter describes the "**two-sample t-test**" method which permits **proof** that one level of a single factor results in a higher average response than another level of one factor. Two-sample t-testing is also used to illustrate randomization and its relationship with proof. There is also "**paired t-test**" in this chapter which generally offers a greater chance of finding significance when the data are paired in some way. Analysis of variance (ANOVA) is introduced in this chapter for multi-level hypothesis testing. Also, "full factorial" designs are introduced to help motivate the so-called "fractional factorial" designs in ▶ Chap. 12.

11

▶ Section 11.2 provides an overview of the different types of DOE and related methods. ▶ Section 11.3 describes two-sample t-testing with examples and a discussion of randomization. ▶ Section 11.4 describes an activity called "randomization", common to all DOE methods and technically required for achieving proof. ▶ Section 11.5 summarizes the material covered. Note that most of the design of experiments method presented here are supported by standard software such as Minitab®, JMP, and DesignExpert® and by Sagata® software. (The author of this book is part owner of Sagata Ltd.)

11.2 Design of Experiments Methods Overview

Five classes of experimentation and analysis methods are described in this book: (1) two-sample t-tests, (2) standard screening using **fractional factorials** (FF), (3) one-shot response surface methods (RSM), (4) sequential response surface methods, and (5) Robust Design based on Profit Maximization (RDPM). A brief summary is offered in ◘ Table 11.1. In addition, two classes of analysis of variance (ANOVA) analysis methods have been provided for determining significance after data has been collected using any experimental plan.

The primary objective is to allow the reader to develop competence in application of methods in each class. Also, decision support information for supporting has been provided for the selection of specific methods of each type, e.g., choosing the number of runs, n, and the parameters used in the analysis. Note that any of these methods could constitute an entire "improvement system".

Table 11.1 Brief summary of methods described in this chapter

Method	Advantage	Disadvantage
Two-sample t-tests	Provide a relatively high level of evidence that a single level of a single factor causes a higher average response	Methods only address one factor-at-a-time (OFAT). Compared with screening using fractional factorials, for comparable total costs the Type I and Type II errors are more likely
Screening using fractional factorials (FF)	Provides an inexpensive way to determine which factors from a long list significantly affect system performance. Sometimes, users apply results to support final engineering design decisions	Compared with response surface methods, the methods generate a relatively inaccurate prediction model. Compared with two-sample t-tests, the level of evidence associated with significance claims is subjectively lower
One-shot Response surface methods (RSM)	Create a relatively accurate prediction model and significance information, permitting identifying of interaction effects	Compared with factor screening methods, these methods require substantially larger numbers of experimental runs for a given number of factors
Sequential Response Surface Methods (RSM)	Generate a relatively accurate prediction model and may require fewer runs than one shot response surface methods	The derived prediction model will, in general, be less accurate than the one from one-shot response surface methods if the method terminates without using all the runs
Robust design based on profit maximization (RDPM)	Builds on RSM to directly maximize the sigma level in a cost-effective manner addressing production noise	Complicated; may require substantial experimental cost
Analysis of variance (ANOVA) followed by multiple t-tests	Offers a standard approach for analyzing significance of factors and/or model terms that addresses the multiplicity of the tests	Compared with Lenth's method and normal probability plots, the Type II errors are generally higher. This is only an analysis method that does not explain which data to collect

Besides randomization, a common aspect of all DOE methods is the importance for the method users in identifying the KIVs and ranges for these factors. The preliminary identification of KIVs derives from engineering judgment. If a poor choice of KIVs and/or ranges is identified, it is unlikely the application of any DOE method will achieve desired results.

Note that all of the methods in Table 11.1 can generate statistical "proof" that changing factors affects average system outputs or responses. In general, derivation of the associated statistical proof relates to the amount and quality of the data collected and not whether the differences detected are important to decision-makers. An important theme in design of experiments is that statistical significance and evidence do not generally translate into "practical" significance.

Example 11.2.1 Method Choices

Question

Which of the following is correct and most complete?

a. FF is sometimes used to give screening information and for final system choices.
b. RSM helps in understanding interaction effects and predicting performance.
c. T-testing can, if applied with randomized experimentation, generate strong proof.
d. All of the above are correct.
e. All are correct except (b) and (d).

Answer

Yes, fractional factorials (FF) are often the last and only design of experiments method used in many projects. Also, modeling the combined effects of factors or "interactions" is possible using response surface methods (RSM). Also, t-testing using randomization can generate proof. Therefore, the correct and most complete answer is (d).

11.3 The Two-Sample T-Test Methodology and the Word "Proven"

The following class of methods is called "two-sample t-testing assuming unequal variances" that can be viewed as the simplest design of experiments methods. Members of this class are distinguished by the initial sample size parameters n_1 and n_2 in *Step 1* and the α level used in *Step 3*.

Roughly speaking, this method is useful for situations in which one is interested in "proving" with a "high level of evidence" that one alternative is better in terms of average response than another. Therefore, there is one factor of interest at two levels. The screening procedure described subsequently can permit several factors to be "proven" significant simultaneously with a comparable number of total tests. However, a subjectively greater level of assumption-making is needed for those screening methods such that the two-sample t-test offers a higher level of evidence.

Definition

The phrase "**blocking factor**" refers to system input variables that are not of primary interest. For example, in a drug study, the names of the people receiving the drug and the placebo are not of primary interest even though their safety is critical.

Algorithm 11.1. Two-sample t-tests

Step 1. *a.* Develop an experimental table or "DOE array" that describes the levels of all blocking factors and the factor of interest for each run. The ordering of the factor levels should exhibit no pattern, i.e., an effort should be made to allocate all blocking factor levels in an unpatterned way. Ideally, experimentation is "**blind**" so that human participants do not know which level they are testing. Unpatterned ordering can be accomplished by putting n_1 As and n_2 Bs in 1 column on a spreadsheet and pseudo-random uniform [0, 1] numbers in the next column. Sorting, we have a "uniformly random" ordering, e.g., 2-1-1-2-2-2-1…

b. Collect $n_1 + n_2$ data, where n_1 of these data are run with factor A at level 1 and n_2 are run with factor A at level 2 following the experimental table.

Step 2. Defining $\bar{y}_1$ as the average of the run responses with factor A at level 1 and s_1^2 as the sample variance of these responses, and making similar definitions for level 2, one then calculates the quantities t_0 and degrees of freedom *(df)* using

$$t_0 = \frac{\bar{y}_1 - \bar{y}_2}{\sqrt{\frac{s_1^2}{n_1} + \frac{s_2^2}{n_2}}} \qquad df = round\left[\frac{\left(\frac{s_1^2}{n_1} + \frac{s_2^2}{n_2}\right)^2}{\frac{\left(s_1^2/n_1\right)^2}{n_1 - 1} + \frac{\left(s_2^2/n_2\right)^2}{n_2 - 1}}\right] \tag{11.1}$$

where "round" means round the number in brackets to the nearest integer.

Step 3. Find $t_{critical}$ using the Excel formula "=TINV(2 * 0.05, df)" or using the critical value from a *t*-table referenced by $t_{\alpha,df}$ (see Table 11.2). If $t_0 > t_{critical}$, then claim that "it has been proven that level 1 of factor A results in a significantly *higher* average or expected value of the response than level 2 of factor A with alpha equal to 0.05".

Step 4. (Optional) Construct two "box plots" of the response data at each of the two level settings (see below). Often, these plots aid in building engineering intuition.

Table 11.2 Values of $t_{critical} = t_{\alpha,df}$

df	α			*df*	α		
	0.01	0.05	0.1		0.01	0.05	0.1
1	31.82	6.31	3.08	7	3.00	1.89	1.41
2	6.96	2.92	1.89	8	2.90	1.86	1.40
3	4.54	2.35	1.64	9	2.82	1.83	1.38
4	3.75	2.13	1.53	10	2.76	1.81	1.37
5	3.36	2.02	1.48	20	2.53	1.72	1.33
6	3.14	1.94	1.44				

Definition

The "**median**" of m numbers is the $[(m+1)/2]$th highest if m is odd. It is the average of the $(m/2)$th highest and the $[(m/2)+1]$th highest if m is even.

Algorithm 11.2. Box and whisker plotting

If the number of data is even, then the 25% (Q1) and 75% (Q3) quartiles are the middle values of the two halves of the data. Otherwise, they are the median including the middle in both halves.

Step 1: Draw horizontal lines at the median, Q1, and Q3.

Step 2: Connect with vertical lines the edges of the Q1 and Q3 lines to form a rectangle or "box".

Step 3: Then, draw a line from the top middle of the rectangle up to the highest data below Q3 + 1.5(Q3 – Q1) and down from the bottom middle of the rectangle to the smallest observation greater than Q1 – 1.5(Q3 – Q1).

Step 4: Any observations above the top of the upper line or below the bottom of the lower line are called "outliers" and labeled with "*" symbols.

Note that, with only 3 data points, software generally does not follow the above exactly. Instead, the ends of the boxes are often the top and bottom observations.

If we were trying to prove that level 1 results in a significantly *lower* average response than level 2, in *Step 3* of Algorithm 11.1, we would test $-t_0 > t_{\text{critical}}$. In general, if the sign of t_0 does not make sense in terms of what we are trying to prove, the above "**one-sided**" testing approach fails to find significance. The phrase "**1-tailed test**" is a synonym for one-sided.

11

To prove there is any difference, either positive or negative, use $\alpha/2$ instead of α and the test becomes "**two-sided**" or "**2-tailed**". A test is called "**double blind**" if it is blind and the people in contact with the human testers also do not know which level is being given to which participant. The effort to become double blind generally increases the subjectively assessed level of evidence. Achieving blindness can require substantial creativity and expense.

The phrase "**Hawthorne effect**" refers to a change in average output values caused by the simple act of studying the system, e.g., if people work harder because they are being watched. To address issues associated with Hawthorne effects and generate a high level of evidence, it can be necessary to include the current system settings as one level in the application of a t-test. The phrase "**control group**" refers to any people in a study who receive the current level settings and are used to generate response data.

Definition

If something is proven using any given α, it is also proven with all higher levels of α. The "**p-value**" in any hypothesis test is the value of α such that the test statistic, e.g., t_0, equals the critical value, e.g., $t_{\alpha,df}$. The phrase "**significance level**" is a synonym for p-value. For example, if the p-value is 0.05, the result is proven with "alpha" equal to 0.05 and the significance level is 0.05. Generally speaking, people trying to prove hypotheses with limited amounts of data are hoping for small p-values.

Using t-testing is one of the ways of achieving evidence such that many people trained in statistics will recognize a claim that you make as having been "proven" with "objective evidence". Note that if t_0 is not greater than $t_{critical}$, then the standard declaration is that "significance has not been established". Then, presumably either the true average of level 1 is not higher than the true average of level 2 or, alternatively, additional data is needed to establish significance.

The phrase "**null hypothesis**" refers to the belief that the factors being studied have no effects, e.g., on the mean response value. Two-sample *t*-testing is not associated with any clear claims about the factors not found to be significant, e.g., these factors are not proven to be "insignificant" under any widely used conventional assumptions. Therefore, failing to find significance can be viewed as accepting the null hypothesis, but it is not associated with proof.

In general, the testing procedures cannot be used to prove that the null hypothesis is true. The Bayesian analysis can provide "posterior probabilities" or chances that factors are associated with negligible average changes in responses after *Step 1* is performed. This nonstandard Bayesian analysis strategy can be used to provide evidence of factors being unimportant.

11.4 T-Test Examples

This section contains two examples, one of which relates to a straightforward application of the t-test method. The second involves answering specific questions based on the concepts. In the first example, an auto company is interested in extending the number of auto bodies that an arc-welding robot can weld without adjustment using a new controller program. The first example is based on the commonly chosen sample size, $n_1 = n_2 = 3$, and selection $\alpha = 0.05$.

If one fails to find significance, that does not mean that the true average difference in responses between the two levels is exactly zero or negative. With additional testing, the test can be re-run and significance might be found. Note that the procedure, if applied multiple times, gives a probability of falsely finding significance (Type I errors) greater than α.

Still, it is common to neglect this difference and still quote the α used in Step 3 as the probability of Type I errors. Therefore, the choice of initial sample size is not critical unless it is wastefully large since additional runs can be added. A rigorous sequential approach would be to pre-plan on performing at most q sets of runs, with tests after each set, stopping if significance is found. Then, the α used for each test could be α/q such that the overall procedure rigorously guarantees an error rate less than α (e.g., 0.05) using the "Bonferroni inequality" which regulates overall errors.

Algorithm 11.3. First t-test example

Step 1: The engineer uses Table 11.3 to determine the run ordering. Pseudo-random uniform numbers were generated and then used to sort the levels for each run. Then, we first input level 1 (the new additive) into the system and observed the response 25. Then, we input level 2 (the current additive) and observed 20 and so on.

Step 2: Responses from welding tests are shown in the right-hand column of Table 11.3. The engineer calculated $\bar{y}_1 = 31.3$, $\bar{y}_2 = 21.3$, $s_1^2 = 30.3$, $s_2^2 = 2.33$, $t_0 = 3.03$, and $df = 2$.

Step 3: The critical value given by Excel "=TINV(0.1, 2)" was $t_{critical} = 2.92$. Since t_0 was greater than $t_{critical}$, we declared, "We have proven that level 1 results in a significantly higher average mean value than level 2 with alpha equal to 0.05." The p-value is 0.047.

Step 4: A box plot from Minitab® software is below which shows that level 1 results in higher number of bodies welded on average. Note that with 3 data Minitab® defines the lowest data point at Q1 and the highest data point as Q3.

Example 11.4.1 Second T-test Application

A work colleague wants to "prove" that his or her software results in shorter times to register the product over the internet on average than the current software. Suppose six people are available for the study: Fred, Suzanne, ... (see Fig. 11.1).

11

Question 1

How many factors, response variables, and levels are involved?

Answer 1

There are two correct answers: (1) two factor (software) at two levels (new and old) and 1 response (time) and (2) two factors (software and people) at two and six levels and 1 response (time). If the same person tested more than one software, people would be a factor.

Table 11.3 One approach to randomize the run order using pseudo-random numbers

Levels	Pseudo-random uniform nos.	Run	Level	Sorted nos.	Response
1	0.583941	1	1	0.210974	$Y_{1,1} = 25$
1	0.920469	2	2	0.448561	$Y_{2,1} = 20$
1	0.210974	3	1	0.583941	$Y_{1,2} = 35$
2	0.448561	4	2	0.589953	$Y_{2,2} = 23$
2	0.692587	5	2	0.692587	$Y_{2,3} = 21$
2	0.589953	6	1	0.920469	$Y_{1,3} = 34$

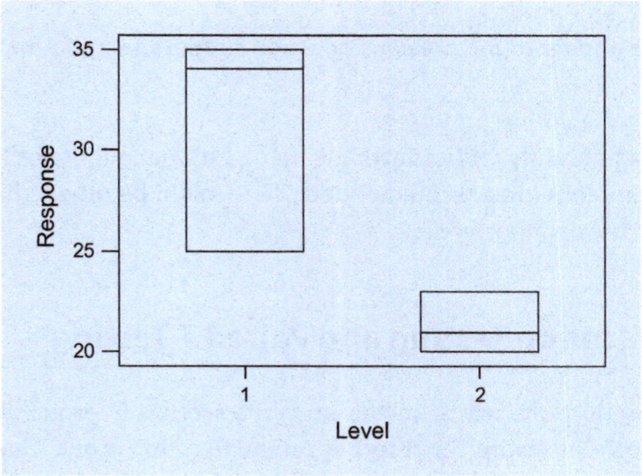

Fig. 11.1 Minitab® box plot and whisker for the autobody welding example

Questions

What **specific** instructions (that a technician can understand) can you give her to maximize the level of evidence that she can obtain?

Answer 2

Assume that we only want one person to test one software. Then, we need to randomly assign people to levels of the factor. Take the names Fred, Suzanne, … and match each to a pseudo-random number, e.g., Fred with 0.82, Suzanne with 0.22, … Sort the names by the numbers and assign the top half to the old and the bottom half to the new software. Then, repeat the process with a new set of pseudo-random numbers to determine the run order. There are other acceptable approaches, but both assignment to groups and run order must be randomized.

Questions 3

In this question, the following data is needed:

New software	Old software
Fred—35.6 s	Juliet—45.2 s
Suzanne—38.2 s	Bob—43.1 s
Jane—29.1 s	Mary—42.7 s

Analyze the above data and draw conclusions that you think are appropriate.

Answer 3

We begin by calculating the following: $\bar{y}_1 = 34.30, \bar{y}_2 = 43.67, s_1^2 = 21.97$, $s_2^2 = 1.80, t_0 = 3.33$, and df = round[2.3] = 2. Note that we are hoping the average time is lower (a better result), therefore the sign of t-critical makes sense and we can ignore it for the calculation. Since 3.33 > 2.92 we have proven that the new software reduces the average registration time with $\alpha = 0.05$.

Questions 4

How might the answer to the previous question support decision-making?

Answer 4

The software significantly reduces average times, but that might not mean that the new software should be recommended. There might be other criteria such as reliability and cost of importance.

11.5 Randomization Testing and Paired T-Testing

The t-testing procedure described in the previous section is probably the most commonly used "hypothesis testing" method. It can be used to "prove" that a factor change "significantly" affects an average response. In this section, we describe two alternative procedures: (1) randomization testing and (2) paired t-testing. Randomization testing is generally relevant when there is strong evidence of a significant difference. It permits the user in such cases to make stronger claims of significance than through ordinary t-testing. This is possible in randomization testing because no assumption is implied about the distribution from which responses come. As described in ► Chap. 19, in standard t-testing, the assumption of normally distributed random errors permits the derivation of the critical values and p-values. Because the randomization derives its p-value without assuming a specific distribution it is called a "nonparametric" hypothesis testing method.

11

Paired t-testing is relevant when there is a natural pairing between observations. For example, we might be testing shoe wear as it depends on spray coating with level (1) being spray 1 and level (2) being spray 2. In our experimental plan, each level might assigned to either the left (L) foot or the right (R) foot of each person. Then, the paired design might be: L1R2, L2R1, L1R2, L1R2, … We would generally apply paired t-testing in such cases because we would generally expect an increased chance of finding significance or higher "power" to find differences. Therefore, randomization testing generally offers lower power than standard t-testing but greater subjective evidence if significance is found. Also, paired t-testing usually offers higher statistical power but is only relevant if there is a natural pairing between observations at different levels.

The first step of randomization test procedure is the same as for standard t-testing procedure in Algorithm 11.1. Steps 2 and 3 are based on calculating all the combinations of possible level assignments to the data and estimating the effects. Consider the example in ◘ Table 11.3. The actual allocation of factor levels to data points is 1-2-1-2-2-1. Yet, if the factor had no effect, we should expect comparable performance for all other $6!/[(3!)(3!)] = 20$ possible allocations of factor levels to effects. ◘ Table 11.4 shows the estimate effects based on all 20 possible allocations. We note that the observed allocation is the most extreme, i.e., it is 1/20. Therefore, we estimate that the nonparametric p-value in this case is 1/20 or 0.05 which is comparable to the parametric p-value of 0.047. The histogram of the estimated effects for all combinations can be called the "reference" distribution which might be highly skewed or "non-normal" (◘ Fig. 11.2).

Table 11.4 The application of the randomization test to the data from the first example in ▶ Sect. 11.4

Combination #	Levels for run						Contribution to the sum						Est. effect
	1	2	3	4	5	6	S1	S2	S3	S4	S5	S6	
1	1	1	1	2	2	2	−25	−20	−35	23	21	34	−0.67
2	1	1	2	1	2	2	−25	−20	35	−23	21	34	7.33
3	1	2	1	1	2	2	−25	20	−35	−23	21	34	−2.67
4	2	1	1	1	2	2	25	−20	−35	−23	21	34	0.67
5	1	1	2	2	1	2	−25	−20	35	23	−21	34	8.67
6	1	2	1	2	1	2	−25	20	−35	23	−21	34	−1.33
7	2	1	1	2	1	2	25	−20	−35	23	−21	34	2.00
8	1	2	2	1	1	2	−25	20	35	−23	−21	34	6.67
9	2	1	2	1	1	2	25	−20	35	−23	−21	34	10.00
10	2	2	1	1	1	2	25	20	−35	−23	−21	34	0.00
11	1	1	2	2	2	1	−25	−20	35	23	21	−34	0.00
12	**1**	**2**	**1**	**2**	**2**	**1**	**−25**	**20**	**−35**	**23**	**21**	**−34**	**−10.00**
13	2	1	1	2	2	1	25	−20	−35	23	21	−34	−6.67
14	1	2	2	1	2	1	−25	20	35	−23	21	−34	−2.00
15	2	1	2	1	2	1	25	−20	35	−23	21	−34	1.33
16	2	2	1	1	2	1	25	20	−35	−23	21	−34	−8.67
17	1	2	2	2	1	1	−25	20	35	23	−21	−34	−0.67
18	2	1	2	2	1	1	25	−20	35	23	−21	−34	2.67
19	2	2	1	2	1	1	25	20	−35	23	−21	−34	−7.33
20	2	2	2	1	1	1	25	20	35	−23	−21	−34	0.67

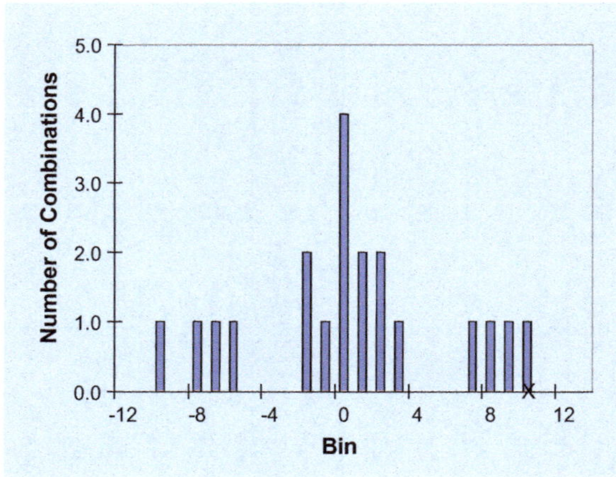

Fig. 11.2 Reference distribution associated with the randomization test

In paired experiments, the additional statistical power can come because much of the unexplained variation is eliminated from consideration by the pairing. For example, the differences between people wearing the shoes are removed from consideration since we only focus on the differences within subjects, i.e., between the shoes on the same person. As another example, consider an experiment done by a golf ball maker would need to the test the equipment used to measure golf balls. The pairing occurs because the same ball is measured on two machines. They would like to prove that one machine gives a higher average reading than another machine. The lack of homogeneity between the golf balls could contribute to compression measurements that could inflate the experimental error and lower the statistical power. Using the paired t-test, the heterogeneity or variability between golf balls would be factored out and the focus would be on differences between the measurements on alternative machines.

The use of hypothesis testing is not emphasized in this book because it can leave the impression that abstract theory is critical to benefit from the methods presented. Here, when describing the somewhat optional subject of paired t-testing, hypothesis testing concepts are also described for completeness. A null hypothesis (H_0) is what one entertains skeptically unless the data proves otherwise. In the golf example, the null hypothesis is that both golf ball presses give the same average values on a given golf ball deflection. Our alternative hypothesis (H_a) often represents what the experimenter hopes to prove is true. In the golf example, the desire would be that the golf ball presses are giving different true average deflection amounts. Therefore, the experimenter generally desires to reject the null hypothesis, e.g., to prove that the presses are different. Such proof is generated when the test statistic value exceeds the critical value, e.g., for t-testing when $t_o > t_{\alpha/2,\, n-1}$.

The test statistic for this hypothesis paired t-test is:

$$t_o = \frac{\bar{d}}{S_d/\sqrt{n}} \tag{11.2}$$

where n is the number of specimen (golf balls) and

$$\bar{d} = \frac{1}{n}\sum_{i=1}^{n} d_i \quad (11.3)$$

is the sample mean of the differences and the sample standard deviation is as follows:

$$S_d = \left[\frac{\sum_{i=1}^{n}\left(d_i - \bar{d}\right)^2}{n-1}\right]^{1/2} \quad (11.4)$$

combining Eqs. (11.3) and (11.4) we get

$$S_d = \left[\frac{\sum_{i=1}^{n} d_i^2 - \frac{1}{n}\left(\sum_{i=1}^{n} d_i\right)^2}{n-1}\right]^{1/2} \quad (11.5)$$

Steps for a paired t-test are as described in Algorithm 11.4.

Algorithm 11.4. Paired T-testing

Step 1: Perform randomized and paired testing similar to ordinary t-testing. Also, optionally set up the null and alternative hypothesis, where μ_d is the value of the difference between press one and press two results. In paired t-testing H_o: $\mu_1 = \mu_2$ is equivalent to testing:

$$H_o: \mu_d = 0$$
$$H_a: \mu_d \neq 0$$

Step 2: Calculate d using the response data and Eq. (11.3).
Step 3: Calculate sample standard deviation using Eq. (11.5).
Step 4: Select the alpha (α) level and identify the relevant critical value.
Step 5: Calculate the t-statistic using Eq. (11.2).
Step 6: Compare the t-value and t-statistic in order to determine if one can reject the null hypothesis. State conclusions

Example 11.5.1 Paired T-test

Question
Use the data from Table 11.5 and a paired t-test to determine if the owner is correct in thinking the presses are giving significantly different average results.

Answer
The relevant hypotheses are:

$$H_o: \mu_d = 0$$
$$H_a: \mu_d \neq 0$$

Table 11.5 Golf ball compression data

Golf ball compression		
Specimen	**Press 1**	**Press 2**
1	90	92
2	110	105
3	80	82
4	75	75
5	92	90
6	105	105
7	99	99
8	108	109
9	85	83

Next, the $\bar{d}$ value can be derived using:

$$\begin{array}{ll} d_1 = 92 - 90 = 2 & d_6 = 105 - 105 = 0 \\ d_2 = 110 - 105 = 5 & d_7 = 99 - 99 = 0 \\ d_3 = 82 - 80 = 2 & d_8 = 109 - 108 = 1 \\ d_4 = 75 - 75 = 0 & d_9 = 83 - 85 = -2 \\ d_5 = 90 - 92 = -2 & \end{array}$$

$$2 + 5 + 2 + 0 - 2 + 0 + 0 + 1 - 2 = 6$$

$$\bar{d} = \frac{1}{n}\sum_{i=1}^{n} d_i = (1/9) * (6) = 0.667$$

The relevant sample standard deviation calculation is:

$$\begin{aligned} S_d &= \left[\frac{\sum_{i=1}^{n} d_i^2 - \frac{1}{n}\left(\sum_{i=1}^{n} d_i\right)^2}{n-1}\right]^{1/2} \\ &= \left[\frac{42 - 0.667}{9-1}\right]^{1/2} \\ &= 2.27 \end{aligned}$$

Selecting $\alpha = 0.05$, the relevant t-critical value is $t_{0.025,8} = 2.306$. We compute the value of the paired t-statistic using:

$$t_o = \frac{\bar{d}}{S_d/\sqrt{n}} = \frac{0.667}{2.227/\sqrt{9}} = 0.88$$

As a result $t_0 < t_{0.025,8}$ and we fail to reject the null. Therefore, we fail to prove anything conclusive. There might be difference that we might find if we had more data. Yet, there is a subjective indication that the presses are not giving very different values. We did apply the relatively powerful paired t-test and failed to find significance.

11.6 ANOVA for Two Sample T-Tests

Analysis of variance (ANOVA) is a set of methods related to the statistical comparison of variances. Perhaps the most common applications of these methods involves using variance estimates to compare response mean differences. These are the methods considered here. By focusing on variance estimate, ANOVA offers the ability to regulate the overall error rate when multiple mean differences are considered. For example, if we are comparing options A, B, and C to find the highest mean, we might perform three tests: A versus B, A versus C, and B versus C. Fisher showed that his ANOVA method offered an adjustable overall error rate, α. This rate was a probability limit of making any wrong conclusion of significance in all of the simultaneous mean tests considered.

The ANOVA procedure is associated with the following assumptions that can be evaluated using data to determine whether they, reasonably, apply to a given situation:

- Response values are normally distributed (or normally distributed to an acceptable approximation) and
- Responses from different experimental runs are independently distributed.

The regression chapter (▶ Chap. 15) describes in relative detail tests of these assumptions which include plots of the so-called "residuals" or estimated errors.

Technically, under the above assumptions the ANOVA generates an F-statistic which can be related to t distributed statistics as follows: $F = t^2$. The F is also a ratio of chi-squared statistics under different assumptions. If the so-called "null" hypothesis (H_0) holds then the ratio is approximately 1.0. On the other hand, if the factor being tested is associated with levels having large mean differences, then the F value is generally much larger than 1.0.

The description of the ANOVA method invariably involves formulas and complicated notation even for single factor experiments. ◘ Table 11.6 describes the notation for experimental data for an experiment at "a" levels written 1, 2, …, a. At each level there are n "replicates" or measurements. In the table (y_{ij}) represents the jth observation taken under the ith factor. The sum of all averages "grand average" including all observations is "$\bar{y}..$".

The ANOVA procedure involves calculating several sums of squares which effectively section or divide the total variation in the data into component parts. The total

◘ **Table 11.6** Data for a single factor ANOVA experiment

Factor level	Observations			Totals	Averages
1	y_{11}	y_{12}	y_{1n}	$y_{1.}$	$\bar{y}_{1\cdot}$
2	y_{21}	y_{22}	y_{2n}	$y_{2.}$	$\bar{y}_{2\cdot}$
⋮	⋮	⋮	⋮	⋮	⋮
A	y_{a1}	y_{a2}	y_{an}	$y_{a.}$	$\bar{y}_{a\cdot}$
				$y_{..}$	$\bar{y}_{..}$

sum of squares is used to quantify all the variability in the response data. The equation for sum of squares total (SS_T) is:

$$SS_T = \sum_{i=1}^{a} \sum_{j=1}^{n} (y_{ij} - \bar{y}_{..})^2 \tag{11.5}$$

The so-called ANOVA decomposition involves dividing the total sum of squares into two parts: treatment and unexplained. The decomposition can be written:

$$SS_T = SSTR = SSE \tag{11.6}$$

where $SSTR$ is the sum of squares due to factors or treatments and SSE is sum of squares of the estimated errors. The specific definitions of both $SSTR$ and SSE are:

$$SSTR = n \sum_{i=1}^{a} (y_{i.} - \bar{y}_{..})^2 \tag{11.7}$$

$$SSE = \sum_{i=1}^{a} \sum_{j=1}^{n} (y_{ij} - \bar{y}_{i.})^2 \tag{11.8}$$

11

The value of this decomposition relates to the fact that, under the abovementioned assumptions, both $SSTR$ and SSE are relatable to the normally distributed noise. Specifically, if the treatment had no effect, then the following quantities should be approximately equal to each other and the variance of repeated responses. These are the mean of squares ($MSTR$) which is simply $SSTR/(a - 1)$ and the mean squared error (MSE) which is $SSE/(N - a)$. The denominators of these quantities are called degrees of freedom which have a geometric interpretation but which are simply numbers that aid in identifying the relevant critical characteristic value. Three different degrees of freedom are relevant: the first is between treatments $(a - 1)$, the second is error $(N - a)$, and the final is total $(N - 1)$. These are then used to find the F-statistic:

$$F_0 = \frac{\text{variance of group means}}{\text{means within group varience}} \tag{11.9}$$

$$= \frac{\text{MSTR}}{\text{MSE}} \tag{11.10}$$

Using an F table or = FINV() in excel we can find $F_{\alpha,\, a-1,\, N-a}$. If $F_0 > F_{\alpha,\, a-1,\, N-a}$ then we can conclude that there is a significant difference in the response means associated with at least two-factor levels. For extension of ANOVA to more than a single factor method please refer to ▶ Chap. 15 of this book.

Question

Below is a chart of data pertaining to the number of seconds a particular flash assembler took to create a single flashlight at different times during the day. Using ANOVA determine if the time of day affects average times, i.e., whether the means differ significantly between two or more times of day (Table 11.7).

Table 11.7 Data for flashlight assembler

Factor level	Observations					Totals	Averages
Time of day	1	2	3	4	5	y_i	$\bar{y}_{..}$
Morning	427	409	411	382	418	2047	409.4
Mid-morning	262	299	294	271	319	1445	289
Earlier afternoon	273	247	235	235	275	1265	253
Late afternoon	309	355	317	338	327	1646	329.2
						$y_{..}$ = 6403	$\bar{y}_{..} = 320.15$

Table 11.8 ANOVA for the flash light assembly study

Source of variance	Sum of squares	Degrees of freedom	Mean squares	F_0
Time of day	67634.55	3	22544.85	59.25059
Error	6088	16	380.5	
Total	73722.55	19		

Answer

Use Eqs. (11.5)–(11.10) to determine whether or not the averages are all statistically the same or different. Some equations were rearranged for convenience.

$$SS_T = \sum_{i=1}^{4}\sum_{j=1}^{5} y_{ij}^2 - \frac{\bar{y}_{..}^2}{N}$$

$$= (427)^2 + (409)^2 + \cdots + (327)^2 - \frac{6403^2}{20} = 73{,}722.55$$

$$SSTR = \frac{1}{n}\sum_{i=1}^{4} y_{i.}^2 - \frac{y_{..}^2}{N}$$

$$= \frac{1}{5}\left[(2047)^2 + \cdots + (1646)^2\right] - \frac{6403^2}{20} = 67{,}634.55$$

$$SS_T = SSTR + SSE$$

$$= 73722.55{-}67634.55 = 6088$$

We now set up these results in an ANOVA table, this is in Table 11.8.

Table 11.8 shows the relevant ANOVA table. Specifically, in deriving the sum or squares values, the degrees of freedom are 3, 16, and 19. Table 11.8 also shows the mean squares derived from the sums of squares divided by the degrees of freedom. F_0 is found by dividing mean squares for time of day by mean squared error. Next, we compute the upper tail of the F-distribution using $F_{0.05,3,16}$ using an F-table or excel to equal 3.24. Since $F_0 > F_{critical}$ we conclude that time of the day does significantly effect the average speed in which the assembler creates the flashlights with $\alpha = 0.05$.

11.7 Full Factorials

In previous sections, we described the methods of t-tests and ANOVA to determine whether changes in the level of a single factor significantly affect the average response values. Next, we focus on an experimental planning method involving multiple factors. The method of "full factorials" is one type of experimental design with intuitive appeal but potentially limited application. Assume there are m factors each with l_i levels, where $i = 1, \ldots, m$ is the factor index. A "full factorial" is a listing of all combinations of the m factors at each of the levels.

For example, consider the design of a paper air plane with m equals two factors. The first, wing length is of interest at $l_1 = 3$ levels. The second, paper type, is of interest at $l_1 = 2$ levels. Table 11.9 shows a full factorial repeated twice, i.e., with $r = 2$ replications. If the experiment were performed following the plan in Table 11.9, the first run would involve constructing an airplane with a wing length of 2 cm based on paper type 1. The second run would involve a wing length of 4 cm and use type 1 paper. In the next section, we describe reasons why performing experiments in standard or "not randomized" order is undesirable.

Even with a single replicate, the number of runs to perform a full factorial can be large. To better understand the relationship between levels, factors and numbers of runs for an experiment consider full factorials in which all factors have the same number of levels, i.e., $l_i = l$ for $i = 1, \ldots, m$. The number of runs is then given by: $n = l^k$. Table 11.10 shows the numbers of runs needed for various m and l combinations from this equation.

Table 11.9 Replicated two factor full factorial

Run (not randomized order)	Wing length (cm)	Paper type
1	2	Type 1
2	4	Type 1
3	6	Type 1
4	2	Type 2
5	4	Type 2
6	6	Type 2
7	2	Type 1
8	4	Type 1
9	6	Type 1
10	2	Type 2
11	4	Type 2
12	6	Type 2

Table 11.10 Runs needed based in number of levels and factors

Number of factors (m)	Number of levels ($l_i = l$)	Number of runs needed (n)
2	2	4
2	4	16
3	2	8
3	4	64
4	2	16
4	4	256
5	2	32
5	4	1,024
6	2	64
6	4	4,096
7	2	128
8	2	256
8	6	1,679,616

Consider that 8 factors at 2 levels requires 256 runs in the experiment. This could be prohibitively costly and time consuming. Also, 8 factors at 6 levels requires 1,679,616 runs, which is likely prohibitively expensive for all but the easiest of experiments. Yet, as described in the theory of experimental design chapter (▶ Chap. 19), the statistical properties of full factorial designs are often undesirable.

If the goal is simply deriving a list of factors with significant effects, the fractional factorials in ▶ Chap. 12 likely offer an economical way to derive the same or similar results. If the goal is to predict average response values accurately as a function of continuous factor settings, then the response surface methods described in ▶ Chap. 13 might actually offer improved prediction accuracy compared with multilevel full factorials and at a fraction of the cost.

11.8 Randomization and Evidence

One activity is common to all of the applications of the design of experiments (DOE) methods in this book. This activity is "**randomization**" which is the allocation of blocking factor levels to runs in a random or unpatterned way in experimental planning. For example, the run order can be considered to be a blocking factor. The act of scrambling the run order is a common example of randomization. Also, the assignment of people and places to factor levels can be randomized.

Philosophically, the application of randomization is critical for proving that certain factor changes affect average response values of interest. Many experts would say that empirical proof is impossible without randomization. Data collection is called an "experiment" if randomization is used and an "observational study" if it is not. Further, many would say experiments are needed for "doing science" although science is also associated with physics-based modeling.

Note that attempts to control usually uncontrollable factors during experimentation can actually work against development of proof, because control can change the system so that proof derived (if any) pertains to a system that is different than the one of interest. Often, the process is aided through the creation of an experimental plan or table showing the levels of the factor and the blocking factors (if any). The use of a planning table is illustrated (poorly) in the next example.

Example 11.8.1 Poor Randomization and Waste

Question 1

Assume that the experimental designer and all testers are watching all trials related to Table 11.4. The goal of the new software is task time reduction. Which is correct and most complete?

a. The data can be used to prove the new software helps with $\alpha = 0.05$.
b. The theory that the people taking the test learned from watching others is roughly equally plausible to the theory that the new software helps.
c. The theory that women are simply better at the tasks than men is roughly equally plausible to the theory that the new software helps.
d. The tests would have been much more valuable if randomization had been used.
e. All of the above are correct except (a).

Answer 1

The experimental plan has multiple problems. The run order is not randomized so learning effects could be causing the observed variation. The assignment of people to levels is not randomized so that gender issues might be causing the variation. The test was run in an unblind fashion, so knowledge of the participants could bias the results. Therefore, the correct answer is (e) (Table 11.11).

Table 11.11 Hypothetical example in which randomization is not used

Run (blocking factor)	Software	Tester (blocking factor)	Average time per task
1	Old	Jim	45.2
2	Old	Harry	38.1
3	Old	George	32.4
4	New	Sue	22.1
5	New	Sally	12.5
6	New	Mary	18.9

Question 2

Which is correct and most complete?

a. Except for randomization issues, t-testing analysis could be reasonably used.
b. $t_0 = 4.45$ for the two sample analysis of software, assuming unequal variances.
c. The experiment would be "blind" if the testers did not know which software they were using and could not watch the other trials.
d. All of the above are correct.
e. None of the above is correct.

Answer 2

Often, in experimentation using t-testing, there are blocking factors that should be considered in planning and yet the t-testing analysis is appropriate. Also, the definition of blind is expressed in part (c). Therefore, the answer is (d).

11.9 Errors from DOE Procedures

Investing in experimentation of any type is intrinsically risky. This follows because if the results were known in advance, experimentation would be unnecessary. Even through competent application of the methods in this book, errors of various types will occur. Probability theory can be used to predict the chances and/or magnitudes of different errors as described in ► Chap. 19. The theory can also aid in the comparison of method alternatives.

In this section, concepts associated with errors in testing hypotheses are described which are relevant to many design of experiments methods. These concepts are helpful for competent application and interpretation of results. They are also helpful for appreciating the benefits associated with standard screening using fractional factorials.

Table 11.12 defines Type I and Type II errors. The definition of these errors involves the concepts of a "true" difference and absence of the true difference in the natural system being studied. Typically, this difference relates to alternative averages in response values corresponding to alternative levels of an input factor. In real situations, the truth is generally unknown. Therefore, Type I and Type II errors are defined in relation to a theoretical construct. In each hypothesis test, the test either results in a declaration of significance or failure to find significance.

Failure to find significance when no difference exists is only a "semi-success" because the declaration is indefinite. Implied in the declaration is that with more runs or slightly different levels, a difference might be found. Therefore, the declaration in the case of no true difference is not as desirable as it could be.

As noted previously, theory can provide some indication of how likely Type I and Type II errors are in different situations. Intuitively, for two-sample t-testing, the chance of errors depend on all of the following:

a. The sample sizes used, n_1 and n_2.
b. The α used in the analysis of results.
c. The magnitude of the actual difference in the system (if any).
d. The sizes of the random errors that influence the test data (caused by uncontrolled factors).

Table 11.12 Definitions of Type I and Type II errors

		Nature or truth	
		No difference exists	Difference exists
Declaration	Significance is found	Type I error	Success
	Failure to find	Semi-success	Type II error

Like many testing procedures, the two-sample t-test method is designed to have the following property. For testing with chosen parameter α and any sample sizes, the chance of Type I error equals α. In one popular "frequentist" philosophy, this can be interpreted in the following way. If a large number of applications occurred, Type I errors would occur in α fraction of these cases. However, the chance of a Type I equaling α is only precisely accurate for specific assumptions about the random errors described in ▶ Chap. 19.

Therefore, fixing α determines the chance of Type I errors. At the same time, the chance of a Type II error can, in general, be reduced through increasing the sample sizes. Also, the larger the difference of interest, the smaller the chance of Type II error. In other words, if the tester is looking for large differences only, the chance of missing these distances and making a Type II error is small, in general.

Example 11.9.1 Testing a New Drug

11

Question

An inventor is interested in testing a new blood pressure medication that she believes will decrease average diastolic pressure by 5 mm Hg. She is required by the FDA to use $\alpha = 0.05$. What advice can you give her?

a. Use a smaller α; the FDA will accept it, and the Type II error chance is lower.
b. Budgeting for the maximum possible sample size will likely help prove her case.
c. She has a larger chance of finding a smaller difference.
d. All of the above are correct.
e. All are correct except (b) and (d).

Answer

As noted previously, if something is proven using any given α, it is also proven with all higher levels of α. Therefore, the FDA would accept proof with a lower level of α. However, generally proving something for a lower α implies an increased chance of Type II error. Generally, the more data, the more chance of proving something that is true. Also, finding smaller differences is generally less likely. Therefore, the correct answer is (b).

11.10 Chapter Summary

This chapter has provided an overview of the design of experiments (DOE) methods in this book. To simplify, fractional factorial methods are useful for screening to find which of many factors significantly affect average response values with only a fraction of the cost of full factorials. Fractional factorials are covered in detail in ▶ Chap. 12. Response surface methods (RSM) are useful for developing relatively accurate surface predictions, including predicting so-called interactions or combined effects of factors on average responses (▶ Chap. 13). Sequential RSM offer a potential advantage in economy in that possibly fewer runs will be used. Robust design methods address the variation of uncontrollable factors and deliver relatively trustworthy system design recommendations (▶ Chap. 14).

The method of t-testing was presented in part to clarify what randomization is and why it matters. Also, t-testing was used to illustrate the use of information to support method related decision-making, e.g., about how many test runs to do at each level. Theoretical information is presented to clarify the chances of different types of errors as a function of method design choices.

Finally, the concept of randomization (a general practice), randomization testing (a specific nonparametric alternative method to standard t-testing) and ANOVA are described. Randomization and ANOVA are relevant to all of the design of the experiments methods in this book. Randomization involves a careful step of planning experimentation that is critical for achieving proof and high levels of evidence. ANOVA is a set of methods including the specific methods covered here that compare means using variance with a controllable Type I error rate.

The following example illustrates how specific key input variables and DOE methods can be related to real-world problems.

Example 11.10.1 Student Retention Study

Question 1

Suppose that you are given a $4,000,000 grant over three years to study retention of students in engineering colleges by the Ohio State Board of Regents. The goal is to provide proven methods that can help increase retention, i.e., cause a higher fraction of students who start as freshmen to graduate in engineering. Describe one possible parameterization of your engineered system including the names of relevant factors and responses.

Answer 1

The system boundaries only include the parts of the Ohio public university network that the Board of Regents could directly influence. These regents can control factors including: (1) the teaching load per faculty (3–7 course per year), (2) the incentives for faculty promotion (weighted towards teaching or research), (3) the class size (relatively small or large), (4) the curriculum taught (standard or hands-on), (5) the level of student services (current or supplemented), and (6) the incentives to honors students at each of the public campuses (current or augmented). Responses of interest include total revenues per college, retention rates of students at various levels, and student satisfaction ratings as expressed through surveys.

Question 2

With regard to the student retention example, how might you spend the money to develop the proof you need?

Answer 2

Changing university policies is expensive. Expenses would be incurred through additions to students' services for selected groups, summer salary for faculty to participate, and additional awards to honors students. Because of the costs, benchmarking and regression analyses techniques applied, using easily obtainable data would be relevant. Still, without randomized experimentation, proof of cause and effect relationships relevant to Ohio realities may be regarded as impossible. Therefore, I would use the bulk of the money to perturb the existing policies. I would begin by dividing the freshman students in colleges across the state into units of approximately the same size in such a way that different units would naturally have minimal interaction. Then, I would assign combinations of the above factor levels to the student groups using random numbers to apply standard screening using fractional factorials with twelve runs ($n = 12$). I would evaluate the responses each year associated with the affected student groups applying the fractional factorial analysis method. As soon as effects appeared significant, I would initiate two-sample t-tests of the recommended settings vs the current using additional groups of students to confirm the results, assuming the remaining budget permits. The fractional factorial and added confirmation runs would likely consume several million dollars. However, it seems likely that the findings would pay for themselves, because the state losses per year associated with poor retention have been estimated in the tens of millions of dollars. These costs do not include additional losses associated with university ratings stemming from poor retention.

11.11 Problems

1. Consider applying DOE to improving your personal health. Which of the following is correct and most complete?
   a. Input factors might include weight, blood pressure, and happiness score.
   b. Output responses might include weight, blood pressure, and happiness score.
   c. Randomly selecting daily walking amount each week could generate proof.
   d. Walking two months 30 min daily followed by two months off can yield proof.
   e. Answers to parts (a) and (d) are both correct.
   f. Answers to parts (b) and (c) are both correct.
2. Which is a benefit of DOE in helping to add definitiveness in design decision-making?
   a. Engineers feel more motivated because proof is not needed for changes.
   b. Tooling costs are reduced since dies must be designed and built only once.
   c. Carefully planned input patterns do not support authoritative proof.
   d. Documentation becomes more difficult since there is no moving target.
   e. Quality likely improves because randomization does not generate rigorous proof.

3. Based on ▶ Chap. 1, which of the following is correct and most complete?
   a. Taguchi contributed to robust design and Box co-invented FF and RSM.
   b. Deming invented FF and invented RSM.
   c. Shewhart invented ANOVA.
   d. All of the above are correct.
   e. All of the above are correct except (a) and (d).
4. Which of the following is correct and most complete?
   a. FF is helpful for finding which factors matter from a long list with little cost.
   b. RSM helps in fine tuning a small number of factor settings.
   c. Robust engineering helps in that it is relatively likely to generate trustworthy settings.
   d. All of the above are relevant advantages.
   e. All of the above are correct except (a) and (d).

Data from ◘ Fig. 11.3 will be used for Questions 5 and 6.

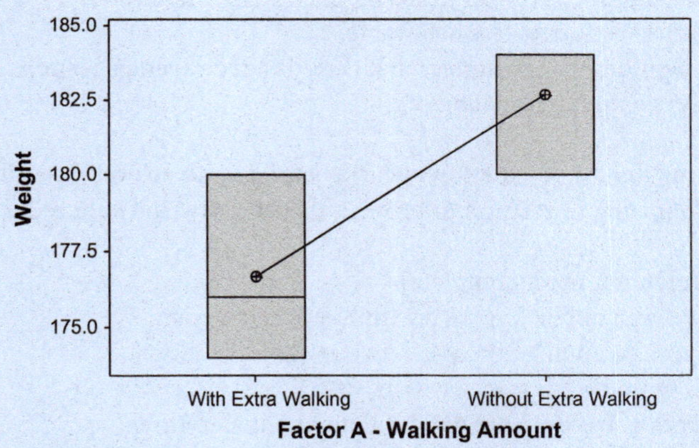

Run	Weight	Factor A – Walking amount
1	184	Without extra walking
2	180	With extra walking
3	184	Without extra walking
4	176	With extra walking
5	174	With extra walking
6	180	Without extra walking

◘ **Fig. 11.3** Data and Minitab® box and whisker plot for weight loss example

5. Which of the following is correct and most complete?
   a. $t_0 = 3.7$, which is ">" the relevant critical value, but nothing is proven.
   b. $t_0 = 2.7$, we fail to find significance with $\alpha = 0.05$, and the plot offers nothing.
   c. $t_0 = 2.7$, which is significant with $\alpha = 0.05$, we can claim proof walking helps.
   d. The run ordering is not random enough for establishing proof.
   e. All of the above are correct except (a).
6. Calculate the degrees of freedom (df) using data from the above example.
7. Consider the Second Two-sample t-test Example 11.4.1 of this chapter. Assume that no more tests were possible. Which is correct and most complete?
   a. Randomly assigning people to treatments *and* run order is essential for proof.
   b. It is likely true that failing to find significance would have been undesirable.
   c. The test statistic indicates a real difference larger than the noise is present.
   d. Significance would also have been found for any value of α larger than 0.05.
   e. Answers to parts (a) and (d) are both correct.
   f. All of the above answers are correct.
8. Assume you t-test with $n_1 = n_2 = 4$. Which is correct and most complete?
   a. Using $n_1 = n_2 = 3$ would likely reduce the chance of Type I and Type II errors.
   b. The chance of finding significance can be estimated using theory.
   c. Random assignment of run ordering makes error rate (both Type I and Type II probabilities) estimates less believable.
   d. Finding significance guarantees that there is a true average response difference.
   e. All answers except (d) are correct.

11

Use the following design of experiments array and data to answer Questions 9 and 10. Consider the following in relation to proving that the new software reduces task times (Table 11.13).

9. Which is correct and most complete?
   a. The above is an application of a within-subjects design.
   b. The above is an application of a between-subjects design.
   c. One fails to find significance with $\alpha = 0.05$.
   d. The degrees of freedom are greater than or equal to three.
   e. All of the above are correct.
   f. All of the above are correct except (a) and (e).

Table 11.13 Software testing data

Run	Software	Tester	Average time per task
1	Old	Mary	45.2
2	New	Harry	38.1
3	Old	George	32.4
4	New	Sue	22.1
5	New	Sally	12.5
6	Old	Phillip	18.9

10. When specimens are heterogeneous and paired, which is correct?
    a. randomization testing likely offers the lowest Type II error rates.
    b. paired t-testing likely offers the lowest Type II error.
    c. you cannot perform any type of test.
    d. Answers to parts (a) and (c) are both correct.
    e. Answers to parts (b) and (c) are both correct.
11. Randomization testing is often preferred because it can be done without
    a. Assumptions about distributions.
    b. Response data.
    c. Randomizing the run order.
    d. Estimating mean differences.
    e. All the above are incorrect.
12. Which of the following is part of an assumption relevant to ANOVA:
    a. Variance of populations are equal.
    b. Independence of successive observations.
    c. Normally distributed random errors.
    d. Randomization.
    e. All of the above are relevant to the proper functioning of ANOVA.
13. Which test statistic does ANOVA use?
    a. T-test.
    b. Z-test.
    c. F-test.
    d. Answers to part (a) and (b) are both correct.
    e. Answers to part (b) and (c) are both correct.
14. A factor relates to a full factorial experimental design in which way?
    a. It is the measured output from the experimental run tests.
    b. Its level settings in the various runs are prescribed by the array.
    c. It is the replication number of the array.
    d. Answers to part (a) and (b) are both correct.
    e. Answer to part (b) and (c) are both correct.
15. A level relates to a full factorial experimental design in which way?
    a. It is the measured output from the experimental run tests.
    b. The settings that factors assume in the runs specified by the array.
    c. Parts of the experiment in which cost estimates are needed.
    d. It is the range of settings described by the array.
    e. There is no such thing as levels in full factorial experiments.
16. How many runs are needed to perform a full factorial experiment with 10 factors at 2 levels?
    a. 1012.
    b. 1024.
    c. 1020.
    d. 100.
    e. 102.

17. Which is correct and most complete?
    a. The blocking factor tester has been randomized over.
    b. $t_0 = 1.45$ for the two sample analysis of software assuming unequal variances.
    c. Harry, Sue, and Sally constitute the control group.
    d. All of the above are correct.
    e. All of the above are correct except (a) and (e).
18. Which is correct and most complete for t-testing or factor screening?
    a. In general, adding more runs to the plan increases many types of error rates.
    b. Type I errors in t-testing include the possibility of missing important factors.
    c. Type II errors in t-testing focus on the possibility of missing important factors.
    d. Standard t-testing can be used to prove the insignificance of factors.
    e. All of the above are correct.
    f. All of the above are correct except (a) and (e).
19. How does paired t-testing differ from two sample t-testing?
20. A bicycle repair shop has two bolt torque wrenches. A torque wrench is used to tighten bolts and clicks when the bolt is tighten to the required amount. One of the bicycle mechanics thinks that one of the torque wrenches (Wrench 2) is reading high on average. They tightened 10 different bolts with both wrenches varying the order in which the wrenches were applied in an unpatterned way. The resulting measured torques are in Table 11.14. Analyze the data using a paired t-test and state conclusions.
21. Find a real-world example where you can perform a paired or unpaired t-test and collect the proper data and perform the analysis.

Table 11.14 Torque data for Problem 20

Torque numbers		
Specimen	Wrench 1 (NM)	Wrench 2 (NM)
1	90	97
2	85	100
3	76	80
4	15	35
5	12	18
6	35	45
7	56	67
8	89	95
9	34	39
10	5	9

22. A car dealership wants to investigate if the size of car effects the length of a lease in years. They collect the following data from their lease department. Use ANOVA to determine if they size of car significantly affects lease time.

Factor level	Observations					Totals	Averages
Size of car	1	2	3	4	5	y_i	$\bar{y}$
Compact	3	2	4	3	2	14	2.8
Mid-size	4	2	3	5	2	16	3.2
Full size	3	4	2	2	3	14	2.8
SUV/Truck	4	3	2	3	3	15	3.0
						$y_{..} = 59$	$\bar{y}_{..} = 2.95$

23. Find a situation in your life in which ANOVA could be used and collect the data then perform ANOVA. If possible, apply randomization. State your conclusions mentioning whether or not results are proven.
24. Plan a hypothetical full factorial experiment involving at least $m=3$ factors and $r=2$ replicates.
25. In your own words define ANOVA and describe when it is applicable.

References

Montgomery DC (2005) Design and analysis of experiments, 6th edn. Wiley

US Patent 4065537 (US Patent Issued on December 27, 1977) Process for producing molded golf balls exhibiting isometric compression. United States Government

DOE: Screening Using Fractional Factorials

T. T. Allen, *Introduction to Engineering Statistics and Lean Six Sigma*,
https://doi.org/10.1007/978-1-4471-7420-2_12

12.1 Introduction

The methods presented in this chapter are primarily relevant when it is desired to determine simultaneously which of many possible changes in system inputs cause average outputs to change. "**Factor screening**" is the process of starting with a long list of possibly influential factors and ending with a usually smaller list of factors believed to affect the average response. More specifically, the methods described in this section permit the simultaneous screening of several (m) factors using a number of runs, n, comparable to but greater than the number of factors ($n \sim m$ and $n > m$).

The methods described here are called "standard screening using fractional factorials" because they are based on the widely used experimental plans proposed by Fisher (1925) and in Plackett and Burman (1946) and Box and Hunter (1961a, b). The term "prototype" refers to a combination of factor levels because each run often involves building a new or prototype system. The experimental plans are called fractional factorials because they are based on building only a fraction of the prototypes that would constitute all combinations of levels for all factors of interest (a full factorial). The analysis methods used were proposed in Lenth (1989) and Ye et al. (2001).

Compared with multiple applications of two-sample t-tests, one for each factor, the standard screening methods based on fractional factorials offer relatively desirable Type I and Type II errors. This assumes that comparable total experimental costs were incurred using the "one-factor-at-a-time" (OFAT) two-sample t-test applications and the standard screening using fractional factorial methods. It also requires additional assumptions that are described in the "decision support" section below. Therefore, the guarantees associated with two-sample t-tests require fewer and less complicated assumptions.

12.2 Standard Screening Using Fractional Factorials

Screening methods are characterized by (1) an "experimental design", **D**, (2) a parameter α used in analysis, (3) vectors that specify the highs, **H**, and lows, **L**, of each factor, and (4) the choice of so-called error rate. The experimental design, **D**, specifies indirectly which prototype systems should be built. The design, **D**, only indirectly specifies the prototypes because it must be scaled or transformed using **H** and **L** to describe the prototypes in an unambiguous fashion that people who are not familiar with DOE methods can understand (see below).

Specifying **D** involves determining the number of "test run" prototypes systems to be built and tested, n, and the number of factors or inputs, m, that are varied, i.e., the dimensions that distinguish specific prototypes. The screening parameter α corresponds to the α in t-testing, i.e., the probability under standard assumptions that significance of at least one factor will be found significant when in actuality no factor influences the output.

The choices for error rate are either the so-called individual error rate (IER) first suggested by Lenth (1989) or the experimentwise error rate (EER) proposed by Ye et al. (2001). The use of IER can be regarded as "liberal" because it implies that the probability of wrongly declaring that at least one factor is significant when actually none has any influence is substantially higher than α. The relatively "conservative" EER guarantees that if no factors influence the response, the probability that one or more is declared significant is approximately α under standard assumptions described in ► Chap. 19. The benefit of IER is the higher "power" or probability of identifying significant factors that do have an effect. Also, sometimes experimenters want to quote some level of proof when proof using the EER is not possible.

Note that Lenth (1989) wrote his method in terms of "effects," which are twice the regression coefficients used in the method below. However, the associated factor of two cancels out so the results are identical with respect to significance.

The method in Algorithm 12.1 is given in terms of only a single response. Often, many responses are measured in *Step 3* for the same prototypes, with the prototypes built from the specifications in the array. Then, *Step 4–7* can be iterated for each of the responses considered critical. Then also, optimization in *Step 6* would make use of the multiple prediction models and evidence relating to any factors and responses might be judged to support performing additional experiments, e.g., using response surface methods (see ► Chap. 13).

Note that the standard screening method generally involves testing only a small fraction of the possible level combinations. Therefore, it is not surprising that the decision-making in *Step 8* is not based on picking the best combination of settings from among the small number tested. Instead, it is based on the prediction model in the main effects plots. Note also that *Step 6* is written in terms of the coefficients, $\beta_{est,2}$, …, $\beta_{est,n}$. Lenth (1989) proposed his analysis in terms of the "effect estimates" that are two times the coefficients. Therefore, the PSE in his definition was also twice what is given. The final significance judgments are unchanged, and the factor of two was omitted for simplicity here.

Algorithm 12.1 Standard screening using fractional factorials

Pre-step. Define the factors and ranges, i.e., the highs, **H**, and lows, **L**, for all factors.

Step 1. Form your experimental array by selecting the first m columns of the array (starting from the left-hand column) in the table below with the selected number of runs n. The remaining $n - m - 1$ columns are unused.

Step 2. For each factor, if it is continuous, scale the experimental design using the ranges selected by the experimenter. $D^s_{i,j} = L_j + 0.5(H_j - L_j)(D_{i,j} + 1)$ for $i = 1, \ldots, n$ and $j = 1, \ldots, m$. Otherwise, if it is categorical simply assign the two levels, the one associated with "low" to −1 and the level with "high" to +1.

Step 3. Build and test the prototypes according to $\mathbf{D}^s$. Record the test measurements for the responses from the n runs in the n dimensional vector **Y**.

Step 4. Form the so-called "design" matrix by adding a column of 1s, **1**, to the left hand side of the entire $n \times (n - 1)$ selected design **D**, i.e., $\mathbf{X} = (\mathbf{1}|\mathbf{D})$. Then, for each of the q responses calculate the regression coefficients $\boldsymbol{\beta}_{est} = \mathbf{AY}$, where **A** is the $(\mathbf{X}'\mathbf{X})^{-1}\mathbf{X}'$ (see the tables below for pre-computed **A**). Always use the same **A** matrix regardless of the number of factors and the ranges.

Step 5. *(Optional)* Plot the prediction model, $y_{est}(\mathbf{x})$, for prototype system output

$$y_{est}(\mathbf{X}) = \beta_{est,1} + \beta_{est,2}x_1 + \cdots + \beta_{est,m}x_m \tag{12.1}$$

as a function of x_j varied from −1 to 1 for $j = 1, \ldots, m$, with the other factors held constant at zero. These are called "**main effects plots**" and can be generated by standard software such as Minitab® or using Sagata® software. A high absolute value of the slope, $\beta_{est,j}$, provides some evidence that the factor, j, has an important effect on the average response in question.

Step 6. Calculate s_0 using

$$s_0 = \text{median}\{|\beta_{est,2}|, \ldots, |\beta_{est,n}|\} \tag{12.2}$$

where the symbols "||" stand for the absolute values. Let S be the set of non-negative numbers $|\beta_{est,2}|, \ldots, |\beta_{est,n}|$ in S with values less than $2.5s_0$ for $r = 1, \ldots, q$. Next, calculate

$$\text{PSE} = 1.5 \times \text{median}\{\text{numbers in } S\} \tag{12.3}$$

and

$$t_{Lenth,j} = |\beta_{est,j+1}|/\text{PSE} \quad \text{for } j = 1, \ldots, m. \tag{12.4}$$

Step 7. If $t_{Lenth,j} > t_{Lenth\ Critical,\alpha,n}$ given in Table 12.1, then declare that factor j has a significant effect for response for $j = 1, \ldots, m$. The critical values, $t_{Lenth\ critical,\alpha,n}$, were provided by Ye et al. (2001). The critical values are designed to control the experimentwise error rate (EER) and the less conservative individual error rate (IER).

Step 8. (Subjective system optimization) If one level has been shown to offer significantly better average performance for at least one criterion of interest, then use that information subjectively in your engineered system optimization. Otherwise, consider adding more data and/or take the fact that evidence does not exist that the level change helps into account in system design.

Table 12.1 Critical values for $t_{Lenth\ critical,\alpha,n}$: (a) EER and (b) IER

(a)				(b)			
n runs				*n* runs			
α	8	12	16	α	8	12	16
0.01	9.715	7.412	6.446	0.01	5.069	4.077	3.629
0.05	4.867	4.438	4.240	0.05	2.297	2.211	2.156
0.10	3.689	3.564	3.507	0.10	1.710	1.710	1.701

12

In general, extreme care should be given for the pre-step, i.e., the parameterization of the engineered system design problem. If the factors are varied over ranges containing only poor prototype system designs, then the information derived from the improvement system will likely be of little value. Also, it is common for engineers to select timidly ranges with the settings too close together. For example, varying wing length from 5 to 6 cm for a paper air plane would likely be a mistake. If the ranges are too narrow, then the method will fail to find significant differences. Also, the chance that good prototype designs are in the experimental region increases as the factor ranges increase.

Further, special names are given for cases in which human subjects constitute an integral part of the system generating the responses of interest. The phrase "**within subjects variable**" refers to a factor in an experiment in which a single subject or group is tested for all levels of that factor. For example, if tests all tests are performed by one person, then all factors are within subject variables.

The phrase "**between subject variables**" refers to factors for which a different group of subjects is used for each level in the experimental plan. For example if each test was performed by a different person, then all factors would be between subject variables. A "**within subjects design**" is an experimental plan involving only within subject variables and a "**between subjects design**" is a plan involving only between subject variables. This terminology is often used in human factors and biological experimentation and can be useful for looking up advanced analysis procedures.

◘ Figure 12.1 provides a practical worksheet following steps similar to the ones in the above method. The worksheet emphasizes the associated system design decision problem and de-emphasizes hypothesis testing. Considering that a single application of fractional factorials can constitute an entire quality project in some instances, it can make sense to write a problem statement or mini-project charter. Also, clarifying with some detail what is meant by key reponses and how they are measured is generally good practice.

Also, small differences shown on main effects plots can provide useful evidence about factors not declared significant. First, if the average differences are small, adjusting the level settings based on other considerations besides the average response might make sense, e.g., to save cost or reduce environmental impacts. Further, recent research suggests that Type II errors may be extremely common and that treating to even small differences on main effect plots (i.e., small "effects") as effective "proof" might be advisable.

Selecting the ranges and the number of runs can be viewed as a major part of the design of the "improvement system". Then, *Steps 2–8* are implementation of the improvement system to develop recommended inputs for the engineered system (◘ Tables 12.2, 12.3, 12.4, 12.5, 12.6, 12.7).

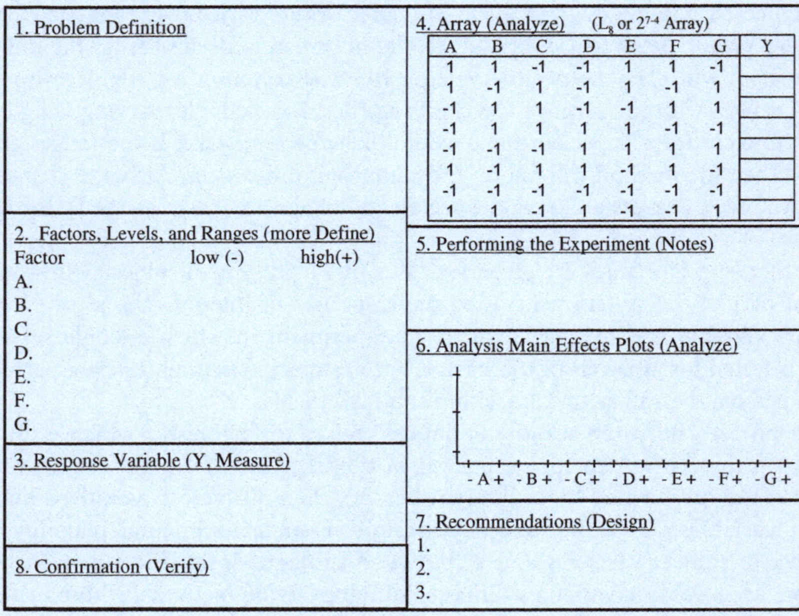

1. Problem Definition

2. Factors, Levels, and Ranges (more Define)

Factor	low (-)	high(+)
A.		
B.		
C.		
D.		
E.		
F.		
G.		

3. Response Variable (Y, Measure)

8. Confirmation (Verify)

4. Array (Analyze) (L_8 or 2^{7-4} Array)

A	B	C	D	E	F	G	Y
1	1	-1	-1	1	1	-1	
1	-1	1	-1	1	-1	1	
-1	1	1	-1	-1	1	1	
-1	1	1	1	1	-1	-1	
-1	-1	-1	1	1	1	1	
1	-1	1	1	-1	1	-1	
-1	-1	-1	-1	-1	-1	-1	
1	1	-1	1	-1	-1	1	

5. Performing the Experiment (Notes)

6. Analysis Main Effects Plots (Analyze)

- A + - B + - C + - D + - E + - F + - G +

7. Recommendations (Design)

1.
2.
3.

Fig. 12.1 Worksheet based on the eight run regular fractional factorial

12

Table 12.2 The n = 8 run regular fractional factorial array

Run	x_1	x_2	x_3	x_4	x_5	x_6	x_7
1	−1	1	−1	1	1	−1	−1
2	1	1	−1	−1	−1	−1	1
3	1	−1	−1	1	−1	1	−1
4	1	−1	1	−1	1	−1	−1
5	−1	1	1	−1	−1	1	−1
6	−1	−1	1	1	−1	−1	1
7	1	1	1	1	1	1	1
8	−1	−1	−1	−1	1	1	1

Table 12.3 (a) The design or **X** matrix and (b) $\mathbf{A} = (\mathbf{X}'\mathbf{X})^{-1}\mathbf{X}'$ for the eight run plan

(a)

1	−1	1	−1	1	1	−1	−1
1	1	1	−1	−1	−1	−1	1
1	1	−1	−1	1	−1	1	−1
1	1	−1	1	−1	1	−1	−1
1	−1	1	1	−1	−1	1	−1
1	−1	−1	1	1	−1	−1	1
1	1	1	1	1	1	1	1
1	−1	−1	−1	−1	1	1	1

(b)

0.125	0.125	0.125	0.125	0.125	0.125	0.125	0.125
−0.125	0.125	0.125	0.125	−0.125	−0.125	0.125	−0.125
0.125	0.125	−0.125	−0.125	0.125	−0.125	0.125	−0.125
−0.125	−0.125	−0.125	0.125	0.125	0.125	0.125	−0.125
0.125	−0.125	0.125	−0.125	−0.125	0.125	0.125	−0.125
0.125	−0.125	−0.125	0.125	−0.125	−0.125	0.125	0.125
−0.125	−0.125	0.125	−0.125	0.125	−0.125	0.125	0.125
−0.125	0.125	−0.125	−0.125	−0.125	0.125	0.125	0.125

Table 12.4 The $n = 12$ run Plackett Burman fractional factorial array

Run	x_1	x_2	x_3	x_4	x_5	x_6	x_7	x_8	x_9	x_{10}	x_{11}
1	1	1	−1	1	1	−1	1	−1	−1	−1	1
2	−1	1	1	1	−1	1	1	−1	1	−1	−1
3	1	1	1	−1	1	1	−1	1	−1	−1	−1
4	−1	1	−1	−1	−1	1	1	1	−1	1	1
5	1	−1	1	1	−1	1	−1	−1	−1	1	1
6	1	1	−1	1	−1	−1	−1	1	1	1	−1
7	−1	1	1	−1	1	−1	−1	−1	1	1	1
8	1	−1	1	−1	−1	−1	1	1	1	−1	1
9	−1	−1	−1	1	1	1	−1	1	1	−1	1
10	1	−1	−1	−1	1	1	1	−1	1	1	−1
11	−1	−1	−1	−1	−1	−1	−1	−1	−1	−1	−1
12	−1	−1	1	1	1	−1	1	1	−1	1	−1

Table 12.5 $A = (X'X)^{-1}X'$ for the 12 run plan

0.083	0.083	0.083	0.083	0.083	0.083	0.083	0.083	0.083	0.083	0.083	0.083
0.083	−0.083	0.083	0.083	−0.083	0.083	0.083	−0.083	−0.083	0.083	−0.083	−0.083
0.083	0.083	−0.083	0.083	0.083	−0.083	0.083	0.083	−0.083	−0.083	−0.083	−0.083
−0.083	0.083	0.083	0.083	−0.083	0.083	−0.083	0.083	−0.083	−0.083	−0.083	0.083
0.083	0.083	−0.083	−0.083	−0.083	0.083	0.083	−0.083	0.083	−0.083	−0.083	0.083
0.083	−0.083	−0.083	0.083	−0.083	−0.083	−0.083	0.083	0.083	0.083	−0.083	0.083
−0.083	0.083	−0.083	0.083	0.083	0.083	−0.083	−0.083	0.083	0.083	−0.083	−0.083
0.083	0.083	0.083	−0.083	0.083	−0.083	−0.083	−0.083	−0.083	0.083	−0.083	0.083
−0.083	−0.083	0.083	0.083	0.083	−0.083	0.083	−0.083	0.083	−0.083	−0.083	0.083
−0.083	0.083	0.083	−0.083	−0.083	−0.083	0.083	0.083	0.083	0.083	−0.083	−0.083
−0.083	−0.083	−0.083	−0.083	0.083	0.083	0.083	0.083	−0.083	0.083	−0.083	0.083
0.083	−0.083	0.083	−0.083	0.083	0.083	−0.083	0.083	0.083	−0.083	−0.083	−0.083

Table 12.6 The $n=16$ run regular fractional factorial array

Run	x_1	x_2	x_3	x_4	x_5	x_6	x_7	x_8	x_9	x_{10}	x_{11}	x_{12}	x_{13}	x_{14}	x_{15}
1	−1	1	1	−1	1	−1	1	1	−1	−1	−1	1	1	−1	−1
2	1	1	1	1	1	1	1	1	1	1	1	1	1	1	1
3	1	1	−1	−1	1	−1	−1	1	1	1	−1	−1	−1	−1	1
4	1	−1	1	1	−1	−1	−1	1	−1	−1	1	1	−1	−1	1
5	−1	−1	−1	1	−1	−1	1	1	1	1	1	−1	1	−1	−1
6	1	−1	−1	1	1	1	−1	−1	1	−1	−1	1	1	−1	−1
7	1	1	−1	1	−1	−1	1	−1	−1	1	−1	1	−1	1	−1
8	1	1	1	−1	−1	1	−1	−1	−1	1	1	−1	1	−1	−1
9	1	−1	1	−1	1	−1	1	−1	1	−1	1	−1	−1	1	−1
10	−1	−1	−1	−1	1	−1	−1	−1	−1	1	1	1	1	1	1
11	−1	1	1	1	−1	−1	−1	−1	1	−1	−1	−1	1	1	1
12	−1	−1	1	−1	−1	1	−1	1	1	1	−1	1	−1	1	−1
13	1	−1	−1	−1	−1	1	1	1	−1	−1	−1	−1	1	1	1
14	−1	1	−1	−1	−1	1	1	−1	1	−1	1	1	−1	−1	1
15	−1	−1	1	1	1	1	1	−1	−1	1	−1	−1	−1	−1	1
16	−1	1	−1	1	1	1	−1	1	−1	−1	1	−1	−1	1	−1

Table 12.7 A = (X′X)$^{-1}$X′ for the 16 run plan

A = 0.0625	1	1	1	1	1	1	1	1	1	1	1	1	1	1	1	1
	–1	1	1	1	–1	1	1	1	1	–1	–1	–1	1	–1	–1	–1
	1	1	1	–1	–1	–1	1	1	–1	–1	1	–1	–1	1	–1	1
	1	1	–1	1	–1	–1	–1	1	1	–1	1	1	–1	–1	1	–1
	–1	1	–1	1	1	1	1	–1	–1	–1	1	–1	–1	–1	1	1
	1	1	1	–1	–1	1	–1	–1	1	1	–1	–1	–1	–1	1	1
	–1	1	–1	–1	–1	1	–1	1	–1	–1	–1	1	1	1	1	1
	1	1	–1	–1	1	–1	1	–1	1	–1	–1	–1	1	1	1	–1
	1	1	1	1	1	–1	–1	–1	–1	–1	–1	1	1	–1	–1	1
	–1	1	1	–1	1	1	–1	–1	1	–1	1	1	–1	1	–1	–1
	–1	1	1	–1	1	–1	1	1	–1	1	–1	1	–1	–1	1	–1
	–1	1	–1	1	1	–1	–1	1	1	1	–1	–1	–1	1	–1	1
	1	1	–1	1	–1	1	1	–1	–1	1	–1	1	–1	1	–1	–1
	1	1	–1	–1	1	1	–1	1	–1	1	1	–1	1	–1	–1	–1
	–1	1	–1	–1	–1	–1	1	–1	1	1	1	1	1	–1	–1	1
	–1	1	1	1	–1	–1	–1	–1	–1	1	1	–1	1	1	1	–1

12.3 Screening Examples

The first example follows the printed circuit board study in Brady and Allen (2002). Note that, in the case study here, the experimentation was done "**on-line**" so that all the units found to conform to specifications in the experimentation were sold to customers. The decision support for choosing the experimental plan is included.

Pre-step. Here, let us assume that the result of "thought experiments" based on "entertained assumptions" was the informed choice of the $n=8$ run design including $m=4$ factors used in the actual study. For ranges, we have $\mathbf{L}=\{$low transistor output, screwed, 0.5 turns, current sink$\}'$ and $\mathbf{H}=\{$high transistor output, soldered, 1.0 turns, alternative sink$\}'$. The factors each came from different people with the last factor coming from rework line operators. Without the ability to study all factors with few runs, the fourth factor might have been dropped from consideration.

Algorithm 12.2 Circuit board example

Step 1. The experimental plan, **D**, was created by selecting the first four columns of the $n=8$ run experimental array above. See part (a) of ◘ Table 12.8. The remaining three columns are unused.

Step 2. All the factors are categorical except for the third factor, screw position, which is continuous. Assigning the factors produced the scaled design, $\mathbf{D}^s$, in ◘ Table 12.8 part (b).

Step 3. 350 units were made and tested based on each combination of process inputs in the experimental plan ($8 \times 350=2800$ units). The single prototype system response values are shown in ◘ Table 12.8 part (c), which are the fraction of the units that conformed to specifications. Note that, in this study, the fidelity of the prototype system was extremely high because perturbations of the engineered system created the prototype system.

Step 4. The relevant design matrix, **X**, and $\mathbf{A}=(\mathbf{X}'\mathbf{X})^{-1}\mathbf{X}'$ matrix are given in ◘ Table 12.3. The derived list of coefficients is (using $\boldsymbol{\beta}_{est}=\mathbf{AY}$):

$$\beta_{est} = \{82.9, -1.125, -0.975, -1.875, 9.825, 0.35, 1.85, 0.55\}'.$$

Step 5. (*Optional*) The prediction model, $y_{est}(\mathbf{x})$, for prototype system output is

$$y_{est}(\mathbf{x}) = 82.9 - 1.125x_1 - 0.975x_2 - 1.875x_3 + 9.825x_4. \quad \textbf{(12.5)}$$

The main effects plot is shown in ◘ Fig. 12.2.

Step 6. We calculated s_0 using

$$s_0 = \text{median}\{|\beta_{est,2}|, \ldots, |\beta_{est,8}|\} = 1.125 \quad \textbf{(12.6)}$$

The set *S* is {1.125, 0.975, 1.875, 0.55, 0.35, 1.85}. Next, calculate

$$\begin{aligned} \text{PSE} &= 1.5 \times \text{ median}\{\text{numbers in } S\} \\ &= (1.5)(1.05) \\ &= 1.58 \end{aligned} \tag{12.7}$$

and

$$\begin{aligned} t_{Lenth,j} &= |\beta_{est,j+1}|/\text{PSE} \\ &= 0.71,\ 0.62, 1.19, 6.24 \ \text{ for } j = 1, \ldots, 4 \text{ respectively.} \end{aligned} \tag{12.8}$$

Step 7. In this case, for many choices of IER versus EER and α, the conclusions about significance were the same. For example, with either $t_{critical} = t_{\text{IER},\alpha=0.1,n=8} = 1.710$ or $t_{critical} = t_{\text{EER},\alpha=0.05,n=8} = 4.876$, the fourth factor "heat sink" had a significant effect on average yield when varied with $\alpha = 0.05$ and using the relatively EER approach. Also, for both choices, the team failed to find significance for the other factors. They might have changed the average response but we could not detect it without more data.

Step 8. Subjectively, the team wanted to maximize the yield and heat sink had a significant effect. It was clear from the main effects plot and the hypothesis testing for heat sink that the high level (alternative heat sink) significantly improved the quality compared with the current heat sink. The team therefore suggested using the alternative heat sink because the material cost increases were negligible compared with the savings associated with yield increases. In fact, the first pass process yield increased to greater than 90% consistently from around 70%. This permitted the company to meet new demand without adding another rework line. The direct savings was estimated to be $2.5 million.

12

In evaluating the cost of poor quality, e.g., low yield, it was relevant to consider costs in addition to the direct cost of rework. This followed in part because production time variability from rework caused the need to quote high lead times to customers, resulting in lost sales.

Example 12.3.1 More Detailed Application

Question 1

Consider the example in Table 12.9. What are D, X, X′X, and A?

Answer 1

D is the entire matrix in Table 12.6 without the column for the runs. Using **X** = (**1**|**D**) one has

X =	1	–1	1	1	–1	1	–1	1	1	–1	–1	–1	1	1	–1	–1
	1	1	1	1	1	1	1	1	1	1	1	1	1	1	1	1
	1	1	1	–1	–1	1	–1	–1	1	1	1	–1	–1	–1	–1	1
	1	1	–1	1	1	–1	–1	–1	1	–1	–1	1	1	–1	–1	1
	1	–1	–1	–1	1	–1	–1	1	1	1	1	1	–1	1	–1	–1
	1	1	–1	–1	1	1	1	–1	–1	1	–1	–1	1	1	–1	–1
	1	1	1	–1	1	–1	–1	1	–1	–1	1	–1	1	–1	1	–1
	1	1	1	1	–1	–1	1	–1	–1	–1	1	1	–1	1	–1	–1
	1	1	–1	1	–1	1	–1	1	–1	1	–1	1	–1	–1	1	–1
	1	–1	–1	–1	–1	1	–1	–1	–1	–1	1	1	1	1	1	1
	1	–1	1	1	1	–1	–1	–1	–1	1	–1	–1	–1	1	1	1
	1	–1	–1	1	–1	–1	1	–1	1	1	1	–	1	–1	1	–1
	1	1	–1	–1	–1	–1	1	1	1	–1	–1	–1	–1	1	1	1
	1	–1	1	–1	–1	–1	1	1	–1	1	–1	1	1	–1	–1	1
	1	–1	–1	1	1	1	1	1	–1	–1	1	–1	–1	–1	–1	1
	1	–1	1	–1	1	1	1	–1	1	–1	–1	1	–1	–1	1	–1

$\mathbf{X}'\mathbf{X} = n \times \mathbf{I}$ (therefore the DOE matrix is "orthogonal"), and **A** is given by ◘ Table 12.7.

◘ **Table 12.8** (a) Design, **D**, (b) scaled design, $\mathbf{D}^s$, and (c) responses, % yield

(a)					(b)					(c)
Run	x_1	x_2	x_3	x_4	Run	x_1	x_2	$x_{3\ (\text{turns})}$	x_4	% Yield
1	–1	1	–1	1	1	Low trans. output	Soldered	0.5	Alternative sink	92.7
2	1	1	–1	–1	2	High trans. output	Soldered	0.5	Current sink	71.2
3	1	–1	–1	1	3	High trans. output	Screwed	0.5	Alternative sink	95.4
4	1	–1	1	–1	4	High trans. output	Screwed	1.0	Current sink	69.0
5	–1	1	1	–1	5	Low trans. output	Soldered	1.0	Current sink	72.3
6	–1	–1	1	1	6	Low trans. output	Screwed	1.0	Alternative sink	91.3
7	1	1	1	1	7	High trans. output	Soldered	1.0	Alternative sink	91.5
8	–1	–1	–1	–1	8	Low trans. output	Screwed	0.5	Current sink	79.8

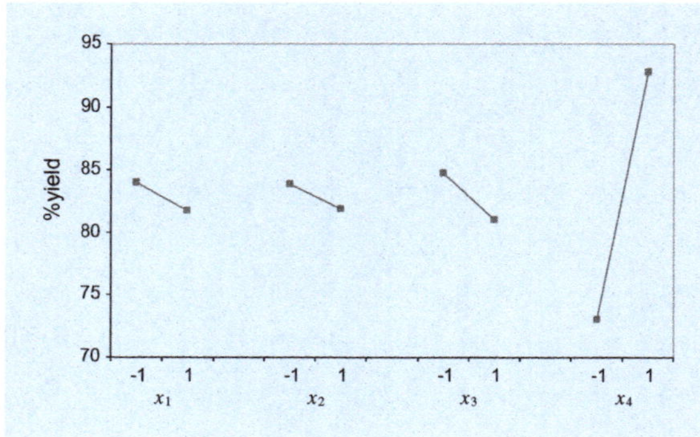

Fig. 12.2 Main effects plot for the printed circuit board example

Table 12.9 The DOE and estimated coefficients in a fictional study

Run (i)	$x_{i,1}$	$x_{i,2}$	$x_{i,3}$	$x_{i,4}$	$x_{i,5}$	Y_1		β_{est} (Coefficients)
1	1	1	−1	1	1	92	β_1 (Constant)	110.42
2	−1	1	1	1	−1	88	β_2 (factor x_1)	−0.58
3	1	1	1	−1	1	135	β_3 (factor x_2)	2.58
4	−1	1	−1	−1	−1	140	β_4 (factor x_3)	−1.42
5	1	−1	1	1	−1	79	β_5 (factor x_4)	−27.25
6	1	1	−1	1	−1	82	β_6 (factor x_5)	0.08
7	−1	1	1	−1	1	141	β_7	−0.42
8	1	−1	1	−1	−1	134	β_8	0.92
9	−1	−1	−1	1	1	81	β_9	−2.25
10	1	−1	−1	−1	1	137	β_{10}	0.08
11	−1	−1	−1	−1	−1	139	β_{11}	−1.08
12	−1	−1	1	1	1	77	β_{12}	0.75

12

Question 2

Analyze the data and draw conclusions about significance. Be standard and conservative (high standard of evidence and low Type I error rate or "α") in your choice of IER versus EER and α.

Answer 2

Lenth's method using the experimentwise error rate (EER) critical characteristic is a standard conservative approach for analyzing fractional factorial data. The individual error rate (IER) is less conservative in the Type I error rate is higher, but the Type II error rate is lower. The needed calculations are as follows:

$$\begin{aligned} s_0 &= \text{median}\{|\beta_{est,2}|, \ldots, |\beta_{est,12}|\} = 0.92, \\ S_1 &\text{ is } \{0.58, 1.42, 0.08, 0.42, 0.92, 2.25, 0.08, 1.08, 0.75\}, \text{ and} \end{aligned} \tag{12.9}$$

$$\begin{aligned} \text{PSE} &= 1.5 \times \text{median}\{\text{numbers in } S_1\} = (1.5) \times (0.75) = 1.125. \\ t_{Lenth,j} &= |\beta_{est,j+1}|/\text{PSE} = 0.516,\ 2.293,\ 1.262,\ 24.222, \text{ and } 0.071 \\ &\quad \text{for } j = 1, \ldots, 5 \text{ respectively.} \end{aligned}$$

Whether $\alpha = 0.01$ or $\alpha = 0.05$, the conclusions are the same in this problem because the critical values are 7.412 and 4.438 respectively. Tests based on both identify that factor 4 has a significant effect and fail to find that the other factors are significant. Factor into system design decision-making that factor 4 has a significant effect on the average response. Therefore, it might be worthwhile to pay more to adjust this factor.

Question 3

Draw a main effects plot and interpret it briefly.

Answer 3

The main effects plot shows the predictions of the regression model when each factor is varied from the low to the high setting, with the other factors held constant at zero. For example, the prediction when the factor x_2 is at the low level is 110.42 – 2.58 = 107.84. Figure 12.3 shows the plot. We can see that factor x_4 has a large negative effect and the other factors have small effects on the average response. If the costs of changing the factors were negligible, the plot would indicate which settings would likely increase or decrease the average response.

Question 4

Suppose that the high level of each factor was associated with a substantial per unit savings for the company, but that demand is assumed to be directly proportional to the customer rating, which is the response. Use the above information to make common-sense recommendations under the assumption that the company will not pay for any more experiments.

Answer 4

Since the high setting of factor x_4 is associated with a significant drop in average response and thus demand, it might not make sense to use that setting to stimulate demand. In the absence of additional information, however, the other factors fail to show any significant effect on average response and demand. Therefore, we tentatively recommend setting these factors at the high level to save cost.

Regular fractional factorials and Plackett Burman designs have a special property in the context of first order regression models. The predicted values for each setting of each factor plotted on the main effects plot are also the averages of the responses associated with that setting in the DOE. In the second example, the prediction when the factor x_2 is at the low level is 110.42 – 2.58 = 107.84. This value is also the average of the six responses when factor x_2 is at the low level.

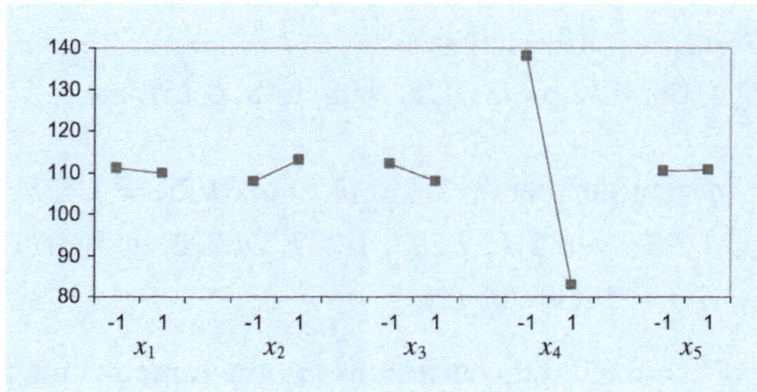

Fig. 12.3 Main effects plot for the fictional example

12.4 Method Origins and Alternatives

In this section, a brief explanation of the origins of the design arrays used in the standard fractional factorial methods is described. Also, some of the most popular alternative methods for planning experiments and analyzing the results are summarized.

12.4.1 Origins of the Arrays

12

It is possible that many researchers from many places in the world independently generated matrices similar to those used in standard screening using fractional factorials. Here, the focus is on the school of research started by the U.K. researcher Sir Ronald Fisher. In the associated terminology, "**full factorials**" are arrays of numbers that include all possible combinations of factor settings for a pre-specified number of levels. For example, a full factorial with three levels and five factors consists of all $3^5 = 243$ possible combinations. Sir Ronald Fisher generated certain fractional factorials by starting with full factorials and removing portions to create half, quarter, eighth, and other fractions.

Box and Hunter (1961a, b) divided fractional factorials into "regular" and "irregular" designs. "**Regular fractional factorials**" are experimental planning matrices that are fractions of full factorials having all of the following property. All columns in the matrix can be formed by multiplying other columns. Irregular designs are all arrays without the above-mentioned multiplicative property. For example, in Table 12.10 (a), it can be verified that column A is equal to the product of columns B times C. Regular designs are only available with numbers of runs given by $n = 2^p$, where p is a whole number. Therefore, possible n equal 4, 8, 16, 32, ….

Consider the three factor full factorial and the regular fractions in Table 12.10. The ordering of the runs in the experimental plan in Table 12.10 (a) suggests one way to generate full factorials by alternating −1s and 1s at different rates for different columns. Note that the experimental plan is not provided in randomized order and

Table 12.10 (a) Full factorial, (b) half fraction, and (c) quarter fraction

(a)				(b)				(c)			
SO	A	B	C	SO	A	B	C	SO	A	B	C
1	−1	−1	−1	1	−1	−1	1	1	−1	−1	1
2	1	−1	−1	2	1	−1	−1	2	1	−1	−1
3	−1	1	−1	3	−1	1	−1				
4	1	1	−1	4	1	1	1				
5	−1	−1	1								
6	1	−1	1								
7	−1	1	1								
8	1	1	1								

should not be used for experimentation in the order given. The phrase "**standard order**" (SO) refers to the not-randomized order presented in the tables.

A "**generator**" is a property of a specific regular fractional factorial array showing how one or more columns may be obtained by multiplying together other columns. For example, Table 12.10 (b) and (c) show selection of runs from the full factorial with a specific property. The entry in column (c) is the product of the entries in columns (a) and (b) giving the generator, (c) = (a)(b). The phrase "**defining relations**" refers a set of generators that are sufficiently complete as to uniquely identify a regular fractional factorial in standard order.

Plackett and Burman (1946) invented a set of alternative irregular fractional factorial matrices available for numbers of runs that are multiples of 4. The Plackett Burman (PB) design was generated using cyclic repetition of a single series. For example, consider the experimental plan used in Table 12.4 and provided in standard order in Table 12.11. In this case, the generation sequence 1, 1, −1, 1, 1, −1, 1, −1, −1, −1, 1 can be used to fill in the entire table. Each successive row is the repetition of this sequence staggered by one each column. The last row is filled in by a row of −1s.

In general, statisticians consider regular designs as preferable to PB designs. Therefore, these designs are recommended for cases in which they are available. This explains why the regular design with eight runs instead of the PB design was provided in Table 12.2. It is interesting to note that computers did not play a key role in generating the regular fractional factorials and Plackett Burman (PB) designs. The creators developed a simple generation approach and verified that the results had apparently desirable statistical properties. Allen and Bernshteyn (2003) and other similar research has used an extremely computational approach to generate new experimental arrays with potentially more desirable properties.

Table 12.11 A Plackett Burman fractional factorial array not in randomized order

Standard order	x_1	x_2	x_3	x_4	x_5	x_6	x_7	x_8	x_9	x_{10}	x_{11}
1	1	1	−1	1	1	−1	1	−1	−1	−1	1
2	1	1	1	−1	1	1	−1	1	−1	−1	−1
3	−1	1	1	1	−1	1	1	−1	1	−1	−1
4	−1	−1	1	1	1	−1	1	1	−1	1	−1
5	−1	−1	−1	1	1	1	−1	1	1	−1	1
6	1	−1	−1	−1	1	1	1	−1	1	1	−1
7	−1	1	−1	−1	−1	1	1	1	−1	1	1
8	1	−1	1	−1	−1	−1	1	1	1	−1	1
9	1	1	−1	1	−1	−1	−1	1	1	1	−1
10	−1	1	1	−1	1	−1	−1	−1	1	1	1
11	1	−1	1	1	−1	1	−1	−1	−1	1	1
12	−1	−1	−1	−1	−1	−1	−1	−1	−1	−1	−1

Example 12.4.2 Experimental Design Generation

Question

Which of the following is correct and most complete?

a. PB generation sequences were chosen carefully to achieve desirable properties.
b. Some columns in regular fractional factorials are not multiples of other columns.
c. Regular fractional factorials are not "orthogonal" because **X′X** is not diagonal.
d. All of the above are correct.
e. All of the above are correct except (a) and (d).

Answer

Plackett and Burman considered many possible sequences and picked the one that achieved desirable properties such as orthogonality. For regular designs, all columns can be achieved as products of other columns, and **X′X** is diagonal for assumptions in this chapter. Therefore, the correct answer is (a).

12.4.2 Alternatives to the Methods in This Chapter

It would probably be more standard to determine significance using a subjective approach based on normal probability plots (see ▸ Chap. 15) or half normal plots (applied to the coefficient estimates and not the residuals). These approaches were proposed by Daniel (1959). The plot-based approaches have the advantage of potentially incorporating personal intuition and engineering judgment into questions about significance. These approaches reveal that the resulting hypothesis tests are based on a higher level of assumption-making and a lower level of evidence than two-sample t-tests.

One disadvantage of probability plot-based analysis is that the subjectivity complicates analysis of screening methods since simulation (see ▶ Chap. 10) of the associated improvement system is difficult or impossible. Also, with the normal probability plots (but not the half normal plots) students can become confused and declare factors significant that have smaller coefficients than other factors that are not declared to be significant.

Another relevant analysis method is so-called "Analysis of Variance" followed by multiple t-tests. This method is described at the end of the chapter. The main benefits of Lenth's method and probability plots are that, under standard assumptions, they have a higher chance of finding significance. The Analysis of Variance method is more conservative and can lead to misleading estimates of Type I and Type II errors.

Also, for reference, the $n=8$ run and $n=16$ run designs in the above plots stem from Box and Hunter (1961a, b) and are called "regular fractional factorials". Regular fractional factorials have the property that all columns can be obtained as the product of some combination of other columns. Researchers call this property the "existence of a defining relation".

The $n=12$ run design does not have this property and is therefore not regular. It is a so-called Plackett-Burman (PB) design because it was proposed in Plackett and Burman (1946). Placket-Burman designs also have the property that each row (except one) has precisely the same sequence of -1's and $+1$'s, except offset.

Table 12.12 The L_{18} orthogonal array used in Taguchi Methods

Run	x_1	x_2	x_3	x_4	x_5	x_6	x_7	x_8
1	2	3	1	3	2	3	1	2
2	1	1	2	2	2	2	2	2
3	1	3	1	2	1	3	2	3
4	2	3	3	2	1	2	3	1
5	1	2	2	2	3	3	1	1
6	1	2	3	3	1	1	2	2
7	2	2	2	3	1	2	1	3
8	1	3	2	3	2	1	3	1
9	2	1	1	3	3	2	2	1
10	2	1	3	2	2	1	1	3
11	1	3	3	1	3	2	1	2
12	2	1	2	1	1	3	3	2
13	1	1	3	3	3	3	3	3
14	1	1	1	1	1	1	1	1
15	2	2	1	2	3	1	3	2
16	1	2	1	1	2	2	3	3
17	2	2	3	1	2	3	2	1
18	2	3	2	1	3	1	2	3

If other regular fractional factorials or PB designs are applied, then the method could still be called "standard". Also, other irregular designs such as the ones discussed in ▶ Chap. 8 can be applied together with the methods from Lenth (1989).

The reader may wonder why only two-level experimental plans are incorporated into the standard screening using fraction factorial. The answer relates to the fact that two-level experimental plans generally offer the best method performance as evaluated by several criteria under reasonable assumptions. ▶ Chapter 8 contains additional relevant criteria and assumptions. Still, there are popular alternatives not based on two-level design including certain so-called Taguchi methods. One such method is based on taking columns from the L_{18}, "orthogonal array" in ◘ Table 12.12 and the following simple analysis methods. Plot the average response for each level of each factor. Then, connect the average responses on the plots. The resulting "marginal plots" roughly predict the response as a function of the inputs.

12.5 Standard Versus One-Factor-at-a-Time Experimentation

The term "standard" has been used to refer to standard screening using fractional factorials. However, other approaches are probably more "standard" or common. The phrase "one-factor-at-a-time" (OFAT) experimentation refers to the common practice of varying one factor over two levels, while holding other factors constant. After determining the importance of a single factor, focus shifts to the next factor. ◘ Table 12.13 (a) shows a standard fractional factorial design and data for a hypothetical example. ◘ Table 12.13 (b) and (c) show OFAT DOE plans for the same problem.

The application of the plan in ◘ Table 12.13 (b) clearly has an advantage in terms of experimental costs compared with the standard method in ◘ Table 12.13 (a). However, with no repeated runs, it would be difficult to assign any level of "proof" to the results and/or to estimate the chances of Type I or Type II errors.

The design in ◘ Table 12.13 (c) represents a relatively extreme attempt to achieve proof using an OFAT approach. Yet, performing two-sample t-test analyses after each set of four tests would likely result in undesirable outcomes. First, using $\alpha = 0.05$ for each test, the chance of at least a single Type I error would be roughly 20%. Advanced readers can use statistical independence to estimate an error rate of 18.6%. With only $n_1 = n_2 = 2$ runs, the chances of identifying effects using this approach are even lower than the probabilities in ◘ Table 18.3 in Chap. 18 with $n_1 = n_2 = 3$. In the next, section information about Type I and II errors suggests that standard screening using fractional factorial methods offers reduced error rates of both types.

The hypothetical response data in ◘ Table 12.13 (a) was generated from the equation $Y = 19 + 6x_1 + 9x_1x_3$ with "+1" random noise added to the first response only. Therefore, there is an effect of factor x_1 in the "true" model. It can be checked (see Problem 18 at the end of this chapter) that $t_{Lenth,1} = 27$ so that the first factor (x_1) is proven using Lenth's method with $\alpha = 0.01$ and the EER convention to affect the average response values significantly. Therefore, the combined effect or "interaction" represented by $9x_1x_3$ does not cause the procedure to fail to find that x_1 has a significant effect.

Table 12.13 (a) Fractional factorial example, (b) low cost OFAT, (c) multiple t-tests

(a)					(b)				(c)			
x_1	x_2	x_3	x_4	Y	x_1	x_2	x_3	x_4	x_1	x_2	x_3	x_4
−1	1	−1	1	24	1	−1	−1	1	1	−1	−1	−1
1	1	−1	−1	15	−1	1	−1	−1	−1	−1	−1	−1
1	−1	−1	1	15	−1	−1	1	−1	1	−1	−1	−1
1	−1	1	−1	33	−1	−1	−1	1	−1	−1	−1	−1
−1	1	1	−1	5					−1	1	−1	−1
−1	−1	1	1	5					−1	−1	−1	−1
1	1	1	1	33					−1	1	−1	−1
−1	−1	−1	−1	23					−1	−1	−1	−1
									−1	−1	1	−1
									−1	−1	−1	−1
									−1	−1	1	−1
									−1	−1	−1	−1
									−1	−1	−1	1
									−1	−1	−1	−1
									−1	−1	−1	1
									−1	−1	−1	−1

It would be inappropriate to apply two-sample t-testing analysis to the data in Table 12.13 (a) focusing on factor x_1. This follows because randomization was not applied with regard to the other factors. Instead, a structured, formal experimental plan was used for these. However, applying one-sided two-sample t-testing (see Problem 19 below) results in $t_0 = 1.3$, which is associated with a failure to find significance with $\alpha = 0.05$. This result provides anecdotal evidence that regular fractional factorial screening based on Lenth's method offers statistical power to find factor significance and avoid Type II errors. Addressing interactions and using all runs for each test evaluation helps in detecting even small effects.

Finally, it is intuitively plausible that standard fractional factorial methods perform poorly when many factors are likely important. This could occur if a relatively high number of factors is used ($n \sim m$) or if the factors are "smart" choices such that changing them does affect the responses.

Advanced readers will observe that Lenth's method is based on assuming that over one half of the relevant main effects and interactions have zero coefficients in the "true" model. If experimenters believe that a large fraction of the factors might have important effects, it can be reasonable to disregard hypothesis testing results and focus on main effects plots. Then, even OFAT approaches might be more reliable.

Example 12.5.1 Printed Circuit Board Related Method Choices

Question

Consider the first "printed circuit board" case study in this chapter. What advice could you provide for the team about errors?

Answer

The chance of false positives (Type I errors) are directly controlled by the selection of the critical parameter in the methods. With only four factors and eight runs, the chances of Type II errors are lower than those typically accepted by method users. Still, only rather large actual differences will likely be found significant unless a larger design of experiments matrix were used, e.g., $n = 12$ or $n = 16$.

12.6 Chapter Summary

A mixed presentation of so-called regular fractional factorials and Plackett Burman designs was presented in this chapter. At present, these DOE arrays are by far the most widely used design of experiments matrices. The Ye, Hamada, Wu modified version of Lenth's method is described as a method to perform simultaneous hypothesis tests on multiple factors. This method is standard enough to be incorporated into popular software such as Minitab®. Together the methods permit users to effectively screen which from a long list of factors, when changed, affects important system outputs of key output variables. There are also discussions of the origins of the fractional factorial experimental matrices and alternative methods. The most widely used design of experiments matrices derive from approaches that are not computational intensive.

12.7 Problems

1. Which is correct and most complete?
   a. Using FF, once an array is chosen, generally only the first m columns are used.
   b. Typically roughly half of the settings change from run to run in applying FF.
   c. Selecting the factors and levels is critical and should be done carefully.
   d. Main effects plots often clarify which factors matter and which do not.
   e. The approved approach for designing systems is to select the DOE array settings that gave the best seeming responses.
   f. All of the above are correct.
   g. All of the above are correct except (e) and (f).
2. Which is correct and most complete?
   a. Plackett Burman designs are not fractional factorials.
   b. Applying standard screening using fractional factorials can generate proof.
   c. A fractional factorial experiment cannot have both a Type I and a Type II error.
   d. All of the above are correct.
   e. All of the above are correct except (a) and (d).

3. Which is correct and most complete?
   a. Adding factors in applying FFs almost always requires additional runs.
   b. Using matrices with smaller numbers of runs helps reduce error rates.
   c. Using the smallest matrix with enough columns is often reasonable for starting.
   d. It is critical to understand where the matrices came from to gain benefits.
   e. All of the above are correct.
   f. All of the above are correct except (a) and (d).
4. Which is correct and most complete?
   a. The EER gives higher critical values than IER and a higher evidence standard.
   b. If significance is not found using IER, it will be found using the EER.
   c. The IER can be useful because its associated higher standard of evidence might still be useful.
   d. All of the above are correct.
   e. All of the above are correct except (a) and (d).

Table 12.14 will be used for Questions 5–7.

5. Which of the following is correct and most complete based on Table 12.14?
   a. There are five factors, and the most standard, conservative analysis uses EER.
   b. Even if four factors had been used, the same **A** matrix would be applied.
   c. The matrix used is part of a matrix that can handle as many as 11 factors.
   d. All of the above are correct.
   e. All of the above are correct except (a) and (d).

Table 12.14 Outputs from a hypothetical fractional factorial application

Run (i)	$x_{i,1}$	$x_{i,2}$	$x_{i,3}$	$x_{i,4}$	$x_{i,5}$	Y_1		β_{est} (Coefficients)
1	1	1	–1	1	1	45	β_1 (Constant)	58.8
2	–1	1	1	1	–1	75	β_2 (factor x_1)	0.4
3	1	1	1	–1	1	80	β_3 (factor x_2)	0.4
4	–1	1	–1	–1	–1	40	β_4 (factor x_3)	16.3
5	1	–1	1	1	–1	75	β_5 (factor x_4)	2.1
6	1	1	–1	1	–1	45	β_6 (factor x_5)	1.3
7	–1	1	1	–1	1	70	β_7	0.4
8	1	–1	1	–1	–1	70	β_8	–0.4
9	–1	–1	–1	1	1	45	β_9	1.3
10	1	–1	–1	–1	1	40	β_{10}	–1.3
11	–1	–1	–1	–1	–1	40	β_{11}	–0.4
12	–1	–1	1	1	1	80	β_{12}	–1.3

6. Which is correct and most complete based on the above table?
   a. Changing factor x_2 over the levels in the experiment can be proven to make a significant difference with IER and $\alpha = 0.05$.
   b. In standard screening, a new set of test prototypes is needed for each response.
   c. Changes in different factors can have significant effects on different responses.
   d. All of the above are correct.
   e. All of the above are correct except (a) and (d).
7. Assume m is the number of factors. Which is correct and most complete?
   a. Regular fractional factorials all have at least one generator.
   b. The model plotted in an optional step is a highly accurate prediction of outputs.
   c. The IER takes into account that multiple tests are being done simultaneously.
   d. All of the above are correct.
   e. All of the above are correct except (a) and (d).

Table 12.15 will be used for Questions 8 and 9.

8. Which of the following is correct (within the applied uncertainty)?
   a. $t_{\text{Lenth},5} = 0.97$, and we fail to find significance with $\alpha = 0.05$ even using IERs.
   b. $t_{\text{Lenth},5} = 1.37$, and we fail to find significance with $\alpha = 0.05$ even using IERs.
   c. $t_{\text{Lenth},2} = 4.57$, which is significant with $\alpha = 0.05$, using the IER.
   d. Lenth's PSE = 0.75 (based on the coefficients not effects) and factor x_4 is associated with a significant effect using the EER and $\alpha = 0.05$.
   e. (a) and (b) are correct.
9. Which is correct and most complete?
   a. It can be reasonable to adjust factor settings of factors not proven to have significant effects using judgment.
   b. Changing factor x_4 does not significantly affect outputs $\alpha = 0.05$ using IERs.
   c. Changing factor x_6 cannot affect any possible responses of the system.
   d. All of the above are correct.
   e. All of the above are correct except (a) and (d).

12

Table 12.15 Outputs from another hypothetical experiment

Run (i)	$x_{i,1}$	$x_{i,2}$	$x_{i,3}$	$x_{i,4}$	$x_{i,5}$	$x_{i,6}$	Y_1		β_{est} (Coefficients)
1	−1	1	–	1	1	−1	20	β_1 (Constant)	35.0
2	1	1	−1	−1	−1	−1	50	β_2 (factor x_1)	1.0
3	1	−1	–	1	−1	1	22	β_3 (factor x_2)	−0.5
4	1	−1	1	−1	1	−1	52	β_4 (factor x_3)	−0.5
5	−1	1	1	–	−1	1	48	β_5 (factor x_4)	−15.0
6	−1	−1	1	1	−1	−1	18	β_6 (factor x_5)	0.5
7	1	1	1	1	1	1	20	β_7(factor x_6)	0.0
8	−1	−1	−1	−1	1	1	50	β_8	−0.5

10. Which is correct and most complete?
    a. Often, the wider the level spacing, the greater the chance of finding significance.
    b. Often, the more data used, the greater the chance of finding significance.
    c. Sometimes finding significance is actually helpful in an important sense.
    d. All of the above are correct.
    e. All of the above are correct except (a) and (d).
11. Which is correct and most complete?
    a. Most software do not print out runs in randomized order.
    b. If no factors have significant effects, main effects plots are rarely (if ever) useful.
    c. Sometimes a response of interest can be an average of three individual responses.
    d. The usage of all runs to make decisions about all factors cannot aid in reducing the effects of measurement errors on the analyses.
    e. All of the above are correct.
    f. All of the above are correct except (a) and (e).
12. Which is correct and most complete?
    a. People almost never make design decisions using FF experiments. Instead, they use results to pick factors for RSM experiments.
    b. Lenth's method is designed to find significance even when interactions exist.
    c. DOE matrices are completely random collections of −1s and 1s.
    d. All of the above are correct.
    e. All of the above are correct except (a) and (d).
13. Standard screening using regular FFs is used and all responses are close together. Which is correct and most complete?
    a. Likely users did not vary the factors over wide enough ranges.
    b. The exercise could be useful because likely none of the factors strongly affect the response. That information could be exploited.
    c. You will possibly discover that none of the factors has a significant effect even using IER and $\alpha = 0.1$.
    d. All of the above are correct.
    e. All of the above are correct except (a) and (d).
14. Which is correct and most complete?
    a. The design in Table 12.14 is a regular fractional factorial.
    b. In applying DOE, factors of interest are controllable during testing.
    c. Sometimes, a good mental test of the factors and levels is the experimenter being honestly unsure about which combination in the DOE matrix will give the most desirable responses.
    d. Usually, one uses the model in the main effects plot to make system design recommendations and does not simply pick the best prototype from the DOE.
    e. All of the above are correct.
    f. All of the above are correct except (a) and (d).

15. Suppose that a person experiments using five factors and the eight run regular array. $\beta_{est} = \{82.9, -1.125, -0.975, -1.875, 9.825, 0.35, 1.85, 0.55\}$.
    a. No factor has a significant effect on the average response with $\alpha = 0.05$.
    b. One fails to find the 4^h factor has a significant effect using IER and $\alpha = 0.05$.
    c. One fails to find the 5^h factor has a significant effect using IER and $\alpha = 0.05$.
    d. The PSE = 2.18.
    e. All of the above are correct.
    f. All of the above are correct except (a) and (d).
16. Assume n is the number of runs and m is the number of factors. Which is correct and most complete?
    a. Normal probability plots are often used instead of Lenth's method for analysis.
    b. A reason not to use 3 level designs for screening might be that 2 level designs give a relatively high chance of finding which factors matter for the same n.
    c. For fixed n, larger m generally means reduced chance of complete correctness.
    d. Adding more factors is often desirable because each factor is a chance of finding a way to improve the system.
    e. All of the above are correct.
    f. All of the above are correct except (a) and (e).
17. Illustrate the application of standard screening using a real example. Experimentation can be based on the paper helicopter design implied by the manufacturing SOP in ▸ Chap. 2 in Example 2.6.1.
    a. Include all the information mentioned in ◘ Fig. 12.1. Define your response explicitly enough such that someone could reproduce your results.
    b. Perform a Lenth's analysis to test whether the estimated coefficients correspond to significant effects with $\alpha = 0.1$ using the IER.
18. Give one generator for the experimental design in ◘ Table 12.15.
19. Which of the following is correct and most complete?
    a. The design in ◘ Table 12.14 is a regular fractional factorial.
    b. The design in ◘ Table 12.15 is a regular fractional factorial.
    c. The design in ◘ Table 12.14 is a full factorial.
    d. All of the above are correct.
    e. All of the above are correct except (a) and (d).
20. Which of the following is correct and most complete?
    a. A regular fractional factorial with 20 runs is available in the literature.
    b. A PB design with 16 runs is available in the literature.
    c. A PB desing with 20 runs is available in the literature.
    d. All of the above are correct.
    e. All of the above are correct except (a) and (d).
21. Consider the columns of the experimental planning matrix as A, B, C, D, E, F, G, and H. Which of the following is correct and complete?
    a. For the design in ◘ Table 12.2, AB = E.
    b. For the design in ◘ Table 12.2, AB = D.
    c. For the design in ◘ Table 12.4, none of the columns can be obtained by multiplying other columns together.
    d. All of the above are correct.
    e. All of the above are correct except (a) and (d).

Short Anwser Section:

1. Consider the example in ◘ Table 12.14. What are **D**, **X**, **X′X**, and **A**?
2. Analyze the data from ◘ Table 12.14 and draw conclusions about significance. Be standard and conservative (high standard of evidence and low Type I error rate or "α") in your choice of IER versus EER and α.
3. Give an example of when you might see a Type I and Type II error along with defining both.
4. Create a fractional factorial experiment and preform it.

References

Allen TT, Bernshteyn M (2003) Supersaturated designs that maximize the probability of finding the active factors. Technometrics 45:1–8

Box GEP, Hunter JS (1961a) The 2^{k-p} fractional factorial designs, part I. Technometrics 3:311–351

Box GEP, Hunter JS (1961b) The 2^{k-p} fractional factorial designs, part II. Technometrics 3:449–458

Brady J, Allen T (2002) Case study based instruction of SPC and DOE. Am Stat 56(4):1–4

Daniel C (1959) Use of half-normal plots in interpreting factorial two-level experiments. Technometrics 1:311–341

Fisher RA (1925) Statistical methods for research workers. Oliver and Boyd, London

Lenth RV (1989) Quick and easy analysis of unreplicated factorials. Technometrics 31:469–473

Plackett RL, Burman JP (1946) The design of optimum multifactorial experiments. Biometrika 33:303–325

Ye K, Hamada M, Wu CFJ (2001) A step-down lenth method for analyzing unreplicated factorial designs. J Qual Technol 33:140–152

DOE: Response Surface Methods

T. T. Allen, *Introduction to Engineering Statistics and Lean Six Sigma*,
https://doi.org/10.1007/978-1-4471-7420-2_13

13.1 Introduction

Response surface methods (RSM) are primarily relevant when the decision-maker desires (1) to create a relatively accurate prediction of engineered system input-output relationships and (2) to "tune" or optimize thoroughly of the system being designed. Since these methods require more runs for a given number of factors than screening using fractional factorials, they are generally reserved for cases in which the importance of all factors is assumed, perhaps because of previous experimentation.

The methods described here are called "standard response surface methods" (RSM) because they are widely used and the prediction models generated by them can yield 3D surface plots. The methods are based on three types of design of experiments (DOE) matrices. First, "central composite designs" (CCDs) are matrices corresponding to (at most) five level experimental plans from Box and Wilson (1951). Second, "Box Behnken designs" (BBDs) are matrices corresponding to three level experimental plans from Box and Behnken (1960). Third, Allen et al. (2003) proposed methods based on so-called "expected integrated mean squared error optimal" (EIMSE-optimal) designs. EIMSE-optimal designs are one type of experimental plan that results from the solution of an optimization problem.

We divide RSM into two classes: (1) "one-shot" methods conducted in one batch and (2) "sequential" methods based on central composite designs from Box and Wilson (1951). This chapter begins with "design matrices" which are used in the model fitting part of response surface methods. Next, one-shot and sequential response surface methods are defined, and examples are provided. Finally, a brief explanation of the origin of the different types of DOE matrices used is given.

13.2 Design Matrices for Fitting RSM Models

13

In the context of RSM, the calculation of regression coefficients without statistics software is more complicated than for screening methods. It practically requires knowledge of matrix transposing and multiplication described in ► Sect. 5.3 and additional concepts. The phrase "functional form" refers to the relationship that constrains and defines the fitted model. For both screening using regular fractional factorials and RSM, the functional forms are polynomials with specific combinations of model terms. Therefore, the model forms can be written:

$$y_{est}(\boldsymbol{\beta}_{est}, \mathbf{x}) = \mathbf{f}(\mathbf{x})'\boldsymbol{\beta}_{est} \tag{13.1}$$

where $\mathbf{f}(\mathbf{x})$ is a vector of functions, $f_{1,j}(\mathbf{x})$, for $j=1, \ldots, k$, only of the system inputs, $\mathbf{x}$. For standard screening methods, the model form relevant to main effects plotting is $f_1(\mathbf{x})=1$ and $f_j(\mathbf{x})=x_{j-1}$ for $j=2, \ldots, m+1$, where m is the number of factors. For example, with three factors the first order fitted model would be: $y_{est}(\boldsymbol{\beta}_{est},\mathbf{x})=\beta_1+\beta_1 x_1+\beta_2 x_2+\beta_3 x_3$. ► Chapter 15 describes general "**linear regression**" models, which all have their structure given by the above equation, in the context of nonlinear models.

The phrase "**model form**" is a synonym for functional form. For one-shot RSM, the model form is

$$\begin{aligned}&\mathrm{f}_1(\mathbf{x}) = 1,\ \mathrm{f}_j(\mathbf{x}) = x_{j-1} \quad \text{for} \quad j = 2, \ldots, m+1,\\ &\mathrm{f}_j(\mathbf{x}) = x_{j-m-1}^2 \quad \text{for} \quad j = m+2, \ldots, 2m+1, \ \text{and}\\ &\mathrm{f}_{2m+2}(\mathbf{x}) = x_1x_2,\ \mathrm{f}_{2m+3}(\mathbf{x}) = x_1x_3,\ \ldots,\ \mathrm{f}_{(m+1)(m+2)/2}(\mathbf{x}) = x_{m-1}x_m.\end{aligned} \tag{13.2}$$

This form is called a "**full quadratic polynomial**".

The functional form permits the concise definition of the $n \times k$ design matrix, **X**. Consider that an experimental plan can itself be written as n vectors, $\mathbf{x}_1, \mathbf{x}_2, \ldots, \mathbf{x}_n$, specifying each of the n runs. Then, the **X** matrix is

$$\mathbf{X} = \begin{bmatrix} \mathbf{f}(\mathbf{x}_1)' \\ \vdots \\ \mathbf{f}(\mathbf{x}_n)' \end{bmatrix} \tag{13.3}$$

Example 13.2.1 Three Factor Full Quadratic

Question

For $m=3$ factors and $n=11$ runs, provide a full quadratic f(x) and an example of D, and the associated X.

Answer

Equation (13.4) contains the requested vector and matrices.

Note that the above form contains quadratic terms, e.g., $\mathrm{f}_5(\mathbf{x}) = x_3^2$. Therefore, the associated linear model is called a "**response surface model**". Terms involving products, e.g., $\mathrm{f}_7(\mathbf{x}) = x_1x_3$, are called **interaction** terms.

$$\mathbf{f}(\mathbf{x}) = \begin{bmatrix} 1 \\ x_1 \\ x_2 \\ x_3 \\ x_1^2 \\ x_2^2 \\ x_3^2 \\ x_1x_2 \\ x_1x_3 \\ x_2x_3 \end{bmatrix} \mathbf{D} = \begin{bmatrix} -1 & -1 & -1 \\ -1 & 1 & -1 \\ 1 & 1 & -1 \\ -1 & 0 & 0 \\ 1 & -1 & 1 \\ 0 & 0 & 0 \\ 1 & 0 & 0 \\ -1 & -1 & 1 \\ -1 & 1 & 1 \\ 1 & 1 & 1 \\ 1 & -1 & -1 \end{bmatrix} \mathbf{D} = \begin{bmatrix} 1 & -1 & -1 & -1 & 1 & 1 & 1 & 1 & 1 & 1 \\ 1 & -1 & 1 & -1 & 1 & 1 & 1 & -1 & 1 & -1 \\ 1 & 1 & 1 & -1 & 1 & 1 & 1 & 1 & -1 & -1 \\ 1 & -1 & 0 & 0 & 1 & 0 & 0 & 0 & 0 & 0 \\ 1 & 1 & -1 & 1 & 1 & 1 & 1 & -1 & 1 & -1 \\ 1 & 0 & 0 & 0 & 0 & 0 & 0 & 0 & 0 & 0 \\ 1 & 0 & 0 & 0 & 0 & 0 & 0 & 0 & 0 & 0 \\ 1 & -1 & -1 & 1 & 1 & 1 & 1 & 1 & -1 & -1 \\ 1 & -1 & 1 & 1 & 1 & 1 & 1 & -1 & -1 & 1 \\ 1 & 1 & 1 & 1 & 1 & 1 & 1 & 1 & 1 & 1 \\ 1 & 1 & -1 & -1 & 1 & 1 & 1 & -1 & -1 & 1 \end{bmatrix} \tag{13.4}$$

Referring back to the first case study in ▶ Chap. 2 with the printed circuit board, the relevant design, **D**, and **X** matrix were:

Table 13.1 A RSM design or "array" and three functional forms

(a)			(b)	(c)	(d)
Run	A	B			
1	−1	−1	$y(\mathbf{x}) = \beta_1 + \beta_2 A + \beta_3 B$	$y(\mathbf{x}) = \beta_1 + \beta_2 A + \beta_3 B + \beta_4 AB$	$y(\mathbf{x})$ = full quadratic polynomial in A and B
2	1	−1			
3	−1	1			
4	0	0			
5	1	1			
6	$-\alpha_C$	0			
7	0	0			
8	0	α_C			
9	α_C	0			
10	0	$-\alpha_C$			

$$\mathbf{D} = \begin{bmatrix} -1 & 1 & -1 & 1 & -1 & 1 & -1 \\ 1 & 1 & -1 & -1 & 1 & -1 & -1 \\ 1 & -1 & -1 & 1 & -1 & -1 & 1 \\ 1 & -1 & 1 & -1 & -1 & 1 & -1 \\ -1 & 1 & 1 & -1 & -1 & -1 & 1 \\ -1 & -1 & 1 & 1 & 1 & -1 & -1 \\ 1 & 1 & 1 & 1 & 1 & 1 & 1 \\ -1 & -1 & -1 & -1 & 1 & 1 & 1 \end{bmatrix} \quad \mathbf{X} = \begin{bmatrix} 1 & -1 & 1 & -1 & 1 & -1 & 1 & -1 \\ 1 & 1 & 1 & -1 & -1 & 1 & -1 & -1 \\ 1 & 1 & -1 & -1 & 1 & -1 & -1 & 1 \\ 1 & 1 & -1 & 1 & -1 & -1 & 1 & -1 \\ 1 & -1 & 1 & 1 & -1 & -1 & -1 & 1 \\ 1 & -1 & -1 & 1 & 1 & 1 & -1 & -1 \\ 1 & 1 & 1 & 1 & 1 & 1 & 1 & 1 \\ 1 & -1 & -1 & -1 & -1 & 1 & 1 & 1 \end{bmatrix} \tag{13.5}$$

so that the last row of the **X** matrix, $\mathbf{f}_1(\mathbf{x}_8)'$, was given by:

$$\mathbf{f}(\mathbf{x}_8)' = \begin{bmatrix} 1 & -1 & -1 & -1 & -1 & 1 & 1 & 1 \end{bmatrix} \tag{13.6}$$

The next example illustrates how design matrices can be constructed based on different combinations of experimental plans and functional forms.

Example 13.2.2 Multiple Functional Forms

In one-shot RSM, the most relevant model form is a full quadratic. However, it is possible that a model fitter might consider more than one functional form. Consider the experimental plan and models in Table 13.1.

Question 1

Which is correct and most complete?

a. A design matrix based on (a) and (b) in Table 13.1 would be 10 × 4.
b. A design matrix based on (a) and (c) in Table 13.1 would be 4 × 10.
c. A design matrix based on (a) and (d) in Table 13.1 would be 10 × 6.
d. All of the above are correct.
e. All of the above are correct except (a) and (d).

Answer 1
Design matrices always have dimensions $n \times k$ so (c) is correct.

Question 2
Which is correct and most complete?
a. The model form in (c) in Table 13.1 contains one interaction term.
b. Using the design matrix and model in (a) and (b) in Table 13.1, **X′X** is diagonal.
c. Linear regression model forms can contain terms like x_1^2.
d. All of the above are correct.
e. All of the above are correct except (c) and (d).

Answer 2
All of the answers in (a), (b), and (c) are correct. Therefore, the answer is (d). Note that, the fact that **X′X** implies the property that this central composite design is "orthogonal" with respect to first order functional forms.

13.3 One-Shot Response Surface Methods

As will be discussed in the context of decision support information below, one-shot RSM is generally preferable to sequential RSM in cases in which the experimenter is confident that the ranges involved are the relevant ranges and a relatively accurate prediction is required. If the decision-maker feels he or she has little knowledge of the system, sequential methods will offer a potentially useful opportunity to stop experimentation having performed relatively few runs but with a somewhat accurate prediction model. Then, one can terminate experimentation having achieved a tangible result or perform additional experiments with revised experimental ranges. In the one-shot experiments described here, the quality of models derived from a fraction of the data has not been well studied. Therefore, the experimenter generally commits to performing all runs when experimentation begins.

One-shot RSM are characterized by (1) an "experimental design", **D** and (2) vectors that specify the highs, **H**, and lows, **L**, of each factor. Note that in all standard RSM approaches, *all factors must be continuous*, e.g., none of the system inputs can be categorical variables such as the type of transistor or country in which units are made. The development of related methods involving categorical factors is an active area for research. Software to address combinations of continuous and categorical factors is available from JMP™ and through ▸ www.sagata.com.

As for screening methods, the experimental design, **D**, specifies indirectly which prototype systems should be built. For RSM, these designs can be selected from any of the ones provided in tables that follow both in this section and in the appendix at the end of this chapter. The design, **D**, only indirectly specifies the prototypes because it must be scaled or transformed using **H** and **L** to describe the prototypes in an unambiguous fashion that people who are not familiar with DOE methods can understand (see below).

Algorithm 13.1 One-Shot Response Surface Methods

Pre-step. Define the factors and ranges, i.e., the highs, **H**, and lows, **L**, for all factors.

Step 1. Select the experimental design from ◘ Tables 13.2 to facilitate the scaling in *Step 2*. Options include the Box Behnken (BBD), central composite (CCD), or EIMSE-optimal designs. If a CCD design is used, then the parameter α_C can be adjusted as desired. If $\alpha_C = 1$, then only three levels are used. The default setting is $\alpha_C = \text{sqrt}\{m\}$, where *m* is the number of factors.

Step 2. Scale the experimental design using the ranges selected by the experimenter. $D^s_{i,j} = L_j + 0.5(H_j - L_j)(D_{i,j} + 1)$ for $i = 1, \ldots, n$ and $j = 1, \ldots, m$.

Step 3. Build and test the prototypes according to $\mathbf{D}^s$. Record the test measurements for the responses for the *n* runs in the *n* dimensional vector **Y**.

Step 4. Form the so-called "design" matrix, **X**, based on the *scaled* design, $\mathbf{D}^s$, following the rules in the above equations. Then, calculate the regression coefficients $\boldsymbol{\beta}_{est} = \mathbf{AY}$, where **A** is the $(\mathbf{X}'\mathbf{X})^{-1}\mathbf{X}'$. Reexamine the approach used to generate the responses to see if any runs were not representative of system responses of interest.

Step 5. (*Optional*) Plot the prediction model, $y_{est}(\mathbf{x})$, for prototype system output

$$
\begin{aligned}
y_{est}(\mathbf{x}) = {} & \beta_{est,1} + \beta_{est,2}x_1 + \cdots + \beta_{est,m+1}x_m + \beta_{est,m+2}x_1^2 + \cdots + \\
& \beta_{est,2m+1}x_m^2 + \beta_{est,2m+2}x_1x_2 + \cdots + \\
& \beta_{est,(m+1)(m+2)/2}x_{m-1}x_m
\end{aligned}
\tag{13.7}
$$

This model is designed to predict average prototype system response for a given set of system inputs, **x**. An example below shows how to make 3D plots using Excel and models of the above form.

Step 6. Apply informal or formal optimization using the prediction model, $y_{est}(\mathbf{x})$, to develop recommended settings. Formal optimization is described in detail in ▶ Chaps. 6 and 19.

13

The above method is given in terms of only a single response. Often, many responses are measured in *Step 3*, derived from the same prototype systems. Then, *Step 4* and *Step 5* can be iterated for each of the responses considered critical. Then also, optimization in *Step 6* would make use of the multiple prediction models.

If EIMSE designs are used, the method is not usually referred to as "standard" RSM. These designs and others are referred to as "optimal designs" or sometimes "computer generated" designs. However, they function in much the same ways as the standard designs and offer additional method options that might be useful.

With scaled units used in the calculation of the **X** matrices, care must be taken to avoid truncation of the coefficients. In certain cases of possible interest, the factor ranges, $(H_j - L_j)$, may be such that even small coefficients, e.g., 10^{-6}, can greatly affect

Table 13.2 RSM designs: (a) BBD, (b) and (c) EIMSE-optimal, and (d) CCD

(a)				(b)				(c)				(d)				
Run	x_1	x_2	x_3	Run	x_1	x_2	x_3	Run	x_1	x_2	x_3	Run	x_1	x_2	x_3	x_4
1	0	−1	1	1	−1	0	1	1	−1	−1	0	1	−1	1	−1	−1
2	0	0	0	2	1	−1	0	2	−1	0	1	2	−1	1	−1	1
3	1	0	1	3	−1	0	−1	3	−1	1	−1	3	0	0	0	0
4	0	−1	−1	4	−1	1	0	4	−1	−1	−1	4	1	−1	−1	−1
5	−1	−1	0	5	0	−1	1	5	−1	−1	1	5	1	−1	1	−1
6	−1	0	1	6	0	1	−1	6	−1	1	1	6	−1	1	1	−1
7	1	1	0	7	0	−1	−1	7	0	−1	1	7	−1	1	1	1
8	1	0	−1	8	0	1	1	8	0	1	−1	8	1	1	−1	1
9	−1	1	0	9	1	0	−1	9	1	−1	−1	9	1	1	1	−1
10	0	0	0	10	0	0	0	10	1	1	−1	10	−1	−1	1	−1
11	1	−1	0	11	−1	−1	0	11	1	1	1	11	−1	−1	−1	−1
12	0	1	−1	12	1	0	1	12	1	−1	1	12	1	−1	1	1
13	0	1	1	13	1	1	0	13	1	0	−1	13	0	0	0	0
14	0	0	0					14	1	1	0	14	0	0	0	0
15	−1	0	−1					15	0	0	0	15	−1	−1	−1	1
								16	0	0	0	16	0	0	0	0

(continued)

Table 13.2 (continued)

(a)				(b)				(c)				(d)				
Run	x_1	x_2	x_3	Run	x_1	x_2	x_3	Run	x_1	x_2	x_3	Run	x_1	x_2	x_3	x_4
												17	1	1	1	1
												18	1	−1	−1	1
												19	1	1	−1	−1
												20	−1	−1	1	1
												21	0	0	0	0
												22	0	0	0	α_C
												23	0	$-\alpha_C$	0	0
												24	0	0	0	$-\alpha_C$
												25	0	0	0	0
												26	$-\alpha_C$	0	0	0
												27	0	α_C	0	0
												28	0	0	$-\alpha_C$	0
												29	0	0	α_C	0
												30	α_C	0	0	0

the predicted response. Therefore, it can be of interest to fit the models based on inputs and design matrices in the original −1 and 1 coding. Another benefit of using "coded" −1, 1 units is that the magnitude of the derived coefficients can be compared to see which factors are more important in their effects on response averages.

The majority of the experimental designs, **D**, associated with RSM have repeated runs, i.e., repeated combinations of the same settings such as $x_1=0$, $x_2=0$, $x_3=0$, $x_4=0$. One benefit of having these repeated runs is that the experimenter can use the sample standard deviation, *s*, of the associated responses as an "assumption free" estimate of the standard deviation of the random error, σ ("sigma"). This can establish the so-called "process capability" of the prototype system and therefore aid in engineered system robust optimization (see below).

In *Step 4*, a reassessment is often made of each response generated in *Step 3*, to see if any of the runs should be considered untrustworthy, i.e., not representative of system performance of interest. ▶ Chapter 15 provides formulas useful for calculating the so-called "**adjusted R-squared**". In practice, this quantity is usually calculated directly by software, e.g., Excel. Roughly speaking, the adjusted R-squared gives the "fraction of the variation explained by changes in the factors". When one is analyzing data collected using EIMSE-optimal, Box Behnken, or central composite designs, one expects adjusted R-squared values in excess of 0.50 or 50%. Otherwise, there is a concern that some or all of the responses are not trustworthy and/or that the most important factors are unknown and uncontrolled during the testing.

Advanced readers might be interested to know that, for certain assumptions, Box Behnken designs are EIMSE designs. Also, an approximate formula to estimate the number of runs, *n*, required by standard response surface methods involving *m* factors is $0.75(m+1)(m+2)$.

13.4 One-Shot RSM Examples

In this section, two examples are considered. The first is based on a student project to tune the design of a paper airplane. The second related to tuning of die casting machine specifications to see if lower weight machines can actually achieve less distorted parts.

The original paper airplane design to be tuned was manufactured using a four step standard operating procedure (SOP). In the first step an A = 11.5 in. long by B = 8.5 in. wide sheet is folded in half lengthwise. In the second step, a point is made by further folding lengthwise the corners inward starting about 2 in. deep from the top. Third, the point is sharpened with another fold, this time starting at the bottom left and right corners so that the plan or top view appears to be an equilateral triangle. In the fourth step, the wing ends are folded C degrees either upwards (positively) or downwards (negatively). Finally, store the airplane carefully without added folds. The goal of the response surface application was to tune factors A, B, and C.

Algorithm 13.2 One shot RSM example

Pre-step. The highs, **H**, and lows, **L**, for all factors are shown in ◘ Table 13.3.

Step 1. The Box Behnken design was selected in ◘ Table 13.4(a) because it was offered a reasonable balance between run economy and prediction accuracy.

Step 2. The scaled $\mathbf{D}^s$ array is shown in ◘ Table 13.4(b).

Step 3. Planes were thrown from shoulder height and the time in air was measured using a stopwatch and recorded in the right-hand column of ◘ Table 13.4(b).

Step 4. The fitted coefficients are written in the regression model:

$$\begin{aligned}\text{Predicted Average Time} &= \\ &-295.14+30.34\,\text{Width}+38.12\,\text{Length}+0.000139\,\text{Angle} \\ &-1.25\,\text{Width}^2-1.48\,\text{Length}^2-0.000127\,\text{Angle}^2 \\ &-1.2(\text{Width})(\text{Length})-0.00778(\text{Width})(\text{Angle}) \\ &+0.00639(\text{Length})(\text{Angle})\end{aligned} \tag{13.8}$$

which has adjusted R^2 of only 0.201. Inspection of the airplanes reveals that the second prototype (Run 2) was not representative of the system being studied because of an added fold. Removing this run did little to change the qualitative shape of the surface but it did increase the adjusted R^2 to 0.55.

Step 5. The 3D surface plot with rudder fixed at 0° is in ◘ Fig. 13.1. This plot was generated using Sagata® Regression.

Step 6. The model indicates that the rudder angle did affect the time but that 0° is approximately the best. Considering that using 0° effectively removes a step in the manufacturing SOP, which is generally desirable, that setting is recommended. Inspection of the surface plot then indicates that the highest times are achieved with width A equal to 7.4 in. and length equal to 9.9 in.

As a second example, consider that researchers at the Ohio State University Die Casting Research Center have conducted a series of physical and computer experiments designed to investigate the relationship of machine dimensions and part distortion. This example is described in Choudhury (1997). Roughly speaking, the objective was to minimize the size and, therefore, cost of the die machine while

◘ **Table 13.3** Example factor ranges

Range\factor	A—width (in.)	B—length (in.)	C—angle (°)
L_j	6.5	9	-90
H_j	8.5	11	90

Table 13.4 Example (a) coded DOE array, **D** and (b) scaled $\mathbf{D}^s$ and response values

(a)			(b)				
A	B	C	Run	A—width	B—length	Angle	Time
0	−1	1	1	7.5	9	90	3.3
0	0	0	2	7.5	10	0	3.1
1	0	1	3	8.5	10	90	2.2
0	−1	−1	4	7.5	9	−90	4.2
−1	−1	0	5	6.5	9	0	1.1
−1	0	1	6	6.5	10	90	5.3
1	1	0	7	8.5	11	0	1.3
1	0	−1	8	8.5	10	−90	1.8
−1	1	0	9	6.5	11	0	3.9
0	0	0	10	7.5	10	0	6.1
1	−1	0	11	8.5	9	0	3.3
0	1	−1	12	7.5	11	−90	0.8
0	1	1	13	7.5	11	90	2.2
0	0	0	14	7.5	10	0	6.2
−1	0	−1	15	6.5	10	−90	2.1

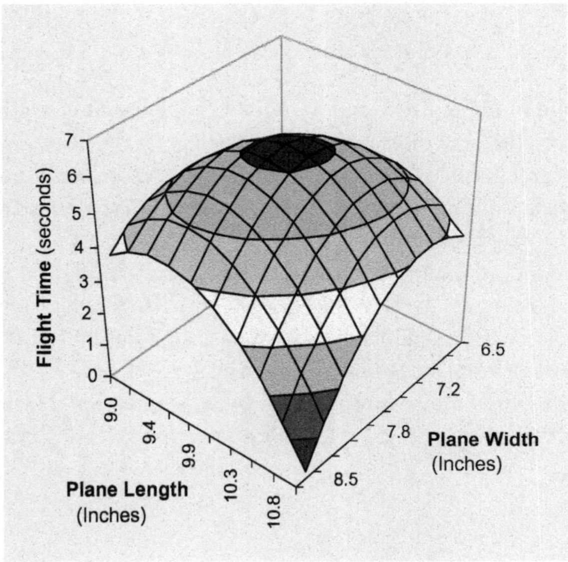

Fig. 13.1 3D surface plot of paper airplane flight time predictions

maintaining acceptable part distortion by manipulating the inputs, x_1, x_2, and x_3 shown in the figure below. These factors and ranges and the selected experimental design are shown in ◘ Fig. 13.2 and ◘ Table 13.5.

Algorithm 13.3 Part distortion example

Step 1. The team prepared the experimental design, **D**, shown in the ◘ Table 13.6 for scaling. Note that this design is not included as a recommended option for the reader because of what may be viewed as a mistake in the experimental design generation process. This design does not even approximately maximize the EIMSE, although it was designed with the EIMSE in mind.

Step 2. The design in ◘ Table 13.6 used the above-mentioned ranges and the formula $D^s_{i,j} = L_j + 0.5(H_j - L_j)(D_{i,j} + 1)$ for $i=1, \ldots, 11$ and $j=1, \ldots, 3$ to produce the experimental plan in ◘ Table 13.7 part (a).

Step 3. The prototypes were built according to $\mathbf{D}^s$ using a type of virtual reality simulation process called finite element analysis (FEA). From these FEA test runs, the measured distortion values $\mathbf{Y}_1, \ldots, \mathbf{Y}_8$ are shown in the ◘ Table 13.7(b). The numbers are maximum part distortion of the part derived from the simulated process in inches.

Step 4. The analyst on the team calculated the so-called "design" matrix, **X**, and $\mathbf{A} = (\mathbf{X}'\mathbf{X})^{-1}\mathbf{X}'$ given by ◘ Tables 13.8 and 13.9. Then, for each of the 8 responses, the team calculated the regression coefficients shown in ◘ Table 13.10 using $\boldsymbol{\beta}_{est,r} = \mathbf{AY}_r$ for $r=1, \ldots, 8$.

Step 5. For example, $y_{est,1}(\mathbf{x})$, is

$$\begin{aligned} y_{est,1}(\mathbf{x}) = {} & 0.0839600 - 0.0169500x_1 - 0.0004868x_2 + 0.0004617x_3 \\ & + 0.0009621x_1^2 + 0.0000323x_2^2 - 0.0000342x_3^2 \\ & - 0.0000373x_1x_2 + 0.0000054x_1x_3 - 0.0000037x_2x_3 \end{aligned} \tag{13.9}$$

The plot in ◘ Fig. 13.3 was developed using Excel. It is relatively easy to generate identical plots using Sagata® software (► www.sagata.com). From these eight prediction models for mean distortion, an additional model was created that was the minimum of the predicted values as a function of **x**. $y_{max}(\mathbf{x}) = \text{Maximum}\,[y_{est,1}(\mathbf{x}), y_{est,2}(\mathbf{x}), \ldots, y_{est,8}(\mathbf{x})]$.

Step 6. In this study, the engineers chose to apply formal optimization. They chose to limit maximum average part distortion to 0.075″ while minimizing the $2.0\,x_1 + x_2$ which is roughly proportional to the machine cost. They included the experimental ranges as constraints, $\mathbf{L} \le \mathbf{x} \le \mathbf{H}$, because they knew that prediction model errors usually become unacceptable outside the prediction ranges. The precise optimization formulation was:

13

$$\begin{aligned}&\text{Minimize: } 2.0\,x_1 + x_2\\&\text{Subject to:}\\&\text{Maximum}\left[y_{est,1}(\mathbf{x}),\ y_{est,2}(\mathbf{x}),\ \ldots,\ y_{est,8}(\mathbf{x})\right] \le 0.075''\\&\qquad\text{and } \mathbf{L} \le \mathbf{x} \le \mathbf{H}\end{aligned} \tag{13.10}$$

which has the solution $x_1 = 6.3$ in., $x_2 = 9.0$ in., and $x_3 = 12.5$ in. This solution was calculated using the Excel solver.

Note that formal optimization is discussed more thoroughly in ▶ Chap. 6. In many cases, the decision-makers will use visual information such as the above plot to weigh subjectively the many considerations involved in engineered system design.

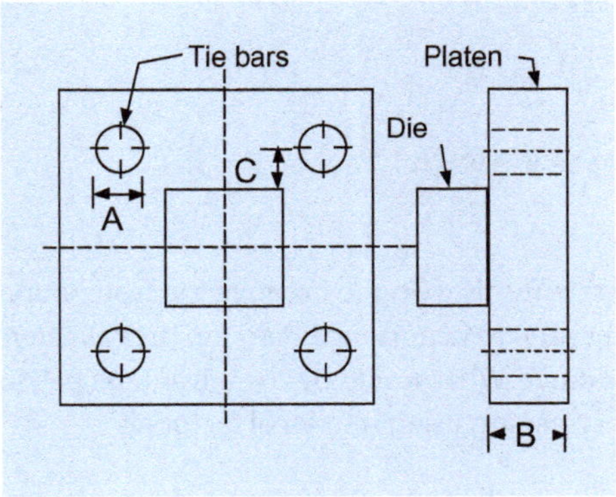

Fig. 13.2 The factor explanation for the one–shot RSM casting example

Table 13.5 The factor and range table for the one–shot RSM casting example

	Factor description	Low (L)	High (H)
X_1	Diameter tie bar (DTB)	5.5″	8.0″
X_2	Platen thickness (PT)	9.0″	14.5″
X_3	Die position (DP)	0.0″	12.5″

Table 13.6 D at 8 part locations in inches

Run	x_1	x_2	x_3
1	−1.0	1.0	1.0
2	1.0	−1.0	1.0
3	1.0	1.0	−1.0
4	1.0	1.0	1.0
5	1.0	0.0	0.0
6	0.0	1.0	0.0
7	0.0	0.0	1.0
8	0.0	0.0	0.0
9	0.5	−1.0	−1.0
10	−1.0	0.5	−1.0
11	−1.0	−1.0	0.5

Example 13.4.1 Food Science Application

Question 1
Assume that a food scientist is trying to improve the taste rating of an ice cream sandwich product by varying factors including the pressure, temperature, and amount of vanilla additive. What would be the advantage of using response surface methods instead of screening using fractional factorials?

Answer 1
Response surface methods generally generate more accurate prediction models than screening methods using fractional factorials, resulting in recommended settings with more desirable engineered system average response values.

Question 2
The scientist is considering varying either three or four factors. What are the advantages of using four factors?

Answer 2
Two representative standard design methods for three factors methods require 15 and 16 runs. Two standard design methods for four factors require 27 and 30 runs. Therefore, using only three factors would save the costs associated with ten or more runs. However, in general, each factor that is varied offers an opportunity to improve the system. Thorough optimization over four factors provably results in more desirable or equivalently desirable settings compared with thorough optimization over three factors.

Table 13.7 (a) D^s and (b) measured distortions at eight part locations in inches

(a)				(b)							
Run	x_1	x_2	x_3	Y_1	Y_2	Y_3	Y_4	Y_5	Y_6	Y_7	Y_8
1	5.50	14.50	12.5	0.0167	0.0185	0.0197	0.0143	0.0113	0.0177	0.0195	0.0153
2	8.00	9.00	12.5	0.006	0.0069	0.0069	0.004	0.0008	0.0078	0.0088	0.0057
3	8.00	14.50	0.00	0.0053	0.0038	0.002	0.0063	0.0069	0.0062	0.0047	0.0072
4	8.00	14.50	12.5	0.0056	0.0067	0.0074	0.0038	0.0016	0.0064	0.0075	0.0046
5	8.00	11.75	6.25	0.0067	0.0066	0.0037	0.0063	0.0051	0.0079	0.0078	0.0076
6	6.75	14.50	6.25	0.0109	0.0109	0.0104	0.0104	0.0093	0.0118	0.0118	0.0113
7	6.75	11.75	12.5	0.0095	0.011	0.0117	0.0072	0.0038	0.0109	0.0123	0.0086
8	6.75	11.75	6.25	0.0106	0.0105	0.0098	0.0100	0.0087	0.0118	0.0118	0.0113
9	7.38	9.00	0.00	0.007	0.0048	0.0013	0.008	0.008	0.0092	0.0069	0.0103
10	5.50	13.13	0.00	0.0163	0.0144	0.012	0.0177	0.0185	0.0175	0.0155	0.0188
11	5.50	9.00	9.38	0.0175	0.0183	0.018	0.0155	0.0122	0.0199	0.0205	0.0178

Table 13.8 Design matrix **X** for Algorithm 13.3

X =

1.00	5.50	14.50	12.50	30.25	210.25	156.25	79.75	68.75	181.25
1.00	8.00	9.00	12.50	64.00	81.00	156.25	72.00	100.00	112.50
1.00	8.00	14.50	0.00	64.00	210.25	0.00	116.00	0.00	0.00
1.00	8.00	14.50	12.50	64.00	210.25	156.25	116.00	100.00	181.25
1.00	8.00	11.75	6.25	64.00	138.06	39.06	94.00	50.00	73.44
1.00	6.75	14.50	6.25	45.56	210.25	39.06	97.88	42.19	90.63
1.00	6.75	11.75	12.50	45.56	138.06	156.25	79.31	84.38	146.88
1.00	6.75	11.75	6.25	45.56	138.06	39.06	79.31	42.19	73.44
1.00	7.38	9.00	0.00	54.39	81.00	0.00	66.38	0.00	0.00
1.00	5.50	13.13	0.00	30.25	172.27	0.00	72.19	0.00	0.00
1.00	5.50	9.00	9.38	30.25	81.00	87.89	49.50	51.56	84.38

Table 13.9 Design matrix **A** for Algorithm 13.3

A =

0.629	−1.439	−0.243	10.565	3.107	−5.165	−14.781	−7.993	−0.296	3.213	13.400
−1.032	0.616	−0.169	−1.601	−3.260	3.332	2.808	1.512	1.318	−1.413	-2.111
0.379	−0.206	−0.101	−0.752	1.254	−1.086	0.934	0.542	−0.520	0.503	−0.947
0.136	0.082	0.245	−0.362	0.115	0.079	0.027	0.053	−0.212	−0.260	0.098
0.138	0.010	0.010	0.052	0.280	−0.232	−0.232	−0.113	−0.085	0.086	0.086
0.002	0.028	0.002	0.011	−0.048	0.058	−0.048	−0.023	0.018	−0.018	0.018
0.000	0.000	0.006	0.002	−0.009	−0.009	0.011	−0.005	0.003	0.003	−0.003
−0.061	−0.061	0.026	0.065	−0.026	−0.026	0.032	0.001	−0.009	−0.009	0.069
−0.027	0.011	−0.027	0.028	−0.011	0.014	−0.011	0.000	−0.004	0.030	−0.004
0.005	−0.012	−0.012	0.013	0.006	−0.005	−0.005	0.000	0.014	−0.002	−0.002

Question 3

The scientist experiments with four factors and develops a second order regression model with an adjusted R-squared of 0.95. What does this adjusted R-squared value imply?

Answer 3

A high R-squared value such as 0.95 implies that the factors varied systematically in experimentation are probably the most influential factors affecting the relevant engineered system average response. The effects of other factors that are not considered most likely have relatively small effects on average response values. The experimenter feel reasonably confident that "what if" analyses using the regression prediction model will lead to correct conclusions about the engineered system.

Table 13.10 The prediction model coefficients for the eight responses

$f_i(x)$	$\beta_{est}(Y_1)$	$\beta_{est}(Y_2)$	$\beta_{est}(Y_3)$	$\beta_{est}(Y_4)$	$\beta_{est}(Y_5)$	$\beta_{est}(Y_6)$	$\beta_{est}(Y_7)$	$\beta_{est}(Y_8)$
1	0.0839600	0.0793900	0.0661300	0.0837000	0.0842100	0.0941700	0.0871700	0.0918800
x_1	−0.0169500	−0.0172800	−0.0111900	−0.0171800	−0.0181000	−0.0173200	−0.0172800	−0.0173000
x_2	−0.0004868	−0.0000179	−0.0018500	−0.0000935	0.0003355	−0.0014730	−0.0008721	−0.0008880
x_3	0.0004617	0.0010780	0.0014900	−0.0000852	−0.0007315	0.0004451	0.0011000	−0.0000978
x_1^2	0.0009621	0.0009889	0.0004559	0.0009975	0.0010850	0.0009564	0.0009792	0.0009821
x_2^2	0.0000323	0.0000144	0.0000891	0.0000255	0.0000214	0.0000550	0.0000388	0.0000400
x_3^2	−0.0000342	−0.0000376	−0.0000221	−0.0000338	−0.0000345	−0.0000321	−0.0000372	−0.0000329
x_1x_2	−0.0000373	−0.0000293	0.0000366	−0.0000673	−0.0000981	−0.0000070	−0.0000205	−0.0000403
x_1x_3	0.0000054	−0.0000294	−0.0000329	0.0000392	0.0000789	−0.0000010	−0.0000353	0.0000275
x_2x_3	−0.0000037	−0.0000097	−0.0000357	−0.0000004	0.0000074	−0.0000014	−0.0000089	0.0000056

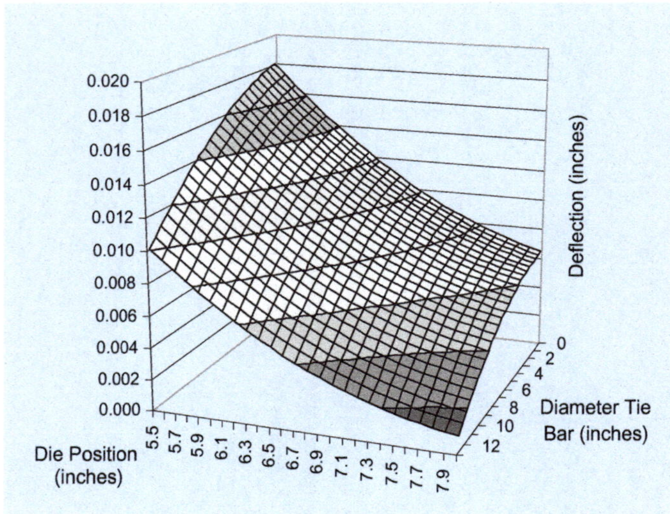

Fig. 13.3 The predicted distortion for the casting example

13.5 Creating 3D Surface Plots in Excel

The most important outcomes of an RSM application are often 3D surface plots. This follows because they are readily interpretable by a wide variety of people and help in building intuition about the system studied. Yet these plots are inappropriate for cases in which only first order terms have large coefficients. For those cases, the simpler main effects plots more concisely summarize predictions.

13 Creating a contour plot in Excel requires manual creation of formulas to generate an array of predictions needed by the Excel 3D charting routine to create the plot. The figure below shows a contour plot of the prediction model for $\mathbf{Y}_1$ in the example above. The second factor, x_2, is fixed at 11 in the plot. The dollar signs are selected such that the formula in cell E8 can be copied to all cells in the range E8:J20, producing correct predictions. Having generated all the predictions and putting the desired axes values in cells, E7:J7 and D8:D20, then the entire region D7:J20 is selected, and the "Chart" utility is called up through the "Insert" menu. An easier way to create identical surface plots is to apply Sagata® software (▸ www.sagata.com), which also includes EIMSE designs (author is part owner) (Fig. 13.4).

13.6 Sequential Response Surface Methods

The method in this section is relevant when the decision-maker would like the opportunity to stop experimentation, having built and tested a relatively small number of prototypes with something tangible. In the version presented here, the

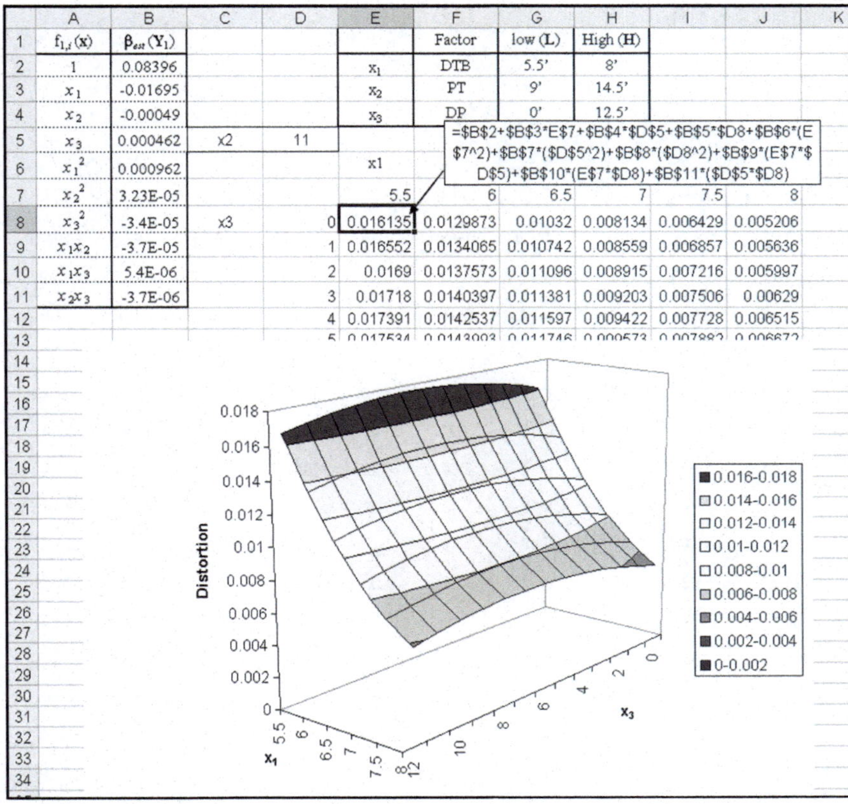

Fig. 13.4 Screen capture showing how to create contour plots

experimenter has performed a fractional factorial experiment as well as derived information pertinent to the question of whether adding pure quadratic terms, e.g., x_1^2, would significantly reduce prediction errors. This information derives from a type of "**lack of fit**" test.

Note that the method described here is not the fully sequential response surface methods of Box and Wilson (1951) and described in textbooks on response surface methods such as Box and Draper (1987). The general methods can be viewed as an optimization under uncertainty method which is a competitor to approaches in Part III of this text. The version described here might be viewed as "two-step" experimentation in the sense that runs are performed in at most two batches. The fully sequential response surface method could conceivably involve tens of batches of experimental test runs.

Two-step RSM are characterized by (1) an "experimental design", **D** (2) vectors that specify the highs, **H**, and lows, **L**, of each factor, and (3) the α parameter used in the lack of fit test in *Step 5* based on the critical values in Table 13.11.

Table 13.11 Critical values of the *F* distribution, $F_{\alpha,\nu1,\nu2}$ **(a)** α = 0.05 and **(b)** α = 0.10

(a)

α = 0.05					v_1					
v_2	1	2	3	4	5	6	7	8	9	10
1	161.45	199.50	215.71	224.58	230.16	233.99	236.77	238.88	240.54	241.88
2	18.51	19.00	19.16	19.25	19.30	19.33	19.35	19.37	19.38	19.40
3	10.13	9.55	9.28	9.12	9.01	8.94	8.89	8.85	8.81	8.79
4	7.71	6.94	6.59	6.39	6.26	6.16	6.09	6.04	6.00	5.96
5	6.61	5.79	5.41	5.19	5.05	4.95	4.88	4.82	4.77	4.74
6	5.99	5.14	4.76	4.53	4.39	4.28	4.21	4.15	4.10	4.06
7	5.59	4.74	4.35	4.12	3.97	3.87	3.79	3.73	3.68	3.64
8	5.32	4.46	4.07	3.84	3.69	3.58	3.50	3.44	3.39	3.35
9	5.12	4.26	3.86	3.63	3.48	3.37	3.29	3.23	3.18	3.14
10	4.96	4.10	3.71	3.48	3.33	3.22	3.14	3.07	3.02	2.98
11	4.84	3.98	3.59	3.36	3.20	3.09	3.01	2.95	2.90	2.85
12	4.75	3.89	3.49	3.26	3.11	3.00	2.91	2.85	2.80	2.75
13	4.67	3.81	3.41	3.18	3.03	2.92	2.83	2.77	2.71	2.67
14	4.60	3.74	3.34	3.11	2.96	2.85	2.76	2.70	2.65	2.60
15	4.54	3.68	3.29	3.06	2.90	2.79	2.71	2.64	2.59	2.54

(b)

α = 0.10					v_1					
v_2	1	2	3	4	5	6	7	8	9	10
1	39.86	49.50	53.59	55.83	57.24	58.20	58.91	59.44	59.86	60.19
2	8.53	9.00	9.16	9.24	9.29	9.33	9.35	9.37	9.38	9.39
3	5.54	5.46	5.39	5.34	5.31	5.28	5.27	5.25	5.24	5.23
4	4.54	4.32	4.19	4.11	4.05	4.01	3.98	3.95	3.94	3.92
5	4.06	3.78	3.62	3.52	3.45	3.40	3.37	3.34	3.32	3.30
6	3.78	3.46	3.29	3.18	3.11	3.05	3.01	2.98	2.96	2.94
7	3.59	3.26	3.07	2.96	2.88	2.83	2.78	2.75	2.72	2.70
8	3.46	3.11	2.92	2.81	2.73	2.67	2.62	2.59	2.56	2.54
9	3.36	3.01	2.81	2.69	2.61	2.55	2.51	2.47	2.44	2.42
10	3.29	2.92	2.73	2.61	2.52	2.46	2.41	2.38	2.35	2.32
11	3.23	2.86	2.66	2.54	2.45	2.39	2.34	2.30	2.27	2.25
12	3.18	2.81	2.61	2.48	2.39	2.33	2.28	2.24	2.21	2.19
13	3.14	2.76	2.56	2.43	2.35	2.28	2.23	2.20	2.16	2.14
14	3.10	2.73	2.52	2.39	2.31	2.24	2.19	2.15	2.12	2.10
15	3.07	2.70	2.49	2.36	2.27	2.21	2.16	2.12	2.09	2.06

Definition

"**Block**" here refers to a batch of experimental runs that are performed at one time. Time here is a blocking factor that we cannot randomize over. Rows of experimental plans associated with blocks are not intended to structure experimentation for usual factors or system inputs. If they are used for usual factors, then prediction performance may degrade substantially.

Definition

A "**center point**" is an experimental run with all of the settings set at their mid-value. For example, if a factor ranges from 15 to 20″ in the region of interest to the experiment, the "center point" would have a value of 17.5″ for the factor.

Definition

Let the symbol, n_c, refer to the number of center points in a central composite experimental design with the so-called block factor having a value of 1. For example, for the $n = 14$ run central composite in part (a) of Table 13.12, $n_C = 3$. Let $y_{average,c}$ and $y_{variance,c}$ be the sample average and sample variance of the rth response for the n_c center point runs in the first block, respectively. Let the symbol, n_f, refer to the number of other runs with the block factor having a value of 1. For the same $n = 14$ central composite, $n_f = 4$. Let $y_{average,f}$ be the average of the response for the n_f other runs in the first block.

An example application of central composite designs is given together with the robust design example in ▶ Chap. 14. In that case the magnitude of the curvature was large enough such that $F_0 > F_{\alpha,1,nC-1}$, for both responses for any α between 0.05 and 0.25.

Note that when $F_0 > F_{\alpha,1,nC-1}$, it is common to say that "the lack of fit test is rejected and more runs are needed." Also, the lack of fit test is a formal hypothesis test like two-sample t-tests. Therefore, if q responses are tested simultaneously, the overall probability of wrongly finding significance is greater than the α used for each test. However, it is less than $q\alpha$ by the Bonferroni inequality. Yet, if accuracy is critically important, a high value of α should be used because that increases the chance that all the experimental runs will be used. With the full amount of runs and the full quadratic model form, prediction accuracy will likely be higher than if experimentation terminates at *Step 5*.

Table 13.12 Central composite designs for (a) 2 factors, (b) 3 factors, and (c) 4 factors

(a)				(b)					(c)					
Run	Block	x_1	x_2	Run	Block	x_1	x_2	x_3	Run	Block	x_1	x_2	x_3	x_4
1	1	0	0	1	1	1	1	1	1	1	−1	1	−1	−1
2	1	1	−1	2	1	1	−1	1	2	1	−1	1	−1	1
3	1	1	1	3	1	0	0	0	3	1	0	0	0	0
4	1	−1	1	4	1	0	0	0	4	1	1	−1	−1	−1
5	1	−1	−1	5	1	−1	−1	−1	5	1	1	−1	1	−1
6	1	0	0	6	1	−1	1	−1	6	1	−1	1	1	−1
7	1	0	0	7	1	−1	−1	1	7	1	−1	1	1	1
8	2	0	−1.41	8	1	−1	1	1	8	1	1	1	−1	1
9	2	−1.41	0	9	1	0	0	0	9	1	1	1	1	−1
10	2	0	0	10	1	1	−1	−1	10	1	−1	−1	1	−1
11	2	0	1.41	11	1	0	0	0	11	1	−1	−1	−1	−1
12	2	0	0	12	1	1	1	−1	12	1	1	−1	1	1
13	2	0	0	13	2	0	$-\alpha_C$	0	13	1	0	0	0	0
14	2	1.41	0	14	2	0	0	0	14	1	0	0	0	0

(continued)

Table 13.12 (continued)

(a)				(b)					(c)					
Run	Block	x_1	x_2	Run	Block	x_1	x_2	x_3	Run	Block	x_1	x_2	x_3	x_4
				15	2	0	0	$-\alpha_C$	15	1	−1	−1	−1	1
				16	2	$-\alpha_C$	0	0	16	1	0	0	0	0
				17	2	0	0	α_C	17	1	1	1	1	1
				18	2	0	α_C	0	18	1	1	−1	−1	1
				19	2	α_C	0	0	19	1	1	1	−1	−1
				20	2	0	0	0	20	1	−1	−1	1	1
									21	2	0	0	0	0
									22	2	α_C	0	0	0
									23	2	0	α_C	0	0
									24	2	0	0	0	α_C
									25	2	0	0	0	0
									26	2	$-\alpha_C$	0	0	0
									27	2	0	$-\alpha_C$	0	0
									28	2	0	0	$-\alpha_C$	0
									29	2	0	0	α_C	0
									30	2	0	0	0	$-\alpha_C$

Algorithm 13.4 Two-step sequential response surface methods

Step 1. Prepare the experimental design selected from the tables below to facilitate the scaling in *Step 2*. The selected design must be one of the central composite designs (CCDs), either immediately below or in the appendix at the end of the chapter.

Step 2 Scale the experimental design using the ranges selected by the experimenter. $D^s_{i,j} = L_j + 0.5(H_j - L_j)(D_{i,j} + 1)$ for $i = 1, \ldots, n$ and $j = 1, \ldots, m$.

Step 3. Build and test the prototypes according to only those runs in $\mathbf{D}^s$ that correspond to the runs with the "block" having a setting of 1. Record the test measurements for the responses for the n_1 runs in the n dimensional vector **Y**.

Step 4. Form the so-called "design" matrix, **X**, based on the *scaled* design, $\mathbf{D}^s$, based on the following model form, **f**(**x**):

$$\begin{aligned} &\mathrm{f}_1(\mathbf{x}) = 1, \mathrm{f}_j(\mathbf{x}) = x_{j-1} \text{ for } j = 2, \ldots, m+1 \\ &\qquad \text{and} \\ &\mathrm{f}_{m+2}(\mathbf{x}) = x_1 x_2,\ \mathrm{f}_{m+3}(\mathbf{x}) = x_1 x_3,\ \ldots,\ \mathrm{f}_{[(m+1)(m+2)/2]-m}(\mathbf{x}) \\ &\qquad = x_{m-1} x_m. \end{aligned} \tag{13.11}$$

(Note that the pure quadratic terms, e.g., x_1^2, are missing.) Then, for each of the q responses calculate the regression coefficients $\boldsymbol{\beta}_{est} = \mathbf{AY}$, where **A** is the $(\mathbf{X'X})^{-1}\mathbf{X'}$.

Step 5. Calculate "**mean squared lack-of-fit**" (*MSLOF*), $y_{variance,c}$, and the F-statistics, F_0 using the following:

$$\begin{aligned} &MSLOF = n_f n_c \left(y_{average,f} - y_{average,c}\right)^2 / \left(n_f + n_c\right) \quad \text{and} \\ &\mathrm{F}_0 = (\mathrm{MSLOF}) / \mathrm{y}_{\mathrm{variance},c} \end{aligned} \tag{13.12}$$

13

where $y_{variance,c}$ is the sample variance of the center point response values for the *r*th response. If $F_0 < F_{\alpha,1,nc-1}$, for all responses for which an accurate model is critical, then go to *Step 8*. Otherwise continue with *Step 6*. The values of $F_{\alpha,1,nc-1}$ are given in ◘ Table 13.9. The available prediction model is $\mathbf{f}_1(\mathbf{x})'\boldsymbol{\beta}_{est}$ based on the above model form with no pure quadratic terms.

Step 6. Build and test the remaining prototypes according to $\mathbf{D}^s$. Record the test measurements for the responses for the n_2 additional runs in the bottom of the $n_1 + n_2$ dimensional vector **Y**.

Step 7. Form the so-called "design" matrix, **X**, based on the *scaled* design, $\mathbf{D}^s$, following the rules for full quadratic model forms, **f**(**x**), as for one-shot methods. (The pure quadratic terms are included.) Then, calculate the regression coefficients $\boldsymbol{\beta}_{est} = \mathbf{AY}$, where **A** is the $(\mathbf{X'X})^{-1}\mathbf{X'}$.

Step 8. *(Optional)* Plot the prediction model, $y_{est}(\mathbf{x}) = \mathbf{f}(\mathbf{x})'\boldsymbol{\beta}_{est}$ for prototype system output to gain intuition about system inputs and output relationships. The example above shows how to make 3D plots using Excel and models of the above form.

Step 9. Apply informal or formal optimization using the prediction models, $y_{est}(\mathbf{x})$, …, $y_{est}(\mathbf{x})$ to develop recommended settings. Formal optimization is described in detail in ► Chap. 6.

Example 13.6.1 Lack of Fit

Question

Suppose that you had performed the first seven runs of a central composite design in two factors, and the average and standard deviation of the only critical response for the three repeated center points are 10.5 and 2.1 respectively. Further, suppose that the average response for the other four runs is 17.5. Perform a lack of fit analysis to determine whether adding additional runs is needed. Note that variance = (standard deviation)^2.

Answer

$MSLOF = [(4)(3)(10.5 - 17.5)^2]/(7) = 84.0$ and $F_0 = 84.0/(2.1^2) = 19.0$. $F_{0.05,1,2} = 18.51$. Since $F_0 > F_{0.05,1,2}$, the lack of fit of the first order model is significant. Therefore, the standard next steps (*Steps 6–9*) would be to perform the additional runs and fit a second order model. Even if we had failed to prove a lack of fit with an F-test, we might choose to add runs and perform an additional analysis to generate a relatively accurate prediction model. Stopping testing saves experimental expense but carries a risk that the derived prediction model may be relatively inaccurate.

13.7 Origin of RSM Designs and Decision-Making

In this section, the origins of the experimental planning matrices used in standard responses surface methods are described. The phrase "**experimental arrays**" is used to describe the relevant planning matrices. Also, information that can aid in decision-making about which array should be used is provided.

13.7.1 Origins of the RSM Experimental Arrays

In this chapter, three types of experimental arrays are presented. The first two types, central composite designs (CCDs) and Box Behnken designs (BBDs), are called standard response surface designs. The third type, EIMSE designs, constitutes one kind of optimal experimental design. Many other types of response surface method experimental arrays are described in Myers and Montgomery (2001).

Box and Wilson (1951) generated CCD arrays by combining three components as indicated by the example in ◘ Table 13.11. For clarity, ◘ Table 13.11 lists the design in standard order (SO), which is not randomized. To achieve proof and avoid problems, the matrix should not be used in this order. The run order should be randomized.

The first CCD component consists of a two level matrix similar or identical to the ones used for screening (▶ Chap. 12). Specifically, this portion is either a full factorials as in ◘ Table 13.11 or a so-called "Resolution V" regular fractional factorial. The phrase "Resolution V" refers to regular fractional factorials with the property that no column can be derived without multiplying at least four other columns together. For example, it can be checked that a 16 run regular fractional factorial with five factors and the generator E = ABCD is Resolution V. Resolution V implies that a model form with all two level interactions, e.g., $\beta_{10}x_2x_3$, can be fitted with accuracy that is often acceptable.

The phrase "center points" refers to experimental runs with all setting set to levels at the midpoint of the factor range. The second CCD component part consists of n_c center points. For example, if factor A ranges from 10 to 15 mm and factor B ranges from 30 to 40 °C, the center point settings would be 12.5 mm and 35 °C. The CCD might have $n_c = 3$ runs with these settings mixed in with the remaining runs. One benefit of performing multiple tests at those central values is that the magnitude of the experimental errors can be measured in a manner similar to measuring process capability in Xbar & R charting (▶ Chap. 4). One can simply take the sample standard deviation, *s*, of the response values from the center point runs.

Advanced readers may realize that the quantity $s \div c_4$ is an "assumption-free" estimate of the random error standard deviation, σ_0. This estimate can be compared with the one derivable from regression (▶ Chap. 15), providing a useful way to evaluate the lack of fit of the fitted model form in addition to the Adjusted R^2. This follows because the regression estimate σ_0 of contains contributions from model misspecification and the random error. The quantity $s \div c_4$ only reflects random or experimental errors and is not effected by the choice of fit model form.

13

The phrase "star points" refers to experimental runs in which a single factor is set to α_C or $-\alpha_C$ while the other factors are set at the mid values. The last CCD component part consists of two star points for every factor. One desirable feature of CCDs is that the value of α_C can be adjusted by the method user. The statistical properties of the CCD based RSM method are often considered acceptable for $0.5 < \alpha_C < \text{sqrt}[m]$, where *m* is the number of factors (◘ Table 13.13).

Box and Behnken (1960) generated BBD arrays by combining two components as shown in ◘ Table 13.12. The first component itself was the combination of two level arrays and sub-columns of zeros. In all the examples in this book, the two level arrays are two-factor full factorials.

In some cases, the sub-columns of zeros were deployed such that each factor was associated with one sub-column as shown in ◘ Table 13.12. Advanced readers may be interested to learn that the general structure of the zero sub-columns itself corresponded to experimental arrays called "partially balanced incomplete blocks" (PBIBs). Of all of the possible experimental arrays that can be generated using the combination of fractional factorials and zero sub-columns, Box and Behnken selected only those arrays that they could rigorously prove minimize the prediction errors caused by "model mis-specification" or bias. Prediction errors associated with bias are described next in the context of EIMSE optimal designs (◘ Table 13.14)

Table 13.13 Two-factor central composite design (CCD) in standard order

Standard order	A	B		
1	−1	−1		
2	1	−1	←	Regular fractional factorial part
3	−1	1		
4	1	1		
5	0	0		
6	0	0	←	Three "center points"
7	0	0		
8	α_C	0		
9	0	α_C	←	"Star" points
10	$-\alpha_C$	0		
11	0	$-\alpha_C$		

Table 13.14 Three factor Box Behnken design (BBD) in standard order

Standard order	A	B	C		
1	−1	−1	0		
2	1	−1	0	←	first repetition
3	−1	1	0		
4	1	1	0		
5	−1	0	−1		
6	1	0	−1	←	second repetition
7	−1	0	1		
8	1	0	1		
9	0	−1	−1		
10	0	1	−1	←	third repetition
11	0	−1	1		
12	0	1	1		
13	0	0	0		
14	0	0	0	←	threecenter points
15	0	0	0		

Allen et al. (2003) proposed "expected integrated mean squared error" (EIMSE) designs as the solution to an optimization problem. To understand their approach, consider that even though experimentation involves uncertainty, much can be predicted before testing begins.

In particular, the following generic sequence of activities can be anticipated in the context of one-shot RSM: tests are performed → a second order polynomial regression model will be fitted → predictions will be requested at settings of future interest. Building on research from Box and Draper (1987), Allen et al. (2003) were able to develop a formula to predict the squared errors that the experimental planner can expect using a given experimental array and generic sequence: perform tests → fit second order polynomial regression model → make predictions.

The assumptions that Allen et al. (2003) used were realistic enough to include contributions from random or "variance" errors from experimentation mistakes and "model bias" cased by limitations of the fitted model form. Bias errors come from a fundamental limitation of the fitted model form in its ability to replicate the twists and turns of the true system input-output relationships.

The formula developed by Allen et al. (2003) suggests that prediction errors are undefined or infinite if the number of runs, n, is less than the number of terms in the fitted model, k. This suggests a lower limit on the possible number of runs that can be used. Fortunately, the number of runs is otherwise unconstrained. The formula predicts that as the number of runs increases, the expected prediction errors decrease. This flexibility in the number of runs that can be used may be considered a major advantage of EIMSE designs over CCDs or BBDs. Advanced readers may realize that BBDs are a subset of the EIMSE designs in the sense that, for specific assumption choices, EIMSE designs also minimize the expected bias.

13.7.2 Decision Support Information (Optional)

13

This section explores concepts from Allen et al. (2003) and, therefore, previews material in ▶ Chap. 18. It is relevant to decisions about which experimental array should be used to achieve the desired prediction accuracy. Response surface methods (RSM) generate prediction models, $y_{est}(\mathbf{x})$ intended to predict accurately the prototype system's input-output relationships. Note that, in analyzing the general method, it is probably not obvious which combinations of settings, $\mathbf{x}$, will require predictions in the subjective optimization in the last step.

The phrase "**prediction point**" refers to a combination of settings, $\mathbf{x}$, at which prediction is of potential interest. The phrase "**region of interest**" refers to a set of prediction points, R. This name derives from the fact that possible settings define a vector space and the settings of interest define a region in that space.

The prediction model, $y_{est}(\mathbf{x})$ with the extra subscript is called an "**empirical model**" since it is derived from data. If there is only one response, then the subscript is omitted. Empirical models can derive from screening methods or standard response surface or from many other procedures including those that involve so-called "neural nets" (see ▶ Chap. 16).

The empirical model, $y_{est}(\mathbf{x})$, is intended to predict the average prototype system response at the prediction point **x**. Ideally, it can predict the engineered system response at **x**. Through the logical construct of a thought experiment, it is possible to develop an expectation of the prediction errors that will result from performing experiments, fitting a model, and using that model to make a prediction. This expectation can be derived even before real testing in an application begins. In a thought experiment, one can assume that one knows the actual average response of the prototype or engineered system would give at the point **x**, $y_{true}(\mathbf{x})$.

The "**true response**" or $y_{true}(\mathbf{x})$ at the point **x** is the imagined actual value of the average response at **x**. In the real world, we will likely never know $y_{true}(\mathbf{x})$, but it can be a convenient construct for thought experiments and decision support for RSM. The "**prediction errors**" at the point **x**, $\varepsilon(\mathbf{x})$, are the difference between the true average response and the empirical model prediction at **x**, i.e., $\varepsilon_r(\mathbf{x}) = y_{true}(\mathbf{x}) - y_{est}(\mathbf{x})$. Since $y_{true}(\mathbf{x})$ will likely be never known in real-world problems, $\varepsilon(\mathbf{x})$ will likely also not be known. Still, it may be useful in thought experiments pertinent to method selection to make assumptions about $\varepsilon(\mathbf{x})$.

Clearly, the prediction errors for a given response will depend on our beliefs about the true system being studied or, equivalently, about the properties of the true model, $y_{est}(\mathbf{x})$. For example, if the true model is very nonlinear or "bumpy", there is no way that a second order polynomial can achieve low prediction errors. ◘ Figure 13.5 illustrates this concept.

Many authors have explored the implications of specific assumptions about $y_{est}(\mathbf{x})$ including Box and Draper (1987) and Myers and Montgomery (2001). The assumptions explored in this section are that the true model is a third order polynomial. Third

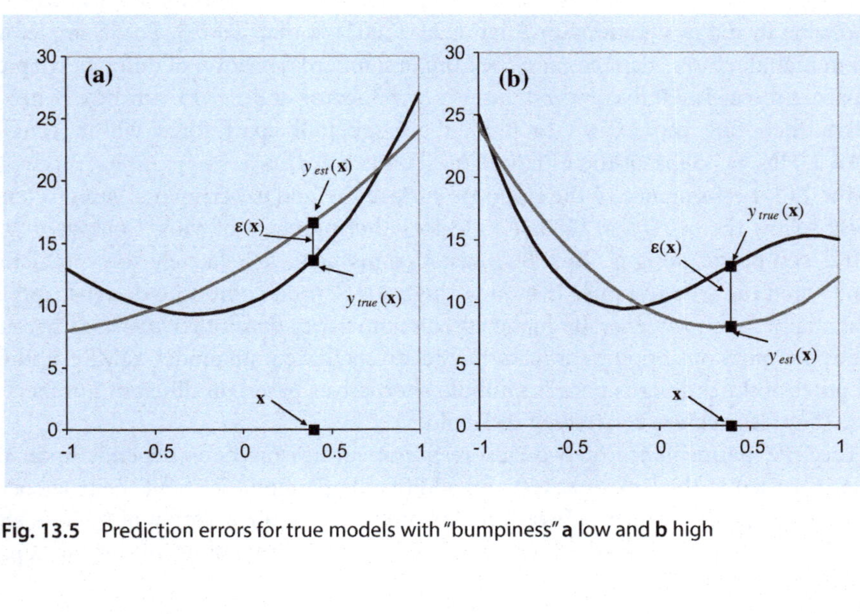

◘ **Fig. 13.5** Prediction errors for true models with "bumpiness" **a** low and **b** high

order polynomials contain all the terms in second order RSM models with the addition of third order terms involving, e.g., x_2^3 and $x_1^2x_2$. Further, it assumes that coefficients are random with standard deviation γ.

Definition

The "**expected prediction errors**" (EPE) are the expected value of the prediction errors, $E[\varepsilon(\mathbf{x})^2]$, with the expectation taken over all the quantities about which the experimenter is uncertain. The EPE is also known as the expected integrated mean squared error (EIMSE). Typically, random quantities involved in the expectation include, the coefficients of the true model, $\boldsymbol{\beta}$, the experimental random errors, ε, and the prediction points, $\mathbf{x}$.

Note that since the prediction errors depend upon the true model and thus $\boldsymbol{\beta}$ under certain assumptions, the expected prediction errors depend upon the standard deviation γ. An interesting result for all linear models is that the expected prediction errors only depend upon the standard deviation of the third order coefficients of the true model in relevant cases. Table 13.15 show the expected prediction errors for alternative RSM experimental designs, **D**. The two method criteria shown are $g_1 = n$, the number of runs and the expected prediction errors, g_2.

The expected prediction errors (EPE) also depend upon the standard deviation of the random errors, σ or "sigma", just as the criteria for screening methods depend on sigma. In practice, one estimates sigma by observing repeated system outputs for the same system input and taking the sample standard deviation. This provides a rough estimate of σ.

To estimate a typical prediction error that one can expect if one uses the RSM method in question, multiply the value in the table by σ. An assumption argued to be reasonable in many situations in Allen et al. (2003) is that $\gamma = 0.5$. For example, if a system manufactures snap tabs and the sample standard deviation of different snap tab
13 pull apart forces is 3.0 lbs. and one uses a $m = 3$ factor and $n = 15$ run Box Behnken design, then one can expect to predict average pull apart force within roughly 0.51×3.0 lbs. $= 1.5$ lbs. or the EPE in natural units is 1.5 lbs.

The EPE performance of the response surface method experimental designs tends to also follow the pattern in Table 13.15 for other numbers of runs. Compared with central composite designs, Box Behnken designs achieve relatively low prediction errors when the true response is bumpy (high γ). Central composite designs, applied sequentially, result in generally higher prediction errors than other methods because of the possibility of stopping earlier with a relatively inaccurate model. EIMSE-optimal and other optimal designs permit multiple alternatives based on different numbers of runs. They also achieve a variety of EPE values.

The EPE performance of two step response surface methods depends upon the value of α used in the lack of fit test and additional assumptions about the true coefficients of the quadratic terms. Therefore, for simplicity, a 40% inflation of the one-shot central composite based method was assumed based on simulations in working papers available from the author.

Table 13.15 Decision support for RSM with three and four factors

	(m)	($g_1=n$)	(g_2=EPE) Expected Prediction Errors			
Design	no. factors	no. runs	γ=0.0	γ=0.5	γ=1	γ=2
Box Behnken	3	15	0.42	0.51	0.88	2.38
EIMSE-optimal	3	11	0.86	0.97	1.38	3.03
EIMSE-optimal	3	16	0.46	0.54	0.82	1.97
Central composite	3	20	0.40	0.56	1.22	3.85
Central composite (two step*)	3	12 or 20	0.55	0.78	1.71	5.40
Box Behnken	4	30	0.48	0.64	1.24	3.67
EIMSE-optimal	4	26	0.43	0.58	1.15	3.46
Central composite	4	30	0.45	0.84	2.37	8.50
Central composite (two step*)	4	20 or 30	0.63	1.17	3.31	11.89

*Two step arrays facilitate stopping fitting a first order model or, alternatively, continuing to fit a full second order model

13.8 Appendix: Additional Response Surface Designs

See Tables 13.16, 13.17, 13.18, 13.19 and 13.20.

13.9 Chapter Summary

This chapter describes the application of so-called response surface methods (RSM). These methods generally result in a relatively accurate prediction of all response variable averages related to quantities measured during experimentation. An important reason why the predictions are relatively accurate is that so-called "interactions" which relate to the combined effects of factors are included explicitly in the predicted models.

Three types of methods were presented. Box Behnken designs (BBDs) were argued to generate relatively accurate predictions because they minimize so-called "bias" errors under certain reasonable assumptions. Central composite designs (CCDs) were presented and explained to offer the advantage that they permit certain level adjust-

Table 13.16 (**a**) 4 factor EIMSE-optimal and (**b**) 5 factor Box Behnken designs

(a)					(b)										
Run	x_1	x_2	x_3	x_4	Run	x_1	x_2	x_3	x_4	x_5	Run	x_1	x_2	x_3	x_4	x_5
1	−1	−1	−1	0	1	−1	−1	1	1	1	23	0	1	0	−1	−1
2	−1	−1	1	0	2	−1	−1	−1	0	−1	24	1	−1	0	0	−1
3	−1	0	0	1	3	−1	−1	0	−1	1	25	1	−1	1	−1	1
4	−1	0	0	−1	4	−1	−1	1	0	−1	26	1	−1	−1	−1	−1
5	−1	1	−1	1	5	−1	1	−1	1	1	27	1	−1	−1	−1	0
6	−1	1	1	0	6	−1	1	−1	−1	1	28	1	−1	0	−1	1
7	−1	1	−1	−1	7	−1	1	1	−1	−1	29	1	−1	1	1	1
8	0	−1	1	1	8	−1	−1	−1	1	−1	30	1	−1	1	1	−1
9	0	−1	1	−1	9	−1	−1	−1	−1	1	31	1	0	−1	1	0
10	0	−1	−1	1	10	−1	−1	0	−1	−1	32	1	0	−1	1	−1
11	0	−1	−1	−1	11	−1	−1	−1	1	1	33	1	1	1	−1	−1
12	0	−1	−1	0	12	−1	−1	1	1	−1	34	1	1	0	1	1
13	0	1	1	0	13	−1	0	1	−1	1	35	1	1	1	−1	1
14	0	1	1	1	14	−1	0	1	−1	0	36	1	1	1	1	−1
15	0	1	1	−1	15	−1	1	−1	−1	−1	37	1	−1	−1	1	1
16	0	1	−1	1	16	−1	1	0	1	-1	38	1	−1	1	−1	−1
17	0	1	−1	−1	17	−1	1	1	1	0	39	1	0	0	1	−1
18	1	−1	1	1	18	−1	1	1	1	1	40	1	1	−1	0	1
19	1	−1	−1	0	19	−1	1	−1	1	−1	41	1	1	1	0	1
20	1	−1	1	−1	20	−1	1	0	0	1	42	1	1	−1	−1	−1
21	1	0	0	1	21	0	−1	0	1	1	43	0	0	0	0	0
22	1	0	0	−1	22	0	1	−1	−1	1	44	0	0	0	0	0
23	1	1	−1	0												
24	1	1	1	0												
25	0	0	0	0												
26	0	0	0	0												

Table 13.17 Central composite designs for five factors

Run	Block	x_1	x_2	x_3	x_4	x_5
1	1	0	0	0	0	0
2	1	−1	−1	1	1	1
3	1	0	0	0	0	0
4	1	1	−1	1	1	−1
5	1	1	−1	−1	1	1
6	1	−1	−1	−1	−1	1
7	1	1	−1	−1	−1	−1
8	1	1	1	1	1	1
9	1	0	0	0	0	0
10	1	−1	−1	−1	1	−1
11	1	0	0	0	0	0
12	1	−1	1	−1	−1	−1
13	1	0	0	0	0	0
14	1	1	1	−1	1	−1
15	1	−1	−1	1	−1	−1
16	1	1	1	1	−1	−1
17	1	−1	1	1	-1	1
18	1	0	0	0	0	0
19	1	−1	1	1	1	−1
20	1	1	−1	1	−1	1
21	1	−1	1	−1	1	1
22	1	1	1	−1	−1	1
23	2	0	0	0	$-\alpha_c$	0
24	2	0	0	0	α_c	0
25	2	$-\alpha_c$	0	0	0	0
26	2	0	0	$-\alpha_c$	0	0
27	2	0	0	0	0	0
28	2	0	0	0	0	$-\alpha_c$
29	2	0	α_c	0	0	0
30	2	α_c	0	0	0	0
31	2	0	0	0	0	α_c
32	2	0	$-\alpha_c$	0	0	0
33	2	0	0	α_c	0	0

Table 13.18 Central composite designs for 6 factors (R = Run, B = Block)

R	B	x_1	x_2	x_3	x_4	x_5	x_6	R	B	x_1	x_2	x_3	x_4	x_5	x_6
1	1	−1	1	1	−1	−1	−1	28	1	1	−1	−1	1	−1	−1
2	1	−1	−1	1	1	1	1	29	1	0	0	0	0	0	0
3	1	−1	1	1	1	−1	1	30	1	1	1	1	1	−1	−1
4	1	0	0	0	0	0	0	31	1	−1	−1	1	−1	−1	1
5	1	−1	1	−1	−1	1	−1	32	1	0	0	0	0	0	0
6	1	1	1	−1	1	−1	1	33	1	0	0	0	0	0	0
7	1	1	−1	−1	−1	1	−1	34	1	1	1	−1	−1	1	1
8	1	1	−1	1	−1	−1	−1	35	1	1	−1	−1	1	1	1
9	1	−1	1	1	1	1	−1	36	1	1	1	1	1	1	1
10	1	0	0	0	0	0	0	37	1	0	0	0	0	0	0
11	1	−1	−1	−1	−1	−1	−1	38	1	1	−1	−1	−1	−1	1
12	1	−1	−1	−1	1	1	−1	39	1	1	1	1	−1	1	−1
13	1	1	−1	1	−1	1	1	40	1	−1	−1	−1	1	−1	1
14	1	−1	1	−1	−1	−1	1	41	2	0	α_C	0	0	0	0
15	1	1	1	−1	−1	−1	−1	42	2	0	0	0	0	0	0
16	1	1	−1	1	1	1	−1	43	2	α_C	0	0	0	0	0
17	1	1	1	−1	1	1	−1	44	2	0	0	0	0	0	$-\alpha_C$
18	1	−1	−1	1	−1	1	−1	45	2	0	0	0	α_C	0	0
19	1	−1	1	1	−1	1	1	46	2	0	$-\alpha_C$	0	0	0	0
20	1	−1	1	−1	1	1	1	47	2	0	0	0	0	α_C	0
21	1	1	1	1	−1	−1	1	48	2	0	0	0	0	0	0
22	1	0	0	0	0	0	0	49	2	0	0	0	0	$-\alpha_C$	0
23	1	0	0	0	0	0	0	50	2	0	0	0	$-\alpha_C$	0	0
24	1	−1	1	−1	1	−1	−1	51	2	0	0	$-\alpha_C$	0	0	0
25	1	−1	−1	1	1	−1	−1	52	2	$-\alpha_C$	0	0	0	0	0
26	1	1	−1	1	1	−1	1	53	2	0	0	α_C	0	0	0
27	1	−1	−1	−1	−1	1	1	54	2	0	0	0	0	0	α_C

Table 13.19 Box Behnken design for 6 factors

Run	x_1	x_2	x_3	x_4	x_5	x_6	Run	x_1	x_2	x_3	x_4	x_5	x_6
1	−1	0	0	−1	−1	0	28	0	0	0	0	0	0
2	0	−1	0	0	−1	−1	29	1	0	0	−1	1	0
3	0	−1	1	0	1	0	30	0	−1	1	0	−1	0
4	1	0	0	−1	−1	0	31	1	1	0	1	0	0
5	0	0	−1	1	0	1	32	0	−1	0	0	1	1
6	−1	0	0	1	−1	0	33	1	0	1	0	0	1
7	0	0	−1	1	0	−1	34	−1	0	−1	0	0	−1
8	0	0	1	1	0	−1	35	1	−1	0	1	0	0
9	0	0	1	−1	0	−1	36	0	1	0	0	−1	1
10	1	1	0	−1	0	0	37	−1	1	0	1	0	0
11	0	0	0	0	0	0	38	0	0	0	0	0	0
12	−1	−1	0	1	0	0	39	0	1	−1	0	1	0
13	1	0	1	0	0	−1	40	−1	0	0	−1	1	0
14	0	1	−1	0	−1	0	41	−1	0	−1	0	0	1
15	1	0	−1	0	0	1	42	0	0	0	0	0	0
16	−1	0	1	0	0	−1	43	1	0	−1	0	0	−1
17	1	0	0	1	1	0	44	0	1	0	0	1	−1
18	0	0	−1	−1	0	−1	45	−1	−1	0	−1	0	0
19	0	0	0	0	0	0	46	−1	0	0	1	1	0
20	0	−1	0	0	−1	1	47	0	0	1	−1	0	1
21	0	−1	−1	0	−1	0	48	0	1	1	0	1	0
22	0	0	0	0	0	0	49	1	−1	0	−1	0	0
23	0	1	0	0	1	1	50	0	−1	0	0	1	−1
24	−1	1	0	−1	0	0	51	0	0	1	1	0	1
25	0	1	1	0	−1	0	52	0	1	0	0	−1	−1
26	1	0	0	1	−1	0	53	−1	0	1	0	0	1
27	0	0	−1	−1	0	1	54	0	−1	−1	0	1	0

Table 13.20 Box Behnken design for 7 factors

Run	x_1	x_2	x_3	x_4	x_5	x_6	x_7	Run	x_1	x_2	x_3	x_4	x_5	x_6	x_7
1	−1	0	−1	0	−1	0	0	32	1	0	−1	0	−1	0	0
2	1	−1	0	1	0	0	0	33	−1	−1	0	−1	0	0	0
3	0	−1	1	0	0	1	0	34	1	0	0	0	0	−1	−1
4	0	0	1	−1	0	0	1	35	1	0	1	0	1	0	0
5	−1	1	0	−1	0	0	0	36	0	0	−1	−1	0	0	1
6	0	1	1	0	0	−1	0	37	0	0	0	1	1	−1	0
7	0	1	0	0	−1	0	1	38	−1	0	1	0	1	0	0
8	0	0	0	0	0	0	0	39	0	0	−1	1	0	0	1
9	1	1	0	1	0	0	0	40	1	1	0	−1	0	0	0
10	−1	0	0	0	0	−1	−1	41	1	0	1	0	−1	0	0
11	0	0	0	1	1	1	0	42	0	−1	1	0	0	−1	0
12	0	1	0	0	−1	0	−1	43	1	0	0	0	0	1	1
13	0	0	0	1	−1	1	0	44	0	0	1	1	0	0	−1
14	−1	0	−1	0	1	0	0	45	0	0	1	−1	0	0	−1
15	0	1	1	0	0	1	0	46	0	−1	−1	0	0	1	0
16	0	0	0	−1	−1	−1	0	47	0	−1	−1	0	0	−1	0
17	0	0	−1	−1	0	0	−1	48	0	−1	0	0	−1	0	1
18	1	0	−1	0	1	0	0	49	0	−1	0	0	1	0	1
19	0	0	0	0	0	0	0	50	1	0	0	0	0	1	−1
20	0	0	0	0	0	0	0	51	1	−1	0	−1	0	0	0
21	0	0	0	−1	1	1	0	52	0	0	0	−1	1	−1	0
22	0	0	0	1	−1	−1	0	53	0	0	0	0	0	0	0
23	0	1	0	0	1	0	−1	54	0	0	1	1	0	0	1
24	−1	0	0	0	0	−1	1	55	−1	−1	0	1	0	0	0
25	−1	0	1	0	−1	0	0	56	0	1	−1	0	0	1	0
26	0	0	0	0	0	0	0	57	−1	0	0	0	0	1	−1
27	0	1	−1	0	0	−1	0	58	−1	0	0	0	0	1	1
28	−1	1	0	1	0	0	0	59	0	−1	0	0	1	0	−1
29	1	0	0	0	0	−1	1	60	0	−1	0	0	−1	0	−1
30	0	1	0	0	1	0	1	61	0	0	−1	1	0	0	−1
31	0	0	0	−1	−1	1	0	62	0	0	0	0	0	0	0

ments and can be used in two-step sequential response surface methods. In these methods, there is a chance that the experimental will stop with relatively few runs and decide his or her prediction model is satisfactory.

The third class of experimental designs presented is the expected integrated mean squared error (EIMSE) designs which are available for a variety of numbers of runs and offer predictive advantages of Box Benken designs. The EIMSE criteria is also used at the end to clarify the relative prediction errors and to help method users decide whether a given experimental design is appropriate for their own prediction accuracy goals.

13.10 Problems

1. Which is correct and most complete?
   a. RSM is mainly relevant for finding which factor changes affect a response.
   b. Central composite designs have at most three distinct levels of each factor.
   c. Sequential response surface methods are based on central composite designs.
   d. All of the above are correct.
   e. All of the above are correct except (a) and (d).
2. Which is correct and most complete?
   a. In a design matrix, there is a row for every run.
   b. Functional forms fitted in RSM are not polynomials.
   c. Linear regression models cannot contain terms like $\beta_1 x_1^2$.
   d. Linear regression models are linear in all the factors (x's).
   e. All of the above are correct.
   f. All of the above are correct except (a) and (e).

Refer to Table 13.21 for Questions 3 and 4.

Table 13.21 (a) A two-factor DOE, (b)—(d) model forms, and (e) ranges

(a)			(b)	(c)	(d)	(e)
Run	A	B				
1	−1	−1	$y(\mathbf{x}) = \beta_1 + \beta_2 A + \beta_3 B$	$y(\mathbf{x}) = \beta_1 + \beta_2 A + \beta_3 B + \beta_4 AB$	$y(\mathbf{x})$ = full quadratic polynomial in A and B	Factor (−1) (+1) A 10.0 N 14.0 N B 2.5 mm 4.5 mm
2	1	−1				
3	−1	1				
4	0	0				
5	1	1				
6	−1.4	0				
7	0	0				
8	0	1.4				
9	0	−1.4				

3. Which is correct and most complete?
   a. A design matrix based on (a) and (b) in ◘ Table 13.19 would be 9 × 4.
   b. A design matrix based on (a) and (c) in ◘ Table 13.19 would be 9 × 4.
   c. A design matrix based on (a) and (d) in ◘ Table 13.19 would be 10 × 6.
   d. All of the above are correct.
   e. All of the above are correct except (a) and (d).
4. Which is correct and most complete?
   a. The model form in (c) in ◘ Table 13.19 contains one interaction term.
   b. Using the design matrix and model in (a) and (b) in ◘ Table 13.19, **X′X** is diagonal.
   c. Using the design matrix and model in (a) and (c) in ◘ Table 13.19, **X′X** is diagonal.
   d. All of the above are correct.
   e. All of the above are correct except (a) and (d).
5. Which is correct and most complete?
   a. Response surface methods cannot model interactions.
   b. In standard RSM, all factors must be continuous.
   c. Pure quadratic terms are contained in full quadratic models.
   d. All of the above are correct.
   e. All of the above are correct except (a) and (d).

   For Question 6, assume
   $f_1(\mathbf{x}) = 1$, $f_j(\mathbf{x}) = x_{j-1}$ for $j = 2, \ldots, m+1$
   and $f_{m+2}(\mathbf{x}) = x_1x_2$, $f_{m+3}(\mathbf{x}) = x_1x_3$, …, $f_{[(m+1)(m+2)/2]-m}(\mathbf{x}) = x_{m-1}x_m$.
6. Which is correct and most complete?
   a. With $m = 3$, $f_7(\mathbf{x}) = x_2x_3$.
   b. This model form contains pure quadratic terms.
   c. With $m = 4$, $f_6(\mathbf{x}) = x_1^2$.
   d. With $m = 5$, $f_2(\mathbf{x}) = 1$.
   e. All of the above are correct.
   f. All of the above are correct except (a) and (e).

13

7. How many factors and levels are involved in the paper airplane example?
8. According to the chapter, which is correct and most complete?
   a. EIMSE designs are an example of optimal or computer generated designs.
   b. Some EIMSE designs are available that have fewer runs than CCDs or BBDs.
   c. Both the choice of DOE matrix and of factor ranges affect design matrices.
   d. One shot RSM generates full quadratic polynomial prediction models.
   e. All of the above are correct.
   f. All of the above are correct except (a) and (e).
9. Which is correct and most complete?
   a. If **X** is $n \times k$ with $n > k$, **X′** (the transpose) is $n \times k$.
   b. If RSM is applied, $(\mathbf{X}'\mathbf{X})^{-1}\mathbf{X}'$ cannot be calculated for quadratic model forms because **X′X** is a singular matrix.
   c. Often, EIMSE designs are not available with fewer runs than CCDs or BBDs.
   d. Central composite designs include fractional factorial, star, and center points.
   e. All of the above are correct.
   f. All of the above are correct except (a) and (e).

10. Which is correct and most complete?
    a. If you only have enough money for a few runs, using screening without RSM might be wise.
    b. In general, factors in a DOE must be uncontrollable during experimentation.
    c. Adjusted R^2 is not relevant for evaluating whether data are reliable.
    d. The number of runs in RSM increases linearly in the number of factors..
    e. All of the above are correct.
    f. All of the above are correct except (a) and (e).
11. Which is correct and most complete?
    a. With three factors, two-step RSM cannot save costs compared with one-shot.
    b. In general, blocks in experimentation are essentially levels of the factor time.
    c. The original sequential RSM can be viewed as an optimization method.
    d. By repeating factor combinations, one can obtain an estimate of sigma.
    e. All of the above are correct.
    f. All of the above are correct except (a) and (e).

For problems 12 and 13, consider the array in Table 13.1(a) and the responses 7, 5, 2, 6, 11, 4, 6, 6, 8, 6 for runs 1, 2, …, 10 respectively. The relevant model form is a full quadratic polynomial.

12. Which is correct and most complete (within the implied uncertainty)?
    a. A full quadratic polynomial cannot be fitted since $(\mathbf{X}'\mathbf{X})^{-1}\mathbf{X}'$ is undefined.
    b. RSM fitted coefficients are 5.88, 1.83, 0.19, 0.25, 0.06, and 2.75.
    c. RSM fitted coefficients are 5.88, 2.83, 0.19, 0.25, 1.06, and 2.75.
    d. All of the above are correct.
    e. All of the above are correct except (a) and (d).
13. Which is correct and most complete (within the implied uncertainty)?
    a. Adjusted R^2 calculated is 0.99 a high fraction of the variation is unexplained.
    b. Adjusted R^2 calculated is 0.99 a high fraction of the variation is explained.
    c. Surface plots are irrelevant since the interaction coefficient is 0.0.
    d. All of the above are correct.
    e. All of the above are correct except (a) and (d).

For Question 14, suppose that you had performed the first seven runs of a central composite design in two factors, and the average and standard deviation of the only critical response for the three repeated center points are 10.5 and 2.1 respectively. Further, suppose that the average response for the other four runs is 17.5.

14. Which is correct and most complete (within the implied uncertainty)?
    a. $F_0 = 29$ and lack of fit is detected.
    b. $F_{0.05,1,nC-1} = 3.29$.
    c. $F_0 = 19$.
    d. All of the above are correct.
    e. All of the above are correct except (a) and (d).
15. Which is correct and most complete?
    a. In two stage RSM, interactions are never in the fitted model form.
    b. In two stage RSM, finding lack of fit indicates more runs should be performed.
    c. Two stage RSM cannot terminate with a full quadratic fitted model in all factors.
    d. In two stage RSM, lack of fit is determined using a t-test.

e. All of the above are correct.
f. All of the above are correct except (c) and (e).

16. Which is correct and most complete based on how the designs are constructed?
 a. Central composite designs do not, in general, contain center points.
 b. A BBD design with seven factors contains the run −1, −1, −1, −1, −1, −1, −1.
 c. Central composite designs contain Resolution V fractional factorials.
 d. CCDs and BBDs were generated originally using a computer.
 e. All of the above are correct.
 f. All of the above are correct except (a) and (e).
17. Which is correct and most complete (according to the text)?
 a. Expected prediction errors cannot be predicted before applying RSM.
 b. Box Behnken designs often foster more accurate prediction models than CCDs.
 c. Predictions about the accuracy of RSM depend on beliefs about the system.
 d. All of the above are correct.
 e. All of the above are correct except (a) and (d).
18. Assume that one had performed the first seven runs of a central composite design in two factors, and the average and standard deviations of the only critical response for the three repeated center points are 9.2 units and 3.1 units respectively. Further, suppose that the average response for the other four runs is 14.6 units. Perform a lack-of-fit analysis to determine whether adding additional runs is needed. Note that variance = (standard deviation)^2.
19. Assume a beverage company is trying to create a top cola by improving upon the taste of the one cola they have already. They plan on varying factors including the pressure, temperature and amount of each ingredient (bottle size could be increased). What method should be used and why?
20. Define the phrase "response surface methods" and include in your answer a description of the two classes of response surface methods described in the chapter.
21. Compare Box Behnken designs and central compose designs using at least two criteria.

References

Allen TT, Yu L, Schmitz J (2003) The expected integrated mean squared error experimental design criterion applied to die casting machine design. J R Stat Soc, Ser C: App Stat 52:1–15

Box GEP, Behnken DW (1960) Some new three-level designs for the study of quantitative variables. Technometrics 30:1–40

Box GEP, Draper NR (1987) Empirical model-building and response surfaces. Wiley, New York

Box GEP, Wilson KB (1951) On the experimental attainment of optimum conditions. J Roy Stat Soc B 13:1–45

Choudhury AK (1997) Study of the effect of die casting machine upon die deflections. Master's thesis, Industrial & Systems Engineering, The Ohio State University, Columbus

Myers R, Montgomery D (2001) Response surface methodology, 5th edn. John Wiley & Sons Inc, Hoboken, NJ

DOE: Robust Design

T. T. Allen, *Introduction to Engineering Statistics and Lean Six Sigma*,
https://doi.org/10.1007/978-1-4471-7420-2_14

14.1 Introduction

In ▶ Chap. 4, it is claimed that perhaps the majority of quality problems are caused by variation in quality characteristics. The evidence is that typically only a small fraction of units fail to conform to specifications. If characteristic values were consistent, then either 100% of units would conform or 0%. Robust design methods seek to reduce the effects of input variation on a system's outputs to improve quality. Therefore, they are relevant when one is interested in designing a system that gives consistent outputs despite the variation of uncontrollable factors.

Taguchi (1993) created several "Taguchi Methods" (TM) and concepts that strongly influenced design of experiments (DOE) method development related to robust design. He defined "**noise factors**" as system inputs, **z**, that are not controllable by decision-makers during normal engineered system operation but which are controllable during experimentation in the prototype system. For example, variation in materials can be controlled during testing by buying expensive materials that are not usually available for production. Let m_n be the number of noise factors so that **z** is an m_n dimensional vector. Taguchi further defined "**control factors**" as system inputs, $\mathbf{x}_c$, that are controllable both during system operation and during experimentation. For example, the voltage setting on a welding robot is fully controllable. Let m_c be the number of control factors so $\mathbf{x}_c$ is an m_c dimensional vector.

Consider that the *r*th quality characteristic can be written as $y_{est,r}$ ($\mathbf{x}_c$, **z**, ε) to emphasize its dependence on control factors, noise factors, and other factors that are completely uncontrollable, ε. Then, the goal of robust engineering is to adjust the settings in $\mathbf{x}_c$ so that the characteristic's value is within its specification limits, LSL_r and USL_r, and all other characteristics are within their limits consistently.

Figure 14.1a shows a case in which there is only one noise factor, *z*, and the control factor combination, $\mathbf{x}_1$, is being considered. For simplicity, it is also assumed that

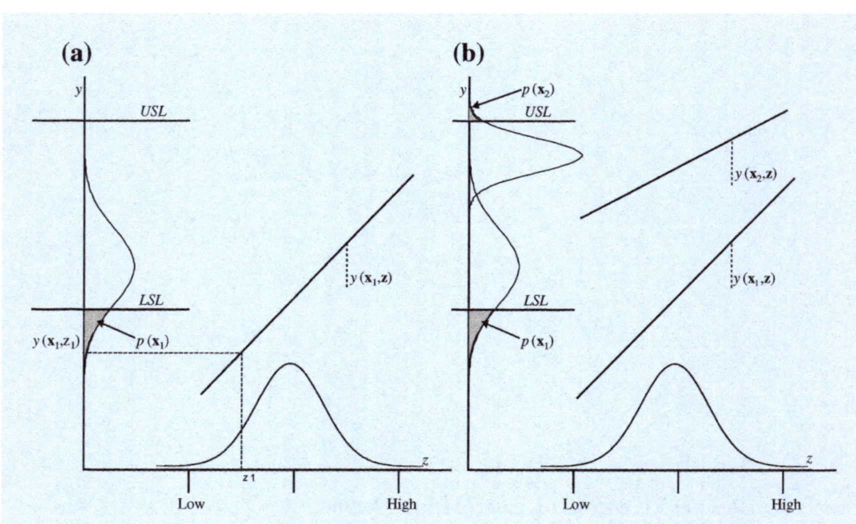

Fig. 14.1 Quality characteristic, *y*, distributions for choices **a** $\mathbf{x}_1$ and **b** $\mathbf{x}_1$ and $\mathbf{x}_2$

there is only one quality characteristic whose subscript is omitted. Also, sources of variation other than z do not exist, i.e., $\varepsilon = 0$, and the relationship between the quality characteristic, $y_{est}(\mathbf{x}_1, z, 0)$, and the z is as shown.

Figure 14.1a focuses on a particular value, $z = z_1$, and the associated quality characteristic value $y_{est}(\mathbf{x}_1, z, 0)$, which is below the specification limit. Also, the Fig. 14.1a shows a distribution for the noise factor under ordinary operations and how this distribution translates into a distribution of the quality characteristic. It also shows the fraction nonconforming, $p(\mathbf{x}_1)$, for this situation.

Figure 14.1b shows how different choices of control factor combinations could result in different quality levels. Because of the nature of the system being studied, the choice $\mathbf{x}_2$ results in less sensitivity of characteristic values than if $\mathbf{x}_1$ is used. As in control charting, sensitivity can be measured by the width of the distribution of the quality characteristic, i.e., the standard deviation, σ, or the process capability. It is also more directly measurable by the fraction nonconforming. It can be said that $\mathbf{x}_2$ settings are more robust than $\mathbf{x}_1$ settings because $p(\mathbf{x}_2) < p(\mathbf{x}_1)$.

In this chapter, multiple methods are presented, each with the goal of deriving robust system settings. First, methods are presented that are an extension of response surface methods (RSM) and are therefore similar to techniques in Lucas (1994) and Myers and Montgomery (2001). These first methods presented here are also based on formal optimization and expected profit maximization such that we refer to them as "robust design based on profit maximization" (RDPM). These methods were first proposed in Allen et al. (2001). Next, commonly used "static" Taguchi Methods are presented, which offer advantages in some cases.

14.2 Expected Profits and Control-by-Noise Interactions

RDPM focuses on the design of engineered systems that produce units. These units could be welded parts in a manufacturing line or patients in a hospital. The goal is to maximize the profit from this activity, which can be calculated as a sum of the revenues produced by the parts minus the cost to repair units that are not acceptable for various reasons.

To develop a realistic estimate of these profits as a function of the variables that the decision-maker can control, a number of quantities must be defined:

1. m is the total number of experimental factors.
2. q is the number of quality characteristics relevant to the system being studied.
3. $\mathbf{x}$ is an m dimensional vector of all experimental inputs which can be divided into two types, control factors, $\mathbf{x}_c$, and noise factors, $\mathbf{z}$, so that $\mathbf{x} = (\mathbf{x}_c'|\mathbf{z}')'$.
4. $\boldsymbol{\mu}_z$ and $\boldsymbol{\sigma}_z$ are m_n dimensional vectors containing the expected values, i.e., $\boldsymbol{\mu}_z = \mathrm{E}[\mathbf{z}]$, and standard deviations, i.e., $\sigma_{z,i} = \mathrm{sqrt}[\mathrm{E}(z_i - \mu_{z,i})^2]$, respectively.
5. $\mathbf{J}$ is a diagonal matrix with the variances of the noise factors under usual operations along the diagonal, i.e., $J_{i,i} = \sigma_{z,i}$ for $i = 1, \ldots, m_n$. (More generally, it is the variance-covariance matrix of the noise factors.)
6. $y_{est,0}(\mathbf{x}_c, \mathbf{z}, \varepsilon)$ is assumed to be the number of parts per year.
7. $y_{est,r}(\mathbf{x}_c, \mathbf{z}, \varepsilon)$ is the rth quality characteristic value function.

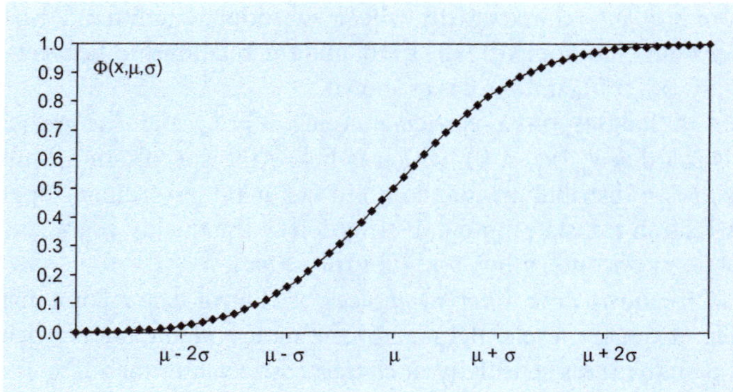

Fig. 14.2 The cumulative normal as a function of parameters μ and σ

8. $p_r(\mathbf{x}_c)$ is the fraction of nonconforming units as a function of the control factors for the rth quality characteristic.
9. w_0 is defined as the profit made per conforming unit.
10. w_r is the cost of the nonconformity associated with the rth characteristic. These failure costs include "rework" (e.g., cost of fixing the unit) and customer "**loss of good will**" (e.g., the cost of warranty costs and lost sales).
11. $\sigma_{total,r}(\mathbf{x}_c)$ is the "**total variation**" at a specific combination of control factors, $\mathbf{x}_c$, i.e., the standard deviation of the rth response taking into account the variation of the noise factors during normal system operation.
12. S_2 is the set of indices associated with responses that are failure probability estimates, and S_1 are all other indices.
13. $\Phi(x, \mu, \sigma)$ is the "**cumulative normal distribution function**", which is the probability that a normally distributed random variable with mean, μ, and standard deviation, σ, is less than x. Values are given by, e.g., Fig. 14.2 or the NORMDIST function in Excel.

14

To calculate $p_r(\mathbf{x}_c)$ under "standard assumptions," it is further necessary to make additional definitions. The full quadratic RSM model is re-written:

$$y_{est,r}(\mathbf{X}_c, \mathbf{Z}, \varepsilon) = \mathbf{f}_1(\mathbf{x})'\boldsymbol{\beta}_{est,r} = b_{0,r} + \mathbf{b}_r'\mathbf{x}_c + \mathbf{x}_c'\mathbf{B}_r\mathbf{x}_c + \mathbf{c}_r'\mathbf{z} + \mathbf{x}_c'\mathbf{C}_r\mathbf{z} + \mathbf{z}'\mathbf{D}_r\mathbf{z} + \varepsilon \quad \text{for all} r \in S_1 \tag{14.1}$$

where, $b_{0,r}$ is the constant coefficient, $\mathbf{b}_r$ is a vector of the first order coefficients of the controllable factors, $\mathbf{B}_r$ is a matrix of the quadratic coefficients of the controllable factors, $\mathbf{c}_r$ is a vector of the first order coefficients of the noise variables, $\mathbf{C}_r$ is a matrix of the coefficients of terms involving control factors and noise factors, and $\mathbf{D}_r$ is a matrix of the quadratic coefficients of the noise factors.

Therefore, the matrix $\mathbf{C}_r$ stores the coefficients of terms such as $x_2 z_1$ for the rth response. Taguchi coined the term "**control-by-noise interactions**" to refer to these terms together with their coefficients. For example, assume that $y_{est,2}(x_1, z_1) = 10.0 + 8.0x_1 + 5.0z_1 + 6.0x_1z_1$. Then, $\mathbf{C}_r$ is a 1×1 matrix given by the number {6.0}. For fixed x_1, it changes the slope of $y_{est,2}$ as a function of z_1.

Figure 14.1b shows how the nonparallelism associated with a control-by-noise interaction can make some control factor combinations more robust. Because of their potential importance in engineering, the phrase "**robustness opportunities**" refer to large control-by-noise factor interaction coefficients, i.e., large values in the $\mathbf{C}_r$ matrices. In some cases, all of these interactions coefficients can be zero and then the system does not offer an opportunity for improving the robustness by reducing the variation of the quality characteristic.

Example 14.2.1 Polynomials in Standard Format

Question

Write out a functional form which is a second order polynomial with two control factors and two noise factors and calculate the related **c** vector and **C** and **D** matrices assuming there is only one quality characteristic, so the index *r* is dropped for the remainder of this chapter.

Answer

The functional form is $y(x_1, x_2, z_1, z_2) = \beta_1 + \beta_2 x_1 + \beta_3 x_2 + \beta_4 z_1 + \beta_5 z_2 + \beta_6 x_1^2 + \beta_7 x_2^2 + \beta_8 z_1^2 + \beta_9 z_2^2 + \beta_{10} x_1 x_2 + \beta_{11} x_1 z_1 + \beta_{12} x_1 z_2 + \beta_{13} x_2 z_1 + \beta_{14} x_2 z_2 + \beta_{15} z_1 z_2$. The matrices are as follows

$$\mathbf{c} = \begin{pmatrix} \beta_4 \\ \beta_5 \end{pmatrix}, \ \mathbf{c} = \begin{pmatrix} \beta_{11} & \beta_{12} \\ \beta_{13} & \beta_{14} \end{pmatrix}, \text{ and } \mathbf{D} = \begin{pmatrix} \beta_8 & 0.5\beta_{15} \\ 0.5\beta_{15} & \beta_9 \end{pmatrix} \quad (14.2)$$

In this chapter, we focus on methods based on $\mathbf{D} = \mathbf{0}$ for convenience. A more complicated and advantageous procedure is in Allen et al. (2001). In general, $\mathbf{D} \neq \mathbf{0}$ and (14.1) is a quadratic form in random variables as described in Johnson and Kotz (1995). With all these definitions and assumptions, the yearly profit, Profit($\mathbf{x}_c$), from running the engineered system can be written

$$\text{Profit}(\mathbf{x}_c) = y_{est,0}(\mathbf{x}_c) \times \{w_0 - \sum_{r \in S1} w_r p_r(\mathbf{x}_c) - \sum_{r \in S2} w_r y_{est,r}(\mathbf{x}_c, \mathbf{z}) \quad (14.3)$$

where

$$p_r(\mathbf{x}_c) = \Phi[LSL_r, \mu_r(\mathbf{x}_c), \sigma_{total,r}(\mathbf{x}_c)] + \{1 - \Phi[USL_r, \mu_r(\mathbf{x}_c), \sigma_{total,r}(\mathbf{x}_c)]\} \quad (14.4)$$

and

$$\mu_r(\mathbf{x}_c, \boldsymbol{\mu}_z) = y_{est,r}(\mathbf{x}_c, \mathbf{z} = \boldsymbol{\mu}_z) \quad (14.5)$$

and

$$\sigma^2_{total,r}(\mathbf{x}_c, \boldsymbol{\sigma}_z, \sigma_\varepsilon) = \sigma_r^2 + (\mathbf{c}_r' + \mathbf{x}_c' \mathbf{C}_r)\mathbf{J}(\mathbf{c}_r' + \mathbf{x}_c' \mathbf{C}_r)' \quad (14.6)$$

where the identity Var[$\mathbf{T}$ $\mathbf{z}$] = $\mathbf{T}$ Var[$\mathbf{z}$] $\mathbf{T}'$ for constant matrix $\mathbf{T}$ has been used. Equations (14.4)–(14.6) all hold under two assumptions. First, $\mathbf{D} = 0$ so that noise-by-noise interactions are ignored. Second, the noise factor values under ordinary operations are normally distributed. Allen et al. (2001) explored more general assumptions that are omitted here for simplicity.

14.3 Robust Design Based on Profit Maximization

Robust Design based on Profit Maximization (RDPM) methods generally require all of the inputs that response surface methods (RSM) require. These include (1) an "experimental design", $\mathbf{D}^s$ and (2) vectors that specify the highs, $\mathbf{H}$, and lows, $\mathbf{L}$, of each factor. In addition, they require (3) the declaration of which factors $\mathbf{x}_c$ are control and which are noise $\mathbf{z}$.

Algorithm 14.1. Robust Design based on Profit Maximization

Step 1: If models of the $p_r(\mathbf{x}_c)$ for all quality characteristics are available, go to *Step 6*. Otherwise continue.
Step 2: For each quality characteristic for which $p_r(\mathbf{x}_c)$ is not available, include the associated response index in the set S_1 if the response is a quality characteristic. Include the response in the set S_2 if the response is the fraction nonconforming with respect to at least one type of nonconformity. Also, identify the specification limits, LSL_k and USL_k, for the responses in the set S_1.
Step 3: Apply a response surface method (all steps except the last, optimization step) to obtain an empirical model of all quality characteristics including the production rate, $y_{est,r}(\mathbf{x}_c, \mathbf{z})$ for $r = 1, \ldots, q$.
Step 4: Estimate the expected value, $\mu_{z,i}$, and standard deviation, $\sigma_{z,i}$, of all the noise factors relevant under normal system operation for all $i = 1, \ldots, m_n$.
Step 5: Estimate the failure probabilities as a function of the control factors, $p_r(\mathbf{x}_c)$ for all quality characteristics, $r \in S_1$, using the formulas in Eqs. (14.3)–(14.5).
Step 6: Obtain cost information in the form of revenue per unit, w_0, and rework and/or scrap costs per defect or nonconformity of type w_r for $r = 1, \ldots, q$.
Step 7: Maximize the profit, Profit($\mathbf{x}_c$), in Eq. (14.3) as a function of $\mathbf{x}_c$

The profit formulation in Eq. (14.3) can be adjusted to the particular situation. For example, sometimes cost information, $w_0, \ldots, w_q$ is not available or other considerations besides the cost of quality are relevant. Alternatively, the control factors might not affect the production rate, e.g., if the process in question is not a manufacturing system or even if it is a system in manufacturing but the related operations are not bottleneck operations. In these cases, it may be useful to adjust the formulation subjectively. Then, the resulting solutions should at least provide insight into which settings result in consistent system outputs. Also, as long as unit specifications are involved, then it is likely that the failure probability functions derived in *Steps* 2–6 will be useful.

14

Example 14.3.1 RDPM and Central Composite Designs

In this section, the proposed methods are illustrated through their application to the design of a robotic gas metal arc-welding (GMAW) cell. This case study is based on a research study at the Ohio State University documented in Allen et al. (2001, 2002).

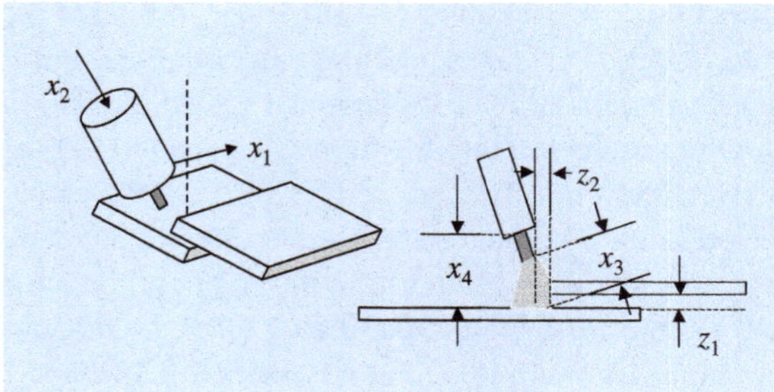

Fig. 14.3 The control and noise factors for the arc welding example

In that study, there were $m_n = 2$ noise factors, z_1 and z_2, $m - m_n = 4$ control factors, $x_1, \ldots, x_{-4}$. These factors are shown in Fig. 14.3. We chose two-step response surface methods because we were not sure that the factor ranges contained the control and noise settings associated with desirable arc welding systems, taking into account the particular power supply and type of material. The two-step approach offered the potentially useful option of performing only 40 tests and stopping with both screening related results and information about two-factor interactions. The central composite design shown with two blocks is given in Table 14.1.

In this study, there were three relevant responses. The rate of producing units was directly proportional to the control factor x_1. Therefore, before doing experiments, we knew that $y_{est,0}(\mathbf{x}_c) = 0.025x_1$ in millions of parts. The other two relevant responses were quality characteristics of the parts produced by the system.

Algorithm 14.2. RDPM and central composite design example

Step 1: In this application, models of $p_1(\mathbf{x}_c)$ and $p_2(\mathbf{x}_c)$ were not readily available. Therefore, it was necessary to go to *Step 2*.

Step 2: In this step, two relevant quality characteristics were identified corresponding to the main ways the units failed inspection or "**failure modes**". In order to save inspection costs and create continuous criteria, the team developed a continuous (1–10) rating system based on visual inspection for each type of described in the first table in ▶ Chap. 10 was utilized. Also, the specification limits $LSL_1 = LSL_2 = 8.0$ and $USL_1 = USL_2 = \infty$ were assigned. Therefore, higher ratings corresponded to better welds.

Step 3: The first 40 experiments shown in Table 14.1 were performed using the central composite design. After the first 40 runs, $n_c =$ center points and $n_f =$ 32 fractional factorial runs. $MSLOF_1 = 19.6$, $y_{variance,c,1} = 0.21$, $F_{0,1} = 91.5 >> F_{0.05,1,7} > 5.59$ so the remainder of the runs in the table below were needed. For, thoroughness we calculated $MSLOF_2 = 24.9$, $y_{variance,c,2} = 0.21$, $F_{0,2} = 116.3 >> F_{0.05,1,7} > 5.59$. Therefore, curvature is significant for both responses. Therefore, also, we performed the remainder of the runs given below. After all of the runs were performed, we estimated coefficients using $\boldsymbol{\beta}_{est,r} = \mathbf{AY}_r$ for $r = 1$ and 2, where $\mathbf{A} = (\mathbf{X}'\mathbf{X})^{-1}\mathbf{X}'$. These multiplications performed using matrix functions in Excel ("Ctrl-Shift-Enter" instead of OK is needed for assigning function values to multiple cells), but the coefficients could have derived using many choices of popular statistical software. Then, we rearranged the coefficients into the form listed in Eq. (14.7), and for the other response related to a quality characteristic in Eq. (14.8) below.

Step 4: The expected value, $\mu_{z,i}$, and the standard deviations, $\sigma_{z,i}$, of the noise factors were based on verbal descriptions from the engineers on our team. Gaps larger than 1.0 mm and offsets larger than ±1.0 of the wire diameters were considered unlikely, where 1.0 WD corresponds to 1.143 mm. Therefore, it was assumed that z_1 was N (mean = 0.25, standard deviation = 0.25) distributed in mm and z_2 was N (mean = 0, standard deviation = 0.5) distributed in wire diameters with zero correlation across runs and between the gaps and offsets. Note that these assumptions gave rise to some negative values of gap, which were physically impossible but were necessary for the analytical formula in Eq. (14.6) to apply. In addition, it was assumed that ε_1 and ε_2 were both N (mean = 0.0, standard deviation = 0.5) based on the sample variances (both roughly equal to 0.25 rating units) of the repeated center points in our experimental design.

Step 5: Based on the "standard assumptions" the failure probability functions were found to be as listed in Eq. (14.9).

Step 6: The team selected (subjectively since there was no real engineered system), w_0 = \$100 revenue per part, w_1 = \$250 per unit and w_2 = \$100 per unit based on rework costs. Burning through the unit was more than twice as expensive to repair since additional metal needed to be added to the part structure as well as the weld. The travel speed was related to the number of parts per minute by the simple relation, $y_{est,0}(\mathbf{x}_c) = 0.025x_1$ in millions of parts, where x_1 was in millimeters per minute.

Step 7: The formulation then became: minimize $0.025x_1[\$100 - p_1(\mathbf{x}_c)\ \$250 - p_2(\mathbf{x}_c)\ \$100]$ where $p_1(\mathbf{x}_c)$ and $p_2(\mathbf{x}_c)$ were given in Eq. (14.3). The following additional constraints, listed in Eq. (14.10) were added because the prediction model was only accurate over the region covered by the experimental design in ◘ Table 14.1.

14

Step 8: This problem was solved using the Excel solver, which uses GRG2 and the solution was x_1 = 1533.3 mm/min, x_2 = 6.83, x_3 = 3.18 mm, and x_4 = 15.2 mm, which achieved an expected profit of \$277.6/min. The derived settings offer a compromise between making units at a high rate and maintaining consistent quality characteristic values despite the variation of the noise factors (gap and offset)

$$b_{0,1} = -179.9$$

$$\mathbf{b}_1' = (0.0 \quad 0.70 \quad 5.11 \quad 23.7)$$

Table 14.1 Welding data from a central composite experimental design

Run	Block	x_1 (TS) (mm/min)	x_2 (R) (–)	x_3 (AL) (mm)	x_4 (CTTW) (mm)	z_1 (Gap) (mm)	z_2 (Offset) (WD)	a_1 (Burn) (0–10)	a_2 (Fusion) (0–10)
1	1	1778.0	6.0	4.8	16.0	1	-0.5	8	3
2	1	1778.0	8.0	4.8	14.0	0	0.5	5	5
3	1	1524.0	7.0	4.0	15.0	0.5	0	10	8
4	1	1270.0	8.0	4.8	14.0	0	−0.5	9	8
5	1	1270.0	6.0	4.8	14.0	0	0.5	10	4
6	1	1270.0	8.0	3.2	16.0	0	−0.5	10	10
7	1	1270.0	6.0	4.8	16.0	1	0.5	9	2
8	1	1524.0	7.0	4.0	15.0	0.5	0	9	8
9	1	1778.0	6.0	3.2	14.0	1	−0.5	8	3
10	1	1778.0	8.0	4.8	16.0	1	0.5	4	4
11	1	1524.0	7.0	4.0	15.0	0.5	0	9	8
12	1	1270.0	6.0	3.2	16.0	0	0.5	10	8
13	1	1270.0	8.0	3.2	14.0	1	−0.5	8	8
14	1	1778.0	6.0	4.8	14.0	0	−0.5	9	8
15	1	1524.0	7.0	4.0	15.0	0.5	0	9	8
16	1	1778.0	8.0	3.2	14.0	1	0.5	2	8
17	1	1778.0	6.0	3.2	16.0	0	−0.5	9	8
18	1	1270.0	6.0	3.2	14.0	1	0.5	9	3

(continued)

14

Table 14.1 (continued)

Run	Block	x_1 (TS) (mm/min)	x_2 (R) (–)	x_3 (AL) (mm)	x_4 (CTTW) (mm)	z_1 (Gap) (mm)	z_2 (Offset) (WD)	a_1 (Burn) (0–10)	a_2 (Fusion) (0–10)
19	1	1270.0	8.0	4.8	16.0	1	−0.5	8	8
20	1	1778.0	8.0	3.2	16.0	0	0.5	5	8
21	1	1270.0	8.0	3.2	16.0	1	0.5	5	7
22	1	1778.0	6.0	3.2	14.0	0	0.5	7	7
23	1	1270.0	6.0	3.2	16.0	1	−0.5	10	8
24	1	1524.0	7.0	4.0	15.0	0.5	0	9	9
25	1	1778.0	6.0	4.8	16.0	0	0.5	8	6
26	1	1778.0	8.0	4.8	16.0	0	−0.5	6	8
27	1	1270.0	8.0	4.8	16.0	0	0.5	8	9
28	1	1778.0	8.0	4.8	14.0	1	−0.5	4	5
29	1	1524.0	7.0	4.0	15.0	0.5	0	9	8
30	1	1778.0	8.0	3.2	14.0	0	−0.5	4	8
31	1	1778.0	8.0	3.2	16.0	1	−0.5	8	8
32	1	1524.0	7.0	4.0	15.0	0.5	0	9	9
33	1	1270.0	6.0	4.8	14.0	1	−0.5	9	2
34	1	1270.0	8.0	4.8	14.0	1	0.5	5	7
35	1	1270.0	6.0	3.2	14.0	0	−0.5	10	9
36	1	1778.0	6.0	4.8	14.0	1	0.5	8	2

(continued)

Table 14.1 (continued)

Run	Block	x_1 (TS) (mm/min)	x_2 (R) (–)	x_3 (AL) (mm)	x_4 (CTTW) (mm)	z_1 (Gap) (mm)	z_2 (Offset) (WD)	a_1 (Burn) (0–10)	a_2 (Fusion) (0–10)
37	1	1778.0	6.0	3.2	16.0	1	0.5	8	2
38	1	1270.0	8.0	3.2	14.0	0	0.5	8	9
39	1	1270.0	6.0	4.8	16.0	0	−0.5	9	7
40	1	1524.0	7.0	4.0	15.0	0.5	0	10	8
41	2	1524.0	7.0	4.0	15.0	0.5	0	10	8
42	2	1524.0	5.0	4.0	15.0	0.5	0	10	8
43	2	1524.0	7.0	4.0	15.0	0.5	−1	10	8
44	2	1524.0	7.0	2.4	15.0	0.5	0	9	8
45	2	1524.0	9.0	4.0	15.0	0.5	0	8	10
46	2	1524.0	7.0	4.0	15.0	0.5	1	8	8
47	2	1016.0	7.0	4.0	15.0	0.5	0	9	8
48	2	1524.0	7.0	4.0	13.0	0.5	0	5	8
49	2	1524.0	7.0	4.0	15.0	−0.5	0	10	8
50	2	1524.0	7.0	4.0	17.0	0.5	0	8	8
51	2	2032.0	7.0	4.0	15.0	0.5	0	8	5
52	2	1524.0	7.0	4.0	15.0	1.5	0	8	2
53	2	1524.0	7.0	4.0	15.0	0.5	0	10	9
54	2	1524.0	7.0	5.6	15.0	0.5	0	10	8

$$\mathbf{B}_1 = \begin{pmatrix} 0.00 & 0.00 & 0.00 & 0.00 \\ 0.00 & -0.22 & 0.00 & 0.25 \\ 0.00 & 0.00 & -0.15 & -0.31 \\ 0.00 & 0.00 & 0.00 & -0.85 \end{pmatrix} \tag{14.7}$$

$$\mathbf{c}_1' = (-2.02 \quad 5.90)$$

$$\mathbf{c}_1 = \begin{pmatrix} 0.00 & 0.00 \\ -0.50 & -0.75 \\ -0.31 & 0.62 \\ 0.25 & -0.25 \end{pmatrix}$$

$$\mathbf{D}_1 = \begin{pmatrix} -0.88 & -1.00 \\ 0.00 & -0.88 \end{pmatrix}$$

$$\mathrm{b}_{0,2} = 88.04$$

$$\mathbf{b}_2' = (\,0.04 \;\; 1.70 \;\; -0.97 \;\; 7.75\,)$$

$$\mathbf{B}_2 \begin{pmatrix} 0.00 & 0.00 & 0.00 & 0.00 \\ 0.00 & 0.06 & 0.08 & -0.06 \\ 0.00 & 0.00 & -0.30 & 0.08 \\ 0.00 & 0.00 & 0.00 & -0.18 \end{pmatrix} \tag{14.8}$$

$$\mathbf{c}_2' = (\,-4.55 \;\; 2.25\,)$$

$$\mathbf{c}_2 = \begin{pmatrix} 0.00 & 0.00 \\ 1.38 & 0.50 \\ -0.16 & 0.00 \\ -0.12 & -0.50 \end{pmatrix}$$

$$\mathbf{D}_2 = \begin{pmatrix} -3.78 & 0.00 \\ 0.00 & -0.78 \end{pmatrix}$$

$$\begin{aligned} p_1(\mathbf{X}) =& 1 + \Phi\left\{\begin{array}{l} 8, \mu_1(x) = y_{est,1}(\mathbf{X}, Z_1 = 0.25, z_2 = 0.0), \\ \sigma_1(\mathbf{x}) = \left[0.25^2 + \right. \\ \left.(\mathbf{c}_1' + \mathbf{X}'\mathbf{C}_1)\mathbf{J}(\mathbf{c}_1' + \mathbf{X}'\mathbf{C}_1)'\right]^{1/2} \end{array}\right\} \\ & - \Phi\left\{\begin{array}{l} \infty, \mu_1(x) = y_{est,1}(\mathbf{X}, Z_1 = 0.25, 0.0), \\ \sigma_1(\mathbf{x}) = \left[0.25^2 + \right. \\ \left.(\mathbf{c}_1' + \mathbf{X}'\mathbf{C}_1)\mathbf{J}(\mathbf{c}_1' + \mathbf{X}'\mathbf{C}_1)'\right]^{1/2} \end{array}\right\} \end{aligned}$$

14

$$p_2(\mathbf{X}) = 1 + \Phi\left\{\begin{array}{l} 8, \mu_2(x) = y_{est,2}(\mathbf{X}, 0.25, 0.0), \\ \sigma_2(\mathbf{x}) = \left[0.25^2 + \right. \\ \left.(\mathbf{c}_2' + \mathbf{X}'\mathbf{C}_2)\mathbf{J}(\mathbf{c}_2' + \mathbf{X}'\mathbf{C}_2)'\right]^{1/2} \end{array}\right\} - \Phi\left\{\begin{array}{l} \infty, \mu_2(x) = y_{est,2}(\mathbf{X}, 0.25, 0.0), \\ \sigma_2(\mathbf{x}) = \left[0.25^2 + \right. \\ \left.(\mathbf{c}_2' + \mathbf{X}'\mathbf{C}_2)\mathbf{J}(\mathbf{c}_2' + \mathbf{X}'\mathbf{C}_2)'\right]^{1/2} \end{array}\right\} \quad (14.9)$$

with

$$\mathbf{J} = \begin{bmatrix} 0.25 & 0 \\ 0 & 0.25 \end{bmatrix}$$

$$\begin{array}{rcl} 1270.0 \le & x_1 & \le 1778.0 \\ 6.0 \le & x_2 & \le 8.0 \\ 3.175 \le & x_3 & \le 4.763 \\ 14.0 \le & x_4 & \le 16.0 \end{array} \quad (14.10)$$

Example 14.3.2 RDPM and Six Sigma

Question
What is the relationship between six sigma and RDPM methods?

Answer
RDPM uses RSM and specific formulas to model directly the standard deviation or "sigma" of responses as a function of factors that can be controlled. Then, it uses these models to derive settings and sigma levels that generate the highest possible system profits. Applying RDPM could a useful component in a six sigma type improvement system.

14.4 Extended Taguchi Methods

The RDPM methods described above have the advantage that they build upon standard response surface methods in ▸ Chap. 13. They also derive an optimal balance between quality and productivity. Next, the original or "static" Taguchi methods are described which offer benefits including relative simplicity.

All design of experiments involve (1) experimental planning, (2) measuring selected responses, (3) fitting models after data is collected, and (4) decision-making. Taguchi refers to his methods as the "Taguchi System" because they consist of innovative, integrated approaches for all of the above. Taguchi Methods approaches for measuring responses and decision-making cannot be used without the application of Taguchi's experimental planning strategies.

The methods described in this section (see Algorithm 14.3) are called "extended" because Taguchi originally focused on approaches to improve single response variable or continuous quality characteristic values. Song et al. (1995) invented the methods

Table 14.2 A Taguchi product array: a in product format and b in standard order

(a)

Outer array			D	-1	1	-1	1
			E	-1	-1	1	1
Inner array			F	1	-1	-1	1
Run	A	B	C				
1	-1	-1	1	y_1	y_2	y_3	y_4
2	1	-1	-1	y_5	y_6	y_7	y_8
3	-1	1	-1	y_9	y_{10}	y_{11}	y_{12}
4	1	1	1	y_{13}	y_{14}	y_{15}	y_{16}

(b)

Run	A	B	C	D	E	F	Y
1	-1	-1	1	-1	-1	1	y_1
2	1	-1	-1	-1	-1	1	y_2
3	-1	1	-1	-1	-1	1	y_3
4	1	1	1	-1	-1	1	y_4
5	-1	-1	1	1	-1	-1	y_5
6	1	-1	-1	1	-1	-1	y_6
7	-1	1	-1	1	-1	-1	y_7
8	1	1	1	1	-1	-1	y_8
9	-1	-1	1	-1	1	-1	y_9
10	1	-1	-1	-1	1	-1	y_{10}
11	-1	1	-1	-1	1	-1	y_{11}
12	1	1	1	-1	1	-1	y_{12}
13	-1	-1	1	1	1	1	y_{13}
14	1	-1	-1	1	1	1	y_{14}
15	-1	1	-1	1	1	1	y_{15}
16	1	1	1	1	1	1	y_{16}

described here to address decision-making involving multiple quality characteristics (as RDPM does). Often, there is more than a single quality characteristic so that there are multiple signal-to-noise ratios and *Step 4* is ambiguous. To address this issue, Song et al. (1995) proposed an "extended Taguchi Method" that involves calculating signal-to-noise ratios for all characteristics and then clarifying which control factor settings are not obviously dominated by other settings. After the obviously poor settings have been removed from consideration, they suggested deciding between the remaining settings based on engineering judgment (Table 14.2).

Algorithm 14.3. Extended Taguchi methods

Step 1. Plan the experiment using so-called "product" arrays. Product arrays are based on all combinations or runs from a "inner array" and an "outer array" which are smaller arrays. Table 14.2 shows an example of a product array based on inner and outer arrays which are four run regular fractional factorials. Taguchi uses many combinations of inner and outer arrays. Often the 18 run array in Table 12.12 is used for the inner array. Taguchi also introduced a terminology such that the regular design in Table 14.2 is called an "L_4" design.

Table 14.2 shows the same experimental plan in two formats. In total, there are 16 runs. The notation implies that there is a single response variable with 16 response data. Taguchi assigns control factors to the inner array, e.g., factors A, B, and C, and the noise factors to the outer array, e.g., factors D, E, and F. In this way, each row in the product format in Table 14.2a describes the consistency and quality associated with a single control factor combination. Note that writing out the experimental plan in "combined array" format as in Table 14.2b can be helpful for ensuring the the runs are performed in a randomized order. The array in Table 14.2 is not randomized to clarify that the experimental plan is the same as the one in Table 14.2a.

Step 2. Once the tests have been completed according to the experimental design, Taguchi based analysis on so-called "signal-to-noise ratio" (SNR) that emphasize consistent performance regardless of noise factor setting for each control factor combination. Probably three most commonly used signal-to-noise are "smaller-the-better" (SNRS), "larger-the-better" (SNRL), and "nominal-is-best" (SNRN). These are appropriate for cases in which high, low, and nominal values of the quality characteristic are most desirable, respectively. Formulas for the characteristic values are:
For example, using the experimental plan in Table 14.2, SNRS value for the first inner array run would equal:

$$\begin{aligned} \text{SNRS} &= -10\,\text{Log}_{10}\left[\text{mean of sum of squares of measured data}\right] \\ \text{SNRL} &= -10\,\text{Log}_{10}\left[\text{mean of sum squares of reciprocal of measured data}\right] \\ \text{SNRN} &= -10\,\text{Log}_{10}\left[\text{mean of sum of squares of \{measured} - \text{ideal\}}\right]. \end{aligned} \tag{14.11}$$

$$-10\,\text{Log}_{10}\left[\left(y_1^2 + y_2^2 + y_3^2 + y_4^2\right) \div 4\right].$$

Similar calculations are then completed for each inner array combination.

Step 3. Create so-called "marginal plots" by graphing the average SNR value for each of control factor settings. For example, the marginal plot for factor A and the design in Table 14.2 would be based on the SNR average of the first and the third control factor combination runs and the second and fourth runs.

Step 4. Pick the factor settings that are most promising according to the marginal plots. For factors that do not appear to strongly influence the SNR, Taguchi suggests using other considerations. In particular, marginal plots based on the average response are often used to break ties using subjective decision-making

Table 14.3 Data for the Taguchi experiment from RSM model predictions

					a_1				a_2					
			L_4	Gap	0.0	1.0	0.0	1.0	0.0	1.0	0.0	1.0		
	L_9			Offset	-0.5	-0.5	0.5	0.5	-0.5	-0.5	0.5	0.5		
TS (mm/min)	Ratio	Arc L. (mm)	CTTW (mm)										SNRL 1	SNRL 2
1270.0	6.0	3.2	14.0		9.7	9.3	9.6	8.2	8.3	4.8	7.2	3.6	19.3	14.2
1270.0	7.0	4.0	15.0		10.7	9.8	10.1	8.2	9.1	6.6	7.9	5.5	19.6	16.8
1270.0	8.0	4.8	16.0		9.2	7.8	8.1	5.7	9.3	7.9	8.1	6.8	17.4	17.9
1524.0	6.0	4.0	16.0		9.5	9.8	9.4	8.7	9.5	5.3	7.5	3.3	19.4	14.1
1524.0	7.0	4.8	15.0		10.0	9.0	9.5	7.9	7.5	4.8	7.0	4.3	19.1	14.8
1524.0	8.0	3.2	15.0		9.1	8.5	7.3	5.6	9.6	8.4	9.1	7.9	17.2	18.8
1778.0	6.0	4.8	15.0		8.9	9.3	9.6	8.9	7.6	3.1	6.2	1.7	19.2	9.2
1778.0	7.0	3.2	16.0		8.3	8.9	7.0	6.6	8.4	5.5	7.0	4.1	17.5	15.0
1778.0	8.0	4.0	14.0		5.5	4.8	4.4	2.7	7.6	6.1	7.7	6.2	11.8	16.6

Example 14.4.1 Welding Process Design Revisited

Question

Without performing any new tests, *sketch* what the application of Taguchi Methods might look like for the problem used to illustrate RPDM.

Answer

To apply the extended Taguchi Methods completely, it would be necessary to perform experiments according to a Taguchi inner and outer array design. Using the L_9 array to determine the combinations of the control factors on the left-hand-side in Table 14.3 below and the L_4 array to determine the noise factors combinations on the right-hand-side in the table below are the standard choices for the Taguchi methods. To avoid prohibitive expense, the response surface models from the actual RSM application were used to simulate the responses shown in the table.

This permitted the use of the standard formulas for bigger-the-better characteristics to calculate the signal-to-noise ratios shown on the right-hand-side of Table 14.3. Random errors were not added to the regression predictions for the response means because random errors might have increased the variability in the comparison and the same regression models were used to evaluate all results.

Because of the way the data were generated, the assumptions of normality, independence and constancy of variance were satisfied so that no transformation of the data was needed to achieve these goals. Transformations to achieve separability and additivity were not investigated because Song et al. (1995) state that their method was not restricted by separability requirements, and the selection of the transformation to

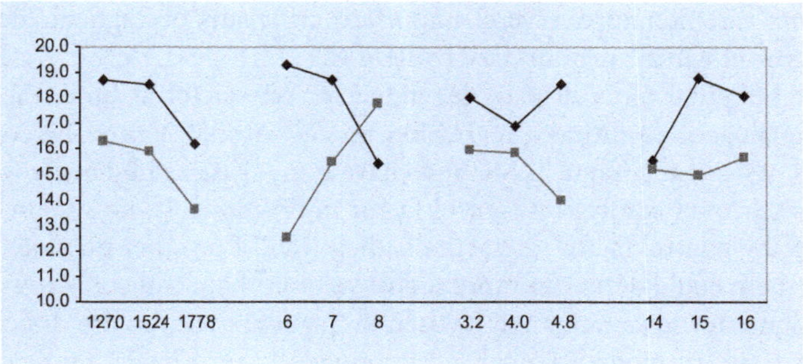

Fig. 14.4 Signal-to-noise ratio marginal plots for the two quality characteristics

achieve additivity involves significant subjective decision-making with no guarantee that a feasible transformation was possible. Control factor settings that were not dominated were identified by inspection of Fig. 14.4. The $f_{i,j}(l)$ refers to the mean values of the *j*th characteristics' average signal-to-noise ratio at level *l* for factor *i*. For example, all combinations of settings having arc length equal to 4.0 mm were dominated since at least one other choice of arc length exists (arc length equal 3.2 mm) for which both signal-to-noise ratio averages are larger.

This first step left 12 combinations of control factors. Subsequently, the formula in Song et al. (1995), which sums across signal-to-noise ratios for different responses, was used to eliminate four additional combinations. The resulting eight combinations included $x_1 = 1270.0$ mm/min, $x_2 = 7.00$, $x_3 = 3.18$ mm, and $x_4 = 15.0$ mm, which yielded the highest expected profit among the group equal to \$90.8/min. The combinations also included $x_1 = 1270.0$ mm/min, $x_2 = 6.0$, $x_3 = 4.80$ mm, and $x_4 = 15.0$ mm, with the lowest expected profit equal to \$–48.7/min. The user was expected to select the final process settings using engineering judgment from the remaining setting combinations, although without the benefit of knowing the expected profit.

Including the parts per minute $Q(x)$, which is proportional to travel speed, as an additional criterion increased by a factor of three the number of solutions that were not dominated. This occurred because setting desirability with respect to the other criteria consistently declined as travel speed increased. The revised process also included several settings which were predicted to result in substantially negative profits, i.e., situations in which expected rework costs would far outweigh sales revenue.

14.5 Literature Review and Methods Comparison

In general, methods in the applied statistics literature relate primarily to modeling the quality losses has been strongly influenced by Taguchi (Nair and Pregibon 1986; Taguchi 1987; Devor et al. 1992; Song et al. 1995; Chet et al. 1995). This concept of variation reduction inside the limits has been influential in the development and instruction of useful quality-control methods (Devor et al. 1992).

From this vast literature, several important criticisms of Taguchi Methods have emerged, some of which are addressed by RDPM:

1. Since marginal plots display dependencies one-factor-at-time, the method is somewhat analogous to fitting a regression model without control-by-control factor interactions. As a result, Taguchi Methods have been criticized by many for their inability to capitalize on control-by-control factor interactions. If the system studied has large control-by-control factor interactions, then RDPM or other methods that model and exploit them could derive far more robust settings than Taguchi Methods.

2. The signal-to-noise ratios (SNR) used in Taguchi methods are difficult to interpret and to relate to monetary goals. This provided the primary motivation for the RDPM approach, which is specifically designed to balance quality improvement needs with revenue issues. As a result, maximizing the SNRs can result in settings that lose money when profitable settings may exist that could be found using RDPM.

3. The extension of Taguchi Methods to address cases involving multiple quality characteristics can involve ambiguities. It can be unclear how to determine a desirable trade-off between different characteristics since the dimensionless signal-to-noise values are hard to interpret and might not relate to profits. For cases in which the system has a large number of quality characteristics, practitioners might even find RDPM simpler to use than Taguchi Methods.

4. In some cases, product arrays can require many more runs than standard response surface methods. This depends on which arrays are selected for the control and noise factors.

5. Taguchi Methods are not related to subjects taught in universities, such as response surface methods (RSM) and formal optimization. Therefore, they might require additional training costs.

Table 14.4 illustrates many of these issues, based on the results for our case study. The table also includes the solutions that engineers on the team thought initially would produce the best welds, which were $x_1 = 1524.0$ mm/min, $x_2 = 7.0$, $x_3 = 4.0$ mm, and

Table 14.4 Summary of the solutions derived from various assumptions

Method/ assumptions	Number runs	Solutions				Expected profit ($/min)
		x_1 (mm/min)	x_2 (−)	x_3 (mm)	x_4 (mm)	
Initial process settings	0	1524.0	7.0	4.0	15.0	165.0
RPDM	54 (or 35*)	1533.3	6.83	3.18	15.2	199.5
Extended Taguchi M.	36					
Highest profit settings		1270.0	7.00	3.18	15.0	201.6
Lowest profit settings		1270.0	6.00	4.80	15.0	18.4

*In case only the fractional factorial and center points are applied

$x_4 = 15.0$ mm, with the expected profit equal to \$14.5/min using the quadratic loss function.

Predictably, the decision-maker has been left with little information to decide between settings offering high profits and low profits. Some of the settings with non-dominated SNR ratios would yield near optimal profits while others would yield near zero profits. In this case, the Taguchi product array actually requires fewer runs than the relevant RSM approach. Note that using an EIMSE optimal design, the RDPM approach could have required only 35 runs (although some expected prediction accuracy loss).

Note that an important issue not yet mentioned can make the Taguchi product array structure highly desirable. The phrase "easy-to-change factors" (ETC) refers to system inputs with the property that if only their settings are changed, the marginal cost of each additional experimental run is small. The phrase "hard-to-change" (HTC) factors refers to system inputs with the property that if any of their settings is changed, the marginal cost of each additional experimental run is large. For cases in which factors divide into ETC and HTC factors, the experimental costs are dominated by the number of distinct combinations of HTC factors, for example, printing off ten plastic cups and then testing each in different environments (ETC factors). Since most of the costs relate to making tooling for distinct cup shapes (HTC factors), printing extra identical cups and testing them differently is easy and costs little.

Taguchi has remarked that noise factors are often all ETC, and control factors are often HTC. For cases in which these conditions hold, the product array structure offers the advantage that the number of distinct HTC combinations in the associated combined array is relatively small compared with the combinations required by a typical application of response surface method arrays.

Finally, Lucas (1994) proposed a class of "mixed resolution" composite designs that can be used in RDPM to save on experimentation costs. The mixed resolution designs achieved lower numbers of runs by using a special class of fractional factorials such that the terms in the matrices $\mathbf{D}_k$ for $k = 1, \ldots, r$ were not estimable. Lucas argued that the terms in $\mathbf{D}_k$ are of less interest than other terms and are not estimable with most Taguchi designs. For our case study, the mixed resolution design (not shown) would have 43 instead of 54 runs. In general, using Lucas mixed resolution composite designs can help make RSM based alternatives to Taguchi Methods like RDPM cost competitive even when all noise factors are ETC.

14.6 Chapter Summary

This chapter describes the goals of robust engineering and two methods to achieve these goals. The objective is to select controllable factor settings so that the effects of uncontrollable factors are not harmful. The first method presented is an extension of standard response surface methods (RSM) called RDPM. This method was originally developed in Allen et al. (2001). The second approach is the so-called static Taguchi Method.

The benefits of the first method include it derives the profit optimal balance between quality and revenues and can easily handle situations involving multiple quality characteristics. Benefits of Taguchi Methods include simplicity and cost advantages in cases when all noise factors are easy-to-change.

Taguchi Methods also have the obvious problem that decision-making is ambiguous if more than a single response or quality characteristic is of interest. For this reason the extension of Taguchi Methods in Song et al. (1995) is described.

14.7 Problems

In general, choose the answer that is correct and most complete.

1. Which is correct and most complete?
   a. Noise factor variation rarely (if ever) causes parts to fail to conform to specifications.
   b. Quality characteristics can be responses in applying RDPM experimentation.
   c. The fraction nonconforming cannot be a function of control factors.
   d. All of the above are correct.
   e. All of the above are correct except (a) and (d).
2. Which is correct and most complete?
   a. Robustness opportunities are always present in systems.
   b. Large control-by-noise factor interactions can cause robustness opportunities.
   c. Noise-by-noise interactions are sometimes neglected in robust engineering.
   d. Total variation of quality characteristics can depend on control factor settings.
   e. All of the above are correct.
   f. All of the above are correct except (a) and (e).
3. Which is correct and most complete?
   a. If there are three control factors and four noise factors, $\mathbf{C}_r$ is 3×4.
   b. Not every quadratic polynomial can be expressed by Eq. (14.1).
   c. The first diagonal element in $\mathbf{B}_r$ is a control-by-noise factor interaction.
   d. The first diagonal element in $\mathbf{B}_r$ is a noise-by-noise factor interaction.
   e. All of the above are correct.
   f. All of the above are correct except (a) and (e).
4. What is the relationship between TOC from ▸ Chap. 2 and RDPM?
5. Which is correct and most complete in relation to extended Taguchi Methods?
   a. These methods are based on central composite designs.
   b. The choice of signal-to-noise ratio depends on the number of control factors.
   c. The methods are called "extended" because they can be used when multiple characteristics are relevant.
   d. All of the above are correct.
   e. All of the above are correct except (a) and (e).
6. Which is correct and most complete in relation to extended Taguchi Methods?
   a. If responses for a control factor combination are: 2, 3, 5, and 3, SNRL = 10.1.
   b. If responses for a control factor combination are: 2, 3, 5, and 3, SNRS = 2.7.
   c. A common goal in applying Taguchi methods is to maximize relevant SNRs.
   d. All of the above are correct.
   e. All of the above are correct except (a) and (e).

7. Which is correct and most complete in relation to extended Taguchi Methods?
   a. Marginal plotting is somewhat similar to regression without control-by-control factor interactions.
   b. Taguchi Methods necessarily involve using formal optimization.
   c. Taguchi product arrays always result in higher costs than standard RSM arrays.
   d. All of the above are correct.
   e. All of the above are correct except (a) and (e).
8. Which is correct and most complete in relation to extended Taguchi Methods?
   a. Taguchi SNR ratios emphasize quality potentially at the expense of profits.
   b. ETC factors generally cost less to change than HTC factors.
   c. Taguchi product arrays call for direct observation of control factor setting combinations tested under a variety of noise factor combinations.
   d. All of the above are correct.
   e. All of the above are correct except (a) and (d).
9. Assume that z_1 and z_2 have means μ_1 and μ_2 and standard deviations σ_1 and σ_2 respectively. Also, assume their covariance is zero. What is

$$\underset{Z}{Var}[(\mathbf{c}' + \mathbf{x}'\mathbf{C})\mathbf{z}] = \underset{Z}{Var}\left[\left[\begin{pmatrix} 2 \\ -1 \end{pmatrix}' + \begin{pmatrix} x_1 \\ x_2 \end{pmatrix}' \begin{pmatrix} 5 & 2 \\ 2 & 8 \end{pmatrix}\right]\mathbf{z}\right],$$

   in terms of μ_1 and μ_2, standard deviations σ_1 and σ_2, and no matrices?
10. List two advantages of RDPM compared with Taguchi Methods.
11. List two advantages of Taguchi Methods compared with RDPM.
12. (Advanced) Extend RDPM to drop the assumption that $\mathbf{D}_r = 0$ for all r.

References

Allen TT, Ittiwattana W, Richardson RW, Maul G (2001) A method for robust process design based on direct minimization of expected loss applied to arc welding. J Manufact Syst 20:329–348

Allen TT, Richardson RW, Tagliabue D, Maul G (2002) Statistical process design for robotic GMA welding of sheet metal. Welding J 81(5):69s–77s

Chen LH, Chen YH (1995) A computer-simulation-oriented design procedure for a robust and feasible job shop manufacturing system. Journal of Manufacturing Systems 14:1–10

Devor R, Chang T, et al (1992) Statistical quality design and control. Macmillan, New York, pp 47–57

Johnson NL, Kotz S et al (1995) Continuous univariate distributions. John Wiley, New York

Lucas JM (1994) How to achieve a robust process using response surface methodology. J Qual Technol 26:248–260

Myers R, Montgomery D (2001) Response surface methodology, 5th edn. Wiley, Hoboken, NJ

Nair VN, Pregibon D (1986) A data analysis strategy for quality engineering experiments. AT&T Technical J 74–84

Song AA, Mathur A et al (1995) Design of process parameters using robust design techniques and multiple criteria optimization. IEEE Trans Syst Man Cybernet 24:1437–1446

Taguchi G (1987) A system for experimental design. UNIPUB, Detroit

Taguchi G (1993) Taguchi methods: research and development. In: Konishi S (ed), Quality engineering series, vol 1. The American Supplier Institute, Livonia, MI

Regression

T. T. Allen, *Introduction to Engineering Statistics and Lean Six Sigma*,
https://doi.org/10.1007/978-1-4471-7420-2_15

15.1 Introduction

Regression is a family of curve-fitting methods for (1) predicting average response performance for new combinations of factors and (2) understanding which factor changes cause changes in average outputs. In this chapter, the uses of regression for prediction and performing hypothesis tests are described. Regression methods are perhaps the most widely used statistics or operations research techniques. Also, even though some people think of regression as merely the "curve fitting method" in Excel, the methods are surprisingly subtle with much potential for misuse (and benefit).

Some might call virtually all curve fitting methods "regression" but, more commonly, the term refers to a relatively small set of "**linear regression**" methods. In linear regression predictions increase like a first order polynomial in the *coefficients*. Models fit with terms like $\beta_{32}\, x_1^2 x_4$ are stilled called "linear" because the term is linear in β_{32}, i.e., if the coefficient β_{32} increases, the predicted response increases proportionally. See ▶ Chap. 16 for a relatively thorough discussion of regression versus alternatives.

Note that standard screening using fractional factorials, response surface methods (RSM), and robust design using profit maximization (RDPM) methods are all based on regression analysis. Yet, regression modeling is relevant whether the response data is collected using a randomized experiment or, alternatively, if it is "**on-hand**" data from an observational study. In addressing on-hand data, primary challenges relate to preparing the data for analysis and determining which terms should be included in the model form.

▶ Section 15.2 focuses on the simplest regression problem involving a single response or system output and a single factor or system input and uses it to illustrate the derivation of the least squares estimation formula. ▶ Section 15.3 describes the challenge of preparing on-hand data for regression analysis including missing data. ▶ Section 15.4 discusses the generic task of evaluating regression models and its relation to design of experiments (DOE) theory. ▶ Section 15.5 describes analysis of variance (ANOVA) followed by multiple t-tests, which is the primary hypothesis-testing approach associated with regression. ▶ Section 15.6 describes approaches for determining a model form either manually or automatically. ▶ Section 15.7 concludes with details about building design matrices for cases involving special types of factors including categorical and mixture variables.

A full understanding of this chapter requires knowledge of functional forms and design matrices from ▶ Chap. 13 and focuses on on-hand data. The chapter also assumes a familiarity with matrix multiplication (see ▶ Sect. 5.3) and inversion. However, when using software such as Sagata® Regression, such knowledge is not critical. For practice-oriented readers supported by software, it may be of interest to study only Sects. 15.3 and 15.5.

15.2 Single Variable Example

Consider the data in ◘ Fig. 15.1a. This example involves a single input factor and a single response variable with five responses or data. In this case, fitting a first order model is equivalent to fitting a line through the data as shown in ◘ Fig. 15.1b. The line shown seems like a good fit in the sense that the (sum squared) distance of the data to the line is minimized. The resulting "best fit" line is $-26 + 32\, x_1$.

(a)

i	$(x_1)_i$	y_i	$y_{est,i}$	$Error_{est,i}$
1	3	70	70	0
2	4	120	102	18
3	5	90	134	–44
4	6	200	166	34
5	7	190	198	–8
			SSE =	3480

(b)

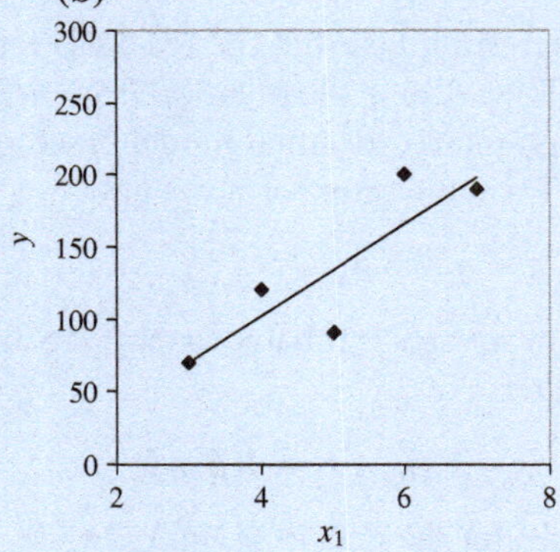

Fig. 15.1 Single factor example **a** data and **b** plot of data and 1st order model

The terms "residual" and "estimated error" refer to the deviation of the prediction given by the fitted model and the actual data value. Let "i" denote a specific row of inputs and outputs. Denoting the response for row i as y_i and the prediction as $y_{est,i}$, the residual is $Error_{est,i} = y_i - y_{est,i}$.

Figure 15.1a also shows the data, predictions, and residuals for the example problem. In a sense, the residuals represent a best guess of how unusual a given observation is believed to be in the context of a given model.

The higher the residual, the more concerned one might be that important factors unexplained by the model are influencing the observation in question. These concerns could lead us to fit a different model form and/or to investigate whether the data in questions constitutes an "outlier" that should be removed or changed because it does not represent the system of interest.

The example in Algorithm 15.1 below illustrates the application of regression modeling to predict future responses. The phrase "**trend analysis**" refers to the application of regression to forecast future occurrences. Such regression modeling constitutes one of the most popular approaches for predicting demand or revenues.

15.2.1 Demand Trend Analysis

Question

A new product is released in two medium-sized cities. The demands in Month 1 were 28 and 32 units and in Month 2 were 55 and 45 units. Estimate the residuals for a first order regression model and use the model to forecast the demand in Month 3.

Answer

The best fit line is $y_{est}(x_1) = 10 + 20x_1$. This clearly minimizes almost any measure of the summed residuals, since it passes through the average responses at the two levels. The resulting residuals are -2, $+2$, $+5$, and -5. The forecast or prediction for Month 3 is $10 + 20 \times 3 = 70$ units.

15.2.2 The Least Squares Formula

It is an interesting fact that the residuals for all observations can be written in vector form as follows. Using the notation from ▶ Sect. 13.2, "**y**" is a column of responses, "**X**" is the design matrix for fitted model based on the data, and "$\mathbf{Error}_{est}$" is a vector of the residuals. Then, in vector form, we have

$$\mathbf{Error}_{est} = \mathbf{y} - \mathbf{X}\boldsymbol{\beta}_{est}. \tag{15.1}$$

The "sum of squares estimated errors" (SSE) is the sum or squared residual values and can be written

$$\mathrm{SSE} = (\mathbf{y} - \mathbf{X}\boldsymbol{\beta}_{est})'(\mathbf{y} - \mathbf{X}\boldsymbol{\beta}_{est}). \tag{15.2}$$

For example, for the data in ◘ Fig. 15.1a, $\beta_{est,1} = -26$, and $\beta_{est,2} = 32$, we have

$$\mathbf{X} = \begin{bmatrix} 1 & 3 \\ 1 & 4 \\ 1 & 5 \\ 1 & 6 \\ 1 & 7 \end{bmatrix} \quad \mathrm{y} = \begin{bmatrix} 70 \\ 120 \\ 90 \\ 200 \\ 190 \end{bmatrix} \qquad \mathrm{y} - \mathbf{x}\beta_{est} = \begin{bmatrix} 0 \\ 18 \\ -44 \\ 34 \\ -8 \end{bmatrix}$$

The example in ◘ Fig. 15.1a is simple enough that the coefficients $\beta_{est,1} = -26$ and $\beta_{est,2} = 32$ can be derived informally by manually plotting the line and then estimating the intercept and slope. A more general formal curve-fitting approach would involve systematically minimizing the SSE to derive the coefficient estimates $\beta_{est,1} = -26$ and $\beta_{est,2} = 32$. The formulation can be written:

$$\begin{aligned} &\text{Minimize:} \qquad \mathrm{SSE} = (\mathbf{y} - \mathbf{X}\boldsymbol{\beta}_{est})'(\mathbf{y} - \mathbf{X}\boldsymbol{\beta}_{est}). \\ &\{\text{by changing } \boldsymbol{\beta}_{est}\} \end{aligned} \tag{15.3}$$

This approach can derive settings for more complicated cases involving multiple factors and/or fitting model forms including second and third order terms.

Minimizing the SSE is much like minimizing $c + b\beta + a\beta^2$ by changing β. Advanced readers will note that the condition that a is non-negative for a unique minimum is analogous to the condition that **X′X** is positive semidefinite. If a is non-negative, the solution to the easier problem is $\beta = -½ \times b \div a$. The solution to the least squares curve fitting problem is

15

$$\boldsymbol{\beta}_{est} = (\mathbf{X}'\mathbf{X})^{-1}\mathbf{X}'\mathbf{y} = \mathbf{A}\mathbf{y} \tag{15.4}$$

where **A** is the "alias" matrix. When the least squares coefficients are used, the sum of squares errors is sometimes written SSE*. For example, using the data in ◘ Table 15.1 (a), we have

$$\mathbf{x} = \begin{bmatrix} 1 & 3 \\ 1 & 4 \\ 1 & 5 \\ 1 & 6 \\ 1 & 7 \end{bmatrix} \qquad \mathbf{x}' = \begin{bmatrix} 1 & 1 & 1 & 1 & 1 \\ 3 & 4 & 5 & 6 & 7 \end{bmatrix} \quad \mathbf{x}'\mathbf{x}\begin{bmatrix} 5 & 25 \\ 25 & 135 \end{bmatrix}$$

$$(\mathbf{x}'\mathbf{x})^{-1}\mathbf{x}' = \begin{bmatrix} 1.2 & 0.7 & 0.3 & -0.3 & -0.8 \\ -0.2 & -0.1 & 0.0 & 0.1 & 0.2 \end{bmatrix}$$

Table 15.1 Two cases involving missing data and regression modeling for forecasting

(a)				(b)			
Point/ Run	x_1 (Month)	x_2 (Interest rate)	y (#sales)	Point/ Run	x_1 (Month)	x_2 (Interest rate)	y (#sales)
1	17	3.5	168	1	15		120
2	18	3.7	140	2	16	3.3	157
3	19	3.5	268	3	17	3.5	168
4	21	3.2	245	4	18	3.5	140
5	22		242	5	19	3.7	268
6	23	3.2	248	6	20	3.5	245
				7	21	3.2	242
				8	22	3.3	248
				9	23	3.2	268

$$\beta_{est} = (\mathbf{X}'\mathbf{X})^{-1}\mathbf{X}'\mathbf{y} = \begin{bmatrix} -26 \\ 32 \end{bmatrix} \quad (15.5)$$

which gives the same prediction model as was estimated by eye and SSE* = 3480.

15.3 Preparing "Flat Files" and Missing Data

Probably the hardest step in data analysis of "on-hand" data is getting the data into a format that regression software can use. The term "field" refers to factors in the database, "points" refer to rows, and "entries" refer to individual field values for specific data points. The term "**flat file**" refers to a database of entries that are formatted well enough to facilitate easy analysis with software. The process of creating flat files often requires over 80% of the analysis time. Also, the process of piecing together a database from multiple sources generates a flat file with many missing entries.

If the missing entries relate to factors not included in the model, then these entries are not relevant. For other cases, many approaches can be considered to address issues relating to the missing entries. The simplest strategy (*Strategy 1*) is to remove all points for which there are missing entries from the database before fitting the model. Many software packages such as Sagata® Regression implement this automatically.

In general, removing the data points with missing entries can be the safest, most conservative approach generating the highest standard of evidence possible. However, in some cases other strategies are of interest and might even increase the believability of results. For these cases, a common strategy is to include an average value for the missing responses and then see how sensitive the final results are to changes in these made-up values (*Strategy 2*). Reasons for adopting this second strategy could be:

1. The missing entries constitute a sizable fraction of entries in the database and many completed entries would be lost if the data points were discarded.
2. The most relevant data to the questions being asked contain missing entries.

Making up data should always be done with caution and clear labeling of what is made up should be emphasized in any relevant documentation.

Example 15.3.1 Handling Missing Data

Question

Consider the data in Table 15.1 related to predicting the number of sales in Month 24 using a first order model using month and interest rate as factors. Evaluate Strategy 2 for addressing the missing data in Table 15.1(a) and (b).

Answer

It would be more tempting to include the average interest rate for the case in Table 15.1(a) than for the case in Table 15.1(b). This follows in part because the missing entry is closer in time to the month for which prediction is needed in Table 15.1(a) than in Table 15.1(b). Also, there is less data overall in Table 15.1(a), so data is more precious. Added justification for making up data for Table 15.1(a) derives from the following sensitivity analysis. Consider forecasts based on second order models and an assumed future interest rate of 3.2. With the fifth point/run removed in Table 15.1(a), the predicted or forecasted sales is 268.9 units. Inserting the average value of 3.42 for the missing entry, the forecast is 252.9 units. Inserting 3.2 for the missing entry, the forecast is 263.4 units. Therefore, there seems to be some converging evidence in favor of lowering the forecast from that predicted with the data removed. This evidence seems to derive from the recent experience in Month 22 which is likely relevant. It can be checked that the results based on Table 15.1(b) are roughly the same regardless of the strategy related to removing the data. Therefore, since removing the data is the simplest and often least objectionable approach, it makes sense to remove point 1 in Table 15.1(b) but not necessarily to remove point 5 in Table 15.1(a).

15.4 Evaluating Models and DOE Theory

Analyzing a flat file using regression is an art, to a great extent. Determining which terms should be included in the functional form is not obvious unless one of the design of experiments (DOE) methods in previous chapters has been applied to planning and data collection. Even if one of the DOE methods and randomization has been applied, several tests are necessary for the derivation of proof.

In general, to be considered trustworthy it is necessary for regression models and their associated model forms to pass several tests. The phrase "**regression diagnostics**" refers to acceptability checks performed to evaluate regression models. Several of these

Table 15.2 Acceptability checks ("✓ guaranteed," "?" unclear, "X" loss unavoidable)

Issue	Measure	DOE	On-hand
Inputs: evidence is believable?	Randomization completed?	✓	×
Inputs: model is supported?	VIFs and correlations	✓	?
Outputs: outliers in the data?	Normal plot of residuals	?	?
Outputs: model is a good fit?	Summary statistics	?	?
Model makes sense?	Subjective assessment	✓	?

diagnostic tests are described in the sections that follow, with the exception of evaluating whether the inputs were derived from a randomized experiment. Material relevant to randomization was described in ▶ Sect. 11.5. The phrase "**input pattern**" refers to the listing of factor levels and runs in the flat file. If standard screening using fractional factorials or response surface methods has been applied, then the input pattern is the relevant experimental array.

Variance Inflation Factors (VIFs) are numbers that permit the assessment of whether reliable predictions and inferences can be derived from the combination of model form and input pattern. A common rule is that VIFs must be less than 10. Note that this rule applies only for formulas involving "standardized" inputs.

Normal Plot of Residuals are graphs that indicate whether the hypothesis tests on coefficients can be trusted and whether specific data points are likely to be representative of systems of interest. Generally, points off the line are outliers.

Summary Statistics are numbers that describe the goodness of fit. For example, R^2 prediction describes the fraction of the variation in the that is explainable by the data. It cannot always be calculated, but when it is available it is relatively reliable.

Table 15.2 shows which issues are solved automatically through the application of randomization and a design of experiments (DOE) methods such as regular fractional factorial or responses surface arrays. In general, models must pass all of the tests including a subjective assessment for the results to be considered critical. If DOEs are performed, the subjective assessment is far less critical because randomization establishes the cause and effect relationship between input changes and response variation.

In an important sense, the main justifications for using design of experiments relate to the creation of acceptable regression models. By using the special experimental arrays and randomizing, much subjectivity is removed from the analysis process. Also, there is the benefit that, if DOE methods are used, it may be possible to properly use the word "proof" in drawing conclusions.

15.4.1 Variance Inflation Factors and Correlation Matrices

This section concerns evaluation of whether a given set of data can be reliably trusted to support fitting a model form of interest. The least squares estimation formula reveals that coefficient estimates can be written as $\boldsymbol{\beta}_{est} = \mathbf{A}\mathbf{y}$ where $\mathbf{A} = (\mathbf{X}'\mathbf{X})^{-1}\mathbf{X}'$ and $\mathbf{A}$ is the "alias" matrix. The alias matrix is a function of the model form fitted and the input factor settings in the data. If the combination is poor, then if any random error, ε_i, influencing a response in **y** occurs, the result will be inflated and greatly change the coefficients.

The term "**input data quality**" refers to the ability of the input pattern to support accurate fitting a modeling of interest. We define the following in relation to quantitative evaluation of input data quality:

1. $\mathbf{D}^s$ is the input pattern in the flat file.
2. **H** and **L** are the highs and lows respectively of the numbers in each column of the input data, $\mathbf{D}^s$.
3. **D** is the input data in coded units that range between −1 and 1.
4. **X** is the design matrix.
5. $\mathbf{X}^s$ is the scaled design matrix (potentially the result of two scalings).
6. n is the number of data points or rows in the flat file.
7. m is the number of factors in the regression model being fitted.
8. k is the number of terms in the regression model being fitted.

The following procedure, in Algorithm 15.1, is useful for quantitative evaluation of the extent to which errors are inflated and coefficient estimates are unstable.

Note that, in *Step 4* the finding of a VIF greater than 10 or a $r_{i,j}$ greater than 0.5 does not imply that the model form does not describe nature. Rather, the conclusions would be that the model form cannot be fitted accurately because of the limitations of the pattern of the input data settings, i.e., the input data quality. More and better quality data would be needed to fit that model.

Note also that most statistical software packages do not include the optional *Step 1* in their automatic calculations. Therefore, they only perform a single scaling. Therefore also, the interpretation of their output in *Step 4* is less credible. In general, the assessment of input data quality is an active area of research, and the above procedure can sometimes prove misleading. In some cases, the procedure might signal that the input data quality is poor while the model has acceptable accuracy. Also, in some cases the procedure might suggest that the input data quality is acceptable, but the model does not predict well and results in poor inference.

Algorithm 15.1. Calculating VIFs and correlations between coefficient estimates

15

Step 1. (Optional) Calculate the scaled input array:

$$D_{i,j} = -1 + 2.0 \times \left(D^s_{i,j} - L_j\right) \div \left(H_j - L_j\right) \text{ for } i = 1,\ldots,n \text{ and } j = 1,\ldots,m.$$

Step 2. Create the design matrix, **X**, associated with **D** if it is available or $\mathbf{D}^s$ and the model form being fitted. Also, create the scaled or "standardized" design matrix $\mathbf{X}^s$ that contains $(k-1)$ columns (no column for the constant term). The entries in $\mathbf{X}^s$ are defined by

$$X^s_{i,j-1} = \left(X_{i,j} - X_{bar,j}\right) \div [s_j \times \text{sqrt}(n-1)] \text{ for } i = 1,\ldots,n \text{ and } j = 2,\ldots,k,$$

where $X_{bar,j}$ is the average of the entries in the jth column and s_j is the standard deviation of the entries in the jth column.

Step 3. Calculate the so-called "**correlation matrix**" which contains the "variance inflation factors" (VIFs) and the correlation between each pair of the *i*th coefficient estimate and the *j*th coefficient estimate ($r_{i,j}$) for $i = 2, \ldots, k$ and $j = 2, \ldots, k$. The matrix, $(\mathbf{X}^{s\prime}\mathbf{X}^{s})^{-1}$, is:

$$\begin{bmatrix} \mathrm{VIF}_2 & r_{23} & \cdots & r_{2k} \\ r_{23} & \mathrm{VIF}_3 & \cdots & \cdots \\ \cdots & \cdots & \cdots & \cdots \\ r_{2k} & \cdots & \cdots & \mathrm{VIF}_k \end{bmatrix} = (\mathbf{X}^{s\prime}\mathbf{X}^{s})^{-1}$$

Step 4. If any of the VIFs is greater than 10 or any of the $r_{i,j}$ are greater than 0.5 declare that the input data quality likely does not permit an accurate fit.

Example 15.4.2 Evaluating Data Quality

Question 1

Consider the data in Table 15.3. Does the data support fitting a quadratic model form?

Answer 1

Following the procedure, in *Step 1*, the **D** matrix in Table 15.3 was calculated using $H_1 = 45$ and $L_1 = 25$. Since there is only a single factor, **D** is a vector. In *Step 2*, the $\mathbf{X}^s$ matrix was calculated using $X_{bar,1} = -0.025$, $s_1 = 1.127$, $X_{bar,2} = 0.953$, and $s_2 = 0.095$. In *Step 3*, the transpose, multiplication, and inverse operations were applied using Excel resulting in the correlation matrix in Table 15.3. *Step 4* results in the conclusion that the input data is likely not of high enough quality to support fitting a quadratic model form since $r_{12} = 0.782 > 0.5$.

Question 2

Interpret visually why a second order model might be unreliable when fitted to the data in Table 15.3(a).

Table 15.3 Example: (a) data and **D**, (b) $\mathbf{X}^s$, and (c) the correlation matrix

(a)			(b)	(c)
$(x_1)_i$	$D_{i,1}$	y_i	$\mathbf{X}^s = \begin{bmatrix} -0.500 & 0.289 \\ -0.500 & 0.289 \\ 0.474 & -0.866 \\ 0.525 & 0.289 \end{bmatrix}$	$\begin{bmatrix} 1.428 & 0.782 \\ 0.782 & 1.428 \end{bmatrix}$
25	−1	110		
25	−1	120		
44	0.9	245		
45	1	260		

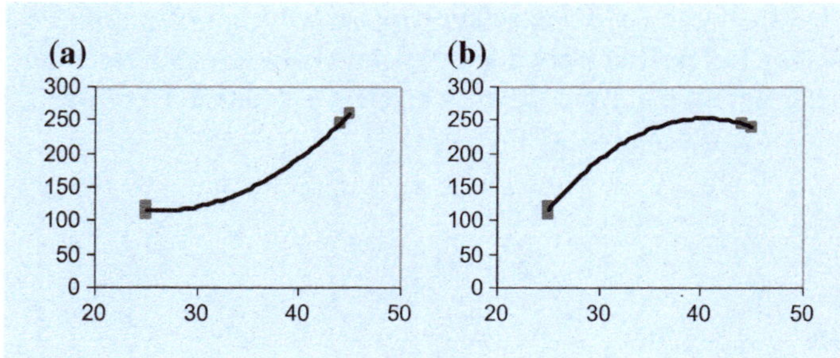

Fig. 15.2 **a** Initial second order model and **b** model from slightly changed data

Answer 2

Figure 15.2a shows the initial fit of the second order model. Figure 15.2b shows the second order fit when the last observation is shifted by 20 downward. The fact that such a small shift compared with the data range causes such a large change in appearance indicates that the input data has low quality and resulting models are unreliable.

15.4.2 Normal Probability Plots and Other "Residual Plots"

Another important regression diagnostic test is based on so-called "**normal probability plots**" of the residuals, $\text{Error}_{est,i}$ for $i = 1, \ldots, n$. Normal probability plots can provide information about whether the model form is adequate. They also aid in identification of response data that are not typical of the system during usual operations. If outliers are detected, this triggers detective work similar to spotting an out-of-control signal in control charting. Data is removed, and the model is refitted only if independent evidence suggests that the data is not representative.

The procedure below shows how to construct normal probability plots of any n numbers, $y_1, \ldots, y_n$, to permit subjective evaluation of the hypothesis that the numbers come from a normal distribution to a good approximation. A rule of thumb is that n must be 7 or greater for the procedure to give reliable results. For regression, if the residuals appear to come from a single normal distribution, then confidence grows in the model form chosen, e.g., one believes that one does not need to include additional terms such as $\beta_{20}\, x_1^2 x_2$ in the model. Also, confidence increases that any hypothesis tests that might be applied provide reliable results within the stated error rates.

Algorithm 15.2. Normal probability plotting

Step 1. Generate an n dimension vector **Z** using the formula

$$Z_i = \Phi^{-1}[(i - 0.5)/n] \text{ for } i = 1, \ldots, n, \tag{15.8}$$

where Φ^{-1} is the cumulative normal distribution with $\mu = 0$ and $\sigma = 1$. The value can be obtained by searching Table 15.4 for the argument and then reading over for the first two digits and reading up for the third digit. Note also, that if $0.5 < s < 1$, then $\Phi^{-1}[s] = 1.0 - \Phi^{-1}[-s]$.

Step 2. Generate $\mathbf{y}_{sorted}$ by sorting in ascending order the numbers in $\mathbf{y}$. Therefore, $y_{sorted,1}$ is the smallest number among $y_1, \ldots, y_n$ (could be the most negative number).

Step 3. Plot the set of ordered pairs $\{y_{sorted,1}, Z_1\}, \ldots, \{y_{sorted,n}, Z_n\}$.

Step 4. Examine the plot. If all numbers appear roughly on a single line then the assumption that all the numbers $y_1, \ldots, y_n$ come from a single normal distribution is reasonable. If the numbers with small absolute values line up but a few with large absolute values are either to the far right-hand-side or to far left hand side, off the (rough) line formed by the others, then we say that the larger (absolute value) numbers probably did not come from the same distribution as the smaller numbers. Probably some factor caused these numbers to have a different origin than the others. These numbers with large absolute values off the line are called "**outliers**".

If outliers are detected, it is generally not desirable to remove automatically the associated data points from the flat file or data set. Instead, detective work must uncover something that makes the associated data not representative of the system of interest before any points are removed. If nothing suspicious is found associated with the outliers, the data points should be retained, and this might suggest that a new model form is needed. In some cases, uncovering the factor whose variation causes outliers can be the most valuable outcome of the regression analysis process.

The term "**heteroscedasticity**" refers to the case in which the residuals do not have constant standard deviation. Heteroscedasticity can be detected using normal probability plots of residuals and observing a relationship that is nonlinear. Heteroscedasticity can be addressed using weighted least squares analysis available in standard software, e.g., Sagata® Regression. Also, it can make sense in some cases to transform the response data, e.g., by taking a natural logarithm of all response data before fitting the model form.

Example 15.4.4 Normal Probability Plotting Residuals

Question

Assume that the residuals are: $\text{Error}_{est,1} = -3.6$, $\text{Error}_{est,2} = -15.1$, $\text{Error}_{est,3} = -1.8$, $\text{Error}_{est,4} = 3.9$, $\text{Error}_{est,5} = -1.4$, $\text{Error}_{est,6} = 4.8$, and $\text{Error}_{est,7} = 2.0$. Use normal probability plotting to assess whether any are outliers.

Answer

Step 1 gives $\mathbf{Z} = \{-1.47, -0.79, -0.37, 0.00, 0.37, 0.79, 1.47\}$. *Step 2* gives $y_{sorted} = \{-15.1, -3.6, -1.8\ -1.4, 2.0, 3.9, 4.8\}$. The plot from *Step 3* is shown in ▪ Fig. 15.3. All numbers appear to line up, i.e., seem to come from the same normal distribution, except for −15.1, which is an outlier. It may be important to investigate the cause of the associated usual response (run 2). For example, there could be something simple and fixable, such as a data entry error. If found and corrected, a mistake might greatly reduce prediction errors.

Table 15.4 If $Z \sim N[0, 1]$, then the table gives $P(Z < z)$. The first column gives firs three digits of z, the top row gives the last digit

	0.00	0.01	0.02	0.03	0.04
−6.0	9.87E−10	9.28E−10	8.72E−10	8.20E−10	7.71E−10
−4.4	5.41E−06	5.17E−06	4.94E−06	4.71E−06	4.50E−06
−3.5	0.00023	0.00022	0.00022	0.00021	0.00020
−3.4	0.00034	0.00032	0.00031	0.00030	0.00029
−3.3	0.00048	0.00047	0.00045	0.00043	0.00042
−3.2	0.00069	0.00066	0.00064	0.00062	0.00060
−3.1	0.00097	0.00094	0.00090	0.00087	0.00084
−3.0	0.00135	0.00131	0.00126	0.00122	0.00118
−2.9	0.00187	0.00181	0.00175	0.00169	0.00164
−2.8	0.00256	0.00248	0.00240	0.00233	0.00226
−2.7	0.00347	0.00336	0.00326	0.00317	0.00307
−2.6	0.00466	0.00453	0.00440	0.00427	0.00415
−2.5	0.00621	0.00604	0.00587	0.00570	0.00554
−2.4	0.00820	0.00798	0.00776	0.00755	0.00734
−2.3	0.01072	0.01044	0.01017	0.00990	0.00964
−2.2	0.01390	0.01355	0.01321	0.01287	0.01255
−2.1	0.01786	0.01743	0.01700	0.01659	0.01618
−2.0	0.02275	0.02222	0.02169	0.02118	0.02068
−1.9	0.02872	0.02807	0.02743	0.02680	0.02619
−1.8	0.03593	0.03515	0.03438	0.03362	0.03288
−1.7	0.04457	0.04363	0.04272	0.04182	0.04093
−1.6	0.05480	0.05370	0.05262	0.05155	0.05050
−1.5	0.06681	0.06552	0.06426	0.06301	0.06178
−1.4	0.08076	0.07927	0.07780	0.07636	0.07493
−1.3	0.09680	0.09510	0.09342	0.09176	0.09012
−1.2	0.11507	0.11314	0.11123	0.10935	0.10749
−1.1	0.13567	0.13350	0.13136	0.12924	0.12714
−1.0	0.15866	0.15625	0.15386	0.15151	0.14917
−0.9	0.18406	0.18141	0.17879	0.17619	0.17361
−0.8	0.21186	0.20897	0.20611	0.20327	0.20045
−0.7	0.24196	0.23885	0.23576	0.23270	0.22965
−0.6	0.27425	0.27093	0.26763	0.26435	0.26109
−0.5	0.30854	0.30503	0.30153	0.29806	0.29460

(continued)

Table 15.4 (continued)

	0.00	0.01	0.02	0.03	0.04
−0.4	0.34458	0.34090	0.33724	0.33360	0.32997
−0.3	0.38209	0.37828	0.37448	0.37070	0.36693
−0.2	0.42074	0.41683	0.41294	0.40905	0.40517
−0.1	0.46017	0.45620	0.45224	0.44828	0.44433
0.0	0.50000	0.49601	0.49202	0.48803	0.48405

	0.05	0.06	0.07	0.08	0.09
−6.0	7.24E−10	6.81E−10	6.40E−10	6.01E−10	5.65E−10
−4.4	4.29E−06	4.10E−06	3.91E−06	3.73E−06	3.56E−06
−3.5	0.00019	0.00019	0.00018	0.00017	0.00017
−3.4	0.00028	0.00027	0.00026	0.00025	0.00024
−3.3	0.00040	0.00039	0.00038	0.00036	0.00035
−3.2	0.00058	0.00056	0.00054	0.00052	0.00050
−3.1	0.00082	0.00079	0.00076	0.00074	0.00071
−3.0	0.00114	0.00111	0.00107	0.00104	0.00100
−2.9	0.00159	0.00154	0.00149	0.00144	0.00139
−2.8	0.00219	0.00212	0.00205	0.00199	0.00193
−2.7	0.00298	0.00289	0.00280	0.00272	0.00264
−2.6	0.00402	0.00391	0.00379	0.00368	0.00357
−2.5	0.00539	0.00523	0.00508	0.00494	0.00480
−2.4	0.00714	0.00695	0.00676	0.00657	0.00639
−2.3	0.00939	0.00914	0.00889	0.00866	0.00842
−2.2	0.01222	0.01191	0.01160	0.01130	0.01101
−2.1	0.01578	0.01539	0.01500	0.01463	0.01426
−2.0	0.02018	0.01970	0.01923	0.01876	0.01831
−1.9	0.02559	0.02500	0.02442	0.02385	0.02330
−1.8	0.03216	0.03144	0.03074	0.03005	0.02938
−1.7	0.04006	0.03920	0.03836	0.03754	0.03673
−1.6	0.04947	0.04846	0.04746	0.04648	0.04551
−1.5	0.06057	0.05938	0.05821	0.05705	0.05592
−1.4	0.07353	0.07215	0.07078	0.06944	0.06811
−1.3	0.08851	0.08691	0.08534	0.08379	0.08226
−1.2	0.10565	0.10383	0.10204	0.10027	0.09853
−1.1	0.12507	0.12302	0.12100	0.11900	0.11702

(continued)

Table 15.4 (continued)

	0.05	0.06	0.07	0.08	0.09
−1.0	0.14686	0.14457	0.14231	0.14007	0.13786
−0.9	0.17106	0.16853	0.16602	0.16354	0.16109
−0.8	0.19766	0.19489	0.19215	0.18943	0.18673
−0.7	0.22663	0.22363	0.22065	0.21770	0.21476
−0.6	0.25785	0.25463	0.25143	0.24825	0.24510
−0.5	0.29116	0.28774	0.28434	0.28096	0.27760
−0.4	0.32636	0.32276	0.31918	0.31561	0.31207
−0.3	0.36317	0.35942	0.35569	0.35197	0.34827
−0.2	0.40129	0.39743	0.39358	0.38974	0.38591
−0.1	0.44038	0.43644	0.43251	0.42858	0.42465
0.0	0.48006	0.47608	0.47210	0.46812	0.46414

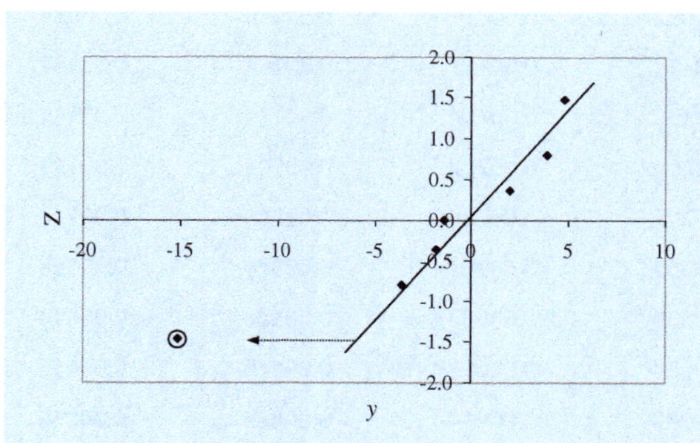

Fig. 15.3 Normal probability plot of the residuals

In the context of screening experiments, analysts might normal probability plot the estimated coefficients instead of applying Lenth's method. The factors judged to be significant (if any) would have coefficients that are outliers.

In addition to normal probability plotting the residuals, it is common to view plots of the $\text{Error}_{est,i}$ plotted versus $y_{est,i}$ and/or the inputs $\mathbf{x}_i$ for each run, $i = 1, \ldots, n$. For the calculations and associated hypothesis tests to be believable, the $\text{Error}_{est,i}$ values should not show an obvious dependence on any other quantities. In general, all of these "**residual plots**" can provide evidence that the functional form needs to be changed and the hypothesis testing results cannot be trusted. Yet, all residual plotting results can also be misleading in cases in which the number of data is comparable to the number of terms in the fitted model.

15.4.3 Summary Statistics

In addition to correlation matrices and residual plots, several numbers called "**summary statistics**" provide often critical information about the adequacy of the model form in question. This section describes four summary statistics: R^2 adjusted, PRESS, R^2 Prediction, and σ_{est}.

Probably the most widely used summary statistic is the "**R^2 adjusted**" that is also written "adjusted R-squared" or R^2_{adj}. This quantity is also sometimes called the "adjusted coefficient of multiple determination". To calculate the adjusted R-squared, it is convenient to use a $n \times n$ matrix, **Q**, with every entry equaling 1.0. This permits calculation of the "sum of squares total" (SST) using

$$\text{SST} = \mathbf{Y}'\mathbf{Y} - \left(\frac{1}{n}\right)\mathbf{Y}'\mathbf{Q}\mathbf{Y} \tag{15.9}$$

Then, the adjusted R-squared (R^2_{adj}) is given by

$$R^2 \text{ adjusted} = 1 - \left(\frac{n-1}{n-k}\right)\left(\frac{\text{SSE}^*}{\text{SST}}\right) \tag{15.10}$$

where k is the number of terms in the fitted model and SSE* is the sum of squares error defined in Eq. (15.3). It is common to interpret R^2_{adj} as the "fraction of the variation in the response data explained by the model".

Example 15.4.6 R^2 Adjusted Calculations

Question

Calculate and interpret R^2 adjusted for the example in Fig. 15.2a.

Answer

The following derive from previous results and definitions:

$$\mathbf{Error}_{est} = \begin{bmatrix} 0 \\ 18 \\ -44 \\ 34 \\ -8 \end{bmatrix}, \ \mathbf{Y} = \begin{bmatrix} 70 \\ 120 \\ 90 \\ 200 \\ 190 \end{bmatrix}, \text{ and } \mathbf{Q} \begin{bmatrix} 1 & 1 & 1 & 1 & 1 \\ 1 & 1 & 1 & 1 & 1 \\ 1 & 1 & 1 & 1 & 1 \\ 1 & 1 & 1 & 1 & 1 \\ 1 & 1 & 1 & 1 & 1 \end{bmatrix} \tag{15.11}$$

Therefore, with $n = 5$ data points, SST = 13720 and R^2 adjusted = 0.662 so that roughly 66% of the observed variation is explained by the first order model in x_1.

If the R^2_{adj} is derived from a formally planned experiment, e.g., standard screening or response surface methods (RSM) have been applied, then one generally expects R^2_{adj} to be greater than 0.75. R^2_{adj} values less than 0.75 generally indicate that important factors are varying uncontrollably, including possibly substantial measurement errors.

Otherwise, if on-hand data is used, the value of R^2_{adj} may be misleading, and limited conclusions can be drawn. Again returning to the first example in this chapter, since the system input (**x**) values do not follow the pattern of a planned experiment, one is skeptical about how much the 0.66 implies.

The next two summary statistics are based on the concept that the SSE can underestimate the errors of regression model predictions on new data. This follows intuitively because the fit might effectively "cheat" by overfitting the data upon which it was based and extrapolate poorly. The phrase "cross-validation" refers to efforts to evaluate prediction errors by using some of the data points only for this purpose, i.e., a set of data points only for testing.

Define $y_{est}(i, \boldsymbol{\beta}_{est}, \mathbf{x})$ as the regression fitted to a training set consisting of all runs except for the ith run. Define the $\mathbf{x}_i$ and y_i as the inputs and response for the ith run respectively. Then, the PRESS statistic is

$$\text{PRESS} = \Sigma_{i,\ldots,n}[y_{est}(i, \boldsymbol{\beta}_{est}, \mathbf{x}) - y_i]^2. \tag{15.12}$$

Because it is based on cross validation, the PRESS is generally more likely to provide an accurate characterization of the errors that the experimenter will face in new situations than the SSE.

The "**R² prediction**" or "R-squared prediction" is

$$R^2 \text{ prediction} = 1 - \left(\frac{n-1}{n-k}\right)\left(\frac{PRESS}{SST}\right). \tag{15.13}$$

As long as the input configuration permits the PRESS to be calculated, the R^2 prediction might be considered preferable to R^2 adjusted. In general, it is easy to identify situations in which a model form would minimize the SSE and/or maximize the R^2_{adj} and yet lead to inaccurate predictions or inferences about the engineered system. It is relatively difficult, however, to imagine a situation in which minimizing the PRESS or maximizing the R^2 prediction would lead to an undesirable model form.

15.4.4 Calculating R² Prediction

15

Question

Calculate and interpret the R^2 prediction for the example in Fig. 15.2a.

Answer

Table 15.5 shows the model coefficients, predictions, and errors in the PRESS sum. Squaring and summing the errors gives PRESS = 6445.41. Then, the R^2 prediction = 0.53. Therefore, the model explains only 53% of the variation and cross validation indicates that some overfitting is occurring.

In ► Chap. 4, the process capability in the context of the Xbar and R charts was defined as $6\sigma_0$. The symbol "σ_0" or "sigma" is the standard deviation of system outputs when inputs are fixed. For establishing the value of σ_0 using Xbar and R charting, it

Table 15.5 Calculations for evaluating the PRESS

Quantity	Data point removed (i)				
	1	2	3	4	5
Constant	–26	–44.000	–15	–11.429	–42
$(x_1)_i$	32	34.571	32	27.143	36
Prediction point	3	4	5	6	7
$y_{est}(i, \boldsymbol{\beta}_{est}, \mathbf{x})$	70	94.286	145	151.429	210
Y	70	120.000	90	200.000	190
Error	0	25.714	–55	48.571	–20

is necessary to remove data associated with any of the 25 periods that is not representative of system performance under usual conditions. This process is similar to the removal of outliers in regression analysis based on normal probability plotting residuals and detective work.

Regression modeling permits estimation of σ_0 without the need to have responses from repetitions of the same system inputs. After not representative data is removed from a process involving residual plots, the model form is refitted. Then, σ_0 can be estimated using

$$\sigma_{est}^2 = \text{SSE}^*/(n-k). \tag{15.14}$$

where SSE* is the sum of squares error for the least squares model, n is the number of runs, and k is the number of terms in the fitted model form. Many software refer to their estimate of "σ_{est}" using "S," including Minitab® and Sagata® Regression.

The value of σ_{est} is useful for at least three reasons. First, it provides a typical difference or error between the regression prediction and actual future values. Differences will often be larger partly because of the regression model predictions are not perfectly accurate with respect to predicting average responses. Second, σ_{est} can be used in robust system optimization, e.g., it can be used as an estimate of σ_r for the formulas in ► Chap. 14.

Third, if the value of σ_{est} is greater by an amount considered subjectively large compared with the standard deviation of repeated response values from the same inputs, then evidence exists that the model form is a poor choice. This is particularly easy to evaluate if repeated runs in the input pattern permit an independent estimate of σ_{est} by taking the standard deviation of these responses. Then, it might be desirable to include higher order terms such as x_1^2 if there were sufficient runs available for their estimation. This type of "**lack of fit**" can be proven formally using hypothesis tests as in two-step response surface methods after the first step in ► Sect. 13.6.

Example 15.4.8 Estimating Sigma Using Regression

Question

Calculate and interpret the value of σ_{est} using the data in Fig. 15.1a.

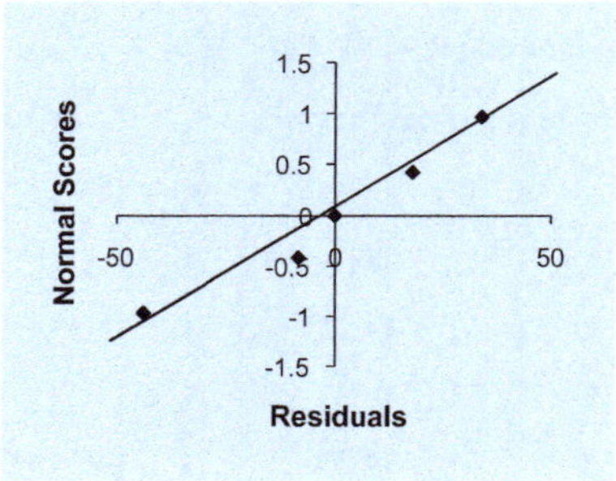

Fig. 15.4 Normal plots of residuals for single factor example

Answer

First, the normal probability plot of residuals in Fig. 15.4 finds no obvious outliers. Therefore, there is no need to remove data and refit the model. From previous problems, the SSE* is 3480 and $\sigma_{est} = \text{sqrt}(3480 \div 3) = 34.1$. Without physical insight about the system of interest or responses from repeated system inputs, there is little ability to assess lack of fit. Typically, outputs from the same system would be within 34.1 units from the mean predicted by the regression model $y_{est}(x_1) = -26 + 32\,x_1$.

15.5 Analysis of Variance Followed by Multiple T-Tests

If all of the acceptability tests are passed, it can be of interest to perform hypothesis tests to prove that model terms are associated with nonzero effects. Even if randomization has not been used in the data collection, it still can be of interest to perform hypothesis testing. In this section, the Analysis of Variance (ANOVA) followed by t-testing method is described in Algorithm 15.5 for hypothesis testing based on regression modeling. This method is perhaps the most common approach used in all standard regression software.

The chief benefit of ANOVA followed by t-tests is that it can detect whether all the data are noise with a regulated Type I error rate regardless of the number of coefficients of interest for testing. Therefore, ANOVA offers the benefits of the Experimentwise Error Rate (EER) in Lenth's methods to cases in which the experimental design is not a regular fractional factorial or Plackett Burman design.

15

As a result, the methods are potentially alternative approaches to Lenth's method described in the context of standard screening using fractional factorials. Generally, the advantage of Lenth's method compared with ANOVA in the context of regular fractional factorials is that Lenth's method offers a higher probability of finding significance under standard assumptions, i.e., a lower Type II error rate. In addition, unlike Lenth's method, standard ANOVA cannot be applied when the number of terms in the fitted regression model equals the number of runs. The reason for this relates to the fact that certain quantities in ANOVA would be zero and, subsequently, ratios based on them would be undefined. Therefore, ANOVA followed by t-tests is probably not relevant for analyzing data from standard screening experiments.

Classic references on ANOVA include Fisher (1925) and other books written by Fisher. Here, only one type of ANOVA method is considered, which might be called "parametric regression based ANOVA". It is called "**parametric**" because the approach involves the potentially "ad hoc" use of the F-distribution which is associated with the assumption that the residuals are normally distributed. The lack-of-fit test in sequential response surface methods is an example of another type of parametric ANOVA. Other types of ANOVA might be relevant for purposes such as comparing the robustness of different methods. Also, nonparametric methods can be useful when a high level of evidence is desired and data is sufficient to offer an acceptable probability of identifying effects, i.e., the nonparametric methods generally require more data to establish significance.

The following are used in the ANOVA method:

1. $\mathbf{D}^s$ is the input pattern in the flat file.
2. $\mathbf{H}$ and $\mathbf{L}$ are the highs and lows respectively of the numbers in each column of the input data, $\mathbf{D}^s$.
3. $\mathbf{D}$ is the input data in coded units that range between -1 and 1.
4. $\mathbf{X}$ is the design matrix.
5. $\mathbf{X}^s$ is the scaled design matrix (potentially the result of two scalings).
6. n is the number of data points or rows in the flat file.
7. m is the number of factors in the regression model being fitted.
8. k is the number of terms in the regression model being fitted.
9. $\mathbf{Y}$ is a vector of response data.
10. $y_{average}$ is the average of all n responses in the n dimensional data vector $\mathbf{Y}$.
11. $\mathbf{J}$ is an n dimensional vector with all entries equal to 1.

It is perhaps most standard to pronounce interactions terms as being significant after the optional scaling in *Step 1* has been performed. Further, it is often reasonable to accept evidence levels associated with p-values greater than 0.05. This follows because the decision-maker may be attempting to determine whether any causal relationship might exist rather than proving that one does exist.

A modified version of the above method is based on an assumption that the standard deviation of the random error, σ, is believed to be known. This could occur, for example, if an Xbar & R chart was used to study this system output and obtain the process capability, 6σ, as described in ▶ Chap. 4. In this approach, one simply substitutes the believed value of σ^2 in place of the MSE in the ANOVA table in *Step 3* and the calculation of the t_i in *Step 5*. Also, the Residuals $df = \infty$, which can be achieved effectively by using the largest number in the F and t tables.

The phrase "random factors" refers to system inputs whose levels are relevant mainly because of their relevance in predicting responses from a large population. For example, the participants are random factors in a drug test because we are not primarily interested in the effects on individuals (the levels) but rather on the effects on a population of people. The formulas in the relevant "**ANOVA Table**" in ◘ Table 15.6 give the same values as formulas in standard textbooks such as Montgomery (2012). If random factors are involved, then modified formulas in Montgomery (2012) should be used to develop more believable inferences about the effects of the factors on the larger population.

Table 15.6 The analysis of variance table for regression-based ANOVA

Source	Sum of squares	df	MS	F value
Regression model	$SSR = (\mathbf{X}\boldsymbol{\beta}_{est} - \mathbf{J}y_{average})'(\mathbf{X}\boldsymbol{\beta}_{est} - \mathbf{J}y_{average})$	$k - 1$	$MSR = SSR/(k - 1)$	$F_0 = MSR/MSE$
Residuals	$SSE = (\mathbf{Y} - \mathbf{X}\beta_{est})'(\mathbf{Y} - \mathbf{X}\beta_{est})$	$n - k$	$MSE = SSE/(n - k)$	

Algorithm 15.3. Analysis of variance followed by multiple t-tests

Step 1. (Optional) Calculate the scaled input array matrix using:

$$D_{i,j} = -1 + 2.0 \times \left(D^s_{i,j} - L_j\right) \div (H_j - L_j) \text{ for } i = 1, \ldots, n \text{ and } j = 1, \ldots, m.$$

Step 2. Create the design matrix, **X**, associated with **D** if it is available or $\mathbf{D}^s$ and the model form being fitted with *k* terms. Calculate the least squares coefficient estimates, $\boldsymbol{\beta}_{est}$, using $\boldsymbol{\beta}_{est} = \mathbf{AY}$ where **A** is the $(\mathbf{X}'\mathbf{X})^{-1}\mathbf{X}'$.

Step 3. Calculate all quantities in the following so-called "**ANOVA table**" in Table 15.6, which includes calculation of the sum of squares regression (SSR), the sum of squares error (SSE), the so-called "**degrees of freedom**" (*df*), the mean squared regression (MSR), and the mean squared error (MSE).

Step 4. If $F_0 < F_{\alpha,k-1,n-k}$ (found using Table 13.9), then stop and declare that "none of the terms in the model has a significant affect on the average response" or, in other words, the data is all noise. Otherwise, go to *Step 5*.

Step 5. Calculate $t_i = (\beta_{est,i})\{[(\mathrm{MSE})(\mathbf{X}'\mathbf{X})^{-1}_{i,i}]^{-1/2}\}$ for $i = 2, \ldots, k$, where $(\mathbf{X}'\mathbf{X})^{-1}_{i,i}$ refers to the *i*th entry on the diagonal of $(\mathbf{X}'\mathbf{X})^{-1}$. If $t_i > t_{\alpha,n-k}$ (found using Table 11.2), then declare, "Term *i* in the regression model is significant for alpha level *α*," and also, "The factors associated with term *i* are significant with alpha level *α*." For the other terms, we conclude only that we "fail to find significance" without additional data.

Example 15.5.1 Single Factor ANOVA Application

15

Question

Calculate and interpret the results of the ANOVA method followed by multiple t-tests based on the data in Fig. 15.1a.

Answer

Table 15.7(a) shows the ANOVA table and Table 15.7(b) shows the calculation of the t-statistic. Note that for a single factor example, the ANOVA p-value is the same as the single factor coefficient p-value, i.e., the chance that the data is all noise

Table 15.7 Single factor (a) ANOVA table and (b) t-test and *p*-values

		df	SS	MS	F	p-value
(a)	Regression	1	SSR = 10,240	10,240	8.83	0.0590
	Residuals	3	SSE = 3480	MSE = 1160		

		Coefficients	Standard error	t stat	p-value
(b)	Constant	−26.00	55.96	−0.46	0.674
	x_1	32.00	10.77	t_1 = 2.97	0.059

can be evaluated with either statistic. With so little data, the p-value of 0.056 can be considered strong evidence that factor x_1 affects the average response values. Note that since it is not clear whether randomized experimentation has been used, it is not proper to declare that the analysis provides proof.

The "**Bonferroni inequality**" establishes that if q tests are made each with an α chance of giving a Type I error, the chance of no false alarm on any test is greater than $1 - q \times \alpha$. Even though additional mathematical results can increase this bounding limit, with even a few tests (e.g., $q = 4$) approaches based on individual testing offer limited overall coverage unless the α values used are very small. ANOVA followed by t-tests can offer the same guarantee while achieving lower Type II error risks than any procedure based on the Bonferroni inequality.

15.6 Regression Modeling Flowchart

The phrase "**stepwise regression**" refers to automatic model form selection procedures. Considering the subjective nature of the acceptability checks in ▶ Sect. 15.4, it is not clear that any automatic procedure can result in an acceptable model. Figure 15.5 gives a reasonably standard semi-automatic approach for establishing regression models to analyze data. This flowchart ties together the diagnostic and analysis of variance (ANOVA) methods described in previous sections.

If a carefully designed, randomized experiment has been performed, the model form may be specified by the DOE method with little ambiguity, e.g., for RSM, the fitted model is generally a second order polynomial. Still, the method in Fig. 15.5 can be used to prune or "edit" the model. Smaller models are simpler and can be more interpretable. For example, there might be other factors besides those purposely varied in experimentation that might be included in the fitted model form. Figure 15.5 is primarily relevant for cases in which data does not come from design of experiments (DOE) applications, i.e., "on-hand" data is being analyzed.

As for calculating variance inflation factors (VIFs) and performing analysis of variance followed by t-tests, starting the flowchart with scaled inputs is generally desirable. Therefore, performing *Step 1* of these procedures is probably a natural first step in all efforts to find a model form. This follows in part because the sizes and significance levels associated with second order terms depend upon the scale of these inputs. Starting

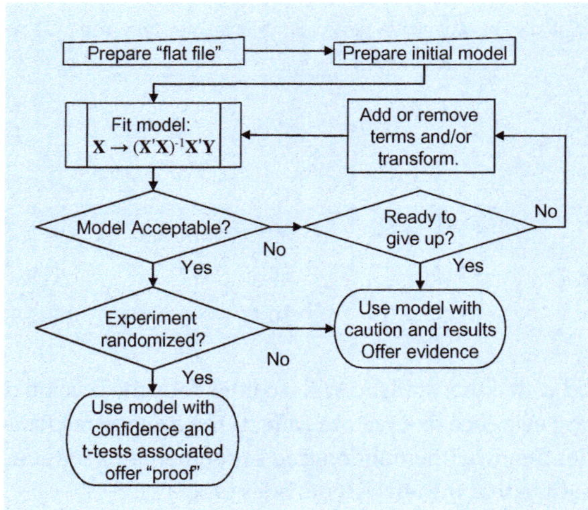

Fig. 15.5 Regression flow chart

with inputs scaled to −1 to 1 provides a natural basis for assessing whether interactions underlie the system performance being studied.

Often, the analysis process in Fig. 15.5 can be completed within a single hour after the flat file is created. A first order model using factors of intuitive importance is a natural starting point. Patterns in the residuals or an intuitive desire to explore additional interactions and curvatures generally provide motivation for adding more terms. Often, adding terms such as x_1^2 is as easy as clicking a button. Therefore, the bottleneck is subjective interpretation of the acceptability of the residual plots and of the model form.

Even though all results from on-hand data should be evaluated with caution, regression analyses often provide a solid foundation for important business decisions. These could include the adjustment of an engineering design factor such as the width of seats on airplanes or the setting aside of addition money in a budget because of a regression forecast of the financial needs.

Example 15.6.1 Method Choices

15

Question

Which of the following is correct and most complete?

a. Even if a randomized experiment has been used, proof might not be achieved.
b. Inspection of residuals could be used to identify unusual observations or outliers.
c. Proof is guaranteed by low *p*-values in regression modeling.
d. All of the above are correct except (c).

Answer

According to the flowchart, a randomized experiment, low *p*-values, and subjectively acceptable residuals are all required for proof. Unusual and untrustworthy data can be identified by observing large values on the plots. For these reasons, the correct answer is (d).

Therefore, "stepwise regression" methods are automatic procedures similar to ◘ Fig. 15.5 except with the automatic assessment based on quantifiable acceptability tests. "**Forward stepwise regression**" involves an initial model that includes only the constant term. "**Backward stepwise regression**" starts with an initial model containing many terms and the removal of terms automatically based on specific diagnostic values. Many stepwise approaches are based on the F-tests which fail to address whether the model form cross-validates well. Sagata® Regression implements a forward stepwise procedure based on the PRESS statistic, which might be considered relatively trustworthy since it is based on cross validation.

The following is an application based on the flowchart, diagnostic plotting, and ANOVA methods. This application involves predicting body fat, which is expensive to measure accurately, as a function of quantities that are inexpensive to measure accurately.

Example 15.6.2 Body Fat Prediction

The following are reproduced with permission from a study by Dr. A. G. Fisher and others and made available through the internet at ▸ http://lib.stat.cmu.edu. Analyze the body fat data in ◘ Table 15.8 and make recommendations to the extent possible for a person in training who is 35 years old, 190 lb, 68 in., with a 42 cm neck, and who wants to lose weight. A good analysis will typically include one or two models, reasons for selecting that model, estimates of the errors of the model, and interpretation for the layperson (◘ Fig. 15.6).

Question 1
What prediction model would you use to predict the body fat of people not in the table such as the person in training and why?

Answer 1
Consider the terms in a full second order model including $f_1(\mathbf{x}) = 1$, $f_2(\mathbf{x}) = \text{Age}, \ldots,$ $f_{15}(\mathbf{x}) = \text{Height} \times \text{Neck}$. The combination of terms up to second order that minimize the PRESS are Age, Age × Weight, and Age × Height. Fitting a model with only these terms using least squares gives %Fat = 3.25 × Age + 0.00699 × Age × Weight – 0.0561 × Age × Height. This model has an $R^2_{adj\_prediction}$ approximately equal to 0.86 so that these few terms explain a high fraction of the variation. The model is also simple and intuitive in that it correctly predicts that older, heavier, and shorter people tend to have relative high body fat percentages. ◘ Figure 15.7 shows the model predictions as a function of height and weight.

Question 2
Does the normal probability plot of residuals support the assumption that the residuals are IID normally distributed?

Answer 2
The normal probability plot provided limited, subjective support for the assumption that the residuals are IID. normally distributed noise. There are no obvious outliers,

Table 15.8 Dimensions and % body fat of 29 people

Age (years)	Weight (lbs.)	Height (in.)	Neck (cm)	% Fat
23	154.25	67.75	36.20	12.3
22	173.25	72.25	38.50	6.1
22	154.00	66.25	34.00	25.3
26	184.75	72.25	37.40	10.4
24	184.25	71.25	34.40	28.7
24	210.25	74.75	39.00	20.9
26	181.00	69.75	36.40	19.2
25	176.00	72.50	37.80	12.4
25	191.00	74.00	38.10	4.1
23	198.25	73.50	42.10	11.7
26	186.25	74.50	38.50	7.1
27	216.00	76.00	39.40	7.8
32	180.50	69.50	38.40	20.8
30	205.25	71.25	39.40	21.2
35	187.75	69.50	40.50	22.1
35	162.75	66.00	36.40	20.9
34	195.75	71.00	38.90	29.0
32	209.25	71.00	42.10	22.9
28	183.75	67.75	38.00	16.0
33	211.75	73.50	40.00	16.5
28	179.00	68.00	39.10	19.1
28	200.50	69.75	41.30	15.2
31	140.25	68.25	33.90	15.6
32	148.75	70.00	35.50	17.7
28	151.25	67.75	34.50	14.0
27	159.25	71.50	35.70	3.7
34	131.50	67.50	36.20	7.9
31	148.00	67.50	38.80	22.9
27	133.25	64.75	36.40	3.7

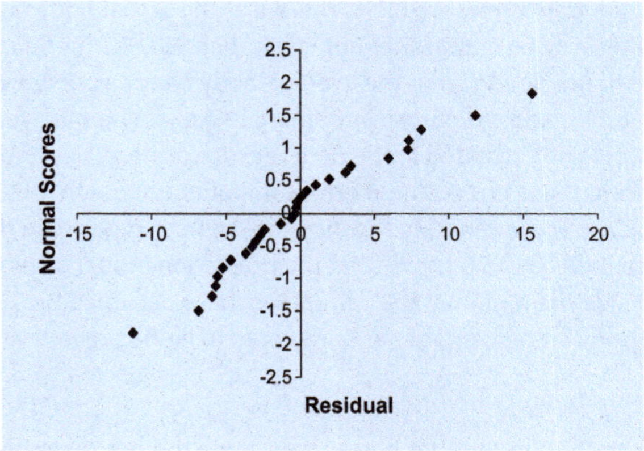

Fig. 15.6 Normal probability plot for the body fat prediction model

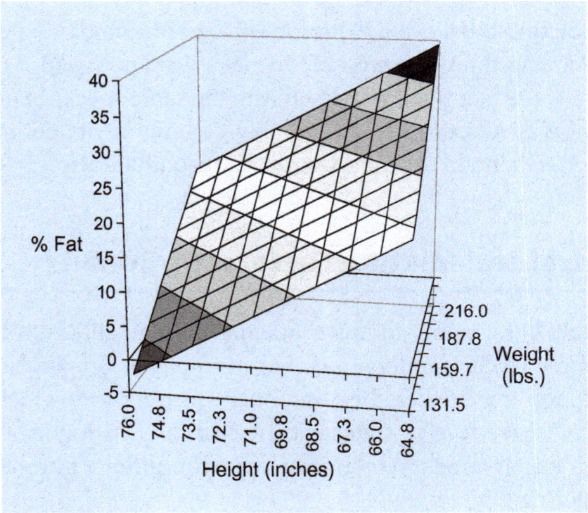

Fig. 15.7 The average body fat percentages predicted for 35-year-old people and plotted using Sagata® regression professional

i.e., points to the far right or left off the line. Since the points do not precisely line up, there could well be missing factors providing systematic errors.

Question 3

What body fat percentage do you predict for the person in training and what are the estimated errors for this prediction?

Answer 3

This model predicts that the average person with $\mathbf{x} = (35, 190, 68, 42)'$ has 25.9% body fat with standard error of the mean $\{\text{Variance}[y_{est}(\boldsymbol{\beta}_{est}, \mathbf{x})]\}^{1/2}$ equal to 2.5%.

The estimated standard errors are 6.1%. Therefore, the actual body fat of the person in training could easily be 6–8% higher or lower than 25.9%. This follows because there are errors in predicting what the average body fat for a person with $\mathbf{x} = (35, 190, 68, 42)'$ (±2.5%), and the person in training is likely to be not average (±6.1%). Presumably, factors not included in the data set such as head size and muscle weight are causing these errors. These error estimates assume that the person in training is similar, in some sense, to the 29 people whose data are in the training set. The surface plot in Fig. 15.6 shows that the prediction model gives nonsensical predictions outside the region of the parameter space occupied by the data, e.g., some average body fat percentages are predicted to be negative.

Question 4

What type of reduction in body fat percentage could the person in training expect by losing 15 lb?

Answer 4

The model predicts that the average person with specifications $\mathbf{x} = (35, 175, 68, 42)'$ would have 22.2% body fat with error of the mean $\{\text{Variance}[y_{est}(\boldsymbol{\beta}_{est}, \mathbf{x})]\}^{1/2}$ equal to 2.5%. Therefore, if the "Joe average" person with the same specifications lost 15 lb, then "Joe average" could expect to lose roughly 4% body fat. It might be reasonable for the person in training to expect losses of this magnitude also.

15.7 Categorical and Mixture Factors (Optional)

"**Categorical factors**" are inputs that can assume only a finite number of levels and the ordering of these levels is ambiguous. For example, a categorical factor might be the supplier company that makes the component in question, which could be Intel, Panasonic, or RCA (three levels). Categorical factors are distinguished from **continuous factors**, which can assume, theoretically, any of an infinite number of levels which have a natural ordering.

15

"Mixture factors" are inputs whose levels are constrained to sum to a constant. For example, these could be the components of a cake such as percent flour, water, and sugar. Percentages must total 100%. Mixture factors require adjustments to response surface methods to make fitting regression models possible. Issues related to categorical factors and mixture factors are described in this section. Analysis of data with categorical outputs is also briefly described with more details in the next chapter.

15.7.1 Regression with Categorical Factors

In general, categorical variables should be avoided as far as possible because their inclusion can greatly increase the number of terms in a model. A general rule is that the number of data or runs needed to fit accurately a model is proportional to the number of terms. Often, engineering insight can permit the experimental team to address the same issue in planning experiments using either a continuous or a categorical factor. For example, color might be considered a categorical factor (e.g., levels might be "green" and "yellow"). At the same time, with suitable equipment it might be possible to address color issues by varying the wavelength of the light, e.g., using a prism. Then wavelength could be the experimental factor resulting in either a savings in experimentation costs or an increase in prediction accuracy or both.

Note tcan only reliably create certain levels of them. For example, imagine that only the temperatures of 20, 25, and 100 °C are available in the laboratory because of experimental limitations. In this case, temperature is continuous and not categorical, since one might be interested in performance at 78 °C (i.e., all "in between" levels are conceivably possible). Also, 20 °C < 25 °C < 100 °C so the level ordering is not ambiguous.

Generally, if categorical factors are at two levels, regression models based on categorical factors can be constructed in the same manner as for continuous factors. However, if three or more levels of one or more categorical factors are involved, the situation is relatively complicated. Then, a mathematical construct called "**contrasts**" are created and treated like "mini-factors" in the analysis. If there are l levels of the categorical factor, then one creates $l-1$ two-level contrasts. These contrasts function in a similar manner in calculations as factors for which experimentation has been conducted at two levels. Therefore, interaction terms can be fitted but pure quadratic or cubic terms cannot.

There are multiple approaches for creating these contrasts that give the same predictions in all situations. The approach described here is to create the ith contrast with values equal to 1 if the categorical factor assumes the corresponding ith level, for $i=1, \ldots, l-1$. Then, in the modeling, no terms involving interactions between these contrasts can be included, although interactions between contrasts and continuous factors can be included.

Table 15.9 shows an example with two factors, both at three levels, with the second being categorical. Part (a) shows the input and output pairs as well as the two

Table 15.9 Example illustrating regression with a three level categorical factor

(a)						(b)							
Run	Temp.	Supplier	x_2	x_3	y	Const.	x_1	x_2	x_3	x_1^2	$x_1 x_2$	$x_1 x_3$	
1	20	Intel	1	0	22	1	20	1	0	400	20	0	
2	80	Panasonic	0	1	33	1	80	0	1	6400	0	80	
3	20	RCA	0	0	21	1	20	0	0	400	0	0	**X** =
4	80	Intel	1	0	3	1	80	1	0	6400	80	0	
5	50	Panasonic	0	1	1	1	50	0	1	2500	0	50	
6	80	RCA	0	0	23	1	80	0	0	6400	0	0	
7	50	Intel	1	0	21	1	50	1	0	2500	50	0	

contrasts. The first of these contrasts can be thought of as associated with the supplier "Intel". Part (b) shows the **X** matrix corresponding to a matrix with all main effects involving x_1, x_2, and x_3, and two-factor interactions involving x_1x_2 and x_1x_3, as well as the pure quadratic term involving x_1^2. Applying Eq. (15.8) to estimate the coefficients gives: $y_{est}(\boldsymbol{\beta}_{est}, \mathbf{x}) = 5.22 + 0.98\, x_1 + 8.00\, x_{2\text{-}9}5.33\, x_3 - 0.01\, x_1^2 - 0.35\, x_1\, x_2 + 1.32\, x_1\, x_{3.}$ To use this model, e.g., to predict the output if Intel were used as the supplier, one would substitute $x_1 = 1.0$ and $x_2 = 0.0$ into Eq. (15.10).

Note that it might make sense to focus on a smaller number of contrasts in an analysis than a complete set. This could aid in intuition-building and leave more degrees of freedom for residuals and/or entertaining other model terms. For example, one might group Intel and Panasonic suppliers together because they have similar quality levels and focus only on the contrast x_2 in the above example. Then, x_3 would not be considered in the analysis.

15.7.2 DOE with Categorical Inputs and Outputs

Many methods have been proposed for planning response surface methods experiments involving categorical factors. Chantarat et al. (2003) offered optimal design of experiments methods with advantages in run economy and prediction accuracy. In this section, we describe what is probably the simplest approach for extending response surface methods, which is based on a product array approach in which a standard response surface array is repeated for all combinations of categorical factor levels. For example, ◘ Table 15.10 shows a product array for two continuous factors and one categorical factor at two levels.

If the product array approach is used, then the fitted model includes: (1) all full quadratic terms for the continuous factors, (2) main effects contrasts for the categorical factors, and (3) interaction terms involving every interaction term contrast and ever one of the continuous factor terms. For example, with two continuous factors and one categorical factor at two levels, the model form is:

$$\begin{aligned} y(x_1, x_2, x_3) = &\beta_1 + \beta_2 x_1 + \beta_3 x_2 + \beta_4 x_3 + \beta_5 x_1^2 + \beta_6 x_2^2 + \beta_7 x_1 x_2 + \beta_8 x_1 x_3 \\ &+ \beta_9 x_2 x_3 + \beta_{10} x_1^2 x_3 + \beta_{11} x_2^2 x_3 + \beta_{12} x_1 x_2 x_3. \end{aligned} \tag{15.15}$$

15 In general, none of the design of experiments and regression methods in this and previous chapters are appropriate if the response is categorical, e.g., conforming or nonconforming to specifications. Logistic regression and neural nets described in the next chapter are relevant when outputs are categorical.

However, if each experimental run is effectively a batch of "b" successes or failures, then the fraction nonconforming can be treated as a continuous response. Moreover, if the batch size and true fraction nonconforming satisfies the following, then it is reasonable to expect that the residuals in regression will be normally distributed:

$$b \times p_0 > 5 \text{ and } b \times (1 - p_0) > 5. \tag{15.16}$$

This is the condition such that binomial distributed random probabilities can be approximated using the "normal approximation" or normal probability distribution

Table 15.10 Product design of a central composite and a two-level categorical factor

(a)								(b)							
SO	A	B	C	SO	A	B	C	Run	A	B	C	Run	A	B	C
1	−1	−1	1	11	−1	−1	2	1	0	−1	2	11	−1	−1	2
2	1	−1	1	12	1	−1	2	2	0	0	1	12	1.4	0	1
3	−1	1	1	13	−1	1	2	3	−1	−1	1	13	0	0	1
4	1	1	1	14	1	1	2	4	1	1	1	14	0	1.4	1
5	0	0	1	15	0	0	2	5	0	0	2	15	1	−1	1
6	0	0	1	16	0	0	2	6	0	1.4	2	16	−1	1	1
7	−1.4	0	1	17	−1.4	0	2	7	−1	0	2	17	0	−1	1
8	1.4	0	1	18	1.4	0	2	8	1	1	2	18	1	−1	2
9	0	−1.4	1	19	0	−1.4	2	9	1.4	0	2	19	0	0	2
10	0	1.4	1	20	0	1.4	2	10	−1	0	1	20	−1	1	2

functions. As for selecting sample sizes in the context of p-charting (in ▶ Chap. 4), a preliminary estimate of a typical fraction nonconforming, p_0, is needed. For example, in the printed circuit board (PCB) described in ▶ Chap. 11, batches of size 350 were used and all estimated fraction nonconforming were between 0.05 and 0.95.

15.7.3 Recipe Factors or "Mixture Components"

In a mixture experiment, the system output is a function of the relative proportion of the q components in a mixture, $x_1, \ldots, x_q$, and not their total amounts. These could be the ingredients in a recipe or the constituents in an alloy or chemical. The components must satisfy

$$\sum_{i=1}^{q} x_i = T \quad \text{and} \quad 0 \le a_i \le x_i \le b_i \le T \quad \text{for} \quad i = 1, \ldots, q \tag{15.17}$$

where T is the sum of all components of interest (often $T = 1$).

The equality constraint on the mixture components in Eq. (15.17) constrains the forms of the models that can be fitted using regression estimation described in Eq. (15.3). For example, if a full quadratic polynomial were fitted to data using any feasible experimental plan satisfying Eq. (15.17), estimation using $\boldsymbol{\beta}_{est} = \mathbf{AY}$ would be impossible because columns of $\mathbf{X}$ would be confounded and $\mathbf{X'X}$ would be singular. Then, all least squares coefficient estimates would be undefined. For this reason, Scheffé (1958) proposed dropping selected terms from full d polynomials with the additional intent of preserving interpretability of the estimated coefficients.

Alternative model schemes developed by Scheffé and other authors have been reviewed in Cornell (2002). An example of the model forms that Scheffé proposed is the so-called "canonical second order" mixture model

$$y(x_1, \ldots, x_q) = \sum_{i=1,\ldots,q} \beta_i x_i + \sum_{i=1,\ldots,q} \sum_{i<j} \beta_{ij} x_i x_j + \varepsilon. \tag{15.18}$$

The model is "canonical first order" if the interaction terms are omitted. Data from Piepel and Cornell (1991) show that models of the form in Eq. (15.18) are by far the most popular in documented case studies. Relevant recent models involving both mixture and process variables are described in Chantarat (2003).

15

Example 15.7.4 Method Choices

Question

Which of the following is correct and most complete?

a. Country of origin is a continuous factor.
b. First order models are particularly relevant when additive effects seem intuitive.
c. Regression methods can establish proof as long as several requirements are met.
d. Predictions outside of the range of the input data are theoretically possible.
e. All of the above are correct except (a).

Answer

Country of origin is a categorical factor because there is no natural ordering. Yes, intuitive additivity is a good justification for first order modeling. Still, intuition often suggests combined effects of factors or interactions are possible. Regression can establish proof and generate predictions which are extrapolations. Both proof and extrapolation are often achieved usefully using regression, but caution is needed.

15.8 Chapter Summary

This chapter begins with a single factor first order regression fitting example and the definitions of residuals and least squares. Next, practical issues are described with respect to preparing inputs for regression software focusing on missing observations. Several criteria are then defined for evaluating the acceptability of a regression model and the associated model form. Analysis of variance followed by t-testing is then described as the primary regression-based hypothesis testing procedure with the benefit that overall Type I errors are regulated. Stepwise regression is then described in the context of a semi-automatic approach for selecting model forms. Finally, issues related to categorical and mixture variables and regressions are discussed.

15.9 Problems

1. A new product is released in two medium-sized cities. The demands in Month 1 were 37 and 43 units and in Month 2 were 55 and 45 units. Which of the following is correct and most complete?
   a. A first order regression forecast for Period 3 is 70 units.
   b. One of the residuals is −2 and another one is 5.
   c. A first order regression forecast for Period 3 is 60 units.
   d. Trend analysis cannot involve regression analysis.
   e. All of the above are correct except (a).
2. Considering the example in ► Sect. 15.2.2, which is correct and most complete?
   a. $\mathbf{X}'\mathbf{X} = \begin{bmatrix} 5 & 25 \\ 25 & 135 \end{bmatrix}$
   b. $\boldsymbol{\beta}_{est}$ is a 2×2 matrix.
   c. $\boldsymbol{\beta}_{est} = [22\ 34]'$ minimizes the sum of squares error.
   d. An optimization solver could not be used to derive least squares estimates.
   e. All of the above are correct.
   f. All of the above are correct except (a) and (d).
3. Consider the data in Table 15.1, which is correct and most complete?
   a. Analysts are almost always given data in a format that make analysis easy.
   b. Making up data can never increase the believability of analysis results.
   c. The missing data point in Table 15.1(a) could be worth saving to produce relatively accurate forecasts.

d. All of the above are correct.
e. All of the above are correct except (a) and (d).

4. Which is correct and most complete with regard to regression diagnostics?
   a. Generally, several acceptability tests must be passed for proof to follow.
   b. VIFs are directly useful for spotting outliers in the response data.
   c. Normal probability plotting does not involve any subjectivity.
   d. Randomization can be achieved after the data has already been collected from an observational study.
   e. All of the above are correct.
   f. All of the above are correct except (a) and (e).
5. Which of the following is correct and most complete?
   a. The optional first step involving scaling can affect VIF values.
   b. Least squares coefficient estimates minimize the sum of squared residuals.
   c. VIFs are not influenced by outliers unless the points are removed.
   d. Changing the last input in Table 15.3(a) to 44 would make the VIFs undefined.
   e. All of the above are correct.
   f. All of the above are correct except (a) and (e).

Table 15.11 is relevant to Questions 6–9. Assume that the model form being fitted is a first order polynomial in factors x_1 and x_2 only, unless otherwise mentioned.

6. Which of the following is correct and most complete (for a model without x_3)?
   a. Including the optional scaling, $\text{VIF}_1 = \text{VIF}_2 = 1.08$.
   b. One of the values on the normal probability plot of residuals is –1.7, –2.4.
   c. In general, $\Phi^{-1}[s] = -\Phi^{-1}[-s]$.

Table 15.11 Data for questions 6–9

x_1—population (thousands of people)	x_2—distance (km)	y—sales ($K/yr)	x_3—service type
2	0.5	75	Dine-in
8	0.1	112	Dine-in
8	1	101	In and deliver
9	0.5	117	In and deliver
12	3	109	Deliver
16	2.5	122	Deliver
20	1	154	In and deliver
23	1.2	156	In and deliver
22	0.75	142	Dine-in
26	1.5	162	Deliver

15

d. The input pattern is clearly not acceptable for fitting a first-order model.
e. All of the above are correct.
f. All of the above are correct except (a) and (e).

7. Which of the following is correct and most complete (for a model without x_3)?
   a. The data derive from an application of standard response surface methods.
   b. No outliers appear in the upper right of the normal probability plot of residuals.
   c. The PRESS value for a first-order model in this case is 3280.
   d. The ANOVA in this case clearly indicates the response data are all noise.
   e. All of the above are correct.
   g. All of the above are correct except (a) and (e).
8. Which of the following is correct and most complete (for a model without x_3)?
   a. The regression estimate for sigma based on a first-order model is 0.944.
   b. The SSR for a first order model is 6637.55.
   c. The SSE for a first order model is 396.45.
   d. There are no obvious outliers on a normal plot of residuals.
   e. All of the above are correct.
   f. All of the above are correct except (a) and (e).
9. Consider a first order regression using all of the factors including x_3 with the sales response. Which is correct and most complete?
   a. Since "Service type" is a three level categorical factor, the coefficient calculations could be aided by creating two "contrasts" or dummy factors.
   b. All of the factors are proven to have a significant effect on sales with $\alpha = 0.05$.
   c. It is unlikely that distance from campus (Distance) really affects sales.
   d. It is impossible that population and service type interact in their effect on sales.
   e. All of the above are correct.
10. Which is correct and most complete with regard to the regression flowchart in Fig. 15.5?
   a. Acceptability checks cannot include residuals plots.
   b. Models of on-hand data must be used with caution.
   c. Adding or removing model terms cannot address outliers.
   d. The initial model must be a first order polynomial.
   e. All of the above are correct.
   f. All of the above are correct except (a) and (e).

Problem 11 is based on the following data set.

11. Calculate the following:
   a. Create a first order regression prediction model from the data.
   b. Create a second order regression prediction model from the data.
   c. Calculate R^2 adjusted for a first order model.
   d. Calculate the PRESS for a first order model.

e. Calculate the R^2 prediction for a first order model.
f. Calculate the estimated t_1 value for the coefficient of x_1 in a first order model.

12. Which is correct and most complete according to the chapter?
 a. Editing models involves adding terms to regression models.
 b. If only a single acceptability check is possible, PRESS might be the best.
 c. If the response is fraction nonconforming, large batch sizes can help make residuals more normally distributed.

Table 15.12 Data for questions 6–9

Run	x_1	x_2	Y	Run	x_1	x_2	Y
1	10	902	2.1	6	10	350	1.4
2	−15	103	0.1	7	−10	920	1.4
3	18	821	2.5	8	23	150	0.9
4	22	100	1.9	9	12	200	1.1
5	22	300	1.7				

Table 15.13 Movie data (A = action, An = animated, C = comedy, D = drama, F = foreign, S = suspense)

Name	Genre	# of stars	Sequel	Critics' rating	5th weekend gross	Cumulative $ (end 5th wk)
Bad Boys II	A	2	1	72	$3,143,914	$128,856,716
Dirty Pretty Things	S	0	0	88	$557,263	$1,986,903
Johnny English	C	1	0	78	$318,985	$26,925,075
Pirates of the Caribbean	A	1	0	82	$13,022,470	$232,750,629
League of Extra…	A	1	0	72	$1,542,272	$62,179,376
Northfork	D	2	0	82	$160,042	$819,913
I Capture the Castle	D	0	0	88	$126,084	$770,491
The Housekeeper	F	0	0	NA	$28,814	$298,818
Madame	F	0	0	NA	$6,493	$88,935
The Cuckoo	F	0	0	NA	$4,437	$66,332
Terminator 3	A	1	1	82	$2,985,446	$142,853,468
Legally Blonde 2	C	1	1	75	$1,408,958	$85,260,859
Swimming Pool	S	0	0	85	$1,008,571	$5,253,781
Sinbad: Legend …	An	2	0	82	$153,993	$25,692,461
28 Days Later	S	0	0	88	$2,341,887	$37,304,321
Charlie's Angels 2	A	3	1	75	$1,460,418	$93,073,452

(continued)

15

Table 15.13 (continued)

Name	Genre	# of stars	Sequel	Critics' rating	5th weekend gross	Cumulative $ (end 5th wk)
On_Line	D	0	0	72	$5,017	$94,894
The Hulk	A	1	0	82	$1,543,240	$128,143,315
Legend of Suri-yothai	F	0	0	NA	$40,956	$277,562
Bonhoeffer	F	0	0	NA	$17,113	$97,623
Rugrats Go Wild!	A	1	0	75	$822,620	$36,801,254
Hollywood Hom-icide	A	1	0	75	$201,872	$29,743,738
Dumb and Dumb-erer 2	C	0	1	68	$123,292	$25,493,066
Jet Lag	C	0	0	78	$50,711	$339,557
The Heart of Me	D	1	0	NA	$8,419	$110,554
Tycoon	F	0	0	NA	$2,417	$74,299
2 Fast 2 Furious	A	0	1	75	$2,641,820	$119,437,965
The Eye	F	0	0	82	$31,867	$339,607
Finding Nemo	An	1	0	92	$13,968,116	$253,991,677
The Italian Job	A	2	0	82	$5,462,902	$76,758,011

d. With mixture factors, some terms in a full polynomial must be dropped.
e. All of the above are correct.
f. All of the above are correct except (a) and (e) (Table 15.12).

13. Analyze the box office data in Table 15.13 and make recommendations to the extent possible for a vice president at a major movie studio. A good analysis will typically include one or two models, reasons for selecting that model, estimates of the errors of the model, and interpretation for the layperson of everything. (Note that this question is intentionally open-ended because that is the way problems are on-the-job). Feel free to supplement with additional real data if you think it helps support your points. (These are from ▶ Yahoo.com)
14. Analyze the real estate data in Table 15.14 and make recommendations to a real estate developer about where to build and what type of house to build for profitability. A good analysis will typically include one or two models, reasons for selecting that model, estimates of the errors of the model, and interpretation of everything for the layperson. (Note that this question is intentionally openended because that is the way problems are on-the-job). Feel free to supplement with additional real data.

Table 15.14 Real estate data

City	#Bedrooms	#Baths	Offering price ($)
Upper Arlington	3	1	154,900
Upper Arlington	2	2	195,000
Upper Arlington	3	2	249,900
Upper Arlington	3	1.5	264,900
Upper Arlington	4	3	279,000
Upper Arlington	3	1.5	290,000
Upper Arlington	4	2.5	312,000
Upper Arlington	4	2.5	357,000
Upper Arlington	5	3.5	375,000
Upper Arlington	3	3	389,000
Upper Arlington	4	2.5	395,900
Upper Arlington	4	2.5	420,000
Upper Arlington	4	3	455,000
Upper Arlington	4	3	499,900
Upper Arlington	4	2.5	499,900
Columbus	3	1.5	150,000
Columbus	3	1	156,900
Columbus	4	2	159,700
Columbus	2	1	159,900
Columbus	4	1.5	167,900
Columbus	4	2.5	181,900
Columbus	2	2	185,000
Columbus	3	3	194,900
Columbus	3	2.5	199,900
Columbus	4	2	213,900
Columbus	3	1	219,900
Columbus	3	1.5	220,000
Columbus	3	2	224,900
Columbus	3	2.5	224,900
Columbus	3	1.5	229,900
Columbus	4	2.5	239,000
Columbus	3	1.5	244,888
Columbus	3	2	244,900

(continued)

Table 15.14 (continued)

City	#Bedrooms	#Baths	Offering price ($)
Columbus	4	1.5	246,900
Columbus	3	2	349900
Columbus	3	2.5	365,000

References

Chantarat N (2003) Modern design of experiments for screening and experimentations with mixture and qualitative variables. PhD dissertation, Industrial & Systems Engineering, The Ohio State University, Columbus

Chantarat N, Zheng N, Allen TT, Huang D (2003) Optimal experimental design for systems involving both quantitative and qualitative factors. In: Ferrin RDM, Sanchez P (eds) Proceedings of the winter simulation conference

Cornell JA (2002) Experiments with mixtures, 3rd edn. New York, Wiley

Fisher RA (1925) Statistical methods for research workers. Oliver and Boyd, London

Montgomery DC (2012) Statistical quality control, 7th edn. New York, Wiley

Piepel GF, Cornell JA (1991) A catalogue of mixture experiments. In: Proceedings of the joint statistical meetings (Aug 19), Atlanta

Scheffé H (1958) Experiments with mixtures. J Roy Stat Soc B 20(2):344–360

Advanced Regression and Alternatives

T. T. Allen, *Introduction to Engineering Statistics and Lean Six Sigma*,
https://doi.org/10.1007/978-1-4471-7420-2_16

16.1 Introduction

Linear regression models are not the only curve-fitting methods in wide use. Also, these methods are not useful for analyzing data for categorical responses. In this chapter, so-called "kriging" models, "artificial neural nets" (ANNs), and logistic regression methods are briefly described. ANNs and logistic regression methods are relevant for categorical responses. Each of the modeling methods described here offers advantages in specific contexts. However, all of these alternatives have a practical disadvantage in that formal optimization must be used in their fitting process.

▶ Section 16.2 discusses generic curve fitting and the role of optimization. ▶ Section 16.3 briefly describes kriging models, which are considered particularly relevant for analyzing deterministic computer experiments and in the context of global optimization methods. In ▶ Sect. 16.3, one type of neural net is presented. ▶ Section 16.4 defines logistic regression models including so-called "discrete choice" models. In ▶ Sect. 16.5, examples illustrate logit and probit discrete choice models.

16.2 Generic Curve Fitting

Many numerical approaches have been proposed for interpolating data points. A subset of these has been developed with the intention of mitigating the effects that random errors have on curve fitting, including linear regression, kriging models, and neural nets. All of these estimate their model parameters, $\boldsymbol{\beta}_{est}$, based on their experimental inputs, $\mathbf{x}_1, \ldots, \mathbf{x}_N$, and outputs, $y_1, \ldots, y_N$, by solving an optimization program of the form

$$\begin{aligned} &\text{Maximize:} && g(\boldsymbol{\beta}_{est}, \mathbf{x}_1, \ldots, \mathbf{x}_N, y_1, \ldots, y_N, h) \\ &\text{Subject to:} && y_{est}(\boldsymbol{\beta}_{est}, \mathbf{x}) = h(\boldsymbol{\beta}_{est}, \mathbf{x}) \\ &&& \boldsymbol{\beta}_{est} \in \mathbf{R}^k \end{aligned} \tag{16.1}$$

where predictions come from $y_{est}(\boldsymbol{\beta}_{est}, \mathbf{x})$, which is based on the so-called functional form, $h(\boldsymbol{\beta}_{est}, \mathbf{x})$, of the fitted model. This "**functional form**" constrains the relationship between the predictions (y_{est}), the inputs ($\mathbf{x}$), and the model parameters ($\boldsymbol{\beta}_{est}$). Generally, $y_{est}(\boldsymbol{\beta}_{est}, \mathbf{x})$ refers to the prediction that will be generated, given $\boldsymbol{\beta}_{est}$ for the mean value of the system response at the point $\mathbf{x}$ in the region of interest. The quantity $\mathbf{R}^k$ refers to the k dimensional vector space of real numbers, i.e., $\beta_{est,i}$ are real numbers $i=1, \ldots, k$. Note the symbol "$\wedge$" can be used interchangeably with "*est*" to indicate that the quantity involved is estimated from the data.

16

For linear regression and many implementations of neural nets, the objective function, g, is the negative of the sum of squares of the estimated errors (SSE). For kriging models, the objective function, g, is the so-called "likelihood" function. Yet, the curve-fitting objective function could conceivably directly account for the expected utility of the decision-maker, instead of reflecting the SSE or likelihood. In the context of regression models, the entries in $\boldsymbol{\beta}_{est}$ are called coefficients. In the context of neural nets, specific $\boldsymbol{\beta}_{est,i}$ refer to so-called weights and numbers of nodes and layers. In the context of kriging models, entries in $\boldsymbol{\beta}_{est}$ are estimated parameters.

16.2.1 Curve Fitting Example

Here we review a curve fitting problem that reveals the special properties of linear regression curve fitting. Consider an example with a single factor involving the five runs, i.e., input, $(x_1)_i$, and output, y_i, combinations, given in Table 16.1. The linear regression optimization problem for estimating the coefficients is given by

$$\text{Maximize:} \quad -\{[y_1 - y_{est}(\boldsymbol{\beta}_{est}, \mathbf{x}_1)]^2 + [y_2 - y_{est}(\boldsymbol{\beta}_{est}, \mathbf{x}_2)]^2 + [y_3 - y_{est}(\boldsymbol{\beta}_{est}, \mathbf{x}_3)]^2 + [y_4 - y_{est}(\boldsymbol{\beta}_{est}, \mathbf{x}_4)]^2 [y_5 - y_{est}(\boldsymbol{\beta}_{est}, \mathbf{x}_5)]^2\} \tag{16.2}$$

$$\text{Subject to:} \quad y_{est}(\boldsymbol{\beta}_{est}, \mathbf{x}_i) = \beta_{est,0}(1) + \beta_{est,1}(x_1)_i \text{ for } i = 1, \ldots, 5 \qquad \boldsymbol{\beta}_{est} \in \mathbf{R}^2 \tag{16.3}$$

Equation (16.3) clarifies that the functional form is a first order polynomial or, in other words, a line. Because Eq. (16.3) is linear in the coefficients and the objective function in Eq. (16.2) is a quadratic polynomial, there is a formula giving the solution.

As described in ► Chap. 15, the solution is given by the formula

$$\boldsymbol{\beta}_{\text{est}} = (\mathbf{X}'\mathbf{X})^{-1}\mathbf{X}'\mathbf{y} = \begin{bmatrix} -26 \\ 32 \end{bmatrix} \tag{16.4}$$

Generally, formula optimization problems are so difficult that there is no formula giving the solution. More commonly, a solution algorithm such as the Excel solver must be applied. Table 16.1 shows the estimated errors and the sum of the squared estimated errors, SSE, for two sets of candidate coefficient solutions. The first set is sub-optimal for the formulation in Eq. (16.2). As shown in Fig. 16.1a, few people would desire this fitted model based on the data compared with the model associated with the coefficients that minimize the sum of squares error are shown in Fig. 16.1b.

Table 16.1 Simple least squares regression example

			Suboptimal β_{est} for (2)		Optimal β_{est} for (2)	
			$\beta_{est,0}$	$\beta_{est,1}$	$\beta_{est,0}$	$\beta_{est,1}$
		Coefficents	200	–10	–26	32
i	$(x_1)_i$	y_i	$y_{est,i}$	$\text{Error}_{est,i}$	$y_{est,i}$	$\text{Error}_{est,i}$
1	3	70	170	–100	70	0
2	4	120	160	–40	102	18
3	5	90	150	–60	134	–44
4	6	200	140	60	166	34
5	7	190	130	60	198	–8
			SSE	22,400		3480

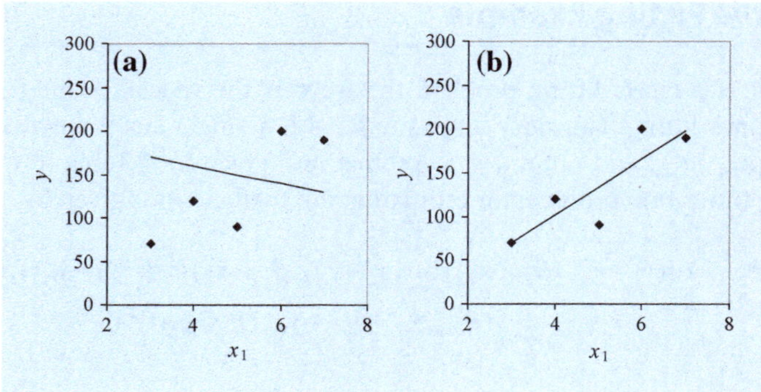

Fig. 16.1 The data and two models **a** sub-optimal and **b** least squares optimal

The coefficients giving the fit in Fig. 16.1b can be derived using the a solver. Using such a solver is necessary for all of the other curve fitting methods in this chapter.

16.3 Kriging Models and Computer Experiments

Matheron (1963) proposed so-called "kriging" meta-models to make predictions in the context of modeling physical, geology-related data. Recently, the application of these same techniques in the context of computer experiments has received significant attention in part because of the above-mentioned advantage that kriging models provide smooth interpolating functions passing through all of the output data, e.g., see Sacks et al. (1989) and Welch et al. (1992). Kriging procedures are relatively difficult to apply because the curve fitting involved requires a nontrivial optimization and, therefore, specialized software. However, as the necessary software becomes increasingly available, there is reason to expect that the methods will enjoy even more widespread application.

Therefore, kriging models under common assumptions provide prediction models, $y_{est}(\boldsymbol{\beta}_{est}, \mathbf{x})$, that pass through all the data points. This is considered to be desirable in the context of certain kinds of experiments that are perfectly repeatable, i.e., the same inputs give the same outputs with $\sigma_0 = 0$. The phrase "**computer experiments**" is often used to refer to finite element method (FEM) and finite difference method (FDM) testing in which prototypes are virtual and no sources of variation are involved in there empirical evaluation.

Kriging models are sufficiently flexible that they can seamlessly extend to situations in which the number of tests grows much higher than those involved in response surface methods. Because kriging models can model input-output relationships with multiple twists and turns, they are considered particularly relevant in the context of optimization.

16.3.1 Design of Experiments for Kriging Models

Deriving desirable experimental plans to foster accurate fitted kriging models is an active area of research. For simplicity, only so-called Latin hypercube designs (LHDs) and space-filling designs for the data collection are considered because these designs have received the most attention in the kriging literature. LHDs have the advantage that they are easy to generate for any number of runs.

The version of LHDs here is based on McKay et al. (1979). For n runs, each of the k factors takes on equally spaced values $-1+1/n$, $-1+3/n$, …, $1-1/n$, in different random orders. Space-filling designs are derived by maximizing the minimum Euclidean distance between all pairs of design points. Table 16.2 shows examples of the LHDs and space-filling designs with the space-filling design generated using the optimization method of Hadj-Alouane and Bean (1997).

16.3.2 Fitting Kriging Models

The following equation offers intuition about how kriging models work:

$$Y(\mathbf{x}) = f(\mathbf{x}) + Z(\mathbf{x}), \tag{16.5}$$

Table 16.2 Example **(a)** latin hypercube and **(b)** space-filling design

(a)				(b)			
Run	A	B	C	Run	A	B	C
1	0.375	0.042	0.875	1	0.040	0.929	0.788
2	0.625	0.458	0.542	2	0.000	0.313	1.000
3	0.042	0.542	0.292	3	1.000	0.010	0.000
4	0.792	0.792	0.125	4	0.253	0.000	0.000
5	0.458	0.375	0.708	5	0.909	0.434	0.475
6	0.208	0.708	0.375	6	0.000	0.586	0.020
7	0.292	0.125	0.458	7	1.000	1.000	0.768
8	0.542	0.292	0.625	8	0.414	1.000	0.273
9	0.125	0.875	0.042	9	1.000	0.909	0.040
10	0.958	0.208	0.792	10	0.980	0.000	0.980
11	0.875	0.625	0.958	11	0.576	0.586	1.000
12	0.708	0.958	0.208	12	0.424	0.000	0.636

where $f(\mathbf{x})$ is a regression model that is potentially the same as a linear regression model and $Z(\mathbf{x})$ is a function that models departures from the regression model. A relevant concept is, therefore, an attempt to more aggressively model the unexplained variation compared with using regression models only.

Attempting to predict the departures, $Z(\mathbf{x})$, from regression is motivated by the fact computer experiments with little or no random error. Sacks et al. (1989) argue that it is reasonable for computer experiments to treat the departures $Z(\mathbf{x})$ as if they can be modeled and not merely considered to be random noise. Following Matheron (1963), Sacks et al. (1989) proposed to model the departures as "realizations" from a Gaussian stochastic process.

Further, they and other authors suggest that the regression component, $f(\mathbf{x})$, should be omitted because of empirical evidence that this gives superior or comparable accuracy. Here also, we focus on the assumption that a constant term only is included in the model instead of a complicated regression model.

The variables used in fitting include:

1. m is the number of factors.
2. $\theta_i \geq 0$ and $0 \leq p_i \leq 2$ for $i = 1, \ldots, m$ are fitted parameters similar to regression coefficients.
3. $R(\mathbf{w}, \mathbf{x})$ is the correlation between the random departures $Z(\mathbf{w})$ and $Z(\mathbf{x})$ for decision vectors $\mathbf{w}$ and $\mathbf{x}$.
4. $\mathbf{R}$ is an $n \times n$ matrix of correlations between the n points in the input array, which is a function of the q_i and p_i.
5. $\boldsymbol{\beta}_{est}$ is the estimated regression coefficient vector. Here, we focus on the assumption that $\boldsymbol{\beta}_{est}$ is just one coefficient, i.e., the constant term.
6. σ_{est} is the estimated standard deviation of the response variation that is roughly proportional to the range of the response.
7. ln L is the "log likelihood" which is the fitting objective analogous to least squares for linear regression.
8. $\mathbf{x}_1, \ldots, \mathbf{x}_n$ are the n input combinations in the input pattern. These could be specified by an experimental design such as a Latin-Hypercube.
9. $x_{i,k}$ refers to the settings of the ith run for the kth factor.
10. $\mathbf{y}$ is the response vector corresponding to the n runs.

Algorithm 16.1. Fitting kriging models

Step 1. Develop a function giving the correlation matrix, $\mathbf{R}$, between the responses at the points, $\mathbf{x}_1, \ldots, \mathbf{x}_n$, using the formula

16

$$\int \mathbf{R}_{i,j}(\boldsymbol{\theta}, \boldsymbol{p}) = \prod_{k=1}^{m} e^{-\theta_k (x_k - x_{i,k})^{pk}} \quad (16.6)$$

where $\mathbf{x}_i$ and $\mathbf{x}_j$ are all pairs of points and $\mathbf{R}$ is a function of $\boldsymbol{\theta}$ and $\boldsymbol{p}$. This matrix is used for calculating and optimizing the likelihood.

Step 2. Calculate $\boldsymbol{\beta}_{est}$ as a function of the fitting parameters using

$$\beta_{est}(\boldsymbol{\theta}, \boldsymbol{p}) = (\mathbf{1}'\mathbf{R}^{-1}\mathbf{1})^{-1}\ \mathbf{1}'\mathbf{R}^{-1}\mathbf{y} \quad (16.7)$$

where **1** is an n dimensional vector of 1s. This coefficient is used for calculating and optimizing the likelihood.

Step 3. Calculate σ_{est} as a function of the fitting parameters using

$$\sigma_{est}^2(\boldsymbol{\theta}, \mathbf{p}) = n^{-1}(\mathbf{y} - \mathbf{1}\beta_{est})'\mathbf{R}^{-1}(\mathbf{y} - \mathbf{1}\beta_{est}). \quad (16.8)$$

Step 4. Calculate ln L as a function of the fitting parameters using:

$$\ln L(\boldsymbol{\theta}, \mathbf{p}) = -\frac{n \ln \sigma_{est}^2 + \ln[\det(\mathbf{R})]}{2}. \quad (16.9)$$

Step 5. Estimate parameters by solving

$$\begin{aligned} &\text{Maximize}: \ \ln L(\boldsymbol{\theta}, \mathbf{p}) \\ &\text{subject to } \theta_i \geq 0 \text{ and } 0 \leq p_i \leq 2 \text{ for } i = 1, \ldots, m. \end{aligned} \quad (16.10)$$

Step 6. Generate predictions at any point of interest, **x**, using

$$y_{est}(\mathbf{x}) = \beta_{est} + \mathbf{r}'(\mathbf{x})\mathbf{R}^{-1}(\mathbf{y} - \mathbf{1}\beta_{est}). \quad (16.11)$$

where $\mathbf{r}(\mathbf{x}) = [R(\mathbf{x}, \mathbf{x}_1), \ldots, R(\mathbf{x}, \mathbf{x}_n)]'$ with $R(\mathbf{w}, \mathbf{x})$ given by

$$R(\mathbf{x}, \mathbf{x}_i) = \prod_{k=1}^{m} e^{-\theta_k (x_{i,k} - x_{j,k})^{pk}}. \quad (16.12)$$

The functions in Eqs. (16.6) and (16.11) represent one of several possible functional forms of interest. With the response of the system viewed as random variables, these equations express possible beliefs about how responses might correlate. The equations imply that repeated experiments at the same points, e.g., $\mathbf{x} = \mathbf{x}_i$ give the same outputs, because the correlations are 1.

Maximization of the likelihood function in *Step 5* can be a difficult problem because there might be multiple extrema in the θ_k and p_k space. Welch et al. (1992) propose a search technique for this purpose that is based on multiple line searches. Commonly, $p_k = 2$ for all k is assumed because this gives rise to often desirable smoothness properties and reduces the difficulty in maximizing the likelihood function.

16.3.3 Kriging Single Variable Example

Question

Fixing $p_1 = 2$, use the data in Table 16.1 to estimate the optimal θ parameter, **R** matrix, and a prediction for $x_1 = 5.5$.

Answer

Equation (16.8) gives an estimate of β equal to 133.18. Next, one derives the log-likelihood as a function of θ_1. Maximizing gives the estimated θ_1 equal to 1.648. The resulting **R** matrix is

$$\mathbf{R} = \begin{bmatrix} 1.0000 & 0.1925 & 0.0014 & 0.0000 & 0.0000 \\ 0.1925 & 1.0000 & 0.1925 & 0.0014 & 0.0000 \\ 0.0014 & 0.1925 & 1.0000 & 0.1925 & 0.0014 \\ 0.0000 & 0.0014 & 0.1925 & 1.0000 & 0.1925 \\ 0.0000 & 0.0000 & 0.0014 & 0.1925 & 1.0000 \end{bmatrix} \tag{16.13}$$

and the prediction is $\boldsymbol{\beta}_{est} + \mathbf{r}'(x_1 = 5.5)\mathbf{R}^{-1}(\mathbf{y} - \mathbf{1}\boldsymbol{\beta}_{est}) = 141.9$ which is somewhat close to the first order linear regression prediction of $-26 + 32 \times 5.5 = 150$.

Allen et al. (2003) compared the prediction accuracy of neural nets with linear models in the context of test functions and response surface methods. The tentative conclusion reached is that kriging models do not offer obvious, substantial prediction advantages despite their desirable property of passing through all points in the data base. However, the cause of the prediction errors was shown to relate to the choice of the likelihood fitting objective and not the bias inherent in the fit model form. Therefore, as additional research generates alternative estimation methods, kriging models will likely become a useful alternative to linear models for cases in which prediction accuracy is important. Also, as noted earlier, kriging models adapt easily to cases in which the number of runs exceeds that in standard response surface methods.

16.4 Neural Nets for Regression Type Problems

Neural nets fascinate many people because their structure is somewhat reminiscent of the way that the human brain generates predictions based on data. These methods also offer potentially great flexibility with respect to the ability to approximate a wide variety of functions. However, this flexibility is also a concern since they offer much potential for misuse. This section focuses on neural nets used for situations involving continuous responses. These neural nets constitute alternatives to linear regression and kriging models.

Here, a simple, spreadsheet-based neural network modeling technique is proposed and illustrated with an example. This discussion is based on results in Ribardo (2000). The principle advantages of the networks described here relate to their pedagogical use, in that they can be completely described. Also, they can be implemented with minimal training and without special software. Note that neural networks are a particularly broad class of modeling techniques. No single implementation could be representative of all of the possible methodologies that have been proposed in the literature. Therefore, the disadvantage of this approach is that other, superior, implementations for similar problems almost surely exist. However, results in Ribardo (2000) with the proposed neural net probably justify a few general comments about training and complexity associated with the existing methods.

Kohonen (1989) and Chambers (2000) review neural net modeling in the context of predicting continuous responses such as undercut in millimeters. Neural nets have also been proposed for classification problems involving discrete responses.

Here also, an attempt is made to avoid using the neural net terminology as much as possible to facilitate the comparison with the other methods. Basically, neural nets are a curve-fitting procedure that, like regression, involves estimating several coefficients (or "weights") by solving an optimization problem to minimize the sum of squares error. In linear regression, this minimization is relatively trivial from the numerical standpoint, since formulas are readily available for the solution, i.e., $\boldsymbol{\beta}_{est} = (\mathbf{X}'\mathbf{X})^{-1}\mathbf{X}'\mathbf{y}$. In neural net modeling, however, this minimization is typically far less trivial and involves using a formal optimization solver. In the implementation here, the Excel solver is used. In more standard treatments, however, the solvers involve methods that are to some degree tailored to the specific functional form (or "architecture") of the model (or "net") being fit. A solver algorithm called "back-propagation" is the most commonly used method for estimating the model parameters (or "training on the training set") for the model forms that were selected. This solver technique and its history is described in Rumelhart and McClelland (1986) and Reed and Marks (1999).

There are many possible functional forms ("architectures") and, unfortunately, little consensus about which of these forms yield the lowest prediction errors for which type of problems, e.g., see Chambers (2000). An architecture called "single hidden layer, feed forward neural network based on sigmoidal transfer functions" with five randomly chosen data in the test set and the simplest "training termination criterion" was arbitrarily selected. One reason for selecting this architecture type is that substantial literature exists on this type of network, e.g., see Chambers (2000) for a review. Also, it has been demonstrated rigorously that, with a large enough number of nodes, this type of network can approximate any continuous function (Cybenko 1989) to any desired degree of accuracy. This fact may be misleading, however, because in practice the possible number of nodes is limited by the amount of data (see below). Also, it may be possible to obtain a relatively accurate net with fewer total nodes using a different type of architecture.

The choice of the number of nodes and the other specific architectural considerations is largely determined by the accepted compromise between the observed variation (high adjusted R^2) and what is referred to as "over-fitting". ◘ Figure 16.2 illustrates the concept of over-fitting. In the selected feed-forward architecture, for each of the l nodes in the hidden layer (not including the constant node, which always equals 1.0), the number of coefficients (or "weights") equals the number of factors, m, plus two. Therefore, the total number of weights is $w = l(m+2)+2$. The additional two weights derive from multiplying the constant node in the final prediction node and the (optional) overall scale factor, which can help in achieving realistic weights.

Several rules of thumb for selecting w and l exist and are discussed in Chambers (2000). If w equals or exceeds the number of data points n, then provably the sum of squares error is zero and the neural net passes through all the points as shown in ◘ Fig. 16.2. If there are random errors in the data, illustrated in ◘ Fig. 16.2 by the ε_is, then prediction of the average response values will be inaccurate since the net has been overly influenced by these "random errors".

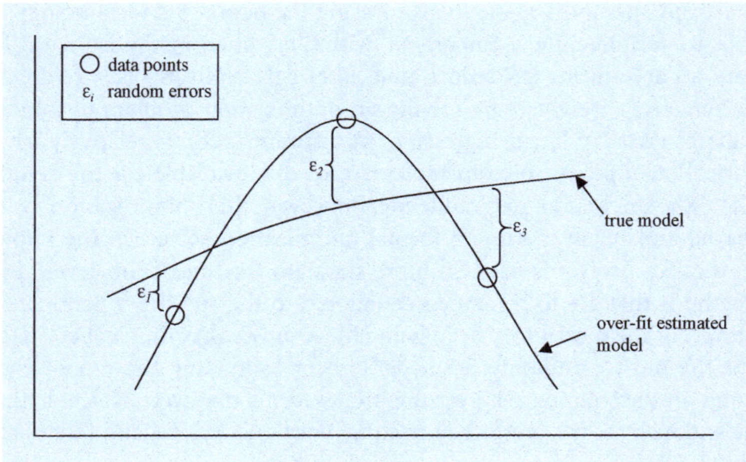

Fig. 16.2 Example of over-fitting

The simple heuristic method for selecting the number of nodes described in Ribardo (2000) will be adopted to address this over-fitting issue in the response surface context. This approach involves choosing the number of nodes so that the number of weights approximately equals the number of terms in a quadratic Taylor series or, equivalently, a response surface model. The number of such terms is $(m+1)(m+2)/2$. In the case study, $m=5$ and the number of terms in the RSM polynomial is 21. This suggests using $l=3$ nodes in the hidden layer so that the number of weights is $3\times(5+2)+2=23$ weights.

One additional complication is the number of runs selected for the so-called "test set". These runs are set aside and not used for estimating the weights in the minimization of the sum of squares error. In the context of welding parameter development from planned experiments, it seems reasonable to assume that the number of runs is typically small by the standards discussed in the neural net literature. Therefore, the ad hoc selection of five random runs for the test set was proposed because this is perhaps the smallest number that could reasonably be expected to provide an independent and reliable estimate of the prediction errors.

A final complication is the so-called "termination criterion" for the minimization of the sum of squares error. In the hopes of avoiding over-fitting inaccuracies illustrated in Fig. 16.3, many neural net users do not attempt to solve the sum of squares minimization problem for the coefficients ("weights") to global optimality. Instead they terminate the minimization algorithm before its completion based on nontrivial rules deriving from inspection of the test set errors. For simplicity, these complications were ignored, and the Excel solver was permitted to attempt to select the weights that globally minimize the sum of squares error.

16

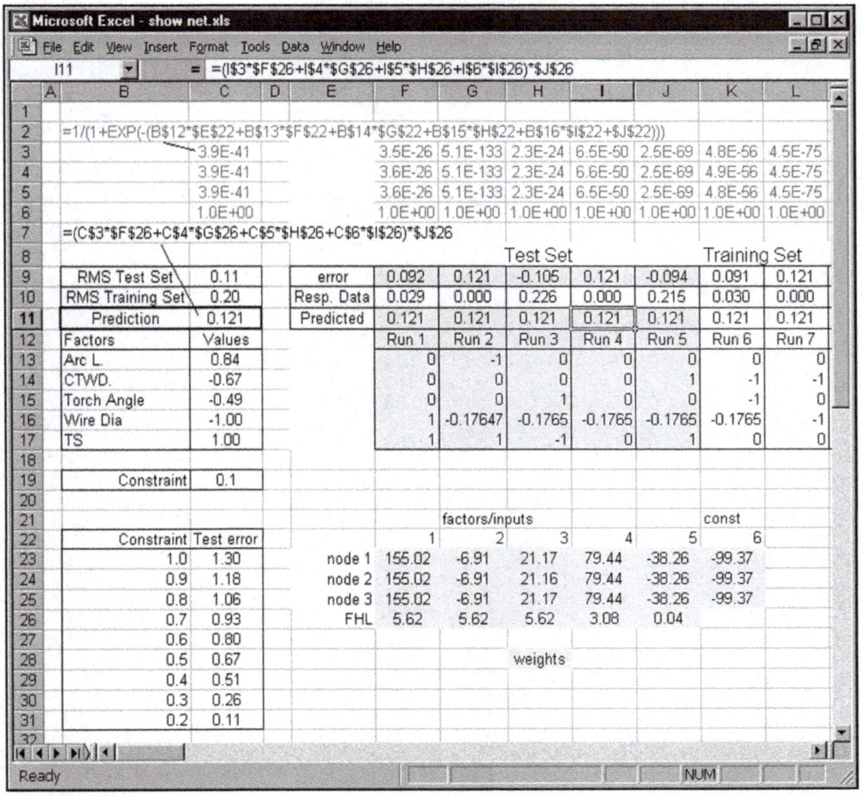

Fig. 16.3 The excel spreadsheet neural net for undercut response

Next the construction of neural nets is illustrated using a welding example from Ribardo (2000). The response data is shown in Table 16.3 from a Box–Behnken experimental design. This data was used to fit ("train") the spreadsheet neural net in Fig. 16.3.

Then 46 identical neural nets were created in Excel, designed to predict each of the test run data based on a common set of weights. A random number generator was used to select five of these runs as the "test set". The Excel solver was used next in order to minimize the training set sum of squares error by optimizing the weights. This procedure resulted in the net shown in Fig. 16.4 for undercut with the weights at the bottom right and the weights for convexity shown in Table 16.4. Figure 16.5 shows a plot of the neural net predictions for the welding example compared with other methods compared in Ribardo (2000).

Table 16.3 Box–Behnken design and data for the neural net from Ribardo (2000)

Run	Std. order	x_1	x_2	x_3	x_4	x_5	Y_1 = undercut	Y_2 = convexity
1	20	0.125	0.750	0	0.062	40	0.0287	0.2258
2	35	0.000	0.750	0	0.052	40	0.0000	1.6837
3	30	0.125	0.750	15	0.052	20	0.2261	−1.0798
4	45	0.125	0.750	0	0.052	30	0.0000	0.4727
5	12	0.125	0.875	0	0.052	40	0.2150	0.9369
6	21	0.125	0.625	−15	0.052	30	0.0300	0.3371
7	37	0.125	0.625	0	0.045	30	0.0000	0.4473
8	42	0.125	0.750	0	0.052	30	0.1572	0.3893
9	1	0.000	0.625	0	0.052	30	0.0000	1.9152
10	41	0.125	0.750	0	0.052	30	0.0000	0.2712
11	40	0.125	0.875	0	0.062	30	0.0000	0.3855
12	28	0.250	0.750	0	0.062	30	0.5174	−1.5789
13	18	0.125	0.750	0	0.062	20	0.8202	−1.8837
14	44	0.125	0.750	0	0.052	30	0.0000	0.4758
15	36	0.250	0.750	0	0.052	40	0.7254	0.3371
16	43	0.125	0.750	0	0.052	30	0.0000	0.3663
17	15	0.000	0.750	15	0.052	30	0.0000	1.2828
18	10	0.125	0.875	0	0.052	20	0.1168	−0.7589
19	17	0.125	0.750	0	0.045	20	0.0153	−0.4141
20	2	0.250	0.625	0	0.052	30	0.7573	−2.1467
21	8	0.125	0.750	15	0.062	30	0.4912	−1.3343
22	22	0.125	0.875	−15	0.052	30	0.0054	0.5323
23	6	0.125	0.750	15	0.045	30	0.2261	0.1176
24	14	0.250	0.750	−15	0.052	30	1.0071	0.1098
25	34	0.250	0.750	0	0.052	20	0.8872	−2.5600
26	46	0.125	0.750	0	0.052	30	0.7066	0.1834
27	29	0.125	0.750	−15	0.052	20	0.4259	0.0248
28	32	0.125	0.750	15	0.052	40	0.0000	0.8287
29	19	0.125	0.750	0	0.045	40	0.0142	0.9482
30	38	0.125	0.875	0	0.045	30	0.0365	0.1249
31	25	0.000	0.750	0	0.045	30	0.0000	0.9082
32	16	0.250	0.750	15	0.052	30	0.6702	−1.4284
33	9	0.125	0.625	0	0.052	20	0.3155	−0.8012

(continued)

Table 16.3 (continued)

Run	Std. order	x_1	x_2	x_3	x_4	x_5	Y_1 = undercut	Y_2 = convexity
34	11	0.125	0.625	0	0.052	40	0.7061	0.9985
35	39	0.125	0.625	0	0.062	30	0.0000	0.1600
36	4	0.250	0.875	0	0.052	30	0.8201	−1.8826
37	27	0.000	0.750	0	0.062	30	0.0000	0.7397
38	26	0.250	0.750	0	0.045	30	0.0000	−0.8550
39	5	0.125	0.750	−15	0.045	30	0.1165	0.1478
40	13	0.000	0.750	−15	0.052	30	0.0000	1.5675
41	7	0.125	0.750	−15	0.062	30	0.0626	0.2740
42	31	0.125	0.750	−15	0.052	40	0.6928	1.0285
43	33	0.000	0.750	0	0.052	20	0.0147	0.7128
44	24	0.125	0.875	15	0.052	30	0.1677	0.1534
45	3	0.000	0.875	0	0.052	30	0.0340	0.9073
46	23	0.125	0.625	15	0.052	30	0.0287	0.2258

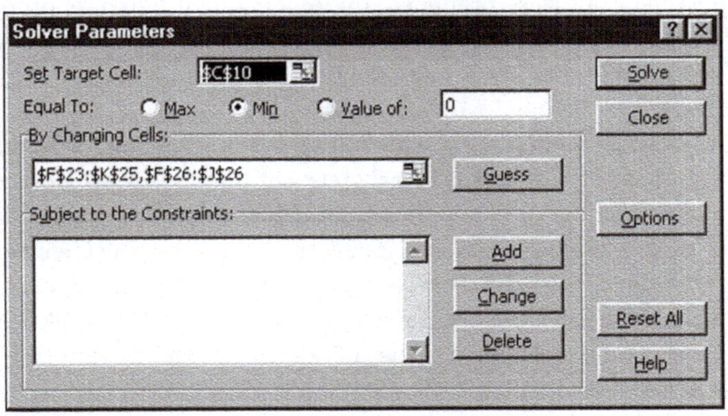

Fig. 16.4 The solver fields for estimating the net coefficients ("weights")

Table 16.4 The weights for the convexity response neural net

	Factors/inputs					Const
	1	2	3	4	5	6
Node 1	7.069	−6.35	−7.12	7.086	−6.2	0.646
Node 2	2.134	2.048	2.899	−2.66	−2.25	−0.26
Node 3	54.01	0.518	37.03	19.98	−38.4	−35.5
FHL	5.26	5.487	5.224	−6.59	−0.28	

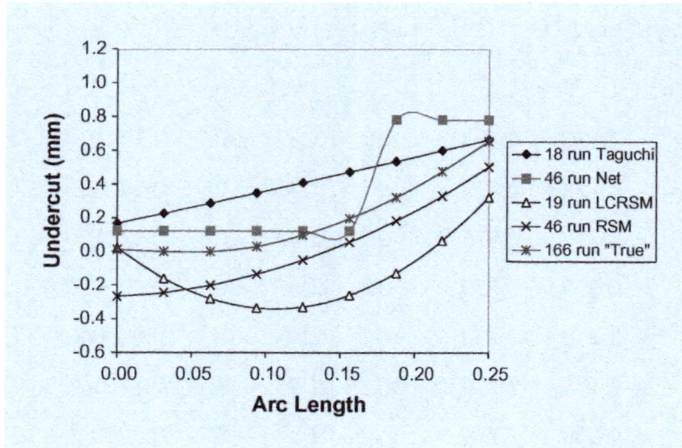

Fig. 16.5 Predictions from the models derived from alternative methodologies

16.5 Logistic Regression and Discrete Choice Models

In many modeling problems of interests, responses are categorical variables, i.e., response levels are discrete and with no natural ordering. Examples might be units either being conforming (level 1) or nonconforming (level 2) to specifications. Also, certain people might purchase product 1 and others might purchase product 2. Then, it maybe of interest to predict the chance that the response will assume any one of categorically different response levels (e.g., see Ben-Akiva and Steven 1985; Hosmer and Lemeshow 1989).

Logistic regression models are a widely used set of modeling procedures for predicting these probabilities. It is particularly relevant for cases in which what might be considered a large number of data points is available. Considering that "**data mining**" is the analysis or very large flat files, logistic regression can be considered an important data mining technique. Also, "discrete choice models" are logistic regression models in which the levels of the categorical variables are options a decision-maker might select. In these situations, the probability is the market share might command when faced with a specified list of competitors.

Logistic regression models including discrete choice models are based on the following concept. Each level of the categorical response is associated with a continuous random variable, which we might call u_i for the "utility" of response level i. If the random variable associated with a given level is highest, that level is response or choice.

16

Figure 16.6a shows a response with two levels, e.g., system options a decision-maker might choose. System 1 random variables have a lower average than system 2 random variables. However, by chance the realization for system 1 (♦) has a higher value than for system 2 (♦). Then, the response would be system 1 but, in general, system 2 would have a higher probability.

Figure 16.6b shows how the distribution means of the two random variables are functions of a controllable input factor x. By adjusting x, it could be possible to tune each system to its optimum resulting in the highest chance that that level (or system) will occur (or be chosen). Note that the input factor levels that tune

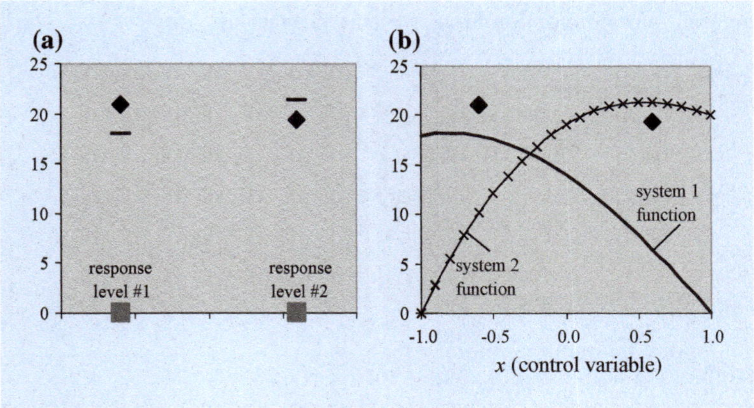

Fig. 16.6 Utilities of **a** two systems each with fixed level and **b** two system functions

one system to its maximum can be different than those that tune another system to its maximum. The goal of experimentation in logistic regression is, therefore, to derive the underlying functions and then to use these functions to predict probabilities.

The specific utilities, u_i, for each level i are random variables. "**Logit models**" are logistic regression models based on the assumption that the random utilities follow a so-called "extreme value" distribution. "**Probit models**" are logistic regression models based on the assumption that the utilities are normally distributed random variables. Sources of randomness in the utilities can be attributed to differences between the average system performance and the actual and/or differences between the individual decision-maker and the average decision-maker.

16.5.1 Design of Experiments for Logistic Regression

Recently, increased attention has been given to the planning of experiments to support logistic regression and discrete choice methoding. Like usual design of experiments, part of planning is selecting which prototype systems should be constructed for testing. An added complication in discrete choice modeling is how to present the prototype system alternatives to decision-makers. "**Choice sets**" refer to combinations of prototypes that are presented from which the people in the experiment select their choices.

For example, if the design of experiment array specifies that short, medium, and tall pens should be made, that is the usual three levels of a continuous factor for a three run DOE. Then, decision-makers are asked to choose between short or medium {choice set #1} and between short and tall {choice set #2}. The remaining combinations, such as showing all three pens (short, medium, and tall) simultaneously, are not necessarily shown.

Table 16.5 An experimental plan satisfying the requirements

Choice set	Height	Price ($)	Person	Choice set	Height	Price ($)	Person
1	Short	10	Maria	3	Short	10	Frank
1	Tall	15	Maria	3	Medium	15	Frank
2	Medium	10	Neville	1	Short	10	Neville
2	Tall	10	Neville	1	Tall	15	Neville

Example 16.5.1 Product Pricing

Question

Develop and experimental plan with the following properties. Three prototypes (short, medium, and tall) are required. Two prices ($10, $15) are tested. Three people are involved in choosing (Frank, Neville, and Maria). People never choose between more than two alternatives at a time.

Answer

There are many possible plans. One solution is shown in Table 16.5. Note that, with the restriction on the choice set size, this becomes a discrete choice problem.

For example, Sandor and Wedel (2001) used the so-called "D_b-error criterion" to generate lists of recommended prototypes and arrangements for their presentation to representative samples of consumers. The D_b-error criterion is analogous to the D-optimality objective (pick the array by maximizing $|\mathbf{X}'\mathbf{X}|$) in response surface contexts because it is based on the maximization of the determinant of the fitted model design matrix. Like D-optimal designs, experimental plans that maximize the D_b-error criterion might, in general, be expected to lead to models that have high prediction errors because the fit model form differs from the true model form, i.e., bias. Therefore, an open research topic is the selection of experimental designs that fosters low prediction errors even if there is bias.

16.5.2 Fitting Logit Models

16

Logit models are probably the most widely used logistic regression and discrete choice models, partly because the associated extreme value distribution makes logit models easy to work with mathematically. The following notation is used in fitting logit models:

1. c is the number of choice sets.
2. $\mathbf{x}_{j,s}$ is an m vector of factor levels (attributes) of response level (alternative system) j in choice set s. In ordinary logistic regression cases, there is only one choice set ($c = 1$), and j is similar to the usual run index in an experimental design array.
3. m_s is the number of response levels (alternative systems) in choice set s.

4. n_s is the number of observations of selections from choice set s.
5. $\boldsymbol{\beta}_{est,j}$ is the estimated coefficient reflecting the average utility of the response level j as a function of the factor levels. Here, the focus is on the assumption that $\boldsymbol{\beta}_{est,j} = \boldsymbol{\beta}_{est}$ for all j.
6. $f_j(\mathbf{x})$ is the functional form of the response j (alternative system) model. Here, the focus is on the assumption that $f_j(\mathbf{x}) = f(\mathbf{x})$ for all j.
7. $p_{j,s}(\mathbf{x},\boldsymbol{\beta}_{est})$ is the probability that the response j with attributes ($\mathbf{x}$) will be selected in the set s.
8. $y_{j,s}$ denotes the number of selections of the alternative j in the choice set s.
9. $ln\ \mathrm{L}(\boldsymbol{\beta}_{est})$ is the log-likelihood which is the fitting objective.

Equation (16.16) can be used to estimate chances that responses will take on specific values. If the responses come from observing peoples' choices, Eq. (16.16) could be used to estimate market shares that a new alternative or product with input values $\mathbf{x}$ might achieve.

The form in Eqs. (16.14) through (16.16) is associated with potentially restrictive "independence from irrelevant alternatives" (IIA) property. This property is that a change in the attributes, $\mathbf{x}$, of one alternative j necessarily results in a change in all other choice probabilities, exactly preserving their relative magnitudes. This property is generally considered not desirable and motivates alternative to logit based logistic regression models.

Note also, some of the attributes associated with specific choices in choice sets could be associated with the decision-makers, e.g., their incomes. This might not require changes to the above formulas as illustrated by the next example. In general, many variations of the above approach are considered in the literation with complications depending on relevant assumptions and the input pattern or design of experiments array.

Algorithm 16.2. Logit model fitting function

Step 1. Observe the n_s selections for $s = 1, \ldots, c$ and document the choice counts $y_{j,s}$ in the context of the input pattern, $\mathbf{x}_{j,s}$.

Step 2. Estimate parameters by maximizing the likelihood and solving

Maximize:
$$ln\mathrm{L}(\beta_{est}) = \sum_{s=1}^{c}\sum_{j=1}^{m_s} y_{j,s}\, ln[p_{j,s}(\mathbf{x}_{j,s}, \beta_{est})] \quad (16.14)$$

where

$$p_{j,s}(\mathbf{x}_{j,s}, \beta_{est}) = \frac{\exp[f(\mathbf{x}_{j,s})'\beta_{est}]}{\Sigma_{j=1,\ldots,ms} \exp[f(\mathbf{x}_{j,s})'\beta_{est}]} \quad (16.15)$$

Step 3. Predict the probability that the response will be level l (alternative l will be chosen), which is associated with factor levels, $\mathbf{x}$, in a choice set with alternatives, $\mathbf{z}_1, \ldots, \mathbf{z}_q$, is the following:

$$p_{j,s}(\mathbf{x}, \beta_{est}) = \frac{\exp[f(\mathbf{x})'\beta_{est}]}{\exp[f(\mathbf{x})'\beta_{est}] + \Sigma_{j=1,\ldots,q} \exp[f(\mathbf{z}_j)'\beta_{est}]} \quad (16.16)$$

Table 16.6 Levels for the design factors

Factor (attribute)	Low (-1)	High (1)
x_1—wing length	8	12
x_2—wing width	3	7
x_3—asking price	26	32
x_4—personal income	0.25	0.7

Example 16.5.3 Paper Helicopter Logistic Regression

Question

Consider an example with a single choice set with $c = 1$, experimental ranges in Table 16.6, and the prototype designs in Table 16.7(a). Assume that there are $n_1 = 20$ people selecting from the prototype designs associated with factor levels x_4 on Table 16.7(b). Use the data to fit a model of the form to predict the average utility:

$$f(\mathbf{x})'\beta = \beta_1 + \beta_2 x_1 + \beta_3 x_2 + \beta_4 x_3 - \beta_5 x_4 + \beta_6 x_1^2 + \beta_7 x_2^2 + \beta_8 x_3^2 + \beta_9 x_4^2 + \beta_{10} x_1 x_2 + \beta_{11} x_1 x_3 + \beta_{12} x_1 x_4 + \beta_{13} x_2 x_3 + \beta_{14} x_2 x_4 + \beta_{15} x_3 x_4. \quad (16.17)$$

Answer

Calculations pertinent to estimating and maximizing the log-likelihood are shown in Table 16.7(b). The Excel solver was used to estimate the coefficients in Table 16.8. These can be used to predict the chance that a customer of a certain income (x_4) would purchase a helicopter with dimensions ($\mathbf{x}$) in a given choice set.

16.6 Software: JMP and Minitab

In recent years, using alternative methods to least squares multiple regression has become much easier. In particular, the JMP® product from SAS® offers an easy-to-use graphical user interface for methods including Kriging or "Gaussian Process" models, artificial neural nets, logistic regression, and other "Generalized Linear Model" methods. Dialogs from JMP® 14 are shown in Figs. 16.7 and 16.8. Fitting logistic regression models with 1M datapoints and 15 regressors may be possible in reasonable times depending on your system specifications. Note that Minitab capabilities are also expanding with impressive routines including for logistic regression of various types and time series modeling.

Table 16.7 (a) Prototypes shown to the people and (b) choices and utility calculations

(a)					(b)				
Response	x_1	x_2	x_3	Choice #	Person	$y_{j,1}$	x_4 – income (×$100K)	Estimated choice prob.	Ln(prob)
$\mathbf{x}_{1,1}$	0	0	0	1	1	11	0.3	5.00E−02	−1.301E+00
$\mathbf{x}_{2,1}$	−1	1	0	2	2	6	0.5	5.00E−02	−1.301E+00
$\mathbf{x}_{3,1}$	0	1	−1	3	3	3	0.4	5.00E−02	−1.301E+00
$\mathbf{x}_{4,1}$	−1	0	−1	4	4	10	0.2	5.00E−02	−1.301E+00
$\mathbf{x}_{5,1}$	1	0	1	5	5	9	0.7	5.01E−02	−1.300E+00
$\mathbf{x}_{6,1}$	−1	0	1	6	6	5	0.3	4.98E−02	−1.303E+00
$\mathbf{x}_{7,1}$	1	0	−1	7	7	1	0.4	5.00E−02	−1.301E+00
$\mathbf{x}_{8,1}$	1	−1	0	8	8	8	0.5	4.99E−02	−1.302E+00
$\mathbf{x}_{9,1}$	0	−1	−1	9	9	6	0.3	4.99E−02	−1.302E+00
$\mathbf{x}_{10,1}$	0	−1	1	10	10	9	0.3	5.00E−02	−1.301E+00
$\mathbf{x}_{11,1}$	0	1	1	11	11	9	0.5	5.00E−02	−1.301E+00
$\mathbf{x}_{12,1}$	−1	−1	0	12	12	6	0.6	5.01E−02	−1.300E+00
$\mathbf{x}_{13,1}$	1	1	0	13	13	1	0.2	5.00E−02	−1.301E+00
					14	7	0.2	5.01E−02	−1.301E+00
					15	7	0.3	4.99E−02	−1.302E+00
					16	5	0.1	5.02E−02	−1.299E+00
					17	8	0.2	5.01E−02	−1.300E+00
					18	13	0.3	5.00E−02	−1.301E+00
					19	5	0.25	4.99E−02	−1.302E+00
					20	2	0.3	5.00E−02	−1.301E+00
								Sum	−2.602E+01

Table 16.8 Beta parameter estimates using maximum likelihood method

Coefficient	Value	Coefficient	Value
β_1	0.2	β_9	0.097652
β_2	0.000399	β_{10}	−3.8E−05
β_3	7.1E−06	β_{11}	0.000899
β_4	−0.00128	β_{12}	−0.0054
β_5	−0.07236	β_{13}	0.003169
β_6	0.000488	β_{14}	0.001148
β_7	0.000749	β_{15}	−0.01027
β_8	−0.00112		

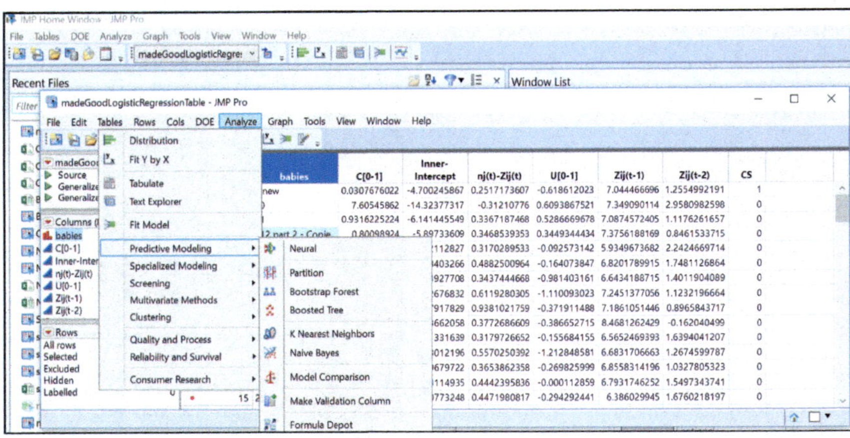

Fig. 16.7 Some of the options available in JMP for modeling

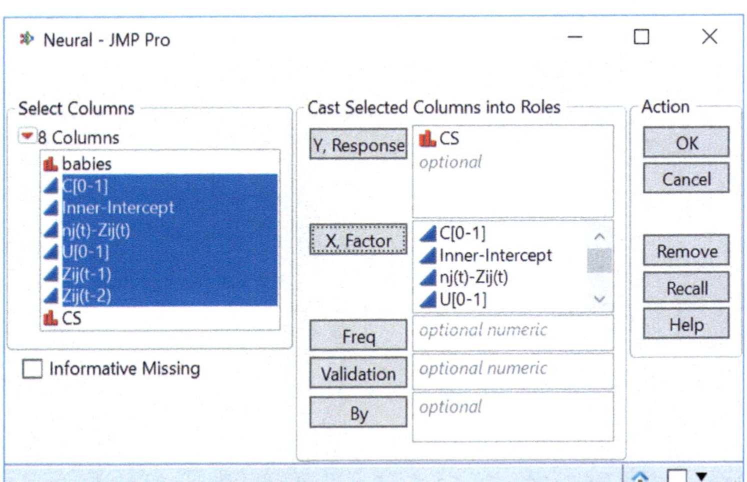

Fig. 16.8 Neural net dialog in JMP

16

16.7 Chapter Summary

This chapter has described three approaches for fitting models to data. All three can be used to predict what might happen in the future if specific input factor settings were chosen (**x**). Kriging modeling is generally considered desirable for deterministic computer experiments such as finite element method (FEM) virtual simulations of physical occurrences (part failures, manufacturing process part quality, contaminant dispersions, …). Artificial neural nets were also described, including an example of so-called "sigmoidal transfer function models" and single hidden layer architectures. The purpose was to show that ANNs can be considered as alternatives to regression and kriging modeling approaches.

Finally, logistic regression models, which can be useful for modeling data with categorical response variables, were briefly described. The resulting models can predict the chance that the system output will assume any given categorical level of interest as a function of input factor settings. More general, discrete choice logistic regression modeling was described, which includes the complication that not all categorical levels might be achievable in any given test. For simplicity, only so-called "logit" logistic regression and discrete choice models were considered. These models are the most analytically tractable logistic regression models.

16.8 Problems

1. Which is correct and most complete based on the data in Table 16.1?
   a. A kriging model prediction would be $y_{est}(\mathbf{x}_1 = 4.5) = 101.5$.
   b. Kriging model predictions could not pass through $x_1 = 3$ and $y_1 = 70$.
   c. FEM experiments necessarily involve random errors.
   d. Kriging models cannot be used for the same problems as regression.
   e. All of the above are correct.
   f. All of the above are correct except (a) and (e).
2. Which is correct and most complete based on the data in Table 16.1 with the last response value changed to 200?
   a. A kriging model prediction would be $y_{est}(\mathbf{x}_1 = 4.5) = 101.5$.
   b. The kriging models in the text are based on the assumption $\sigma_0 = 0.0$.
   c. The maximum log-likelihood value for θ_1 is less than 1.5.
   d. **R** in this case is an $n \times n$ matrix.
   e. All of the above are correct.
   f. All of the above are correct except (a) and (e).
3. Which is correct and most complete?
   a. Artificial neural nets are only relevant for predicting categorical responses.
   b. At least two hidden layers are generally needed for accurate prediction.
   c. The fittable parameters in neural nets are often called weights.
   d. The fitting objective in neural nets is usually the likelihood function.
   e. All of the above are correct.
   f. All of the above are correct except (a) and (e).

4. Which is correct and most complete in relation to artificial neural nets (ANNs)?
   a. RSM arrays cannot be used in neural net fitting.
   b. The Excel solver can be used in fitting ANNs.
   c. Often, so-called "back propagation" algorithms are used in fitting ANNs.
   d. Rules of thumb exist for estimating the most derable number of nodes.
   e. All of the above are correct.
   f. All of the above are correct except (a) and (e).
5. Which is correct and most complete (according to the chapter)?
   a. Discrete choice modeling is effectively a subset of logistic regression.
   b. Linear regression is a natural alternative to discrete choice modeling.
   c. The functions that predict the underlying utility must be first order.
   d. Probit models assume that utilities follow an extreme value distribution.
   e. All of the above are correct.
   f. All of the above are correct except (a) and (e).
6. Which is correct and most complete (according to the chapter)?
   a. The paper helicopter example involves four controllable design parameters.
   b. People might choose different alternatives because of personal differences.
   c. Maximum likelihood estimation can be used in discrete choice modeling.
   d. Uncovering the underlying utility surface might help for predicting market share of new products.
   e. All of the above are correct.
   f. All of the above are correct except (a) and (e).

References

Allen T, Bernshteyn M, Kabiri K, Yu L (2003) A comparison of alternative methods for constructing meta-models for computer experiments. J Qual Technol 35(2):1–17

Ben-Akiva M, Steven RL (1985) Discrete choice analysis. MIT Press, Cambridge, Mass

Chambers M (2000) Queuing network construction using artificial neural networks. PhD dissertation, The Ohio State University, Columbus

Cybenko G (1989) Approximations by superpositions of a sigmoidal function. Mathematics of control, signals, and systems. Springer, New York

Hadj-Alouane AB, Bean JC (1997) A genetic algorithm for the multiple-choice integer program. Oper Res 45:92–101

Hosmer DW, Lemeshow S (1989) Applied logistic regression. Wiley, New York

Kohonen T (1989) Self-organization and associative memory. Springer series in information sciences, vol 8, 3rd edn. Springer, London

Matheron G (1963) Principles of geostatistics. Econ Geol 58:1246–1266

McKay MD, Conover WJ, Beckman RJ (1979) A comparison of three methods for selection values of input variables in the analysis of output from a computer code. Technometrics 21:239–245

Reed RD, Marks RJ (1999) Neural smithing: supervised learning and feed-forward artificial neural net. MIT Press, Cambridge, Mass

Ribardo C (2000) Desirability functions for comparing parameter optimization methods and for addressing process variability under six sigma assumptions. PhD dissertation, Industrial & Systems Engineering, The Ohio State University, Columbus

Rumelhart DE, McClelland JL (eds) (1986) Parallel distributed processing: exploration in the microstructure of cognition. Foundations, vol 1. MIT Press, Cambridge, Mass

Sacks J, Welch W, Mitchell T, Wynn H (1989) Design and analysis of computer experiment. Stat Sci 4:409–435

Sandor Z, Wedel M (2001) Designing conjoint choice experiments using managers' prior beliefs. J Mark Res XXXVIII:430–444

Welch WJ, Buck, RJ, Sacks J, Wynn HP, Mitchell TJ, Morris MD (1992) Screening, predicting, and computer experiments. Technometrics 34:15–25

DOE and Regression Case Studies

T. T. Allen, *Introduction to Engineering Statistics and Lean Six Sigma*,
https://doi.org/10.1007/978-1-4471-7420-2_17

17.1 Introduction

In this chapter, two additional case studies illustrate design of experiments (DOE) and regression being applied in real-world manufacturing. The first study involved the application of screening methods for identifying the cause of a major quality problem and resolving that problem. The second derives from Allen et al. (2000) and relates to the application of a type of response surface method. In this second study, the design of an automotive part was tuned to greatly improve its mechanical performance characteristics.

Note that ▶ Chap. 13 contains a student project description illustrating standard response surface methods and what might realistically be achieved in the course of a university project. Also, ▶ Chap. 14 reviews an application of sequential response surface methods to improve the robustness and profitability of a manufacturing process.

17.2 Case Study: The Rubber Machine

In this section, the so-called "Rubber Machine" case study is presented. This study is similar to the printed circuit board (PCB) study from an earlier chapter and from Brady and Allen (2002) described in ▶ Chap. 12. In this rubber machine study, a machine was essentially broken for several months, and the techniques permitted resolution of the related quality problems, greatly increasing return on investment. The study also illustrates the dangers and inefficiencies of one-factor-at-a-time (OFAT) approaches to experimentation described in ▶ Chap. 12.

17.2.1 The Situation

A Midwestern factory makes a small component used in air conditioning compressors for in-home applications, shown in ◘ Fig. 17.1. The company has established itself as the low cost leader in its sector and has maintained over 50% of the world market for the type of component produced. For confidentiality reasons, we will refer to the part as a "bottle cap" which is the informal name sometimes used within the company. In the years prior to the study, the company had been highly successful in reducing production costs and improving profits through the intelligent application of lean manufacturing including value stream mapping (▶ Chap. 5) and other industrial and quality engineering-related techniques. Therefore, the management of the company was generally receptive to the application of formal procedures for quality and process improvement.

17

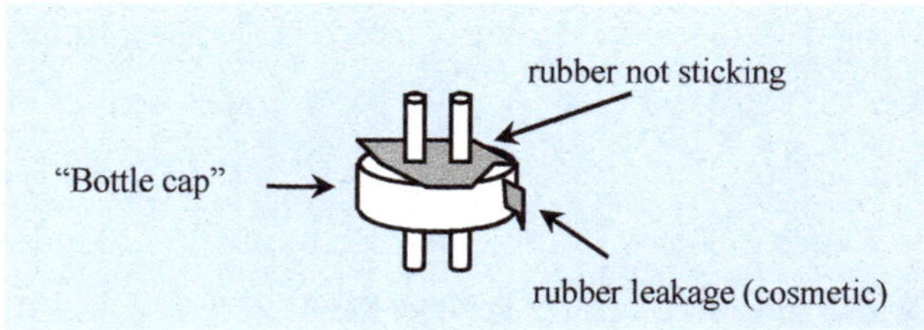

◘ Fig. 17.1 "Bottle cap" part

In its desire to maintain momentum in cost-cutting and quality improvement, the company decided to purchase two new machines for applying rubber to the nickel-plated steel cap and hardening the rubber into place. The machines cost between \$250,000 and \$500,000 each in direct costs. These new machines required less labor content than the previous machines and promised to achieve the same results more consistently. Unfortunately, soon after the single production line was converted to using the new machine, the rubber stopped sticking on roughly 10% of the bottle caps produced. Because this failure type required expensive rework as well as 100% inspection and sorting, the company reverted to its old process.

17.2.2 Background Information

A large fraction of the engineering and management resources of the small company were deployed as an initial team to fix the new machines. During a period of roughly 3 months, engineers disrupted production in order to test their theories by running many units with one factor adjusted and then adjusting the settings back (one-factor-at-a-time, OFAT). Unfortunately, all of the tests results were inconclusive. In addition, at least one polymer expert was flown into inspect the problem and give opinions. Several months after the machines had been installed in the plant, the company was still unable to run them. An additional series of OFAT experiments were conducted to investigate the effects of seven factors on the yields. Again, the results were inconclusive.

17.2.3 The Problem Statement

Because of the unexpected need to use the old process, the company was rapidly losing money due to overtime and disruptions in the product flow through the plant. Therefore, the problem was to adjust the process inputs (**x**) to make the rubber stick onto the nickel plating consistently using the newer machines. Unfortunately, the engineering and technicians had many theories about which factors should be adjusted to which levels, with little convincing evidence supporting the claims of each person (because of the application of OFAT). Seven candidate input factors were identified whose possible adjustment could solve the problem. Also, considering the volume of parts produced and the ease of inspection, it was possible to entertain the use of reasonably large batch sizes, i.e., $b = 500$ was possible.

Example 17.2.1 Rubber Machine Initial Results

Question
Which of the following could the first team most safely be accused of?

a. Leaders stifled creativity by adopting an overly formal decision-making approach.
b. The team forfeited the ability to achieve statistical proof by using a nonrandom run order.

c. The team failed to apply engineering principles and relied too much on statistical methods.
d. The team failed to devote substantial resources to solve the problem.

Answer
This answer is virtually identical to the one in the printed circuit-board study. Compared with many of the methods described in this book, team one has adopted a fairly "organic" or creative decision style. Also, while it is usually possible to gain additional insights through recourse to engineering principles, it is likely that these principles were consulted in selecting factors for OFAT experimentation to a reasonable extent. In addition, the first team did provide enough data to determine the usual yields prior to implementing recommendations. Therefore, the criticisms in (a), (c), and (d) are probably not fair. According to ▶ Chap. 11, random run ordering is essential to establishing statistical proof. Therefore, (b) is correct.

17.3 The Application of Formal Improvement Systems Technology

A team of two people trained in design of experiments (including the author) persuaded the engineering supervisor in charge of fixing the machine to apply design of experiments methods. A team was created for planning the experiment, conducting the tests, and analyzing the results. Drawing on the engineering talent of the team, the factors described in Table 17.1 were chosen. The output or response selected was the fraction of $b = 500$ parts for which the rubber would not stick. Typical fractions nonconforming were expected to be greater than $p_0 = 0.05$. An initial budget allocation for 8 test runs, each involving 500 parts, was allocated. All 4000 parts could be made and tested in a single day.

Example 17.3.1 Rubber Machine DOE Plan

Question
Which is correct and most complete (according to previous chapters)?
a. The fraction nonconforming in this case should not be treated as a continuous response.

Table 17.1 Factors and the ranges decided by the engineering team

Factor	Low (−1)	High (+1)
A. Floor delay	0–15 min	12+ h
B. Temp. at priming	Warm	Hot
C. Oven temp.	300 °F	380 °F
D. Primer thickness	Thin color	Thick color
E. Chamber humidity	Ambient	90 °F 90%
F. Shot size	−0.75 turns	Full shot
G. Extra oven time	<2 min	15 min

17

b. Response surface methods are a good fit because the important factors are known.
c. A fractional factorial screening experiment could be applied with up to seven factors.
d. The relevant response is categorical, so regression cannot be applied.
e. All of the above are correct.

Answer

According to ▶ Chap. 15, the response can be treated as categorical because $b \times p_0 > 5$, i.e., more than five units are expected to be nonconforming in all test runs. Therefore, (a) and (d) are false. There was a long list of potential candidates. Also, the budget allocation was for only eight runs. Therefore, (b) is false and response surface methods would not be a good fit. ▶ Chapter 12 describes methods permitting an eight run experiment involving seven factors. Therefore, (c) is correct.

The improvement team selected the eight run fractional factorial in Table 17.2 to structure experimentation. The resulting fractions nonconforming are also described in the right-hand column. Interestingly, all fractions were lower than expected perhaps because of a Hawthorne effect, i.e., the act of watching the process carefully seems to have improved the quality.

One of the factors involved a policy decision about how long parts could wait in queue in front of the rubber machine before they would need to be "reprimed" using an upstream "priming" machine. This factor was called "floor delay". If the results had suggested that floor delay was important, the team would have issued recommendations relating to the redesign of engineering **policies** about production scheduling to the plant management. It was recognized that we probably could not directly control the time parts waited. With 4000 parts involved in the experiment, complete control of the times would have cost too much time.

Therefore, the team could only control decisions within its sphere of influence. Implicitly, therefore, the "**system boundaries**" were defined to correspond to what could be controlled, e.g., a maximum time of 15 min recommended for parts to sit without being re-primed in our recommended guidelines. This was the **control factor**. To simulate the impacts of possible decisions the team would make on this issue, parts

Table 17.2 The experimental design and the results from the rubber machine study

Run	A	B	C	D	E	F	G	Y_1
1	1	−1	1	−1	1	−1	1	4.4
2	−1	−1	−1	1	1	1	1	0
3	1	1	−1	−1	1	1	−1	0
4	−1	1	1	1	1	−1	−1	3.8
5	−1	1	1	−1	−1	1	1	0
6	1	1	−1	1	−1	−1	1	0.6
7	−1	−1	−1	−1	−1	−1	−1	2.8
8	1	−1	1	1	−1	1	−1	0

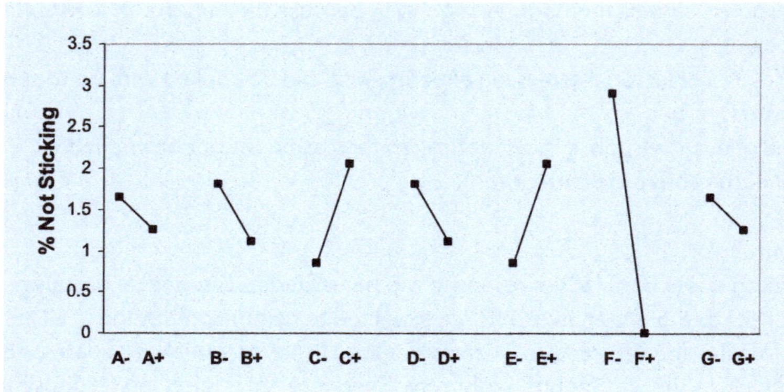

Fig. 17.2 Main effects plots derived from the fractional factorial experiment

were either re-primed in the experiment if they waited longer than 15 min or they were constrained to wait at least 12 h.

The main effects plot in Fig. 17.2 and the results of applying Lenth's analysis method both indicated that Factor F likely affected the fraction nonconforming. Because the statistic called "t_{Lenth}" for this factor is greater than the "critical value" $t_{\text{IER},0.05,8} = 2.297$ and the order of experimentation was determined using randomization, many people would say that "this factor was proven to be significant with $\alpha = 0.05$ using the individual error rate."

Note that this conclusion is associated with a lower standard of evidence because it was based on a fractional factorial array-based method rather than a two-sample t-test and the so-called individual error rate (IER) was used (see ▶ Chap. 12). Yet, most importantly, it was immediately confirmed that adjusting shot size to the high level effectively eliminated the sticking problem. The effect was proven and confirmed.

The other potentially important factors included E (chamber humidity). The result for chamber humidity was surprising. This factor is not found to be significant using the Length hypothesis test but the posterior probability suggests that it might be important. The expensive "hydrolyzer" machine had been bought precisely to help eliminate the sticking problem. It created another step in the process before the injection of the rubber. Yet, the results indicated that there was a non-negligible probability that the hydrolyzer was actually making the problem worse, i.e., increasing the % not sticking (Table 17.3).

After the experiment and analysis, it was discovered that some of the maintenance technicians in the plant had been adjusting the shot size intermittently based on their intuition about how to correct another less serious problem that related to the "leaker" cosmetic defect. This problem was less serious because the rework operation needed to fix the parts for this defect involved only scraping off the parts, instead of pulling off all the rubber, cleaning the part, and starting over. The maintenance staff involved had documented their changes in a notebook, but no one had thought to try to correlate the changes with the incidence of defective parts.

Table 17.3 Analysis results for the rubber machine screening experiment

Factor	Estimated coefficients (β_{est})	t_{Lenth}
A	−0.2	0.48
B	−0.35	0.85
C	0.6	1.45
D	−0.35	0.85
E	0.6	1.45
F	−1.45	3.52
G	−0.2	0.48

A policy was instituted and documented in the standard operating procedures (SOPs) that the shot size should never be changed without direct permission from the engineering supervisor, and the "hydrolyzer" machine was removed from the process. The non-sticking problem effectively disappeared, and production shifted over to the new machines, saving roughly $15K/month in direct supplemental labor costs. The change also effectively eliminated costs associated with production disruption and having five engineers billing their time to an unproductive activity. Another team was created to address the less important problem of eliminating the fraction of parts exhibiting the cosmetic defect.

17.4 Case Study: Snap Tab Design Improvement

A major automotive manufacturer was attempting to save assembly cost by using plastic fasteners instead of screws to hold together its air conditioning cases. Since plastic fasteners are molded into the plastic case itself, all fasteners can be engaged in a single operation with minimal assembly time. Alternatively, screws must be inserted and engaged singly, requiring higher assembly cost. Since the engineers were unable to find any acceptable existing snap tab designs, the question was whether snap tabs of sufficient strength and acceptable size could be developed in time for the launch of a new vehicle program whose budget was paying for the development. A major concern was whether the expected cost savings would justify the development cost.

The selected snap fit design concept is shown in Fig. 17.3. The four design factors were identified through the application of a cause and effect matrix (altered to protect confidential information) shown in Table 17.4. This "pre-screening" clearly identified four factors as being much more relevant than the others. For this "loop-hook" topology, accurate engineering models were not available to predict the pull-apart force (force at time of joint failure) and insertion force as a function of the four design parameters in Fig. 17.3. Even virtual prototypes using finite element analysis cost at least $3K each for testing. The allocated budget permitted only 12 virtual prototypes to be built and tested.

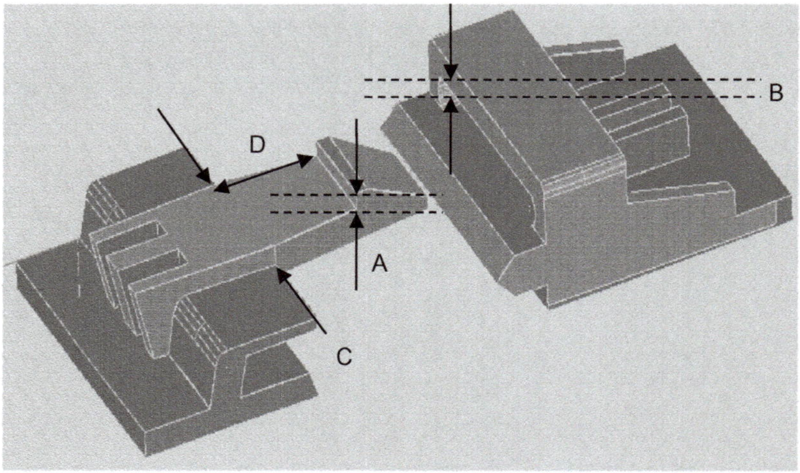

Fig. 17.3 The snap tab design concept optimized in our case study

Example 17.4.1 Snap Fit DOE Plan

Question

Which is correct and most complete (according to previous chapters)?

a. Central composite designs are available with 4 factors and 12 runs.
b. Screening experiments generally do not permit fine tuning parameters.
c. The number of runs is less than the number of terms in a quadratic polynomial.
d. Non-standard methods were required to address the tuning and cost objectives.
e. All of the above are correct.
f. All of the above are correct except (a) and (e).

Answer

Even with only a single centerpoint, the smallest standard central composite design with four factors has 25 runs (► Chap. 13). With their goal of finding which factors affect responses, screening methods generally have only two levels and do not permit fine tuning taking into account quadratic terms and/or interaction terms. According Eq. (13.2), the number of terms is $0.5 \times (4+1) \times (4+2) = 15$, which is greater than the budgeted number of runs. Therefore, a nonstandard method must be used, since fitting at least some quadratic curvatures and interactions was desirable for tuning. Therefore, the correct answer is (f).

The constraint on test runs followed from the fact that each test to evaluate pull-apart and insertion forces required roughly 3 days of two people working to create and analyze a finite element method simulation. Since management was only willing to guarantee enough resources to perform 12 experimental runs, application of the standard central composite design, which required at least 25 runs, was impossible.

Table 17.4 Cause and effect (C&E) matrix used for "pre-screening" factors

Issue	Manufacturing engineer rating	Material type	(A) Ledge width	Flat length	(D) Tab height	(B) Loop thickness	(C) Loop thickness	Entry angle	Loop radius
Easy to assemble	4.5	3.5	9.0	2.0	2.5	8.0	8.0	7.5	4.0
Strong enough to replace screws	10.0	4.5	10.0	3.5	8.0	7.5	7.5	1.0	3.0
Factor rating number (F')		60.8	140.5	44.0	91.3	111.0	111.0	43.8	48.0

Table 17.5 LCRSM: **(a)** initial design **(b)** the model forms, and **(c)** the additional runs

(a)					Form	(b)	(c)				
Run	A	B	C	D	#1:	$\beta_0+\beta_A A+\beta_B B+\beta_C C+\beta_D D+$	Run	A	B	C	D
1	−0.5	−1	−0.5	1		$\beta_A 2A^2+\beta_B 2B^2+\beta_C 2C^2+$	A1	−1	1	−1	1
2	1	1	−1	1		$\beta_{AB} AB+\beta_{AC} AC+\beta_{BC} BC$	A2	−1	−1	−1	−1
3	−1	1	1	1			A3	−1	1	1	−1
4	1	−1	−0.5	−0.5	#2:	$\beta_0+\beta_A A+\beta_B B+\beta_C C+\beta_D D+$	A4	1	1	−1	−1
5	0	0	−1	0		$\beta_A 2A^2+\beta_B 2B^2+\beta_D 2D^2+$					
6	0	1	0	0		$\beta_{AB} AB+\beta_{AD} AD+\beta_{BD} BD$					
7	−0.5	−1	1	−0.5							
8	−1	0	0	0	#3:	$\beta_0+\beta_A A+\beta_B B+\beta_C C+\beta_D D+$					
9	1	1	1	−1		$\beta_A 2A^2+\beta_C 2C^2+\beta_D 2D^2+$					
10	−1	1	−1	−1		$\beta_{AC} AC+\beta_{AD} AD+\beta_{CD} CD$					
11	0	0	0	−1							
12	0.5	−0.5	0.5	0.5	#4:	$\beta_0+\beta_A A+\beta_B B+\beta_C C+\beta_D D+$					
13	0.5	−0.5	0.5	0.5		$\beta_B 2B^2+\beta_C 2C^2+\beta_D 2D^2+$					
14	0.5	−0.5	0.5	0.5		$\beta_{BC} BC+\beta_{BD} BD+\beta_{CD} CD$					

Even a 2 level design that permits accurate estimation of interactions contains 16 runs, so just the first experiment in two-step RSM could not be applied. Even the small central composite design, which had at least 17 runs was practically impossible (Myers and Montgomery 2001). Note that two similar optimization projects were actually performed using different materials. These necessities led to the use of a non-standard response surface methods.

The majority of standard design of experiments (DOE) methods were presented initially in the *Journal of the Royal Statistical Society: Series B* and *Technometrics*. These and other journals including the *Journal of Quality Technology*, the *Journal of Royal Statistical Society: Series C*, *Quality & Reliability Engineering International*, and *Quality Engineering* contain many innovative DOE methods. These methods can address non-standard situations, such as those involving categorical and mixture factors (▶ Chap. 15), and/or potentially result in more accurate predictions and declarations for cases in which standard methods can be applied.

In this study, the team chose to apply so-called "low cost response surface methods" (LCRSM) from Allen et al. (2000) and Allen and Yu (2002). Those papers provide tabulated, general-purpose experimental designs for three, four, and five factors each with roughly half the number of runs of the corresponding central composite designs and comparable expected prediction errors. Table 17.5 shows the design of experiments (DOE) arrays and model forms relevant to LCRSM. Table 17.6 shows the actual DOE array used in the case study. Note that no repeated tests were needed because finite element method (FEM) computer experiments have little or no random error, as described in ▶ Chap. 16.

Table 17.6 Experimental runs and the measured pull-apart and insertion forces

Run	A	B	C	D	Y_1	Y_2
1	1.25	1.7	12.5	10.00	55.95	15.39
2	2.00	2.1	10.0	10.00	101.76	19.92
3	1.00	2.1	20.0	10.00	101.23	21.02
4	2.00	1.7	12.5	6.25	52.93	18.55
5	1.50	1.9	10.0	7.50	59.93	13.42
6	1.50	2.1	15.0	7.50	80.54	15.90
7	1.25	1.7	20.0	6.25	60.87	14.70
8	1.00	1.9	15.0	7.50	72.02	13.51
9	2.00	2.1	20.0	5.00	102.70	22.81
10	1.00	2.1	10.0	5.00	51.36	23.79
11	1.50	1.9	15.0	5.00	59.42	26.33
12	1.75	1.8	8.8	8.75	81.94	13.50

17.5 The Selection of the Factors

Using only four parameters to specify such a complex topology such as loop hook snap tabs leads to inevitable ambiguities. For example, should the loop width vary along with the tab width (factor C)? It was arbitrarily selected to vary the loop width linearly as a function of C. Similarly, if factor C is the width at the base of the snap tab, how should that relate to the width at the end of the tab (C′)? Again, it was decided somewhat arbitrarily that $C' = C - 7$ mm. Also, as the tab width changes, at what levels do we change the integer number of support brackets?

It was decided again somewhat arbitrarily to add a separate bracket for every 7 mm of tab width. Therefore, a change to factor C implied changes to the tab width at the base and end, a change in the loop width, and, potentially, a change in the number of support brackets. In this "**parameterization**" or framing of the problem, one could not "dial up" a wide loop and a narrow tab. Figure 17.4 shows the selected primary factors (A, B, C, and D) and the sub-factors that depended on them (C′, D′, and D″).

Note that any ambiguity in the choice of parameterization could actually increase interest by the practitioner in the methods described in this book. This follows because these technologies permit more factors to be studied, modeled, and optimized over with generally higher probabilities of achieving desired outcomes than alternatives such as one-factor-at-a-time (OFAT). With more factors, one has substantially greater freedom to investigate parameterizations that permit effects to be separated and better understood. As we will discover in the case study, the guessed parameterization helped in the achievement of remarkable performance improvements.

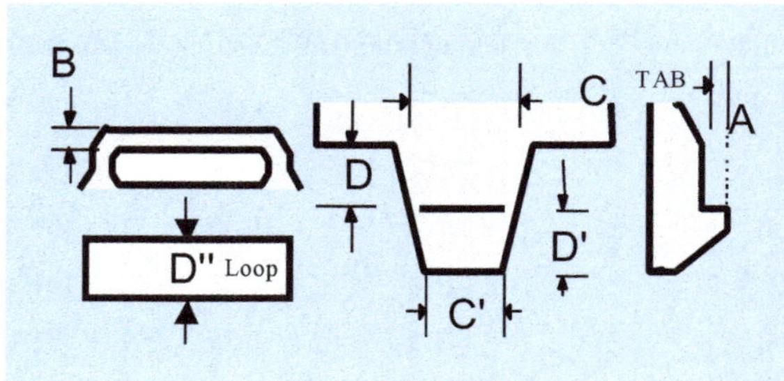

■ **Fig. 17.4** Experimental factors in the parameterization chosen

17.6 General Procedure for Low Cost Response Surface Methods

The application of low cost response surface methods is similar to that of standard two-step response surface methods except multiple models are fit instead of one and the diagnostic test is different. Additional details are available in Allen and Yu (2002). The major steps as described in Algorithm 17.1 are experimental set-up and testing, modeling, diagnostics, and additional testing if needed. The case study is described in the next section.

17.7 The Engineering Design of Snap Fits

This section describes the application of LCRSM to derive the empirical prediction models of the pull-apart and insertion forces for the snap tab project. Results are modified slightly to preserve confidentiality. In the real study, a similar method was applied and achieved similar results.

The steps in the development of the model for the snap tab pull apart force and insertion forces were as follows. In *Step 1*, the team used the factors shown in ■ Table 17.6. The 12 sets of responses are also shown. The engineering ranges for the factors A, B, C, and D were 1.0 mm to 2.0 mm, 1.7 mm to 2.1 mm, 10.0 mm to 20.0 mm, and 5.0 mm to 10.0 mm, respectively. The response Y_1 was the pull-apart force in pounds (lb), and Y_2 was the insertion force in lb. The data derived from the 12 finite element analyses are also shown in ■ Table 17.6 in the right-hand columns. ■ Figure 17.5 illustrates finite element method (FEM) runs, showing the stresses placed on each element of the snap fit during a simulated pull apart at the point of breakage.

In *Step 2*, the four linear regression model forms in ■ Table 17.5(b) were fitted to each of the responses and selected the one with the lowest sum or squares error. The selected models for each response were:

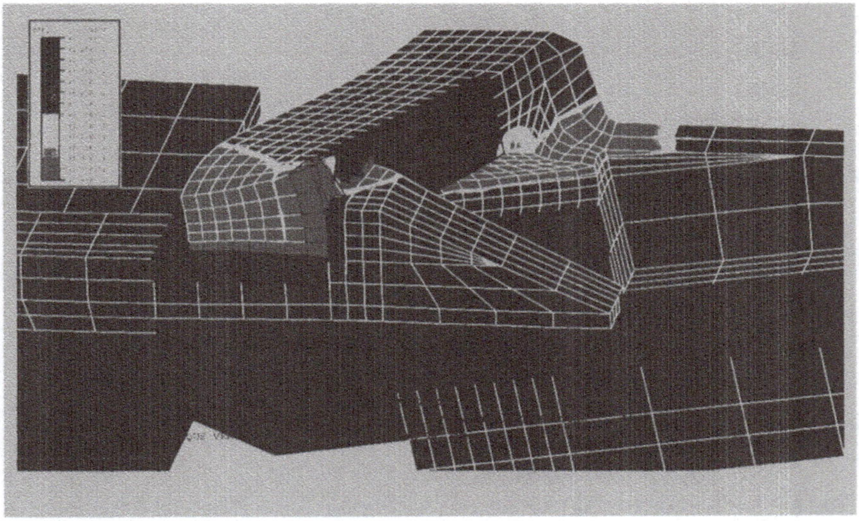

Fig. 17.5 Finite element analysis (FEA) simulation of the snap tab

$$y_{est,1} = 72.06 + 8.98A + 14.12B + 13.41C + 11.85D + 8.52A^2 - 6.16B^2 + 0.86C^2 + 3.93AB - 0.44AC - 0.76BC$$

and

$$y_{est,2} = 14.62 + 0.80A + 1.50B - 0.32C - 3.68D - 0.45A^2 - 1.66C^2 + 7.89D^2 - 2.24AC - 0.33AD + 1.35CD. \quad (17.1)$$

Algorithm 17.1. Low cost response surface methods

Step 1: Set up the experiment by taking the experimental design appropriate for the relevant number of factors from the appropriate table. Here only the four factor design in Table 17.5(a) is given, which is given in scaled (–1,1) units. Scale to engineering units, e.g., see Table 17.6, perform the experiments, and record the responses.

Step 2: Create the regression model(s) of each response by fitting the appropriate set of candidate model forms from Allen and Yu (2002). For the design in Table 17.5(a), this is the set in Table 17.5(b). The model fitting uses least squares linear regression. Select the fit model form with the lowest sum of squares error.

Step 3: (The Least Squares Coefficient Based Diagnostic) Calculate

$$\beta_{q,est} = (\sum_{i}^{q} \beta_{i,est}^2)^{1/2}(q-1)^{-1/2} \tag{17.2}$$

where $\beta_{i,est}$ are the least squares estimates of the q second order coefficients in the model chosen in *Step 2*. Include coefficients of terms like A^2 and BC, but not first order terms such as A and D. Estimate the maximum acceptable standard error of prediction or "plus or minus" accuracy goal, $\sigma_{prediction}$. If $\beta_{q,est} \le 1.0\sigma_{prediction}$, refit the model form in the engineering units. Stop. Otherwise, or if there is any special concern with the accuracy, continue to *Step 4*. Special concerns might include mid-experiment changes to the experimental design. The default assumption for $\sigma_{prediction}$ is that it equals 2.0 times the estimated standard error, because then the achieved expected "plus or minus" accuracy approximately equals the error that would be expected if the experimenter applied substantially more expensive methods based on composite designs.

Step 4: Perform additional experiments specified in the appropriate if needed, e.g., ▪ Table 17.5(c). After the experiment, fit a full quadratic polynomial regression model as in ordinary response surface methods. The resulting model is expected to have comparable prediction errors (within 0.2σ) as if the full central composite with 27 runs had been applied.

In the modified *Step 3*, the choice was made to set the desired accuracy to be $\sigma_{prediction} = 3.0$ lb or ± 3 lb accuracy for the pull-apart force and $\sigma_{prediction} = 3.0$ lb for the insertion force. The square roots of the sum of squares of the quadratic coefficients divided by the number of quadratic coefficients, 6, for the two responses were $\beta_{q,est} = 4.6$ lb and 3.5 lb respectively. Since these were less than their respective cutoffs, 2.0×3.0 lb $= 6.0$ lb for the pull force and 2.0×3.0 lb $= 6.0$ lb for the insertion force, we stopped. No more experiments were needed. The expected average errors that resulted from this procedure were estimated to roughly equal their desired values.

Compared with central composite designs using 25 distinct runs, there was a savings of 13 runs, which was approximately half the experimental expense. The expected average errors that resulted from this procedure were as small or smaller than desired, i.e., within ± 3 lb for both pull apart and insertion forces averaged over the region of interest. This **prediction** accuracy oriented experiment was likely considerably more accurate than what could be obtained from a **screening** experiment such as either of the first two case studies. Also, the project was finished on time and within the budget.

The models obtained from the low cost response surface methods procedure were then optimized to yield the recommended engineering design. The parameters were constrained to the experimental region both because of size restrictions and to assure good accuracy of the models. An additional constraint was that the insertion force of the snap tab needed to be less than 12 lb to guarantee easy assembly. The formal optimization program that we used was

$$\begin{aligned} &\text{Maximize:} && y_{est,1}(\mathrm{A,B,C,D}) \\ &\text{Subject to:} && y_{est,2}(\mathrm{A,B,C,D}) \le 12.0\,\text{lb} \\ & && -1.0 \le \mathrm{A,B,C,D} \le 1.0 \end{aligned}$$

where we expressed the variables in coded experimental units.

Using a standard spreadsheet solver, the optimal design was A = 1.0, B = 0.85, C = 1.0, and D = 0.33. ◘ Figure 17.5 shows the region of the parameter space near the optimal. The insertion force constraint is overlaid on the contours of the pull force. Forces are in pounds. In engineering units, the optimal engineering design was A = 2.0 mm, B = 2.07 mm, C = 20 mm, and D = 8.3 mm, with predicted pull-apart force equal to $y_{est,1}$(A, B, C, D) = 118 lb.

Note that all factors have at least one associated model term that is large in either or both of the models derived from the model selection for the insertion and pull-apart forces. If the team had used fewer factors to economize, then important opportunities to improve the quality would likely have been lost because the effects of the missing factors would not have been understood. These missing factors would likely have been set to sub-optimal values.

The results of the snap fit case study are summarized in ◘ Fig. 17.7. The "current" model derived from existing standard operating procedures (SOPs) in the corporate design guide. Results associated with the "best guess" design, chosen after run 1 was completed, and the final recommended design, are also shown in ◘ Fig. 17.7. Neither the best guess design nor the current model designs were strong and small enough to replace screws. The size increase was deemed acceptable by the engineers because the improved strength made replacing screws feasible (◘ Fig. 17.6).

Note that there was a remarkable agreement between the predicted and the actual pull-apart forces (within 3%), which validates both the low cost response surface method errors and our procedure for finite element simulation. The resulting optimized design was put into production and into the standard operating procedures. Some savings was achieved, but unanticipated issues caused the retention on screws on many product lines.

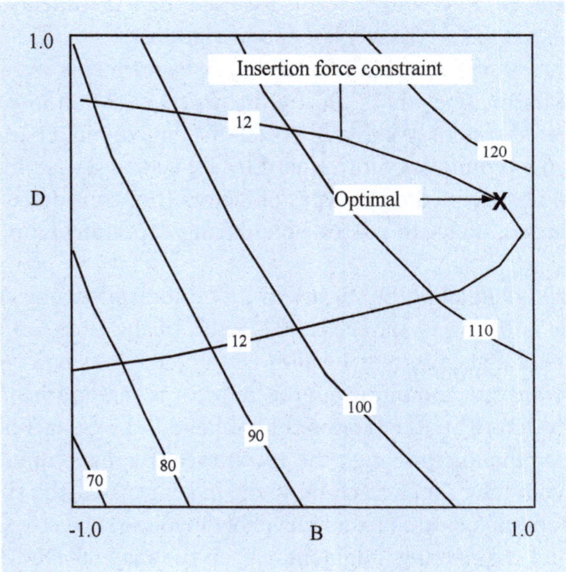

◘ **Fig. 17.6** Insertion force constraint on pull force contours with A = 1 and C = 1

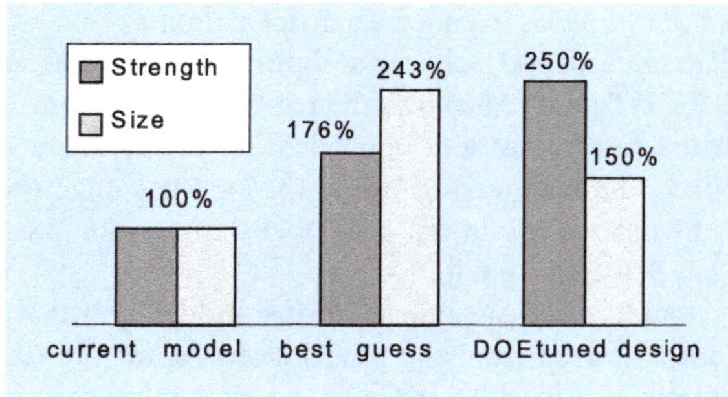

Fig. 17.7 Improvement of the snap fit achieved in the case study

17.8 Concept Review

Two additional case studies have been described in which the participants all believed that formal improvement systems related technology helped them to more than recoup their investment in experimentation and analysis. Reviewing the common features of these studies may help the reader to evaluate better whether formal improvement systems might achieve similar successes for a given new system design problem.

In all three studies, the participants had **sufficient authority** and resources **to experiment** on either the actual physical system that they were designing (e.g., the PCB and rubber machine studies) or, at least, a similar "surrogate" system (e.g., virtual FEA simulation in the snap tab study). The experimental outputs were assumed to relate identically to engineered system outputs for a given combination of inputs. Confirmation experiments were performed on physical prototype systems for the snap tab study, but otherwise in all cases the teams assumed that the fidelity of the prototype systems was high enough that **fidelity issues** were ignored.

Also, the factors selected as inputs in all of the experiments were all directly controllable by the team members both during the experiment and in subsequent operation of the engineered system. We therefore call this type of input parameter a "**control factor**" following the terminology introduced by Taguchi and described in ▶ Chap. 14. Note that Taguchi also defined other types of factors (not considered in these studies) including **noise factors** that are controllable during experimentation but not during system operation.

Because the factors in all of the studies could be controlled, one can think of them as "dials" that one is trying to tune, e.g., the width of the snap tab is a "**continuous** factor". Some of these dials may only be allowed to point to a small number of discrete settings, e.g., the transistor mounting approach factor is "**categorical**" or, equivalently, "**qualitative**" or "**discrete**" (either screwed or soldered). Some factors are parameters in policies or recommendations, e.g., the recommended maximum waiting time for parts after priming in the rubber machine study. In some cases, the specification of the precise definitions of the control factors inevitably involves subjective decision-making and, hopefully, good engineering judgment (e.g., in the snap tab study).

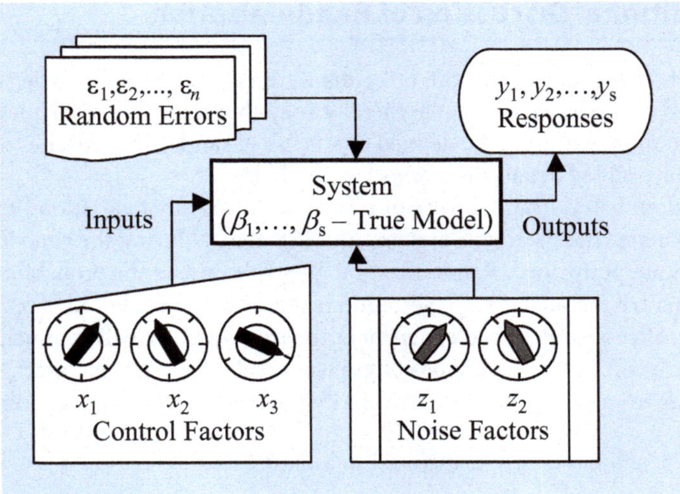

Fig. 17.8 The relationship of terminology associated with system design

In this view of system design pictured in the Fig. 17.8, the decision-makers are asked to determine settings of the control factor dials so that the outputs, the y_is, consistently achieve some desired values. One of the challenges for formal data collection and analysis methods is to facilitate accurate estimation of the system true performance as a function of the control factors, despite changes in the outputs during experimentation because of the random errors, the ε_is. These errors presumably occur because other, usually unknown, factors are changing. Armed with the estimates of the true model functions, β_is, one can then attempt to optimize the control factors to achieve desirable system outputs during normal operations.

Example 17.8.1 Experimentation in Hospitals

Question
Assume you are an administrator working in a hospital to reduce cost and improve customer service. What might your control factors and responses be for a 3-month improvement project?

Answer
As an administrator, you cannot control the manner in which surgeries are performed nor which drugs are prescribed. You can, however, recommend and experiment with factors including the numbers of different types of nurses on call during the week, details of the insurance documentation process, and the numbers of beds in the different wards. Responses might include the times until patients see medical personnel, customer satisfaction ratings, and monthly personnel costs.

17.9 Additional Discussion of Randomization

Note that, in each experiment, the test runs were performed in an order determined by a random number generator. We therefore say that these experiments were "**randomized**". Randomization can be defined by the use of random approaches to specify all otherwise unspecified details of the experimental plan.

The wisdom behind randomization relates to the way that variation of factors that both (1) influence the prototype systems and (2) change during the time in which the experiments are performed. Randomization greatly increases the probability that these factors will enter the analysis as the random noise that the methods are designed to address, i.e., after a randomized experiment the experimenter will be much more confident that control factors that appear significant really do affect the average response. The following examples are designed to clarify the practical value of randomization.

Example 17.9.1 Rubber Machine Example Revisited

Imagine that the rubber machine experiment had been performed in an order not specified by pseudo-random numbers. Table 17.7 shows the same experimental plan and data from the rubber machine study except the run order is given in an order that displays some of the special properties of the experimental matrix. For example, the columns corresponding to factors E, F, and G have an "elegant" structure. This is a run order that Box and Hunter (1961) might have first generated in their derivation of the matrix from combinatorial manipulations.

As in the real study, all of the runs with high fractions of nonconforming units correspond to prototype systems in which the shot size was low. However, without randomization, another simple explanation for the data confuses the issue of whether shot size causes nonconforming units. The people performing the study might simply have improved in their ability to operate the system, i.e., a "**learning effect**". Notice that only the first four runs are associated with poor results. The absence of randomization in this imagined experiment would greatly diminish the value of the collected data.

Table 17.7 Hypothetical rubber machine study performed in a nonrandomized order

Run	A	B	C	D	E	F	G	Y_1
1	1	−1	1	−1	1	−1	1	4.4
2	−1	1	1	1	1	−1	−1	3.8
3	1	1	−1	1	−1	−1	1	0.6
4	−1	−1	−1	−1	−1	−1	−1	2.8
5	−1	−1	−1	1	1	1	1	0
6	1	1	−1	−1	1	1	−1	0
7	−1	1	1	−1	−1	1	1	0
8	1	−1	1	1	−1	1	−1	0

17

Example 17.9.2 Drug Testing Example

Consider a simple experiment in which a drug is given along with a placebo to a test group and a control group. ▶ Chapter 11 shows one way to use random approaches to assign people to groups. However, suppose that the experimenter does not use pseudo-random numbers to assign which subjects to the test and control groups and instead permits the subjects to divide themselves. It seems likely that smokers, who generally have poorer health, might naturally group together because of shared interests.

If they concentrated into the control group, any positive benefits associated with the drug might be suspect. This follows because the negative health outcomes for the control group could easily be caused by smoking and not the absence of the drug. Using pseudo-random numbers makes this type of confusion or "**confounding**" extremely unlikely. For example, if there are 10 smokers in a group of 30, the chance that all 10 would be randomly assigned to a test group of 15 is less than 0.000001.

Because of the desirable characteristics from randomization, researchers in multiple fields associate the word "**proof**" with the application of randomized experimental plans. Generally, researchers draw an important distinction between inferences drawn from "**on-hand data**", i.e., data not from randomized experimental plans, which they call **observational studies**, and the results from randomized experimental plans. In language that I personally advocate, one can only claim a hypothesis is "**proven**" if one has a mathematical proof with stated assumptions or "axioms" derivation of the hypothesis from the standard model in physics, or evidence from hypothesis testing, based on randomized experimental plans.

The issue of fidelity further complicates the use of the word proof. As noted earlier, in all of the studies, the stakeholders were comfortable with the assumption that the prototype systems used for experimentation were acceptable surrogates for the engineered systems that people cared about, i.e., that made money for the stakeholders. Still, it might be more proper to say that causality was proven in the randomized experiments for the prototype systems and not necessarily for the engineered systems. Conceivably, one could prove a claim pertinent to a low fidelity prototype system in the laboratory but not be able to generalize that claim to the important, highest fidelity, real-world system in production. Although methods to address concerns associated with fidelity are a subject of ongoing research, fidelity issues, while extremely important, continue to be largely outside the scope of formal statistical methods.

Note that randomization benefits are associated with the effects of factors that are not controlled. Since these factors are often overlooked, the experimenter may not have the option of controlling and fixing them. Yet, it is also not clear that controlling these factors would be desirable (even if it were possible) since their variation might constitute an important feature of the engineered system. Therefore, a tightly controlled prototype system might be a low fidelity surrogate for the engineered system. This explains why proof is generally associated with randomization and not control.

17.10 Chapter Summary

This chapter contains two case studies. In the first, two rubber machines were malfunctioning and causing a production bottleneck. Standard screening using fractional factorial methods were applied to identify the cause and suggest a prompt and successful remedy. One of the associated factors used in the study was not a setting on a machine but rather a way of stating policy to employees. In the second study, an innovative design of experiments methods called low cost response surface methods (LCRSM) was applied to develop a surface prediction of strength and insertion effort for snap tabs. Formal optimization of the resulting surface models permitted the doubling of the strength with small increase in size.

17.11 Problems

Use the following information to answer Questions 1–3:

PCB Study Revisited: A company assigns a team of electrical engineers to improve the low first-pass yield on a printed circuit-board (PCB) line. The nonconforming units are reworked and shipped and the final yield is much higher (99%). The company lead time is not world class and sales are being lost. The electrical engineers have come up with 10 possibly important factors, without consulting with rework operators. Then, they performed OFAT with small sample sizes (batches of 20 units at each level, each a success or failure). They did not use t-testing or any formal test and simply implemented the settings that seemed most promising. After the OFAT testing, only three of their factors seemed to make a big difference. None of the setting before or after their changes corresponded to those in any corporate SOP.

1. Which is correct and most complete (according to this book)?
   a. The engineers have a high level of evidence, and the settings should work.
   b. Randomness and interactions could have confused them. Yield might be worse.
   c. The cost of low first-pass yield is almost all direct payments to operators.
   d. They certainly elicited factors from those with the most process knowledge.
   e. All of the above are correct.
   f. All of the above are correct except (a) and (e).
2. Which is correct and most complete (according to this book)?
   a. With experts in-the-loop, it is rarely (if ever) critical to consult SOP settings.
   b. They likely made some Type I errors by assuming that all factors mattered.
   c. They could have used 12 batches based on a Plackett Burman (PB) array and likely avoided errors.
   d. Up to 3 additional factors could have been used with only 16 batches structured according to a regular fractional factorial (FF).
   e. All of the above are correct.
   f. All of the above are correct except (a) and (e).

17

The above approach resulted in a disastrous drop in the yield (to 40%), and an IE "DOE expert" was called into plan new experiments. Someone other than an electrical engineer then suggested an additional factor to consider.

3. Which is correct and most complete?
   a. A reasonable first step is to return the process to the documented settings.
   b. They could study four factors with eight batches of units according to a regular FF.
   c. A list of factors to study should come from engineers and operators.
   d. Lenth's method with EER for the analysis will likely cause few Type I errors.
   e. All of the above are correct.
   f. All of the above are correct except (a) and (e).

The IE let team found that the operator suggested factor was critical, adjusted only it and increased the yield to 95%.

Use the following scenario for Questions 4.

Furniture Study: A furniture manufacturer will lose an important Japanese customer until they can fix an elusive surface finish problem. They are convinced that the cause of the small fraction of unacceptable units relates to an interaction between controllable factors. They are considering studying one categorical (natural or composite wood) and three continuous factors that they are pretty sure all matter including the noise factor, humidity.

4. Which is correct and most complete?
   a. Starting with standard screening using fractional factorials is natural because interactions are modeled.
   b. They could reasonably use a single, standard Box Behnken array with four factors.
   c. A reasonable recommendation is to perform two Box Behnken DOEs, one for each level of the categorical factor.
   d. There is no way to use any RSM method since they have a categorical factor.
   e. All of the above are correct.
   f. All of the above are correct except (a) and (e).

Answer Questions 5 and 6 based only on the following information.

Snap Tab Study Variant: An automotive company wanted to replace screws on its air conditioner cases with snap tabs. The problem was that their snap tabs were less than half as strong as what was needed. They performed an RSM study to tune the four factor settings that they considered.

5. Which is correct and most complete?
   a. As is usual in DOE recommendations, they should recommend the settings corresponding to the best run in their DOE.
   b. If RSM was applied to computer simulations, errors from the DOE modeling process and the simulation could cause poor real-world confirmation results.
   c. Before testing, selecting a range of factor settings that contain good values inside places less demand on expert judgment than directly selecting the final settings.
   d. It is possible that no feasible snap tab design could be found from the analysis.
   e. All of the above are correct.
   f. All of the above are correct except (a) and (e).

6. Which is correct and most complete?
   a. Using an EIMSE optimal design, the snap tab study could have been done with 20 runs or fewer.
   b. The LCRSM approach used could have generated a less accurate prediction model than any of the standard RSM approaches in ► Chap. 13.
   c. By dropping fixing factor C, it would have been possible to using an EIMSE optimal design and 16 or fewer runs.
   d. All of the above are correct.
   e. All of the above are correct except (a) and (d).
7. Which of the following is correct and most complete?
   a. RSM should have been used in the rubber experiment to focus on shot size and humidity since everyone knew these mattered most.
   b. A regression analysis of on-hand data could conceivably have suggested that shot size was causing the sticking problem.
   c. Regression of on-hand data could have proven that shot size was the problem.
   d. All of the above are correct.
   e. All of the above are correct except (a) and (d).
8. Perform an experiment involving four factors and one or more responses using standard screening using fractional factorials or responses surface methods. The experimental system studies should permit building and testing individual prototypes requiring less than $5 and 10 min time.

References

Allen TT, Yu L (2002) Low cost response surface methods for and from simulation optimization. Qual Reliab Eng Int 18:5–17

Allen TT, Yu L, Bernshteyn M (2000) Low cost response surface methods applied to the design of plastic snap fits. Qual Eng 12:583–591

Box GE, Hunter JS (1961) The 2k—p fractional factorial designs. Technometrics 3(3):311–351

Brady J, Allen T (2002) Case study based instruction of SPC and DOE. Am Stat 56(4):1–4

Myers RH, Montgomery DA (2001) Response surface methodology, 5th edn. Wiley, Hoboken, NJ

DOE and Regression Theory

T. T. Allen, *Introduction to Engineering Statistics and Lean Six Sigma*,
https://doi.org/10.1007/978-1-4471-7420-2_18

18.1 Introduction

As is the case for other six sigma-related methods, practitioners of six sigma have demonstrated that it is possible to derive value from design of experiments (DOE) and regression with little or no knowledge of statistical theory. However, understanding the implications of probability theory can be intellectually satisfying and enhance the chances of successful implementations.

Also, in some situations, theory can be practically necessary. For example, in cases involving mixture or categorical variables (▶ Chap. 15), it is necessary to go beyond the standard methods and an understanding of theory is needed for planning experiments and analyzing results. This chapter focuses attention on three of the most valuable roles that theory can play in enhancing DOE and regression applications. For a review of basic probability theory, refer to ▶ Chap. 10.

First, applying t-testing theory can aid in decision-making about the numbers of samples and the α level to use in analysis. Associated choices have implications about the chances that different types of errors will occur. Under potentially relevant assumptions, the chance of wrongly declaring significance (a Type I error) might not be the α level used. Also, if the number of runs is not large enough, a lost opportunity for developing statistical evidence is likely (a Type II error).

Second, theory can aid in the many decisions associated with standard screening using fractional factorials. Decisions include which DOE array to use, which alpha level to use in analysis, and whether to use the individual error rate (IER) or experiment-wise error rate (EER) critical values. With multiple factors being tested simultaneously, many Type I and Type II errors are possible in the same experiment.

Third, in applying responses surface methods (RSM) and regression in general, the resulting prediction models will unavoidably result in some inaccuracy or prediction errors. Theory can aid in predicting what those errors will be and aid in the selection of the design of experiment (DOE) array. In general, design of experiment arrays (DOE) can be selected from a pre-tabulated set or custom designed. "**Optimal design of experiments**" is the activity of using theory and optimization to select custom designed experimental arrays. The huge number of possible arrays to select from explains in part why many people consider DOE the most complicated of six sigma related methods.

▶ Section 18.2 describes general concepts associated with design of experiments and regression theory. The discussion introduces the need for pseudo-random number generation which is described in ▶ Sect. 18.3. Sections 18.4 and 18.5 describe the use of pseudo-random numbers for supporting t-test and fractional factorial decision-making respectively. In ▶ Sect. 18.6, the assumptions underlying the theory of linear polynomial regression are described. ▶ Section 18.7 describes the evaluation of a simple response surface methods (RSM). ▶ Section 18.8 describes formulas useful for supporting RSM decision-making and calculating efficiently the so-called EIMSE criterion.

18.2 Design of Experiments Criteria

It might seem surprising that the chances of errors can usefully be estimated quantitatively even before experimentation begins. Yet such predictions are possible using probability theory, including those that relate to Type I and II and prediction inaccuracy. Evaluations prior to experimentations should not be entirely surprising since it is widely known that t-testing with $\alpha = 0.05$ is associated with a 0.05 estimated chance of Type I errors, under at least some assumptions.

In general, the phrase "**DOE criteria**" refers to evaluations of method quality available for making method choices, e.g., which array to use, before experiments are performed. Criteria comprise the objective such as minimizing expected squared prediction error and the assumptions needed to calculate criteria values. ◘ Table 18.1 previews the criteria used in this chapter to support method related decision-making. Other criteria include so-called "resolution" described in ► Chaps. 13 and 14 and so-called "**D-efficiency**" described later in this chapter.

The term "**simulation**" refers to the use of pseudo-random numbers to evaluate criteria. Simulation is not always needed because in some cases criteria can be evaluated using calculus or in other ways that do not require pseudo-random numbers. Even when simulation is unneeded, the concept of evaluating method selection choices through testing scenarios shown in ◘ Fig. 18.1 is perhaps central to all applications of probability theory to support method selection.

◘ Figure 18.1 also shows a decision-maker trying to select which levels to use for three test runs. The right-hand-side shows one possible scenario. In this scenario, response data are made up for the purposes of a "**thought experiment**" in which the method user imagines what might happen if three distinct, evenly spaced levels are applied. Also, a hypothetical "prediction point" is imagined where prediction will be requested after the experiment and analysis. From the model that would be fitted, a prediction follows. Also, regression t-testing would suggest significance of factor x_1 for affecting the average response.

Clearly, the made-up data in a thought experiment is not associated with any real evidence of whether x_1 affects the response, nor does it help in making predictions. However, such hypothetical data can be useful in careful comparisons of alternative method options. Using simulation, it is possible to test millions of possible scenarios and use them to calculate estimates of DOE criteria for rating method options.

◘ **Table 18.1** Preview of the design of experiments criteria explored in this chapter

Method	Criterion		
	Objective	Assumptions	Relevance
T-testing	Type I and II errors probabilities	Responses are normally distributed with selected means	Correct declarations during analysis
Standard screening	Type I error and Type II error probabilities	Hierarchical assumptions based on normality and unknown true models	Correct declarations during analysis
One-shot RSM	Expected squared prediction errors or the "EIMSE"	Random, independent true model coefficients, errors, and prediction points	Accuracy of predictions after experimentation

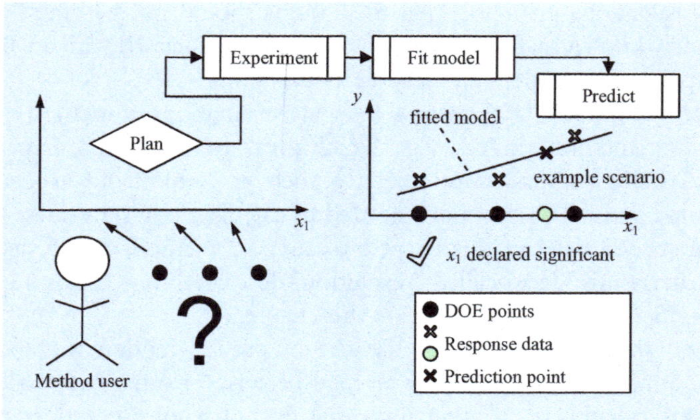

Fig. 18.1 An example DOE design problem with one simulation run or scenario

18.3 Generating "Pseudo-Random" Numbers

Pseudo-random numbers are needed for simulating method performance and simulation-based estimation of criteria. In this section, practical ways to generate approximately random numbers or "pseudo-random" numbers are described. Results from Press et al. (1993, pp. 275–286) are used throughout. We begin with the definition of the uniformly distributed random variables, U, over the interval $[a, b]$. The notation that we will use is $U \sim \mathrm{U}[a, b]$. Uniform random variables have the distribution function $f_u(x) = (a - b)^{-1}$ for $a \leq x \leq b$ and $f_u(x) = 0$ otherwise.

The initial starting point of most simulations are approximately independent identically (IID) distributed random numbers from a uniform distribution between $a = 0$ and $b = 1$, written U[0, 1]. As noted in ▶ Chap. 10, "**independent**" means that one is comfortable with the assumption that the next random variable's distribution is not a function of the value taken by any other random variables for which the independence is believed to apply.

For example, if a person is very forgetful, one might be comfortable assuming that this person's arrival times to class on two occasions are independent. Under that assumption, even though the person might be late on one occasion (and feel bad) the person would not modify his/her behavior and the chance of being late the next time would be the same as always. Formally, if $f(x_1, x_2)$ is the "joint probability density function", then independence implies that it can be written $f(x_1, x_2) = f(x_1)f(x_2)$. Also, the phrase "**identically distributed**" means that all of the relevant variables are assumed to come from exactly the same distribution.

Consider the sequences of numbers $Q_1, Q_2, \ldots, Q_n$ and $U_1, U_2, \ldots, U_n$ given by

18

$$\begin{aligned} Q_i &= \mathrm{mod}\left(1{,}664{,}525 Q_{i-1} + 1{,}013{,}904{,}223, 2^{32}\right) \\ U_i &= \frac{Q_i}{2^{32}} \quad \text{for } i = 1, \ldots, \infty \text{ with } Q_0 = 1 \end{aligned} \tag{18.1}$$

where the function "mod" returns the remainder of the first quantity in the brackets when divided by the second quantity. For example 14 mod 3 is 14 – 4(3) = 2. The phrase "**random seed**" refers to any of the numbers Q_1, …, Q_n, which starts a sequence.

Then, starting with $Q_0 = 3$, the first eight values $i = 1$, …, 8 of the Q_i sequence are 1,018,897,798, 2,365,144,877, 3,752,335,016, 3,345,418,727, 1,647,017,498, 3,714,889,393, 2,735,194,204, and 1,668,571,147. Also, the associated U_i are 0.23723063, 0.550678204, 0.873658577, 0.778915995, 0.383476144, 0.864940088, 0.636837027, and 0.388494494. We know that these numbers are *not random* since they follow the above sequence, and all values can be predicted precisely at time of planning. In fact, the sequence repeats every 4,294,967,296 digits so that there is necessarily a perfect correlation between each element and the element 4,294,967,296 after it (they are identical). Therefore, the numbers are *not independent*, even if they appear random, if small strings are considered. Still, considering the histogram of the first 5000 numbers in ◘ Fig. 18.2, it might be of interest to *pretend* that they are IID U[0, 1].

For the computations in subsequent chapters, numbers are used based on different, more complicated sequences of pseudo-random numbers given by the function "ran2" in *Numerical Recipes* on pp. 282–283. Yet, the concept is the same. The sequence that will be used also repeats but only after 2.3×10^{18} numbers. Therefore, when ran2 is used one can comfortably entertain the assumption that these are perfect IID uniform random variables.

18.3.1 Other Distributions

Generally, pseudo-random numbers for distributions other than uniform are created starting with uniformly distributed pseudo-random numbers. The "univariate transformation method" refers to one popular way to create these random numbers, illustrated in ◘ Fig. 18.3. An initial pseudo-random U[0, 1] number U is transformed to

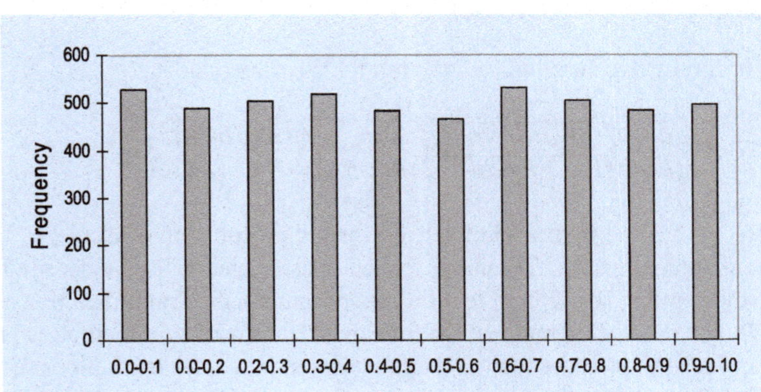

◘ **Fig. 18.2** Histogram of 5000 numbers from a sequence of pseudo-random numbers

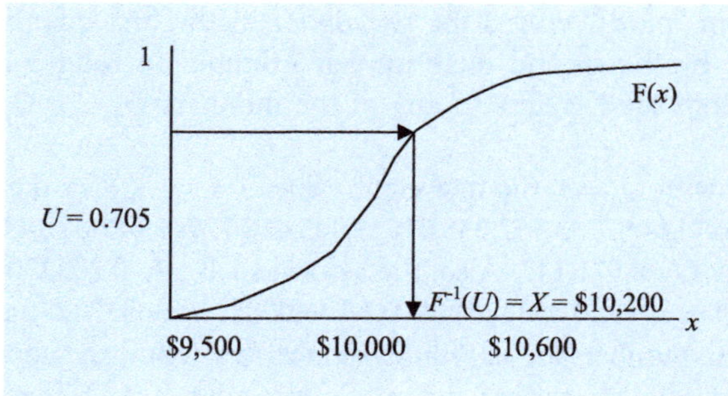

Fig. 18.3 One way to derive a pseudo-random number, *X*

another number, *X*, using the so-called "inverse cumulative distribution" or *F* function associated with the distribution of interest.

Since the *U* has a roughly equal chance of hitting anywhere along the vertical axis, the chance that *X* will lie in any interval on the horizontal axis is proportional to the slope of the curve at that point. One can write this slope (d/d*x*)*F*(*x*). From the "Liebniz rule" in calculus (see the Glossary), we can see that (d/d*x*)*F*(*x*) = *f*(*x*) if and only if

$$F(x) = \int_{-\infty}^{x} f(x)\mathrm{d}x \quad (18.2)$$

which is the definition of the "**cumulative distribution function**" (CDF) associated with the density function *f*(*x*).

For example, the cumulative distribution function, *F*(*x*), for the triangular distribution function with $a = \$9{,}500$ and $b = \$10{,}600$, with $c = \$10{,}000$ is

$$F(x) = \begin{cases} 0 & \text{if } x \le a \\ \frac{(x-a)^2}{(b-a)(c-a)} & \text{if } a < x \le c \\ 1 - \frac{(b-x)^2}{(b-a)(b-c)} & \text{if } c < x < b \\ 1 & \text{if } x \ge b \end{cases} \quad (18.3)$$

The inverse cumulative distribution function for the triangular is

$$F^{-1}(u) = \begin{cases} a + \sqrt{(b-a)(c-a)u} & \text{if } u < (c-a)/(b-a) \\ b - \sqrt{(1-u)(b-a)(b-c)} & \text{otherwise} \end{cases} \quad (18.4)$$

As long as one has the inverse cumulative distribution function available, $F^{-1}(u)$, one can generate approximately IID random variables associated with any density function *f*(*x*) by first generating IID U[0, 1] pseudo-random numbers, *U*, and then transforming, $X = F^{-1}(U)$. For many distributions, the univariate transformation method is built into standard spreadsheet software such as Excel. However, for some distributions such as the so-called "triangular" distribution described in the next example, it can be necessary to calculate the inverse cumulative distribution and perform all steps by hand.

18

Example 18.3.1 Simulating Future Revenues

Suppose someone tells you that she believes that revenues for her product line will be between \$1.2M and \$3.0M next year, with the most likely value equal to \$2.7M. She says that \$2.8M is much more likely than \$1.5M.

Question 1

Define a distribution function consistent with her beliefs.

Answer 1

One distribution function satisfying these conditions is a triangular distribution with $a=\$1.2M$, $b=\$3.0$, and $c=\$2.7M$ in Fig. 18.4.

Question 2

Use your own distribution function from Question 1 to estimate the probability, according to her beliefs, that revenue will be greater than \$2.6M.

Answer 2

$$\begin{aligned} \mathrm{P}(X > 2.6) &= \text{the shaded area above} \\ &= 1-\mathrm{P}(X \leq 2.6) \text{ where } \mathrm{P}(X \leq 2.6) \text{ is the CDF for 2.6 and} \end{aligned} \quad \textbf{(18.5)}$$

$$\mathrm{P}(X \leq 2.6) = \frac{(2.6-1.2)^2}{(3.0-1.2)(2.7-1.2)} = 0.73 \Rightarrow \mathrm{P}(X > 2.6) = 0.27\,.$$

Question 3

Generate or show in detail how to generate three pseudo-random samples from the distribution defined in Question 2. Start with the pseudo-random uniform numbers 0.23, 0.78, and 0.51.

Answer 3

$$F^{-1}(u) = \begin{cases} \sqrt{(3.0-1.2)(2.7-1.2)u+a} & \text{if } u < (2.7-1.2)/(3.0-1.2) \\ \sqrt{3.0-(1-u)(3.0-1.2)(3.0-2.7)} & \text{otherwise} \end{cases} \quad \textbf{(18.6)}$$

Plugging in and marking the units we obtain: \$1.9M, \$2.65M, and \$2.37M.

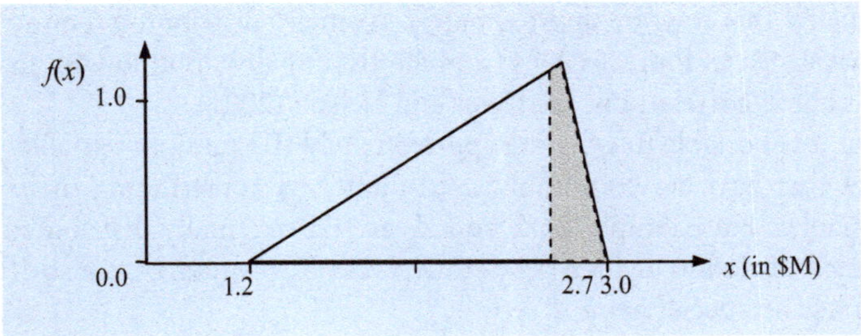

Fig. 18.4 A proper distribution function consistent with the stated beliefs

18.3.2 Correlated Random Variables

Sometimes, one is interested in investigating assumptions about random variables that include correlations between them, i.e., the random variables are not independently distributed. An example might be the prices of X_1—automotive, X_2—oil stocks, and X_3—natural gas stocks with assumed means μ_1, μ_2, and μ_3 respectively. From past data and/or expert opinion one might want to entertain the assumptions that

$$\begin{aligned}
&\mathrm{E}[(X_1-\mu_1)(X_2-\mu_2)] = -13(\$/\text{share})^2\\
&\mathrm{E}[(X_1-\mu_1)(X_3-\mu_3)] = -10.5(\$/\text{share})^2\\
&\mathrm{E}[(X_2-\mu_2)(X_3-\mu_3)] = 8.5(\$/\text{share})^2\\
\text{and}\\
&\mathrm{E}[(X_1-\mu_1)^2] = 28.25(\$/\text{share})^2\\
&\mathrm{E}[(X_2-\mu_2)^2] = 12.25(\$/\text{share})^2\\
&\mathrm{E}[(X_3-\mu_3)^2] = 18(\$/\text{share})^2.
\end{aligned} \tag{18.7}$$

Then, one would like our pseudo-random numbers to reflect these "correlations" and, e.g., have similar sample correlations.

Suppose that we have $F^{-1}(u)$ available for a normal distribution with mean $\mu=0$ and standard deviation $\sigma=1$. Then, one can generate, Z_1, Z_2, Z_3 approximately IID standard normal random variables. It is a fact verifiable by linear algebra and calculus that if we form the matrices **V** and **T**:

$$\mathbf{V} = \begin{bmatrix} 28.25 & -13 & -10.5 \\ -13 & 12.25 & 8.5 \end{bmatrix} = \mathbf{T'T} \text{ with } \mathbf{T} = \begin{bmatrix} 5 & -1.5 & -1 \\ -1.5 & 3 & 1 \end{bmatrix} \tag{18.8}$$

where **T** is called the "Cholesky decomposition" or "square root" of **V**, then we can generate pseudo-random numbers (in this case stock prices), X_1, X_2, X_3, with the desired correlations and means using the formula:

$$\begin{bmatrix} X_1 \\ X_2 \\ X_3 \end{bmatrix} = \mathbf{T} \begin{bmatrix} Z_1 \\ Z_2 \\ Z_3 \end{bmatrix} + \begin{bmatrix} \mu_1 \\ \mu_2 \\ \mu_3 \end{bmatrix} \tag{18.9}$$

Because of the unusual properties of the normal distribution, one can also say that the X_i calculated this way are approximately normally distributed. For a recent reference on generating random variables from almost *any* distribution with many possible assumptions about correlations, see Deler and Nelson (2001).

Note that it is possible to generate approximately IID random variables from many distributions that have no commonly used names by constructing them from other random variables. For example, if Z_1 and Z_2 are IID normally distributed with mean 0 and standard deviation 1, then $X_1=\sin(Z_1)$ and $X_2=\sin(Z_2)$ are also IID, but their distribution has no special name.

18

18.3.3 Monte Carlo Simulation (Review)

The central limit theorem provides a mathematical framework with which to evaluate the averages and standard deviations of simulated numbers. The results of that theorem are repeated from ▶ Chap. 10 using the following symbols:

1. $X_1, X_2, \ldots, X_n$ are random variables assumed to be independent identically distributed (IID). These could be quality characteristic values outputted from a process with only common causes operating. They could also be a series of outputs from some type of numerical simulation.
2. $f(x)$ is the common density function of the identically distributed $X_1, X_2, \ldots, X_n$.
3. Xbar_n is the sample average of $X_1, X_2, \ldots, X_n$. Xbar_n is effectively the same as Xbar from Xbar charts with the "n" added to call attention to the sample size.
4. σ is the standard deviation the $X_1, X_2, \ldots, X_n$, which do not need to be normally distributed.

The CLT focuses on the properties of the sample averages, Xbar_n.

If $X_1, X_2, \ldots, X_n$ are independent, identically distributed (IID) random variables from a distribution function with any density function $f(x)$ with finite mean and standard deviation, then the following can be said about the average, Xbar_n, of the random variables.

Defining

$$\text{Xbar}_n = \frac{(X_1 + X_2 + \cdots + X_n)}{n} \text{ and } Z_n = \frac{\text{Xbar}_n - \int_{-\infty}^{\infty} u f(u) \mathrm{d}u}{(\sigma/\sqrt{n})}, \tag{18.10}$$

it follows that

$$\lim_{n\to\infty} \Pr(Z_n \le x) = \int_{-\infty}^{x} \frac{1}{\sqrt{2\pi}} e^{-\frac{1}{2}u^2} du. \tag{18.11}$$

In words, averages of n random variables, Xbar_n, are approximately characterized by a normal probability density function. The approximation improves as the number of quantities in the average increases. A reasonably understandable proof of this theorem, i.e., the above assumptions are equivalent to the latter assumption, is given in Grimmet and Stirzaker (2001, Chap. 5).

To review, the expected value of a random variable is:

$$E[X] = \int_{-\infty}^{\infty} u f(u) du \tag{18.12}$$

Then, the CLT implies that the sample average converges, Xbar_n, converges to the true mean $E[X]$ as the number of random variables averaged goes to infinity. Therefore, the CLT can be effectively rewritten as

$$E[X] = \text{Xbar}_n + e_{\text{MC}}, \tag{18.13}$$

where e_{MC} is normally distributed with mean 0.000 and standard deviation σ ÷ sqrt[n] for "large enough" n. It is standard to refer to Xbar_n as the "Monte Carlo simulation estimate" of the mean, $E[X]$. There, with only common causes operating, the Xbar chart user is charting Monte Carlo estimates of the mean.

Since σ is often not known, it is sometimes of interest to use the sample standard deviation, s:

$$s = \sqrt{\frac{\sum_{i=1}^{n} (X_i - Xbar_n)^2}{n-1}} \quad (18.14)$$

Then, it is common to use

$$\sigma_{estimate} = s \div c_4 \quad (18.15)$$

where c_4 comes from Table 10.3. Therefore, the central limit theorem provides us with an estimate of the errors of Monte Carlo estimates.

18.3.4 The Law of the Unconscious Statistician

A result from integration theory broadens the applicability of expected value Monte Carlo. It is called the law of the unconscious statistician:

$$\mathrm{E}\left[g(X)\right] = \int_{-\infty}^{\infty} g(x)f(x)\,dx \quad (18.16)$$

where $f(x)$ is the distribution function the random variable x. Then, to calculate $\mathrm{E}[g(X)]$, we can generate IID $g(X)$ using IID X from the distribution function $f(x)$. In this way, Monte Carlo simulation can evaluate a wide variety of expected values. For example, if $g(X)$ is an "**indicator function**" which is a 1 if an event A occurs and 0 otherwise, then $\mathrm{E}[g(X)] = \Pr\{A\}$.

This law can be proven using some of the basic definitions associated with integrals. Intuitively, if the probability that $\{X = x\}$ is proportional to $f(x)$, the probability that $\{g(X) = g(x)\}$ is also proportional to $f(x)$.

Example 18.3.2 Unconscious Statistician Example

Question

Estimate $\int_1^3 e^{x^2} x^2\,dx$.

Answer

Rewriting, we have

$$\int_1^3 e^{x^2} x^2\,dx = \int_{-\infty}^{\infty} \left[(3-1)\left(e^{x^2} x^2\right)\right] f(x)dx$$
$$= E\left[g(X)\right] \text{ where } g(x) = 2e^{x^2} x^2 \quad (18.17)$$

18

and where $f(x)$ is the density function for a uniform distribution with $a=1$ and $b=3$.
Also, $X \sim U[1, 3]$, i.e., X is uniformly distributed with $a=1$ and $b=3$.
Therefore, the pseudo-random U[0, 1] numbers 0.23723063, 0.550678204, 0.873658577, 0.778915995, 0.383476144, 0.864940088, 0.636837027, 0.388494494, and 0.033923503 can be used to construct the pseudo-random sequence 1.474461, 2.101356, 2.747317, 2.557832, 1.766952, 2.72988, 2.273674, which pretend to be IID U[1, 3]. Using the inverse cumulative is equivalent to multiplying by $(b-a)$ and then adding a.
From this sequence, one constructs the sequence 38.24, 730.71, 28628.23, 9081.29, 141.71, 25691.32, 1818.08, 148.51, and 7.13, which we pretend are IID samples of $2e^{X^2}X^2$. The average of these numbers is 7365.0 and the standard deviation is 11607. Therefore, the Monte Carlo estimate for the original integral is 7365.0 with estimated error 11607/3 = 3869.0. Using 10,000 pseudo-random numbers the estimate is 10949.51 with standard error 255.9. Therefore the true integral value is very likely within 768 of 10949.5 (three standard deviations or $3.0 \times \sigma_{estimate}$).

18.4 Simulating T-Testing

In this section, simulation is used to study the decision to invest in the applying the "**two-sample t-test** assuming unequal variances" method described in ▶ Chap. 11. It is perhaps true that the primary objective of the t-test procedure is the following. People must be stopped from claiming that their product, service, or idea (level 1) causes a more desirable average response then level 2, when it either does nothing or makes things worse. For example, a salesman might be selling snake oil as something that makes hair grow when it does not.

The admission associated with t-testing is that even if changing levels does nothing and the procedure is applied correctly, there is some low probability significance will be established. Therefore, a criterion that can be used to evaluate the t-test strategy is the probability that the test will wrongly indicate significance, i.e., a "Type I error" is made and the snake oil salesman fools us.

The following assumptions can be used to create and/or verify the t-critical values used in all standard t-test procedure:

1. When level 1 is inputted, responses are IID normally distributed with mean, μ_1, and standard deviation, σ_1.
2. When level 2 is inputted, responses are IID normally distributed with mean μ_2 and standard deviation, σ_2.
3. $\mu_1 = \mu_2 + \Delta$ and $\Delta = 0.0$ if Type I errors are being simulated.

Under these assumptions and when $\alpha = 0.05$, the probability of wrongly finding significance is well known to be 0.05 independent of μ_1, σ_1, μ_2, and σ_2. This is the defining property of the t-test strategy. As an example of evaluating a procedure using Monte

Carlo, we next show how this probability (0.05) can be estimated for the case in example 1 in the proceeding section.

Example 18.4.1 Simulation of Type I Errors

Suppose we are interested in entertaining assumptions of the standard type with $\mu_1 = \mu_2 = 0$, and $\sigma_1 = 5$ and $\sigma_2 = 1$. Then, any conclusion of significance is a mistake since the average true responses are the same independent of the level, i.e., a snake oil salesman is at work.

Question 1

How can we write the probability of Type I error as a expectation assuming that a two-sample t-test procedure with n_1 and n_2 is applied?

Answer 1

The probability of wrongly indicating significance can be written in terms of the random indicator function, $I(Y_{1,1}, Y_{1,2}, Y_{1,3}, Y_{2,1}, Y_{2,2}, Y_{2,3})$, which is a function of the six random responses, $Y_{1,1}, \ldots, Y_{2,3}$. The function "$I()$" equals 1.0 if the procedure indicates significance and 0.0 otherwise. With these definitions, the mistake probability is $E[I(Y_{1,1}, Y_{1,2}, Y_{1,3}, Y_{2,1}, Y_{2,2}, Y_{2,3})]$.

Question 2

How can we estimate this probability numerically?

Answer 2

To estimate this probability, we can sample pseudo-random IID normally distributed $Y_{1,1}, \ldots, Y_{2,3}$ according to the appropriate distributions and derive from these numbers pseudo-random $I(\mathbf{Y})$. The central limit theorem says that if we average enough $I(\mathbf{Y})$, the result will converge to the true value. The simulation in our example uses the following randomly generated numbers $Y_{1,1} = -1.501$, $Y_{1,2} = -6.388$, $Y_{1,3} = 1.221$, $Y_{2,1} = -0.818$, $Y_{2,2} = 0.661$, and $Y_{2,3} = -0.760$. Then, functional relationships are used to calculate $t_0 = -0.842$, $df = 2$, and $I(\mathbf{Y}) = 0$. Performing these operations 10,000 times and averaging the derived probability estimate is 0.049 with estimated standard error 0.002 We can see that the Monte Carlo is trying to estimate the number 0.05 which is the exact Type I error associated with the test strategy described above under standard assumptions. Table 18.2 illustrates results from applying a spreadsheet solver to estimate the Type I error rate.

In practice, one does not need to use simulation since the critical values are already tabulated to give a pre-specified Type I error probability. Still, it is interesting to realize that the error rates can be reproduced. Similar methods can be used to estimate Type I error rates based on assumptions other than normally distributed responses. Also, simulation can also be used to evaluate other properties of this strategy including Type II error as described in the next example.

Table 18.2 Simulations used to estimate the probability of Type I error (α)

No.	$Y_{1,1}$	$Y_{1,2}$	$Y_{1,3}$	$Y_{2,1}$	$Y_{2,2}$	$Y_{2,3}$	$\bar{y}_1$	$\bar{y}_2$	s_1^2	s_2^2	t_0	df	$t_{critical}$	I(Y)
1	−1.501	−6.388	1.221	−0.818	0.661	−0.760	−2.223	−0.306	14.867	0.702	−0.842	2	2.920	0
2	6.382	5.992	8.666	0.179	−0.031	−0.116	7.013	0.011	2.086	0.023	8.352	2	2.920	1
3	−10.918	−1.171	5.475	−1.137	0.610	0.092	−2.205	−0.145	67.984	0.805	−0.430	2	2.920	0
4	−5.434	−3.451	−8.452	−0.425	0.285	−0.680	−5.779	−0.273	6.342	0.250	−3.714	2	2.920	0
5	−9.235	−4.888	−3.868	2.008	−0.617	−0.564	−5.997	0.276	8.123	2.251	−3.373	3	2.353	0
6	−10.590	−2.840	−2.020	−1.457	−0.985	−1.044	−5.150	−1.162	22.362	0.066	−1.459	2	2.920	0
7	0.674	−1.827	−1.635	−1.163	0.895	−0.973	−0.929	−0.414	1.938	1.293	−0.497	3	2.353	0
8	−1.851	6.713	−0.426	0.044	0.483	0.498	1.479	0.342	21.059	0.067	0.428	2	2.920	0
9	−0.931	−2.566	9.861	−1.296	−0.650	−0.867	2.121	−0.938	45.595	0.108	0.784	2	2.920	0
10	4.328	11.878	−3.275	1.904	1.218	1.097	4.311	1.406	57.402	0.189	0.663	2	2.920	0
⋮	⋮	⋮	⋮	⋮	⋮	⋮	⋮	⋮	⋮	⋮	⋮	⋮	⋮	⋮
10^4	3.806	1.534	3.202	−0.669	−1.538	1.773	2.847	−0.145	1.385	2.948	2.490	3	2.353	1

Example 18.4.2 Simulation of Type II Errors

Suppose you are thinking about using a t-test to "analyze" experimental data in which one factor was varied at two levels with two runs at each of the two levels. Suppose that you are interested in entertaining the assumption that the true average response at the two levels differs by $\Delta = 0.5$ s and that the random errors always have standard deviations $\sigma_1 = \sigma_2 = 0.3$ s.

Question 1

What additional assumptions are needed for estimation of the power, i.e., the probability that the t-test will correctly find significance?

Answer 1

Many acceptable answers could be given. The assumed mean difference must be 0.5. For example, assume the level 1 values are IID N($\mu_1 = 0$, $\sigma_1 = 0.3$) and the level 2 values are IID N($\mu_2 = 0.5$, $\sigma_2 = 0.3$). Note that power equals 1 – probability of Type II errors so that it is higher if we find significance more often.

Question 2

What would one Monte Carlo run for the estimation of the power under your assumptions from Question 1 look like? Arbitrary random-seeming numbers are acceptable for this purpose.

Answer 2

The responses were generated arbitrarily, being mindful that the second level responses should be roughly higher by 0.5 than the first. Then, the other numbers were calculated: $Y_{1,1} = 0.60$, $Y_{1,2} = -1.20$, $Y_{2,1} = 2.10$, $Y_{2,2} = 0.10$, $\bar{y}_1 = -0.30$, $\bar{y}_2 = 1.10$, $s_1^2 = 1.62$, $s_2^2 = 2.00$, $t_0 = -1.274$, df = 2, $t_{critical} = 2.92$, $I() = 0$, because we failed to find significance in this simulation or though experiment.

Question 3

How might the Type II error probability be derived through averaging the indicator function values from many simulation runs influence your decision-making?

Answer 3

If one feels that the estimated Type II error probability for a given true effect size is too high (subjectively), then we might re-plan the experiment to have more runs. With more runs, we can generally expect the probabilities of Type I and Type II errors to decrease and the probability to correctly detect effects of any given size to increase.

18.4.1 Sample Size Determination for T-Testing

18

Next, the implications of simulation results are explored related to the choices of the initial sample sizes n_1 and n_2. Table 18.3 provides information in support of decisions about the method parameters n_1, n_2, and α. Table 18.3 shows the chance that

Table 18.3 The probability of a t-test's finding significance

$n_1 = n_2$	$\Delta \div \sigma$				
	0.001	0.5	1	2	5
$\alpha = 0.01$					
3	1.0%	2.3%	4.8%	15.6%	67.6%
6	1.0%	5.4%	19.4%	71.5%	100.0%
$\alpha = 0.05$					
3	5.0%	11.4%	22.1%	52.2%	98.1%
6	5.0%	11.9%	47.7%	93.8%	100.0%

significance will be declared under the standard assumptions described in the last section. If $\Delta = 0$, then the table probabilities are the Type I error rates. "**Power**" (β) is often used to refer to the probability of finding significance when there is a true difference, i.e., $\Delta \neq 0$. Therefore, the probability of a Type II error is $1 - \beta$. Interpolating or extrapolating linearly to other sample sizes might give some insights.

To use the decision support information in Table 18.3, it is necessary to entertain assumptions relating to the size of the true prototype system to the average response change that it is desirable to detect, Δ. Also, it is necessary to estimate the typical difference, σ, between responses from prototype systems with identical inputs. These numbers must be guessed, and then the implications of various decisions about the methods to use can be explored as illustrated in the following example. Note that the quantity, $\Delta \div \sigma$, is sometimes called the "signal to noise ratio" even though it is not related to the "SN" ratio in Taguchi Methods (from ▶ Chap. 15).

Example 18.4.3 Sample Sizes for Fuel Testing

Question 1

An auto racer is interested to know if a new oil additive reduces her race time by 10.0 s, i.e., $\Delta = 10.0$ s. Also, the racer may know that, with no changes in his or her vehicle or strategy, times typically vary ± 5.0 s. What is a reasonable estimate of $\Delta \div \sigma$?

Answer 1

A reasonable signal-to-noise ratio estimate is $\Delta \div \sigma = 2.0$.

Question 2

Assume that the cost of the fuel additive is not astronomically high. Therefore, the racer is willing to tolerate a 5% risk, wrongly concluding that the additive helps when it does not. What a level makes sense for this case?

Answer 2

Clearly, $\alpha = 0.05$ is by definition appropriate.

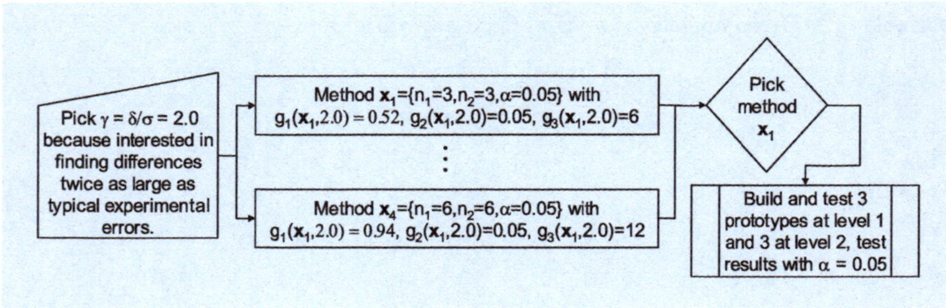

Fig. 18.5 Example t-test method (initial sample size and α) selection

Question 3

The racer is considering using $n_1 = n_2 = 6$ test runs. Would this offer an high chance of detecting the effect of interest?

Answer 3

Yes, Table 18.3 indicates that this approach would give greater than or equal to 93.8% probability of finding average differences significant if the true benefit of the additive is a reduction on average greater than 10.0 s. Under standard assumptions, the Type I error rate would be 0.05 and the Type II error rate would be 0.062. In other words, if the effect of the additive is strong, starting with 6 runs gives an excellent chance of proving statistically that the average difference is nonzero.

Question 4

Flow chart a decision process resulting in the selections $\alpha = 0.05$ and $n_1 = n_2 = 6$ using criteria power (g_1), Type I error rate (g_2), and number of runs (g_3).

Answer 4

See Fig. 18.5.

18.5 Simulating Standard Screening Methods

Before the tests are performed, the experimenter must select the experimental plan or "design", **D**, which includes selecting the number of runs, *n*, and factors, *m*. Also, during analysis one must select IER or EER critical values and the value of α. All these choices and the properties of the prototype system have implications for the success criteria (g_1, g_2,…) associated with the methods.

The following definitions can be use to generate simulation results:

18

1. The actual change in the average response cause by a change in the factors, τ, is called the "**effect**" of that factor.
2. "**Important factors**" are factors that, when changed from one to another predefined level, result in an actual change, τ, that is greater than a pre-specified amount, Δ. Therefore, important factors satisfy $\tau > \Delta$.

3. As in ▶ Chap. 4, "σ" refers to the standard deviation of the "**random errors**", ε_i. This can be estimated using control charting, system knowledge, or experience.
4. p_0 is the expected fraction of factors that are important or, in other words, the believed probability that any given factor will have an important effect.

In terms of the assumed values of Δ, σ, p_0, and a few additional assumptions described in Allen and Bernshteyn (2003), the criteria in ◻ Tables 18.4 and 18.5 can be calculated using simulation. The additional assumptions relate to the possibility of interactions in the true model and potentially nonzero values for unimportant factors.

◻ **Table 18.4** Probability of finding a given important factor significant (the power)

Assumptions	*n*	Factors (*m*)						
		3	4	5	6	7	8	9
Liberal ($\Delta = 2.0\sigma$, $p_0 = 0.25$, IER, $\alpha = 0.05$)	8	0.95	0.90	0.82	0.74	0.73	–	–
Conservative ($\Delta = 1.0\sigma$, $p_0 = 0.5$, EER, $\alpha = 0.10$)		0.69	0.61	0.45	0.36	0.33	–	–
Liberal ($\Delta = 2.0\sigma$, $p_0 = 0.25$, IER, $\alpha = 0.05$)	16	0.96	0.99	1.00	0.98	0.97	0.96	0.93
Conservative ($\Delta = 1.0\sigma$, $p_0 = 0.5$, EER, $\alpha = 0.10$)		0.74	0.79	0.99	0.77	0.93	0.87	0.84

◻ **Table 18.5** Probability of complete correctness identifying important factors (p_{CS})

Assumptions	*n*	Factors (*m*)						
		3	4	5	6	7	8	9
Liberal ($\Delta = 2.0\sigma$, $p_0 = 0.25$, IER, $\alpha = 0.05$)	8	0.79	0.73	0.57	0.44	0.36	–	–
Conservative ($\Delta = 1.0\sigma$, $p_0 = 0.5$, EER, $\alpha = 0.10$)		0.45	0.13	0.17	0.09	0.07	–	–
Liberal ($\Delta = 2.0\sigma$, $p_0 = 0.25$, IER, $\alpha = 0.05$)	16	0.79	0.76	0.57	0.76	0.45	0.53	0.30
Conservative ($\Delta = 1.0\sigma$, $p_0 = 0.5$, EER, $\alpha = 0.10$)		0.52	0.58	0.35	0.48	0.36	0.31	0.16

Table 18.4 shows the probability the method will find any given important factor to be significant (g_1), i.e., the power. The "**probability of correct selection**" is the chance that the list of factors declared to be significant and factors not declared to be significant matches the lists of factors that are actually important and not important. This second criterion (g_2) is written p_{CS}. Looking at the tables, it is possible to decide whether eight runs and the choice of IER offers acceptable risks.

Example 18.5.1 Rubber Machine Problem Revisited

Question 1

What assumptions about Δ, σ, and p_0 might have seemed appropriate to planners of the rubber machine study (▶ Chap. 17) before their experiment?

Answer 1

The team would likely have been happy with factors reducing the fraction nonconforming by $\Delta = 2\%$ or more. Also, they would likely agree that only $p_0 = 50\%$ of the factors were important (but they did not know which ones, of course). A reasonable estimate for σ based on 500 samples would be 0.02 or 2%.

Question 2

Use Table 18.4 to estimate the power and p_{CS}. Interpret this information.

Answer 2

Under conservative assumptions, the power might have been as low as 0.33 and the p_{CS} as low as 0.07. However, looking back on the results it seems that p_0 was actually $1 \div 7 = 0.14$. This means that the hypothetical planners would have likely overestimated their own abilities to identify important factors in experimental planning. Further, such overestimation might have wrongly made them believe that they needed more runs. With the sparsity present in the actual system (small true p_0), the chance of the method finding the important factor was probably closer to 0.73.

In general, Tables 18.4 and 18.5 provide information pertinent to selecting specific design of experiments arrays associated with different numbers of factors, m, and the assumptions use in the analysis. These analysis assumptions include EER or IER and the specific α used. The tables summarize criteria values for two combinations of assumptions about Δ, σ, p_0, EER or IER, and α. Note that the combination $\alpha = 0.10$ with the EER might not be viewed as conservative since $\alpha = 0.10$ is higher than $\alpha = 0.05$. However, using lower values of α might not yield acceptable g_1 (power) criterion value because of the inherent conservatism of the EER assumptions.

Simulation results in Tables 18.4 and 18.5 support the following general insights about standard screening using fractional factorials. First, using the IER increases the power (but also the chance of Type I errors) compared with using the EER. Second, using more factors generally reduces the probabilities of correct selection. This corresponds to common sense in part because, with more factors, more opportunities for errors are possible. Also, more interactions in the true model are possible that can reduce the effectiveness of the screening analysis. Finally, the better a job engineers or other team members do in selecting factors, the higher the p_0.

Unfortunately, high values of p_0 actually decreases the method performance. In technical jargon, the chance of correct selection shrinks because the methods are based on the assumption of "sparsity" or small p_0.

18.6 Evaluating Response Surface Methods

▶ Chapter 13 includes a definition of the expected prediction errors for different assumptions about the system being studied and choices of response surface method (RSM) designs. To the extent that the goal of experimentation is to produce accurate prediction models, the criterion here is central to DOE theory. In this section, the details of the "expected prediction error" calculations are explained. First, the rationale for the associated assumptions is given in terms of Taylor series expansions. Second, a simulation approach for evaluating the expected prediction errors is given. Third, a formulaic approach that is more computationally efficient than simulation is provided. This formulaic approach also is used to suggest insights about the theory of RSM and regression.

18.6.1 Taylor Series and Reasonable Assumptions

Assumptions about the true model, $y_{true}(\mathbf{x})$, are critical to the theory of experimental design. Clearly, if one knew the exact true model before experimentation and the only goal was accurate prediction of the mean response values, then experimentation would not be needed. At the same time, it is of interest to explore assumptions about the true model and the robustness of our method choices to the aspects of these assumptions that are uncertain.

The discussion begins by concentrating on the single factor (x_1) case for simplicity. Generalizations to more than one factor are considered afterward. Taylor's theorem applies under the often reasonable condition that $y_{true}(\mathbf{x})$ is "infinitely differentiable" (i.e., "smooth" with no spikes) over the region of interest. Under these conditions, the theorem gives that whatever the true model is, it can be expressed uniquely and with perfect accuracy as (see, e.g., Simmons 1996, p. 500)

$$y_{true}(x_1) = \sum_{i=0}^{\infty} \frac{y_{true}^{(i)}(a)}{i!}(x_1 - a)^i \tag{18.18}$$

over the same interval where $y_{est}^{(d)}(a)$ is the dth derivative of $y_{est}(x_1)$ with respect to x_1 evaluated at a. Another result is based on a Taylor Series truncated at order "d" and Lagrange's error formula. This formula states that (Simmons 1996, p. 500):

$$y_{true}(x_1) = \left[\sum_{i=0}^{d} \frac{y_{true}^{(i)}(a)}{i!}(x_1 - a)^i\right] + \frac{y_{true}^{(d+1)}(c)}{(n+1)!}(x_1 - a)^{d+1} \tag{18.19}$$

for some c satisfying $a < c < x_1$. Therefore, as long at $y_{true}^{(d+1)}(c)$ is small compared with $(n+1)!$ and the other terms, one can truncate at order d with small errors. Also, notice that the truncated expansion is simply a polynomial in x_1 of order d.

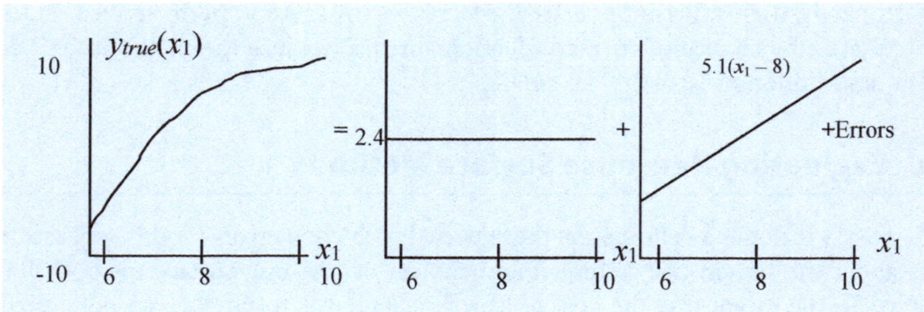

Fig. 18.6 Shows how the Taylor series approximates a function over an interval

It is perhaps helpful to consider the following question. In which practical situations it might be reasonable to assume that $y_{true}^{(d+1)}(c)$ is small enough such that the Lagrange error can be neglected? Figure 18.6 investigates the very approximate expansion with $d = 1$ to aid in intuition building related to Taylor series. The implication is that whenever the response is "somewhat smooth" because changing the inputs is likely to only gradually affect average outputs, then it is reasonable to neglect expansion errors.

These considerations motivate the assumption used in the next factor that the true model can be well approximated by a third order polynomial. They also provide a hint about the situations in which prediction performance of RSM and regression in general might be poor. Poor performance generally occurs when the true model is "bumpy" or third and higher order terms in the Taylor series approximation are needed to provide an accurate approximation. Since Taylor's theorem holds for cases involving more than a single factor, the assumption of a third order true model is often reasonable for those cases as well.

Unfortunately, Taylor's theorem provides little guidance about the values of the coefficients in the expansion. For example, $2.1x_1 + 0.5x_1^3$ is a third order model. However, prior to experimentation, one has little guidance about whether this model is somehow more relevant for thought experiments than other third order models. The assumption of IID N(0, γ^2) assumptions is often entertained where γ is an adjustable assumption parameter. This assumption has the property that positive and negative values are equally likely which might be appropriate for certain cases of interest.

Example 18.6.1 Example Expansion

Question
What is the $d = 2$ Taylor series expansion of $y_{true}(x_1) = \exp(x_1)$ around the point $x_1 = 0$?

Answer
In this case, all derivatives equal 1 when evaluated at $x_1 = 0$. Therefore, the expansion is $y_{true}(x_1) = 1 + x_1 + 0.5\,x_1^2$.

18.6.2 Regression and Expected Prediction Errors

In this section, simulation of the expected prediction errors is illustrated. Knowing an estimate of the prediction accuracy that one is likely to achieve prior to experimentation can help in evaluating whether a different experimental array should be used which might involve more test runs. The simulation-based estimation is mainly important because it can help clarify concepts related to all experimental design criteria. Simulation here is based on the following assumptions:

1. "**Prediction points**" ($\mathbf{x}_p$) are input combinations where prediction will likely be requested after the experimentation and analysis. The assumption considered here is that these are uniformly distributed over a region of interest. For example, likely settings of interest might be anywhere between the test levels.
2. As in ► Chap. 4, "**random**" or "repeatability" errors in experiments, ε_i for $i = 1, \ldots, n$, are IID normally distributed with mean 0 and standard deviation σ.
3. "**True model coefficients**" (β_i) are the hypothetical coefficients in the unknown system performance model. Here, it is assumed that these are IID N(0, γ^2).

Example 18.6.2 Illustration of an Error Simulation Run

Question

Develop an example illustrating the relationship of design of experiment arrays, prediction points, random errors, and a true model.

Answer

Figure 18.7 illustrates the concepts associated with simulating an application of regression. A simulation run begins with a hypothetical true model known to the method tester, not the imagined method user. The method user performs tests on the system and gets the response data. Then, a model is fitted, significance is found through analysis of variance followed by multiple t-tests, and a prediction is made at the prediction point.

The method tester knows that there is a true difference caused by the factor, so therefore the declaration of significance is desirable. The tester, knowing the true model, also knows the prediction errors at the prediction point. Averaging the squared errors from multiple simulation runs gives an estimate of the expected squared prediction errors, which is referred to in the next section as the "expected integrated mean squared error" (EIMSE).

Next, two example simulation runs useful for estimating the EIMSE quantitatively are illustrated. The first, a simulation run, starts with assumptions and generates an $n = 4$ dimensional simulated data vector, **Y**. Assume that the experimental plan allocates test units at the points $x_1 = -1.0$ mm, $x_1 = 0.0$ mm, $x_1 = -0.5$ mm, and $x_1 = 1.0$ mm. The assumed true model form is $\beta_0 + \beta_1 x_1 + \beta_2 x_1^2 + \beta_3 x_1^3 + \varepsilon$. One starts with the pseudo-random numbers 0.236455525, 0.369270674, 0.504242032, 0.704883264, 0.050543629, 0.369518354.0.774762962, 0.556188571, 0.016493236. We use the first

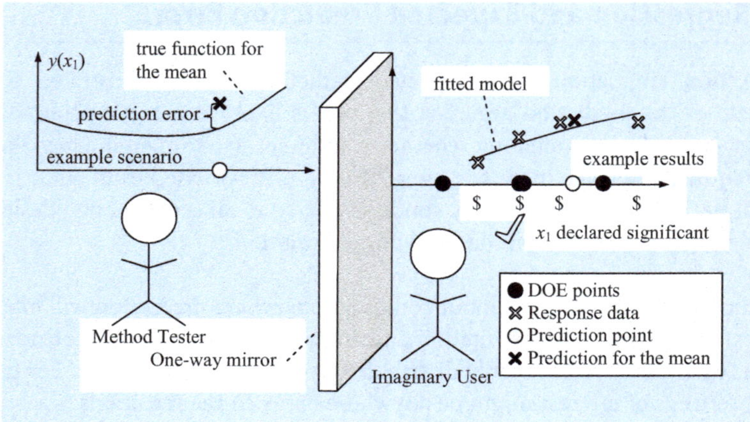

Fig. 18.7 Illustrates a simulation of an experimental design application

four to generate pseudo-random true model coefficients from a N(0, γ^2). Then, we use the next four numbers to generate four random errors.

The Excel function "NORMINV" can be used to generate pseudo-random normally distributed random numbers from pseudo-random uniformly distributed random numbers. Note that this is not needed since Excel also has the ability to generate normally distributed numbers directly; however, it is good practice to generate all random numbers from the same sequence. Combining all this information gives the four simulated random experimental response values **Y** = (–1.275717523, –0.474320672, 0.018436594, –0.180546067)′.

The next assumption needed to estimate the expected prediction errors is that a second order polynomial will be fitted after the experiment using least squares regression. Associated with this model form is the design matrix $\mathbf{X}_1$. The construction of design matrices is described in ► Chap. 13. In this case, the design matrix, $\mathbf{X}_1$, and the estimated coefficients, $\boldsymbol{\beta}_{est}$, are

$$\mathbf{X}_1 = \begin{bmatrix} 1 & -1 & 1 \\ 1 & -0.5 & 0.25 \\ 1 & 0 & 0 \end{bmatrix} \qquad \text{so that } \boldsymbol{\beta}_{est} = \begin{bmatrix} 0.004332 \\ 0.552287 \\ -0.73481 \end{bmatrix} \tag{18.20}$$

where we have used the following formula to estimate the coefficients:

$$\boldsymbol{\beta}_{est} = (\mathbf{X}_1'\mathbf{X}_1)^{-1}\mathbf{X}_1'\mathbf{Y} \tag{18.21}$$

Note that coefficients derived from this equation automatically minimize the sum of the squared estimated prediction errors.

From the above assumptions, x_p is sampled from a uniform distribution. The predicted and actual value key point give the simulated pseudo-random error ε_p. For the first simulation run, the ninth random number to generate the prediction

18

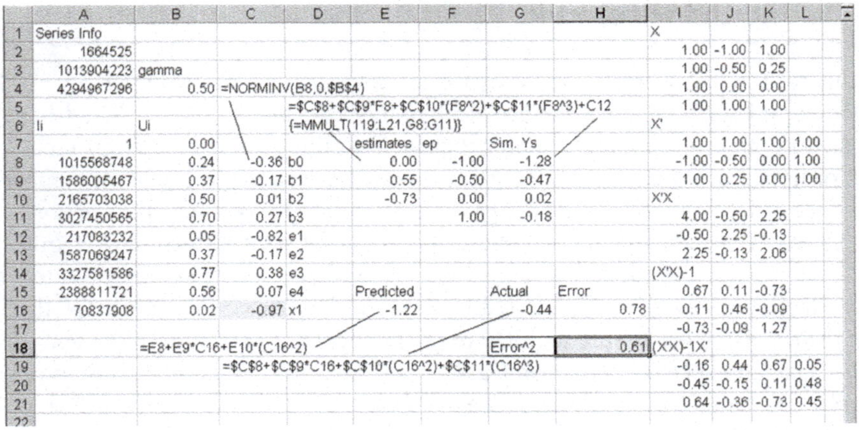

Fig. 18.8 A single simulation run in the MC estimation of the EIMSE

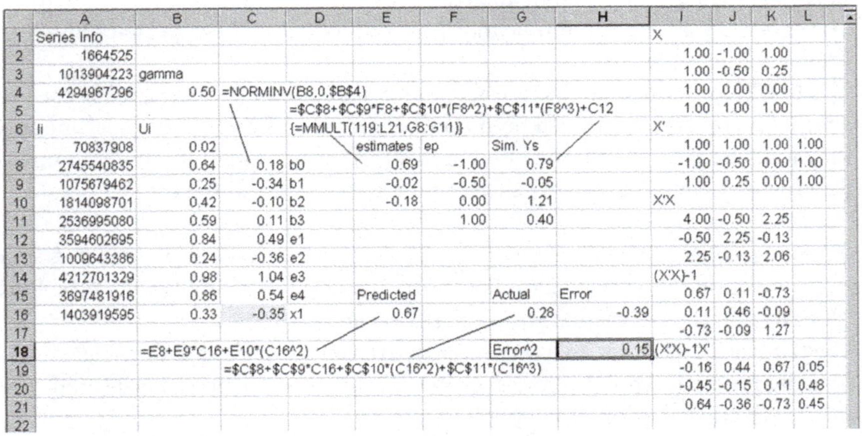

Fig. 18.9 A second run with the only difference being the random seed

point $x_1 = -0.97$ mm from a U[–1 mm, 1 mm]. Then, we calculate the pseudo-random values for $y_{est}(\mathbf{x})$ and $y_{true}(\mathbf{x})$ and the error ε_p and the value $(\varepsilon_p)^2$. For this example, we have $\varepsilon_p = (-1.22) - (-0.44) = 0.78$ and $(\varepsilon_p)^2 = 0.61$. Figure 18.8 shows how Microsoft® Excel can be used to perform the simulation run.

To estimate the EIMSE with negligible errors using simulation, potentially thousands of simulations would be needed. Figure 18.9 shows a second simulation run in which the entire process is repeated using the next nine numbers from the pseudo-random numbers, i.e., starting with the last Q_i used as the final random seed. The resulting $\varepsilon_p = 0.16$. Thus, the $n = 2$ Monte Carlo estimate is the sample average 0.38 with estimated error stdev(0.61, 0.16)/sqrt(2) = 0.25.

18.6.3 The EIMSE Formula

In this section, the formula for the expected integrated mean squared error (EIMSE) criterion is described. As for the last section, the concepts are potentially relevant for predicting the errors of any "**empirical model**" in the context of a given input pattern or design of experiments (DOE) array. Also, this formula is useful for comparing response surface method (RSM) designs and generating them using optimization.

The parts of the name include the "**mean squared error**" which derives from the fact that empirical models generally predict "mean" or average response values. The term "integrated" was originally used by Box and Draper (1959) to refer to the fact that the experimenter is not interested in the prediction errors at one point and would rather take an expected value or integration of these areas of all prediction points of interest. The term "expected" was added by Allen et al. (2003) who derived the formula presented here. It was included to emphasize the additional expectation taken over the unknown true system model.

Important advantages of the EIMSE compared with many other RSM design criteria such as so-called "D-efficiency" include:

1. The sqrt(EIMSE) has the simple interpretation of being the expected plus or minus prediction errors.
2. The EIMSE criteria offers a more accurate evaluation of performance because it addresses contributions from both random errors and "**bias**" or model-misspecification, i.e., the fact that the fitted model form is limited in its ability to mimic the true input-output performance of the system being studied.

An advantage of the EIMSE compared with some other criteria is that it does not require simulation for its evaluation. The primary reason that simulation of the EIMSE was described in the last section was to clarify related concepts.

The following quantities are used in the derivation of the EIMSE formula:

1. $\mathbf{x}_p$ is the prediction point in the decision space where prediction is desired.
2. $\rho(\mathbf{x}_p)$ is the distribution of the prediction points.
3. R is the region of interest which describes the area in which $\rho(\mathbf{x}_p)$ is nonzero.
4. $\boldsymbol{\beta}_{true}$ is the vector of true model coefficients.
5. $\boldsymbol{\varepsilon}$ is a vector of random or repeatability errors.
6. σ is the standard deviation of the random or repeatability errors.
7. $y_{true}(\mathbf{x}_p, \boldsymbol{\beta}_{true})$ is the true average system response at the point $\mathbf{x}_p$.
8. $y_{est}(\mathbf{x}_p, \boldsymbol{\beta}_{true}, \boldsymbol{\varepsilon}, \text{DOE})$ is the predicted average from the empirical model.
9. $\mathbf{f}_1(\mathbf{x})$ is the model form to be fitted after the testing, e.g., a second order polynomial.
10. $\mathbf{f}_2(\mathbf{x})$ contains terms in the true model not in $\mathbf{f}_1(\mathbf{x})$, e.g., all third order terms.
11. $\boldsymbol{\beta}_1$ is a k_1 dimensional vector including the true coefficients corresponding to those terms in $\mathbf{f}_1(\mathbf{x})$ that the experimenter is planning to estimate.
12. $\boldsymbol{\beta}_2$ is a k_2 dimensional vector including the true coefficients corresponding to those terms in $\mathbf{f}_2(\mathbf{x})$ that the experimenter is hoping equal 0 but might not. These are the source of bias or model mis-specification related errors.

13. $\mathbf{X}_1$ is the design matrix made using $\mathbf{f}_1(\mathbf{x})$ and the DOE array.
14. $\mathbf{X}_2$ is the design matrix made using $\mathbf{f}_2(\mathbf{x})$ and the DOE array.
15. R is the "region of interest" or all points where prediction might be desired.
16. $\boldsymbol{\mu}_{11}$, $\boldsymbol{\mu}_{12}$, and $\boldsymbol{\mu}_{22}$ are "moment matrices" which depend only on the distribution of the prediction points and the model forms $\mathbf{f}_1(\mathbf{x})$ and $\mathbf{f}_2(\mathbf{x})$.
17. "E" indicates the statistical expectation operation which is here taken over a large number of random variables, $\mathbf{x}_p$, $\boldsymbol{\beta}_{true}$, $\boldsymbol{\varepsilon}$.
18. $\mathbf{X}_{N,1}$ is the design matrix made using $\mathbf{f}_1(\mathbf{x})$ and all the points in the candidate set.
19. $\mathbf{X}_{N,2}$ is the design matrix made using $\mathbf{f}_2(\mathbf{x})$ and all the points in the candidate set.

Example 18.6.3 Hydroforming Press Design

Question

A consultant is working with a manufacturer who wants to use design of experiments to tune its process settings. The consultant observes the press and finds that the thinnest point on the manufactured part is around 5.0 mm ± 2.0 mm. Also, the process engineer is curious about which pressure to use (2000–2300 psi), which radius is used on the design (3.0–6.0 mm), and which thickness of input tube is used (2–3 in.). The company is willing to do 20 or more test runs. Use this information to develop reasonable assumptions for σ and R and choices of $\mathbf{f}_1$ and k_1.

Answer

The assumptions $\sigma = 2.0$ mm and R is the cube defined by the ranges 2000–2300 psi, 3.0–6.0 mm, and 2–3 in. seem reasonable. With the goal of tuning, the choices $\mathbf{f}_1(\mathbf{x})' = [1\ x_1\ x_2\ x_3\ x_1^2\ x_2^2\ x_3^2\ x_1x_2\ x_1x_3\ x_2x_3]$ and, therefore, $k_1 = 10$ seem appropriate which can easily be estimated with $n = 20$ runs.

With these definitions, the general formula for the expected integrated mean squared error is:

$$\text{EIMSE(DOE)} = \underset{\mathbf{x}_p,\boldsymbol{\beta}_{true},\boldsymbol{\varepsilon}}{E} \{[y_{true}(\mathbf{x}_p,\boldsymbol{\beta}_{true}) - y_{est}(\mathbf{x}_p,\boldsymbol{\beta}_{true},\boldsymbol{\varepsilon},\text{DOE})]^2\}\ . \quad (18.22)$$

Note that this formula could conceivably apply to any type of empirical or fitted model, e.g., linear models, kriging models, or neural nets. This section focuses on linear models of the form

$$y_{true}(\mathbf{x}_p,\boldsymbol{\beta}_{true}) = \mathbf{f}_1(\mathbf{x})\boldsymbol{\beta}_1 + \mathbf{f}_2(\mathbf{x})\boldsymbol{\beta}_2. \quad (18.23)$$

For properly constructed design matrices $\mathbf{X}_1$ and $\mathbf{X}_2$ based on the DOE and model forms (see ▶ Chap. 13), the response vector, $\mathbf{Y}$, describing all n experiments is

$$\mathbf{Y} = \mathbf{X}_1\boldsymbol{\beta}_1 + \mathbf{X}_2\boldsymbol{\beta}_2 + \boldsymbol{\varepsilon}\ . \quad (18.24)$$

It is perhaps remarkable that, for linear models, the above assumptions imply:

$$\text{EIMSE(DOE)} = \sigma^2\text{Tr}[\mu_{11}(\mathbf{X}_1'\mathbf{X}_1)^{-1}] + \text{Tr}[\mathbf{B}_2\Delta] \quad (18.25)$$

where "Tr[ ]" is the trace operator, i.e., gives the sum of the diagonal elements, and

$$\mathbf{B}_2 = \underset{\boldsymbol{\beta}_2}{E}[\boldsymbol{\beta}_2\boldsymbol{\beta}_2'], \Delta = \mathbf{A}'\mu_{11}\mathbf{A} - \mu_{12}'\mathbf{A} + \mathbf{A}'\mu_{12} + \mu_{22}, \text{ and}$$
$$\mathbf{A} = (\mathbf{X}_1'\mathbf{X}_1)^{-1}\mathbf{X}_1'\mathbf{X}_2 \tag{18.26}$$

and

$$\mu_{ij} = \int_R \rho(\mathbf{x}_p)\mathbf{f}_i(\mathbf{x}_p)\mathbf{f}_j(\mathbf{x}_p)' d\mathbf{x}_p \text{ for } i = 1 \text{ or } 2 \text{ and } i \le j \ge 2. \tag{18.27}$$

Note that we have assumed that the random variables $\mathbf{x}_p$, $\boldsymbol{\beta}_{true}$, and $\boldsymbol{\varepsilon}$ are independently distributed. If this assumption is not believable, then the formulas might not give relevant estimates of the expected squared prediction errors. However, simulation based approaches similar to those described in the last section can be applied directly to the definition in Eq. (18.23). This was the approach taken in Allen et al. (2000) and Allen and Yu (2002).

Example 18.6.4 EIMSE Basics

Question 1

$\mathbf{f}_1(\mathbf{x})' = [1\ x_1\ x_1^2\ x_1^3]$ and $\rho(x_1) = 0.5$ for $-1 \le x_1 \le 1$. What is μ_{11}?

Answer 1

The results below follow from Eq. (18.27):

$$\mu_{11} = \int_R \rho(\mathbf{x}_p)\mathbf{f}_1(\mathbf{x}_p)\mathbf{f}_1(\mathbf{x}_p)' d\mathbf{x}_p$$
$$= \int_R 0.5 \begin{pmatrix} 1 & x_1 & x_1^2 & x_1^3 \\ x_1 & x_1^2 & x_1^3 & x_1^4 \\ x_1^2 & x_1^3 & x_1^4 & x_1^5 \\ x_1^3 & x_1^4 & x_1^5 & x_1^6 \end{pmatrix} dx_1 = \begin{pmatrix} 1 & 0 & \frac{1}{3} & 0 \\ 0 & \frac{1}{3} & 0 & \frac{1}{5} \\ \frac{1}{3} & 0 & \frac{1}{5} & 0 \\ 0 & \frac{1}{5} & 0 & \frac{1}{7} \end{pmatrix} \tag{18.28}$$

Question 2

Clearly, a large number of assumptions are needed to evaluate the EIMSE. How valuable can the formula-outputted numbers be?

Answer 2

The EIMSE is a rationalization in an important sense. Its value primarily relates to the criterion's use to identify undesirable input patterns and to compare different design of experiment arrays. For example, in ▶ Chap. 13, the sqrt(EIMSE) is used to compare standard response surface methods (RSM) designs. Also, a decision-maker can use the EIMSE to decide how many runs are needed for their array.

It is perhaps true that the EIMSE in Eq. (18.25) is of unavoidable importance in the theory of design of experiments (DOE). The phrase "**integrated variance**" refers to the first term in the EIMSE formula which is proportional to the random errors believed to be associated with the experimental system. If the system is perfectly repeatable (as in certain computer experiments), this term is zero. The phrase "**expected bias**" refers to the second term in the EIMSE formula, which is proportional to quantities associated with the magnitudes of the expected bias term. The equation reveals that the expected prediction errors do not depend on the unknown true coefficients $\boldsymbol{\beta}_1$ but only on the expected outer product represented by $\mathbf{B}_2$.

The importance of the EIMSE follows despite the challenges involved with its calculation. These challenges include developing reasonable assumptions and performing the needed calculations. The challenges associated with developing reasonable assumptions have caused many researchers to attempt clever ways to work around the problem, e.g., concentrating only on the expected bias or similar constructs (e.g., see Box and Draper 1987). However, it is not clear whether any alternative criterion can be substituted, and the computational challenge has been made easier by modern computers (Allen et al. 2003).

Related to developing the needed assumptions, challenges divide into assumptions about (1) $\mathbf{x}_p$, (2) $\boldsymbol{\beta}_2$, and (3) $\boldsymbol{\varepsilon}$:

1. Assumptions about the prediction point, $\mathbf{x}_p$, are needed to calculate the moment matrices $\boldsymbol{\mu}_{11}$, $\boldsymbol{\mu}_{12}$, and $\boldsymbol{\mu}_{22}$. The term "**candidate set**" refers to a very large design of experiments array with N runs that includes many if not all of the input combinations of potential interest in the design region. These points could be generated randomly according to the distribution function $\rho(\mathbf{x}_p)$. By default, the distribution of interest is uniform over the region of interest and the region is defined by the ranges or the factors in the input pattern or DOE array. For example, with $m = 3$ factors, the default region of interest would be a cube in the design space.
   The design matrices associated with the candidate points are written $\mathbf{X}_{N,1}$ and $\mathbf{X}_{N,2}$. If the candidate points are a random sample from the region of interest and N is large enough, then

$$\boldsymbol{\mu}_{ij} = (N^{-1}) \times \mathbf{X}_{N,i}\mathbf{X}_{N,j}' + \boldsymbol{\varepsilon}_{\mathrm{MC}} \text{ for } i = 1 \text{ or } 2 \text{ and } i \leq j \geq 2, \quad \textbf{(18.29)}$$

   which is established by the central limit theorem and $\boldsymbol{\varepsilon}_{MC}$ is the Monte Carlo error.

2. During the experimentation process, the experimenter will include only some terms in the fitted model and thus effectively assume that the other terms equal zero. Therefore, the experimenter hopes that the true system coefficients of the terms assumed to equal zero, $\boldsymbol{\beta}_2$, actually are equal zero. Yet, what to assume about these errors is unclear. Clearly, it is unwise to assume that all the $\boldsymbol{\beta}_2$ are all zero or $\mathbf{B}_2 = E[\boldsymbol{\beta}_2\boldsymbol{\beta}_2'] = \mathbf{0}$. This type of wishful thinking is embodied by criteria such as the integrated variance and D-efficiency. These criteria lead to optimistic views about the prediction accuracy and poor decision-making.
   Here, two kinds of assumptions about $\mathbf{B}_2$ are considered. The first is $\mathbf{B}_2 = \gamma^2 \times \mathbf{I}$, where γ is an adjustable parameter that permits studying of sensitivity of a DOE and model form to bias errors. For example, $\gamma = 0$ represents the assumption that $\boldsymbol{\beta}_2$ is zero and the EIMSE is the integrated variance. The second derives from DuMouchel and Jones (1994). It can be shown that assumptions in that paper imply

$$\mathbf{B}_2 = \gamma^2 \times \mathbf{C}^2, \tag{18.30}$$

where **C** is a diagonal matrix, i.e., the off-diagonal entries equal 0.0. The diagonal entries equal the ranges of the columns, i.e., Max[ ] – Min[ ], of the matrix a given by

$$\alpha = \mathbf{X}_{N,2} - \mathbf{X}_{N,1}(\mathbf{X}_{N,1}'\mathbf{X}_{N,1})^{-1}\mathbf{X}_{N,1}'\mathbf{X}_{N,2}. \tag{18.31}$$

The DuMouchel and Jones (1994) default assumption is $\gamma = 1$. Their choices are also considered the default here because they can be subjectively more reasonable for cases in which the region of interest has an usual shape. For example, in experiments involving mixture variables (see ► Chap. 15), certain factors might have much more narrow ranges than other factors. Then, the assumption $\mathbf{B}_2 = \gamma^2 \times \mathbf{I}$ could imply a belief that certain terms in $\mathbf{f}_2(\mathbf{x})\boldsymbol{\beta}_2$ have far more impact on errors than other terms. Fortunately, for many regions of interest, including many cases with cuboidal regions of interest, the two types of assumptions are equivalent.

3. The EIMSE formula above is based on the assumptions that the random errors in $\boldsymbol{\varepsilon}$ are independent of each other and have equal variance. Huang and Allen (2005) proposed a formula for cases in which these assumptions do not apply. Calculation of the EIMSE formula here requires an estimate of the standard deviation of the random errors, σ. This is the same sigma described in ► Chap. 4 which characteristics the common cause variability of the system.

Example 18.6.5 Single Factor Design Comparison

Question

Consider two experimental plans for a single variable problem: $\text{DOE}_1 = [-1\ 0\ 1]'$ and $\text{DOE}_2 = [-1\ 0.95\ 1]'$. Assume that $\gamma = 0.4$ and the fitted model form will be $\mathbf{f}(\mathbf{x})' = [1\ x_1]$. What type of model form is this? Use default assumptions to estimate the expected prediction errors associated with the process of experimenting with each experimental design, writing down the data, and fitting the model form $\mathbf{f}(\mathbf{x})$.

Answer

The fitted model is a first order polynomial. One assumes that $\sigma = 1$, $\gamma = 1$ (true response has a default level of bumpiness), $\mathbf{f}_2(\mathbf{x}) = [x_1^2]$, and the x_1 input is equally likely to be between –1 and 1 so $x_1 \sim U[-1, 1]$. First, using evenly spaced candidate points on the line [–1, 1], one derives $\mathbf{C} = 1$ and $\mathbf{B}_2 = \gamma^2\mathbf{C}^2 = 0.2$. The calculations are

$$\mathbf{X}_1 = \begin{bmatrix} 1 & -1 \\ 1 & 0 \\ 1 & 1 \end{bmatrix}, \mathbf{X}_2 = \begin{bmatrix} 1 \\ 0 \\ 1 \end{bmatrix}, \mathbf{A} = (\mathbf{X}_1'\mathbf{X}_1)^{-1}\mathbf{X}_1'\mathbf{X}_2 = \begin{bmatrix} 0.6667 \\ 0 \end{bmatrix},$$
$$\boldsymbol{\mu}_{11} = \int_{-1}^{1} \rho(x)\mathbf{f}_1(x)\mathbf{f}_1'(x)dx = \begin{bmatrix} 1 & 0 \\ 0 & 0.33 \end{bmatrix}, \tag{18.32}$$

$$\mu_{12} = \int_{-1}^{1} \rho(x)\mathbf{f}_1(x)\mathbf{f}_2{}'(x)dx = \begin{bmatrix} 0.33 \\ 0 \end{bmatrix}, \mu_{22} = \int_{-1}^{1} \rho(x)\mathbf{f}_2(x)\mathbf{f}_2{}'(x)dx = [0.2],$$

$$\Delta = \mathbf{A}'\mu_{11}\mathbf{A} - \mu_{12}'\mathbf{A} + \mathbf{A}'\mu_{12} + \mu_{22} = 0.2\text{, and}$$
$$\text{EIMSE}(\text{DOE}_1) = \sigma^2\text{Tr}[\mu_{11}(\mathbf{X}_1'\mathbf{X}_1)^{-1}] + \text{Tr}[\mathbf{B}_2\Delta] = 0.1667 + 0.032 = 0.2\,.$$

For DOE_2, the matrices μ_{11}, μ_{12}, and μ_{22} are the same and

$$\mathbf{X}_1 = \begin{bmatrix} 1 & -1 \\ 1 & 0.95 \\ 1 & 1 \end{bmatrix}, \mathbf{X}_2 = \begin{bmatrix} 1 \\ 0.9025 \\ 1 \end{bmatrix}, \mathbf{A} = \begin{bmatrix} 0.9750 \\ -0.0237 \end{bmatrix}, \tag{18.33}$$

$$\text{EIMSE}(\text{DOE}_2) = \sigma^2\text{Tr}[\mu_{11}(\mathbf{X}_1'\mathbf{X}_1)^{-1}] + \text{Tr}[\mathbf{B}_2\Delta] = 0.2438 + 0.1987 = 0.3.$$

The higher EIMSE for the DOE_2 correctly reflects the obvious fact that the second design is undesirable. Using DOE_2, experimenters can expect roughly 50% higher prediction squared errors. It could be said that DOE_2 causes higher errors.
It should be noted that the formula derivation originally required two steps. First, in general

$$\boldsymbol{\beta}_2'\Delta\boldsymbol{\beta}_2 = \text{Tr}[(\boldsymbol{\beta}_2\boldsymbol{\beta}_2'\Delta)], \tag{18.34}$$

which can be proven by writing out the terms on both sides of the equality and showing they are equal. Also, for constant matrix $\boldsymbol{\Delta}$ and matrix of random variables $(\boldsymbol{\beta}_2\boldsymbol{\beta}_2')$, it is generally true that:

$$E\{\text{Tr}[(\boldsymbol{\beta}_2\boldsymbol{\beta}_2')\Delta]\} = Tr[E[\boldsymbol{\beta}_2\boldsymbol{\beta}_2']\Delta]\} = Tr[\mathbf{B}_2\Delta]. \tag{18.35}$$

Finally, note that the above EIMSE criteria have limitations which offer opportunities for future research. These include that the criterion has not been usefully developed for sequential applications relative to linear models. After some data is available, it seems reasonable that this data could be useful for updating beliefs about the prediction errors after additional experimentation. In addition, efficient ways to minimize the EIMSE to generate optimal experimental designs have not been identified.

18.7 Chapter Summary

In this chapter, design of experiments (DOE) criteria are defined. These criteria provide information before experimentation begins about what can be expected afterwards. This information aids in the selection about which design of experiments array best fit the objectives of the experimenter.

Table 18.6 Design of experiments planning, analysis, and decision-making summary

Method	Plan experiment	Analyze/Fit	Decide/ Design
Two-sample t-testing	Two levels of a single factor performed in random order	T-testing method	Common sense approach
Standard screening	Regular fractional factorials and Plackett-Burman arrays	Lenth's method and main effects plots	Common sense approach
One-shot response surface methods	Central composite designs (CCDs), Box Behnken designs (BBDs), and EIMSE optimal designs	Linear regression, second order fitted model	Formal optimization
Robust design using profit maximization	Same as RSM	Same as RSM	Formal maximization of the expected profit

The evaluation of DOE criteria requires assumptions about what will happen after data is collected and, potentially, statistical simulation. Fortunately, for most DOE methods, what happens after experimentation is largely predictable prior to testing. Table 18.6 overviews planning, analysis, and decision-making associated with the methods described in this text.

Statistical simulation involves pseudo-random numbers and the central limit theorem. The uses of simulation to evaluate Type I and Type II error-rate criteria are described. Then, simulation is applied to estimated the expected squared prediction errors or, equivalently, the expected integrated means square error (EIMSE) criterion. The final section describes a formula that can more efficiently evaluate the EIMSE under specific assumptions and the details of its calculation.

18.8 Problems

1. Which of the following is correct and most complete?
   a. Random variables are numbers whose values are known at time of planning.
   b. Probabilities can generally be written as expected values of indicator functions.
   c. Probability theory and simulation can generate information of interest to people considering which methods or strategies to apply.
   d. mod(9,7) = 2.
   e. All of the above are correct.
   f. All of the above are correct except (a) and (e).

2. Which of the following is correct and most complete?
   a. Method evaluation criteria can include subjective judgments about ease of use.
   b. Many quantitative properties of methods including Type I error rates can be calculated using the assumption that there is a true average difference and simulation.
   c. The assumptions used to calculate Type I and Type II error rates are the same.
   d. All of the above are correct.
   e. All of the above are correct except (d) and (e).
3. Which of the following is correct and most complete?
   a. Recursive sequences of numbers can seem random to the untrained eye.
   b. The pseudo-random number generation procedure in the text involves generating two sequences, integer seeds and approximately continuous numbers.
   c. Slopes of the cumulative distribution functions are proportional to probability density functions.
   d. All of the above are correct.
   e. All of the above are correct except (a) and (d).
4. Which of the following is correct and most complete?
   a. If X is uniform[10, 12], then $\Pr\{X<11.5\}=0.25$ or 25%.
   b. If X is uniform[10, 12], then the inverse cumulative for X is $F^{-1}(u)=10+2u$.
   c. Generating random numbers rarely starts with generating U[0, 1] deviates.
   d. All of the above are correct.
   e. All of the above are correct except (a) and (d).
5. Which of the following is correct and most complete?
   a. If applied correctly, t-testing cannot result in undesirable declarations.
   b. If a hypothesis testing method derives appropriate results 95 times out of 100 simulated tests, then an MC estimate for the error probability is 0.05.
   c. If random variables (RVs) are combinations of RVs, the CLT does not apply.
   d. All of the above are correct.
   e. All of the above are correct except (a) and (d).
6. Which of the following is correct and most complete?
   a. Data from a thought experiment proves a factor affects system average outputs.
   b. If there is no true effect, finding significance in t-testing is a Type I error.
   c. Type II errors are only possible if there is no true difference.
   d. All of the above are correct.
   e. All of the above are correct except (a) and (d).
7. Which of the following is correct and most complete?
   a. Using the central limit theorem, one can estimate the error of estimates.
   b. Monte Carlo generally gives expected values with no errors.
   c. If four simulation runs give, 0, 1, 0, and 0. The $\sigma_{estimate}=0.5$.
   d. All of the above are correct.
   e. All of the above are correct except (a) and (d).
8. Which is correct and most complete?
   a. In calculating the EIMSE, one must know the true values of fitted coefficients.
   b. The true level of bumpiness likely affects derived prediction errors.
   c. In calculating the EIMSE, one assumes something about the random errors.
   d. All of the above are correct.
   e. All of the above are correct except (a) and (d).

9. Which is correct and most complete?
   a. The EIMSE cannot compare the expected accuracies associated with RSM designs prior to experimentation.
   b. The moment matrices ($\mu_{i,j}$) can only be calculated knowing DOE array.
   c. $\mathbf{X}_1$ and $\mathbf{X}_2$ could be the design matrices associated with $\mathbf{f}_1$ and $\mathbf{f}_2$ respectively.
   d. All of the above are correct.
   e. All of the above are correct except (a) and (d).
10. Which is correct and most complete?
   a. The EIMSE is always equal to or larger than the integrated variance.
   b. The bias depends on assumptions about the coefficients not in the fitted model.
   c. Moment matrices only depend on the models forms and assumptions about where prediction will be requested.
   d. All of the above are correct.
   e. All of the above are correct except (a) and (d).
11. Which is correct and most complete?
   a. In planning experiments, one generally does not know the true model.
   b. Without added information, one must assume $\mathrm{E}[\boldsymbol{\beta}_2\boldsymbol{\beta}_2'] = \mathbf{0}$.
   c. In general, $\mathbf{x}'\mathbf{A}\mathbf{x} = \mathrm{Tr}[\mathbf{A}]\ \mathbf{x}\mathbf{x}'$.
   d. All of the above are correct.
   e. All of the above are correct except (a) and (d).

12. Use the assumption, $\mathbf{B}_2 = (1.25)^2\,\mathbf{I}$ and standard assumptions to calculate the EIMSE for the following DOE array.

Run	A	B	C
1	-0.5	-1.0	-0.5
2	-0.5	0.5	-1.0
3	0.5	1.0	-0.5
4	-1.0	0.0	0.0
5	0.5	-1.0	0.5
6	0.0	0.0	0.0
7	1.0	0.0	0.0
8	-0.5	-0.5	1.0
9	-0.5	1.0	0.5
10	0.5	0.5	1.0
11	0.5	-0.5	-1.0

18

References

Allen TT, Bernshteyn M (2003) Supersaturated designs that maximize the probability of finding the active factors. Technometrics 45:1–8

Allen TT, Yu L (2002) Low cost response surface methods for and from simulation optimization. Qual Reliab Eng Int 18:5–17

References

Allen TT, Yu L, Bernshteyn M (2000) Low cost response surface methods applied to the design of plastic snap fits. Qual Eng 12:583–591

Allen TT, Yu L, Schmitz J (2003) The expected integrated mean squared error experimental design criterion applied to die casting machine design. J Roy Stat Soc Ser C: Appl Stat 52:1–15

Box GEP, Draper NR (1959) A basis for the selection of a response surface design. J Am Stat Assoc 54:622–654

Box GEP, Draper NR (1987) Empirical model-building and response surfaces. Wiley, New York

Deler B, Nelson BL (2001) Modeling and generating multivariate time series with arbitrary marginals using a vector autoregressive technique. In: Proceedings of the 2001 Winter Simulation Conference.

DuMouchel W, Jones B (1994) A simple bayesian modification of D-optimal designs to reduce dependence on a assumed model. Technometrics 36:37–47

Huang D, Allen T (2005) Design and analysis of variable fidelity experimentation applied to engine valve heat treatment process design. J Roy Stat Soc (Ser C) 54(2):1–21

Grimmet GR, Stirzaker DR (2001) Probability and random processes, 3rd edn. Oxford University Press, Oxford

Press WH, Flannery BP, Teukolsky SA, Vetterling WT (1993) Numerical recipes in C: the art of scientific computing, 2nd edn. Cambridge University Press, New York. ► www.nr.com

Simmons GF (1996) Calculus with analytic geometry, 2nd edn. McGraw Hill, New York

Optimization and Management

Optimization and Strategy

T. T. Allen, *Introduction to Engineering Statistics and Lean Six Sigma*,
https://doi.org/10.1007/978-1-4471-7420-2_19

19.1 Introduction

The selection of confirmed key system input (KIV) settings is the main outcome of a six sigma project. The term "**optimization problem**" refers to the selection of settings to derive to formally maximize or minimize a quantitative objective. ► Chapter 6 described how formal optimization methods are sometimes applied in the assumption phase of projects to develop recommended settings to be evaluated in the control or verify phases.

Even if the decision-making approach used in practice is informal, it still can be useful (particularly for theorists) to imagine a quantitative optimization problem underlying the associated project. This imagined optimization problem could conceivably offer the opportunity to quantitatively evaluate whether the project results were the best possible or the project could be viewed as a lost opportunity to push the system to its true potential. The phrase "**project decision problem**" refers to the optimization problem underlying a given six sigma project.

In this part of the book, "**strategy**" refers to decision-making about a project including the selection of methods to be used in the different phases. The strategic question of whether to use the six sigma method or "adopt" six sigma on a companywide basis is briefly discussed in ► Chap. 21, but the focus is on project decision problems. Therefore, strategy here is qualitatively different than design of systems that are not methods. A second optimization problem associated with six sigma projects involves the selections of techniques to derive most efficiently the solution of the underlying project decision problem. For example, in some cases benchmarking can almost immediately result in settings that push a system to its potential. Then, benchmarking could itself constitute a nearly optimal strategy because it aided in the achievement of desirable settings with low cost.

In this chapter, optimization problems and formal methods for solving them are described in greater detail. This discussion includes optimization problems taking into account uncertainty. For example, the robust design optimization described in ► Chap. 14, uncontrollable "noise" factors constitute random variables that can affect the consistency of system quality. Formal approaches to design strategy almost necessarily involve uncertainty. They also include but are not limited to optimal design of experiment. This chapter also includes so-called "black box simulation optimization" methods relevant for solving optimization under uncertainty problems.

"**Tolerance design**" refers to the selection of specifications for individual components using formal optimization. ► Chapter 20 describes the application of decision-making under uncertainty to tolerance design. ► Chapter 21 closes with a discussion of six sigma strategy focusing six sigma as an approach for optimization under uncertainty. Also, opportunities for future research are described.

19.2 Formal Optimization

As described in ► Chap. 6, formal optimization is associated with a process of precisely defining the elements of a decision problem into a "mathematical program" and using an automatic procedure to derive recommended settings. Let **x** refer to the *m* dimensional vector including all the KIV settings to be selected. Let $g(\mathbf{x})$ be the precisely defined objective to be maximized, and let the set of $\mathbf{x}$ values of interest in the decision space be $\boldsymbol{M}$, which is defined by q "**constraints**" that limit feasible or possible solutions.

It is standard to refer to $g(\mathbf{x})$ as the "**objective function**" which quantifies the decision-makers goals for the system being designed. In terms of these definitions, the general mathematical program can be defined as

$$\begin{aligned} &\text{Maximize: } g(\mathbf{x}) \\ &\text{Subject to: } \mathbf{x} \in M \end{aligned} \tag{19.1}$$

"**Operations researchers**" translate or "formulate" problems into forms identical or equivalent to Eq. (19.1). Operations researchers also develop automatic procedures to solve Eq. (19.1). As long as $\boldsymbol{M}$ is bounded, there is at least one solution, $\mathbf{x}_{\text{optimal}}$, to the above program for each function $g(\mathbf{x})$. In general, the objective function might constitute an accurate key output variable (KOV) prediction for the system and the $\mathbf{x}_{\text{optimal}}$ would then be the best possible key input variable (KIV) settings.

Consider the following single factor optimization example. A decision-maker has $\mathbf{x} = [x_1]$, where one is maximizing a quadratic polynomial, $g(x_1) = -11 + 12x_1 - 2x_1^2$. Also, imagine that one can only control x_1 over the range from $x_1 = 2$ to $x_1 = 5$. This defines the region $\boldsymbol{M}$. The associated mathematical program is

$$\begin{aligned} \text{Maximize: } g(x_1) &= -11 + 12x_1 - 2x_1^2 \\ x_1 &\in [2, 5] \end{aligned} \tag{19.2}$$

▫ Figure 19.1 plots $g(x_1)$ as a function of x_1. From this plot, it is obvious that the solution to Eq. (19.2) is $\mathbf{x}_{\text{optimal}} = [x_{1,\text{optimal}}] = 3$. The implied solution method can be called

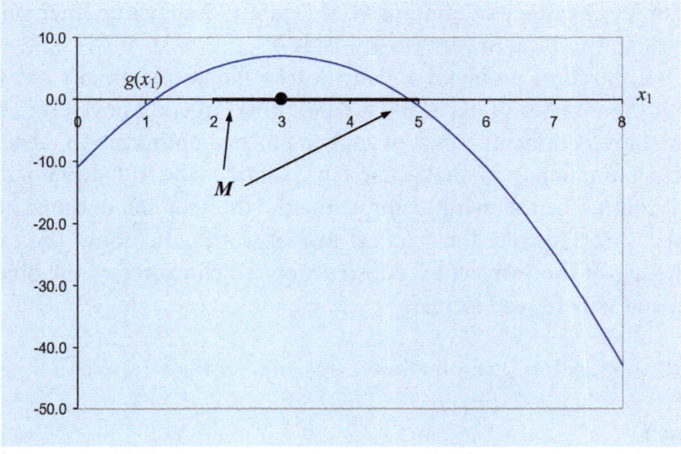

▫ **Fig. 19.1** Illustration of the optimization region, $\boldsymbol{M}$, and the solution to (2), $\mathbf{x}_{\text{optimal}} = 3$

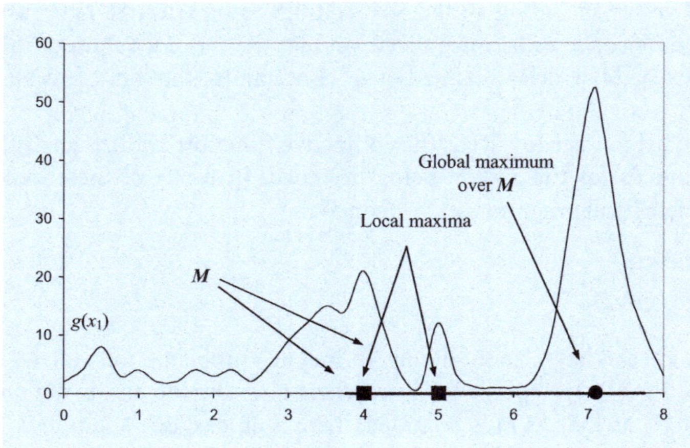

Fig. 19.2 A formulation with two local maxima and one global maximum

"complete enumeration" over a fine grid, i.e., inputting effectively all possible inputs and picking the solution with the highest value of $g(x_1)$. Then, the recommended system design is to set $\mathbf{x}_{optimal}$ to 3 for normal system optimization.

The numbers −11, 12, and −2 in Eq. (19.2) might have been derived from experimentation and regression, perhaps, but this first example is really a "toy" problem for the purposes of illustration. This problem is not representative of actual problems that decision-makers might encounter. This follows because (probably) most formal optimization problems of interest involve considerably larger decision spaces, $\boldsymbol{M}$, i.e., more decision variables and/or ranges that contain so-called **local maxima**.

A point is a local maximum if all neighboring points in $\boldsymbol{M}$ have lower objective values but there exists at least one solution, $\mathbf{x}_{optimal}$, in $\boldsymbol{M}$ that has a higher objective value. Solutions that have the highest objective values in $\boldsymbol{M}$ are called "**global maxima**" (or global optima in general) because their associated objective value is higher (or more extreme) than for any other solution in $\boldsymbol{M}$. Figure 19.2 shows another single variable problem in which two local maxima exist inside $\boldsymbol{M}$.

Solving optimization problems to find a true global maximum can be difficult, particularly if the number of decision variables m is large, e.g., over 100. However, in practice the relatively difficult aspect of applying formal optimization relates to obtaining an objective function, $g(\mathbf{x})$, that accurately quantifies the truly key input variable in a given application. The following example revisits the snap tab optimization problem from ▶ Chap. 17 to illustrate formulation in real-world situations. The example also illustrates the use of the "subject to" construction which expresses the objective function, $g(\mathbf{x})$, in an easier to read format.

Example 19.2.1 Snap Tab Optimization Problem Revisited

Question 1
Provide an example of formulation in a real-world problem.

Answer 2
In the snap tab case study, a large number of factors were considered and strategy limitations did not permit the creation of accurate models of the KOVs as a function of all KIVs. For this reason, cause and effect matrices were used to shorten then KIV list. Then, an innovative design of experiments (DOE) method was used to quantify input-output relationships.

Question 2
Rewrite the snap tab formulation from ▶ Chap. 17 into the form in Eq. (19.1).

Answer 2
The revised formulation follows.

Maximize: $g(\mathbf{x}) = y_{1,est}(x_1, x_2, x_3, x_4) - \infty \text{Maximum}[y_{2,est}(x_1, x_2, x_3, x_4) - 12.0, 0]$

Subject to:

$$
\begin{aligned}
&y_{1,est} - (72.06 + 8.98x_1 + 14.12x_2 + 13.41x_3 + 11.85\text{D} + 8.52x_1^2 \\
&\qquad - 6.16x_2^2 + 0.86x_3^2 + 3.93x_1x_2 - 0.44x_1x_3 - 0.76x_2x_3) = 0, \\
&y_{2,est} - (14.62 + 0.80\,x_1 + 1.50\,x_2 - 0.32\,x_3 - 3.68\,x_4 - 0.45\,x_1^2 \\
&\qquad -1.66\,x_3^2 + 7.89\,x_4^2 - 2.24\,x_1x_3 - 0.33\,x_1x_4 + 1.35\,x_3x_4) = 0, \text{ and} \\
&-1 \le x_1, x_2, x_3, x_4 \le 1.
\end{aligned}
\tag{19.3}
$$

In the real snap tab study, the Excel solver was used with multiple starting points to derive the recommended settings, $x_1 = 1.0$, $x_2 = 0.85$, $x_3 = 1.0$, and $x_4 = 0.33$. An exercise at the end of the chapter involves using the Excel solver to derive these settings by coding and solving Eq. (19.3). To solve this problem one needs to activate the "Solver" option under the "Tools" menu. It may be necessary to make the solver option available in Excel because it might not have been installed. Do this using the "Add-Ins" option, also under the "Tools" menu.

Example 19.2.2 Die Casting Machine Design

Question 1
If part distortion causes \$0.6M per year per mm in average distortion in rework costs and the current gate position is 9 mm, suppose any change cost \$0.2M. Formulate decision-making about gate position as an optimization problem. Suppose a die casting engineer has the following prediction model for average part distortion y (in mm), as a function of gate position, x_1 (in mm): $y(x_1) = 5.2 - 4.1x_1 + 1.5\,x_1^2$.

Answer 1
Calculating $y(x_1 = 9 \text{ mm}) \times (\$0.6\text{M}) = \$53.8\text{M}$, a relevant formulation is:

$$
\underset{x_1}{\text{Maximize}}\ g(x_1) = -\text{Minimum}\left[y(x_1)(\$0.6\text{M}) + \$0.2\text{M},\ \$53.8\text{M}\right] \tag{19.4}
$$

Question 2

Solve the problem in Question 1 and make recommendations.

Answer 2

Assuming the current setting is not optimal, $d/dx_1\ [g(x_1)] = 0 = 0.6[-4.1 + 2(1.5)\ x_{1,opt}]$, $x_{1,opt} = 1.36$ mm $\Rightarrow g(x_1) = \$1.6\text{M} < \$53.8\text{M}$, so the assumption is valid. Therefore, unless conditions other than gate position are more important, the casting engineer should seriously consider moving the gate position to 1.36 mm.

19.2.1 Heuristics and Rigorous Methods

Often, enumerating or testing all feasible solutions is not possible in a reasonable amount of time. Therefore, more careful study is needed to find a global maximum or even just obtain a solution of reasonable quality. Thousands of types of optimization problems have been identified and studied. Table 19.1 lists a small sampling of possible types of problems. For many problems of the last two types in the table, no procedures exist that can guarantee the attainment of global optimal solutions for large problems, e.g., $m > 1000$ factors or decision variables and $q > 1000$ constraints.

For added insight, consider classic optimization problems in which the objectives are linear and the constraints form a "**convex**" set, i.e., all between other feasible points are also feasible. If a linear program is being solved, a competent operations researcher should be able to propose a solution method that can guarantee the attainment of a global optimum in a time that might be considered reasonable.

"**Polynomial time**" refers to types of problems that, when the size parameters increase large, a global maximum can be guaranteed in computing times that increase relatively slowly compared with some polynomial in these size parameters. For example, times are always less than a polynomial function of m and q for some finite coefficients. See Papadimitriou (1994) for more information. The ability to generate a

Table 19.1 Overview of several types of optimization problems

$g(x)$	M	Size parameters	Name	Type
Linear is the x_i	Convex set	m and q	Linear program	Polynomial
Linear plus a term $\mathbf{x}'\mathbf{B}\mathbf{x}$ with positive semidefinite (PSD) $\mathbf{B}$	Convex set	m and q	Quadratic program	Polynomial
Linear plus a term $\mathbf{x}'\mathbf{B}\mathbf{x}$ with non-PSD $\mathbf{B}$	Convex set	m and q	Quadratic program	Non-polynomial
Nonlinear with integer constraints on $\mathbf{x}$	Nonconvex	m, q, objective function size	General integer program (IP)	Non-polynomial

global maxima efficiently cannot be guaranteed for some quadratic programs, many types of integer programs, and for perhaps most problems of interest. The class of other, more challenging problems is "**non-polynomial time**" or "**np-hard**" problems.

"**Heuristics**" are procedures that do not guarantee to find a global maxima in polynomial time. By contrast, "**rigorous algorithms**" are procedures associated with a mathematically proven claim about the objective values of the solutions produced in polynomial time. Sometimes, one also uses rigorous algorithms to refer to methods that eventually converge to a global maxima.

Generally, rigorous algorithms are only available for polynomial time optimization problems. Much, perhaps most, of the historical contributions to the study of operations research relates to exploiting the properties of specific formulations to produce methods that guarantee the attainment of global optimal solutions in reasonable time periods. Yet the properties of the objective function, $g(\mathbf{x})$, only permit operations researchers and computer scientist to apply heuristic solution methods. For example, the Excel solver has some difficulty solving Eq. (19.3) even though the optimization is over only four decision variables, because the quadratic program does not have a convex **B**.

In practice, most problems of interest cannot be solved in polynomial time. This is particularly true for those that relate to deriving optimal strategies including optimal design of experiment (DOE) arrays. However, a competent operations researcher always checks, in case a rigorous method could be applied and the attainment of a global maximum can be guaranteed in reasonable time.

Also, users of formal optimization must balance computational performance of the procedures (the time the computer takes to generate a solution) against guarantees of achieving optimal solutions and against the human time required to "code up" or acquire the software used for the automatic solution. Partly motivated by considerations of reducing the amount of human time required to solve optimization problems, interest in both academics and industry continues to grow in so-called "general-purpose" heuristics. These general-purpose heuristics include methods such as genetic algorithms (GAs), simulated annealing, and taboo searches. This chapter focuses on genetic algorithms because of their popularity and subjective elegance.

Conceivably, a general-purpose heuristic might even be used for a polynomial time problem. This could follow because it might require less human time. Also, in some cases, a general-purpose heuristic might even find an acceptable solution in a shorter computing time. For example, the simplex method is a nonpolynomial time algorithm used to solve convex linear programs even though polynomial time methods are available. It is used because of its high average efficiency and the fact that it does offer a rigorous conformation of optimality when it terminates.

Example 19.2.3 Generic Optimization

Question

Which is correct and most complete?

a. Linear programs with convex constraints are np-hard.
b. Facing an np hard problem often gives an excuse to use a heuristic like Gas.
c. Generating DOEs by maximizing a criterion is typically an np-hard problem.
d. All of the above are correct.
e. All of the above are correct except (a) and (d).

Answer

According to Table 19.1, linear programs with convex constraints are solvable in polynomial time (not np-hard). Facing an np-hard problem means that even a competent operations research cannot, in general, guarantee the achievement of a global maximum so that a heuristic can be a reasonable approach. Yes, generally strategy related optimization problems such as optimal DOE generation are np-hard. Therefore, the correct and most complete answer is (e).

19.3 Stochastic Optimization

An important special case of the general optimization program in Eq. (19.1) occurs if the function $g(\mathbf{x})$ is an expected value taken over some random variables, $\mathbf{Z}$. Problems of this type are called stochastic optimization problems. A general formulation of "**stochastic optimization**" problems can be written

$$\begin{aligned} &\text{Maximize: } g(\mathbf{x}) = \underset{\mathbf{Z}}{E}\left[g_2(\mathbf{x}, \mathbf{Z})\right] \\ &\text{Subject to: } \mathbf{x} \in \boldsymbol{M} \end{aligned} \tag{19.5}$$

and where $\mathbf{Z} = [Z_1, Z_2, \ldots, Z_q]'$ where the Z_i are random variables with known distribution functions.

The semantic distinction between problems of the form in Eq. (19.5) and those of the form in Eq. (19.1) is blurred by the realization that every problem of the form in Eq. (19.1) could include a term $+E[0]$ in the objective function and would therefore might be called a stochastic optimization problem. To create a practically useful distinction, therefore, the phrase "**stochastic optimization**" here refers to the study of problems in which the decision-maker believes that it is necessary to estimate the objective function, $g(\mathbf{x})$, using some form of numerical integration, e.g., Monte Carlo simulation. Therefore, a problem may be a stochastic optimization problem for one person. For another person who knows more about statistics and calculus, the problem might not be stochastic since numerical integration might not be needed.

For example, consider a simple version of the well-studied "**newsvendor**" problem. Assume that a newsvendor is deciding how many papers to purchase on a given day for resale, x_1. Suppose the purchase price to the newsvendor is \$0.20 and the selling price is \$0.50. Further, suppose that the newsvendor would like to entertain the assumption that the number of units that will be sold the next day, Z_1, is given by a normal distribution, $N(\mu = 100, \sigma = 25)$, rounding the sales numbers down to the nearest integer. Also, assume that the vendor must throw away any unsold items, losing all the money spent on them and that he or she is interested in maximizing the expected profit for the next day's sales.

The problem can be written

$$\begin{aligned}&\underset{Z_1}{\text{Maximize: }} g(x_1) = E\left[\$0.50\ \text{Minimum}\,(x_1, Z_1) - \$0.20x_1\right]\\&x_1 \in \{\text{whole numbers}\}\end{aligned} \quad (19.6)$$

where $Z_1 \sim N(\mu = 100, \sigma = 25)$. Therefore, the first term in the expectation is the revenue from sales and the second term is the upfront cost.

A person knowledgeable about statistics and calculus might recognize that the objective function, $g(x_1)$, in Eq. (19.6) can be expressed in terms of the mean value of a truncated normal distribution for which interpolation functions might be used. To that person, numerical integration would not be needed, probably permitting more efficient code to be developed for the problem solution. Then, he or she might not refer to Eq. (19.6) as a stochastic optimization problem. Instead, he or she might call this problem "**deterministic**" or not requiring numerical integration.

Still, many people might use Monte Carlo to estimate $g(x_1)$ in Eq. (19.6) in the context of their optimization method. For them, Eq. (19.6) would be a stochastic optimization problem. Whichever way one uses to solve Eq. (19.6), the solutions is $\mathbf{x}_{\text{optimal}} = 106$ newspapers. Also, it is acceptable for anyone to say that Eq. (19.6) is a stochastic optimization problem because of the ambiguity. Next, we define a general-purpose heuristic for solving stochastic optimization problems, which could be used to derive $\mathbf{x}_{\text{optimal}} = 106$ newspapers.

Example 19.3.1 Die Casting Gate Design

Question 1

If part distortion causes \$0.6M per year per mm in average distortion in rework costs and the current gate position is 9 mm. Suppose any change cost \$0.2M. Formulate decision-making about gate position as an optimization problem. Suppose a die casting engineer has the following prediction model for average part distortion, y (in mm), as a function of gate position, x_1 (in mm): $y(x_1) = 5.2 - 4.1x_1 + 1.5\,x_1^2$.

Answer 1

Calculating y(9 mm) × (\$0.6M) = \$53.8M, a relevant formulation is

$$\underset{x_1}{\text{Maximize:}}\ \ \text{g}(x_1) = -\text{Minimum}\left[y(x_1)(\$0.6\text{M}) + \$0.2\text{M},\ \$53.8\text{M}\right] \quad (19.7)$$

Question 2

Solve the problem in Question 1 and make recommendations.

Answer 2

Assuming the current setting is not optimal $d/dx_1\,[g(x_1)] = 0 = 0.6[-4.1 + 2(1.5)x_{1,\text{opt}}]$, $x_{1,\text{opt}} = 1.36$ mm ⇒ $g(x_1) = \$1.6\text{M} < \$53.8\text{M}$ so the assumption is valid. Therefore, unless conditions other than gate position are more important, the casting engineer should seriously consider moving the gate position to 1.36 mm.

The phrase "**simulation optimization**" refers to stochastic optimization in which Monte Carlo or a similar simulation approach is used for evaluating solutions inside the optimization procedures. If Monte Carlo is used, then t-tests and the Bonferroni inequality can be incorporated to make simultaneous solution comparisons and to ensure that solution quality is not degrading as the procedures progress. In general, the primary guarantees of solution quality derive from hypothesis testing type sub-procedures incorporated into the heuristics. Andradóttir (1998) and Swisher et al. (2000) survey many simulation optimization methods and discuss the related issues.

In general, the key difference related to solving simulation optimization and deterministic problems is that, in solving simulation optimization problems, the solution method must determine which solutions should be investigated *and* how thoroughly they should be evaluated. At each evaluation, the optimization method must determine the number of samples to be used which implies a Monte Carlo random error from a distribution with a certain standard deviation.

19.4 Genetic Algorithms

In this section, one of many possible general-purpose heuristics or "optimization solver" is described, a genetic algorithm (GA). This GA itself represents only one of many types of genetic algorithms. Further, other classes of general-purpose heuristics include simulated annealing and taboo searches. No claims are made about the GA provided here except that it is reasonably easy to code and understand.

Also, perhaps because it is easy to code GAs and adapt them to new problems, GAs in general constitute one of the most popular types of methods in operations research-related industries for solving optimization problems. The GAs presented in this chapter are potentially useful for stochastic optimization problems as well as other "deterministic" optimization problems.

19.4.1 Genetic Algorithms for Stochastic Optimization

Genetic algorithms were first proposed in Rechenberg (1964) and further developed in Holland (1975). De Jong (1975) proposed so-called "elitist" genetic algorithms for general-purpose simulation optimization. Elitist genetic algorithms copy a fraction of the solutions considered best from generation to generation. De Jong reasoned that with several solutions being copied instead of merely the best as in other types of GAs, the random error associated with simulation would be less likely to cause the best solution to be lost.

Aizawa and Wah (1994) introduced the concept of changing the numbers of samples dynamically in the context of using GAs for simulation optimization to reduce computational effort compared with fixed sample size approaches including those proposed by De Jong. However, the Bayesian sequential sampling-based procedures Aizawa and Wah proposed required significant problem specific knowledge and did not guarantee convergence to an acceptable solution with any specifiable probability. Chelouah and Siarry (2000) introduced the flexibility of GAs based on continuous numbers that we also include. More recently there has also been considerable interest in linking genetic algorithms with statistical selection and ranking methods, e.g., Bernshteyn (2001), Boesel (1999), and Yu (2000).

Our own most recent research, documented in Ittiwattana (2002), provides a procedure, based on elitist genetic algorithms, that guarantees long run convergence result for simulation optimization problems building on Rudolph (1996). The results prove long-run attainment of a solution within a value of a global optimum solution that can be pre-specified by the method user with a probability that can also be pre-specified. Therefore, one of the motivations for presenting the elitist genetic algorithm described next is that it is helpful for understanding our more computationally efficient, rigorous method.

19.4.2 Populations, Cross-over, and Mutation

Genetic algorithms are motivated by the process of natural evolution and the observation that the nature is successful in creating fit organisms adapted to their environments. GAs differ from the majority of optimization methods which iterate from a single solution to another single solution. Instead, GAs iterate between whole sets of solutions or "**populations**". The population being considered is called the current generation. Individuals in this population are called chromosomes and are each associated with one possible solution to the mathematical program. Iteration is based on the natural processes including probabilistic mating, crossover based reproduction, and mutation based on the fitness.

To understand how these natural processes relate to solving mathematical programs, it is helpful to understand first how an individual chromosome can be interpreted as a solution to a mathematical program. The chromosome is typically stored in coded form, e.g., as a vector of real numbers between 0 and 1. Each of these numbers is called a "gene" with reference to natural selection. This form must be decoded to be interpreted as a system design option, **x**. Often, the number of genes is the dimension of **x**. After decoding, the objective function can be calculated. If the feasible region is a hypercube, then this decoding could be as simple as a linear transformation. Once the associated solution x is decoded, the value $g(\mathbf{x})$ is calculated and called the "fitness" of the member of the population involved.

As an example of decoding, consider the deterministic optimization problem

$$\text{Maximize: } g(\mathbf{x}) = \text{determinant}\begin{bmatrix} x_1 & x_2 \\ x_3 & x_4 \end{bmatrix} \qquad (19.8)$$
$$2.0 \le x_1, x_2, x_3, x_4 \le 5.0$$

In ◘ Fig. 19.3, in part (a) an example chromosome is represented by a four-vector of numbers between 0 and 1. According to the specifics of a problem, this chromosome

(a)	(b)	(c)
	Rescale → (3.8 2.6 2.6 3.5)	$g(\mathbf{x}) = \det \begin{bmatrix} 3.8 & 2.6 \\ 2.6 & 3.5 \end{bmatrix}$
(0.6 0.2 0.2 0.5)	Regroup → $\begin{bmatrix} 3.8 & 2.6 \\ 2.6 & 3.5 \end{bmatrix}$	= 6.5 = the "fitness"

Fig. 19.3 **a** A sample chromosome, **b** the decoded solution, **c** the fitness

can be decoded into a matrix with elements from 2 to 5 by multiplying each number by 3 adding 2 to each number and, i.e., a linear transformation. The fitness can be evaluated as a function of the derived matrix.

19.4.3 An Elitist Genetic Algorithm with Immigration

The following method is proposed for student use. It is coded in the Appendix to this chapter, and it is called "toycoolga". The initial population of N chromosomes is generated using U[0, 1] pseudo-random numbers. Then, the method is iterated for a pre-specified number of generations. In each iteration, the chromosomes associated with the top e estimated fitness values are copied from the current generation to the next generation. If Monte Carlo (MC) is used for the fitness evaluation, then n simulations are used for all evaluations. The putatively top e solutions are called the elitist subset. The next c chromosomes in the new generation are created through so-called "**one-point crossovers**" that mix two solutions. The word "putatively" refers to the fact that we do not know for certain which solutions have the highest mean values because of MC errors.

In one-point crossovers, two chromosomes or "parents" are selected from the current generation. Each chromosome has an equal probability of being selected for parenting. Then, a random integer, I, between zero and the number of genes, m, is selected. The chromosome entered into the next population contains the same first I genes as the first parent and the remaining $m - I$ genes come from the second parent. The remaining $N - e - c$ chromosomes in the new generation are generated using U[0, 1] pseudo-random numbers. The term "**immigrants**" refers to these remaining chromosomes. They are generated using pseudo-random U[0, 1] random numbers.

Table 19.2 illustrates one iteration of the proposed genetic algorithm assuming $N=10$, $e=2$, and $c=6$. The fitness is calculated using the objective value in Eq. (19.6), which does not require Monte Carlo. It seems plausible that this procedure will continue to improve. The immigrants and the crossovers derived from them will cause "new blood" to continually enter the population, reducing the probability of all solutions becoming identical and associated with a sub-optimal objective value.

Table 19.2 **(a)** The population at generation t; (***b**) the population at generation $t+1$

(a)						(b)					
Chromosome					Fitness	Chromosome					Fitness
1	0.6	0.2	0.2	0.5	6.5	Elitist	0.7	0.2	0.0	0.4	7.9
2	0.0	0.5	0.9	0.0	−12.5	Elitist	0.9	0.1	0.9	0.6	7.1
3	0.9	0.1	0.9	0.6	7.1	Crossover 2 and 7	0.0	0.5	0.4	0.5	−4.2
4	0.5	0.4	0.1	0.3	2.8	Crossover 7 and 1	0.7	0.2	0.2	0.5	7.6
5	0.5	0.0	0.2	0.1	2.9	Crossover 10 and 10	0.5	0.3	0.6	0.9	5.4
6	0.1	0.2	0.3	0.0	−2.9	Crossover 10 and 2	0.5	0.5	0.9	0.0	−9.5
7	0.7	0.6	0.4	0.5	2.2	Crossover 5 and 7	0.5	0.0	0.2	0.5	7.1
8	0.7	0.2	0.0	0.4	7.9	Crossover 1 and 7	0.6	0.2	0.2	0.5	6.5
9	0.5	0.0	0.9	0.3	−2.1	Immigrant 9	0.1	0.2	0.3	0.0	−2.9
10	0.5	0.3	0.6	0.9	5.4	Immigrant 10	0.1	0.5	0.9	0.9	−5.9

19.4.4 Test Stochastic Optimization Problems

The first problem is originally from De Jong (1975), but also incorporates the modification of Aizawa and Wah (1994):

$$\underset{x}{\text{Maximize:}}\ g(\boldsymbol{x}) = -\underset{\omega}{E}\left(\sum_{i=1}^{30} i \cdot x_i^4 + 64\omega\right) \tag{19.9}$$

where $-1.28 \leq x_i \leq 1.28\ \forall\ i=1, \ldots, 30\ \omega \sim N(\mu=0, \sigma=1)$. This is the problem coded into the fitness function in the Appendix.

The second problem is from Mühlenbein et al. (1991)

$$\underset{x}{\text{Maximize:}}\ g(\boldsymbol{x}) = -\underset{\omega}{E}\left[(20 + \sum_{i=1}^{20}(x_i^2 - \cos(2\pi x_i)) + \omega)\right] \tag{19.10}$$

where $-5.12 \leq x_i \leq 5.12\ \forall\ i=1, \ldots, 20\ \omega \sim N(\mu=0, \sigma=35)$. These problems provide a nontrivial challenge to the proposed optimization method. Optimal design of experiments (DOE) problems such as the simulation optimization ones in Allen and Bernshteyn (2003) provide substantially more computationally intensive challenges.

19.5 Variants on the Proposed Methods

Several aspects of commonly used GAs are described in this section including variant rules for crossover and mutation. Commonly, the generations are of a constant size as in "toycoolga" in the Appendix. Also, very generally the iterative process of going from one generation to another is Markovian in the sense that the content of each generation depends explicitly only on the previous generation.

GAs are usually classified by a type of selection mechanism that is used for deciding which pairs of chromosomes should be chosen for crossing over. Common types include fitness proportionate, rank, tournament, and elitist selections. Assume that the maximization problem is being considered. Let N be the size of the generation and f_i the fitness value of the ith solution. Note that the elements that would most closely mirror natural processes would include fitness proportional selection, tournament selection, and an absence of elitist selection.

19.5.1 Fitness Proportionate Selection

Additionally assume the objective function is positive for all solutions. The probability p_i that the individual i will be selected for mating is defined as

$$p_i = [\sum_{i=1,\ldots,N} f_i]^{-1}(f_i) \tag{19.11}$$

This selection is vulnerable to converging to a local optimum since, once found, the local optimum will be assigned a high probability to be selected and its components will multiply in the next generation. If the search does not identify a better solution soon enough, this local optimum will fill up the whole generation. Once all solutions are alike, only mutation will be able to produce different solutions, thus deteriorating the efficiency of the algorithm.

19.5.2 Ranking Selection

In ranking selection, the f_i are ranked for $i=1, \ldots, N$. Then, the probabilities of selection are functions of the ranks only. This approach prevents a much better, but a local solution from being excessively successful in the selection process and eventually dominating the whole generations.

19.5.3 Tournament Selection

To select a solution for mating, a group of size $q \geq 2$ (tournament size) is drawn from the generation with replacement. The solution of this group with the highest objective function value passes to the mating. The process continues until enough crossed over solutions are produced.

19.5.4 Elitist Selection

Under previous schemes of selection not involving elitist subsets it is possible that the best solution will not be passed to the next generation. A popular method to ensure monotonicity of the objective value of the best candidate from generation to generation is to copy a subset of $e \geq 1$ of the putatively best solutions from generation to generation. Elitist selection, in general, is an additional selection feature that can be combined with any of the previously described.

19.5.5 Statistical Selection of the Elitist Subset

Ittiwattana (2002) describes the application of statistical selection and ranking methods to guarantee that the putatively optimal e have a bounded probability of containing a "good" solution in the context of stochastic optimization problems.

19.6 Appendix: C Code for "Toycoolga"

The program below is written in ANSI C. It is designed to maximize the toy problem given in Eq. (19.7). For this toy problem, one can easily evaluate the true fitness of each solution without noise for a check. In real problems, the true mean could probably only be estimated precisely through using an extremely large number of simulations, "n_evaluations".

```
#include <stdio.h>
#include <stdlib.h>
#include <math.h>

#define sizeOfGeneration 100
long noOfGenerations=100;
double eliteFraction=0.1;
/* fraction of chromosomes to be copied to the next generation */
double randomFraction=0.1;
/* how many chromosomes will be created randomly in each generation*/
long n_evaluations=100; /* Number of evaluations of the objective functions to do
for each solution */
unsigned long seed=1;
#define SIZE 30 /* The chromosome size, i.e., the number of decision variables. */

void evaluate(struct CANDIDATE *x, long noOfEvals,
int objective, int noise);
void GAsearch(struct CANDIDATE *result);
void generateCandidate(struct CANDIDATE *newborn);
double fitness(struct CANDIDATE *x,long noise);
long lmax(long a,long b);
int compare(const void *vp, const void *vq);
double ranS(unsigned long *iseed);
double gasdev(unsigned long *iseed);
long getRandomNumber(unsigned long *seed, long levels);

struct CANDIDATE {
        double fitness;
        double vector[SIZE];
        double stdev;
        long evals_used;};

/******** Here the program actually begins *********/
void main() {
        struct CANDIDATE solution,exactsolution;
        printf("GA starts ... \n");
        GAsearch(&solution);
        exactsolution=solution;
printf("The program is completed. You can see the results in outputfile.txt \n");}
/*************************************/

/**** This search performs a GA for minimization */
void GAsearch(struct CANDIDATE *result) {
```

```
        long gen_iter,i;
        long nelite,nrandom;
        int parent1,parent2,coord,cut;
struct                  CANDIDATE                  generation[sizeOfGeneration],
newgeneration[sizeOfGeneration], currentbest;
        FILE *output,*bestsolution;

        output=fopen("outputfile.txt","w");
        /* Convert the eliteFraction and randomFraction
from fractions to numbers */
        nelite=(long)(sizeOfGeneration*eliteFraction);
        nrandom=lmax(1,(long)(sizeOfGeneration*randomFraction));

        /***** Initialize the 0th generation ******/
        for (i=0;i<sizeOfGeneration;i++)
                generateCandidate(&generation[i]);
        /* The chromosomes have been filled now,
but the fitness still contains garbage */

        /**** Here is the big GA cycle ******/
for (gen_iter = 0; gen_iter < noOfGenerations; gen_iter++)
        {/* Going over generations starts now */
                /* Evaluate all fitnesses */
for (i=0;i<sizeOfGeneration;i++) evaluate(&generation[i],n_evaluations,0,0);

                /* Ranking the solutions (smallest first) */
                /* qsort is a standard "c" function.
                qsort(generation,sizeOfGeneration,
sizeof(struct CANDIDATE),compare);

                /* STEP I */
                /* Cloning (copying) of the top number of solutions
specified by nelite */
                for (i=0; i<nelite; i++) newgeneration[i]=generation[i];

                /* STEP II */
                /* Filling the new generation with crossovers */
                for (i=nelite;  i < (sizeOfGeneration-nrandom);  i++)
                {
                        /* Randomly picking parents */
        parent1=getRandomNumber(&seed,(long)(sizeOfGeneration/2));
                        /* picking 1st one from a better part */
```

```
                    parent2=getRandomNumber(&seed,sizeOfGeneration);
                    /* One-point crossover is implemented */
                    cut=1+getRandomNumber(&seed,SIZE-1); /* this
decides after which coordinate
                    to cut the solutions */
                    for (coord=0;coord<cut;coord++)

newgeneration[i].vector[coord]=generation[parent1].vector[coord];
                    for (coord=cut;coord<SIZE;coord++)
newgeneration[i].vector[coord]=generation[parent2].vector[coord];
              } //Crossovers are in.

              /* STEP III */
/* Filling the rest of the new generation with random solutions */
              for (i=(sizeOfGeneration-nrandom);i<sizeOfGeneration;i++)
                    generateCandidate(&newgeneration[i]);

/* The new generation is created. Make it a current generation now */
for (i=0;i<sizeOfGeneration;i++) generation[i]=newgeneration[i];

/* Printing into the file the top solution's actual value */
              currentbest=generation[0];
              evaluate(&currentbest,1,0,0);
/* Evaluate without the noise */
              printf("Generation N %ld  achieved %8.3lf\n",gen_iter,
currentbest.fitness);
              fprintf(output,"%6.2lf\n",currentbest.fitness);
        } /* Done going over generations */

        *result=currentbest;
        bestsolution=fopen("bestsolution.txt","w");
        int ind; for ( ind=0;ind<SIZE;ind++)
              fprintf(bestsolution,"%5.2lf ",currentbest.vector[ind]);
        fclose(output);
        fclose(bestsolution);}

/********* evaluate  *********/
/* noOfEvals - number of evaluations to be used for this candidate
noise = 1 the fitness includes noise, 0 - no noise
In most cases, there is noise and no way to turn it off.*/
void evaluate(struct CANDIDATE *x,long noOfEvals,int objective,
int noise) {
        long i; double sum; sum=0;
```

```
        if (noise==0) noOfEvals=1;
        if (objective==0)
        {       for (i=0;i<noOfEvals;i++) sum+=fitness(x,noise);
                sum=sum/noOfEvals;
                x->fitness=sum;
                x->evals_used=noOfEvals;        }
else {printf("Only one objective is defined at this point\n"); exit(1);} }

/****************/
void generateCandidate(struct CANDIDATE *newborn) {
        long i; for (i=0;i<SIZE;i++) newborn->vector[i]=ranS(&seed);
        newborn->fitness=666; /* just to set it to something */
        newborn->evals_used=-1; /* that's nothing was used */ }

/*********************************************/
long lmax(long a,long b) {return (a>b)?a:b;}

/**************************************/
int compare(const void *vp, const void *vq) {
        const struct CANDIDATE *p;
        const struct CANDIDATE *q;
        p=(struct CANDIDATE *) vp;
        q=(struct CANDIDATE *) vq;
        if ((*p).fitness<(*q).fitness) return -1;else
                if ((*p).fitness>(*q).fitness) return 1;
                else return 0;                  }

//What follows is a quick and dirty way to generate approximately
//uniform [0,1] random numbers.
double ranS(unsigned long *iseed) {
        *iseed = *iseed*1664525L + 1013904223L;
        return  *iseed/4294967296.0;}

//What follows is a quick and dirty way to generate approximately
//normally distributed numbers exploiting the central limit theorem.
double gasdev(unsigned long *iseed)
{       long i; double sum=0.0;
for(i=0;i<12;i++) sum+=ranS(iseed);
return (sum - 6.0);}

long getRandomNumber(unsigned long *seed,long levels) {
// Returns a uniform number from 0 to upBound [0,upBound-1]
        return (long)(ranS(seed)*(levels) ); }
```

```
/***************/
//The fitness function is the part that is tailored to each problem.
//The program will minimize the expected fitness value.
double fitness(struct CANDIDATE *x, long noise) {
        int j; double s,vect[SIZE];
        /* We'll have to decode the chromosome first */
        for (j=0;j<SIZE;j++) vect[j]=-1.28+(x->vector[j])*1.28*2;
        /* now calculate the fitness */
        s=0;
        for (j=0;j<SIZE;j++) s+=(j+1)*pow(vect[j],4.0);
        if (noise) s+=64*gasdev(&seed);
        return s; }
```

19.7 Chapter Summary

This chapter has defined general optimization concepts including the idea that two types of optimization problems are associated with each six sigma project. One of these problems relates to the most desirable selection of methods to efficiently derive desirable settings. The chapter focuses on "stochastic optimization" with the solution methods using Monte Carlo for evaluating and comparing solutions. A simple genetic algorithm is provided useful for a wide variety of stochastic optimization problems including many robust system design and strategy design problems.

19.8 Problems

1. Which is correct and most complete?
    a. Optimal strategy cannot involve method selection.
    b. Tolerance design cannot involve optimal selection of components.
    c. Constraints can be used to specify the feasible region, ***M***.
    d. All of the above are correct.
    e. All of the above are correct except (a) and (d).
2. Which is correct and most complete?
    a. Objective functions rarely (if ever) correspond to key output variables.
    b. Formulation is an activity that operations researchers often attempt.
    c. In a maximization problem, a local maximum objective value can be higher than global maxima objective value.
    d. All of the above are correct.
    e. All of the above are correct except (a) and (d).
3. Which is correct and most complete?
    a. Complete enumeration generally involves testing a small fraction of solutions.
    b. Facing an np hard problem often gives an excuse to use a heuristic like GAs.
    c. Many optimal DOE generation problems are np hard.
    d. All of the above are correct.
    e. All of the above are correct except (a) and (d).

4. Which is correct and most complete?
   a. Usual genetic algorithms maintain in memory only a single solution at a time.
   b. Elitist genetic algorithms involve copying no solution to the next generation.
   c. Immigration can be important to avoiding convergence to a local minimum.
   d. All of the above are correct.
   e. All of the above are correct except (a) and (d).
5. Which is correct and most complete?
   a. Quadratic programs with convex constraints might or might not be np hard.
   b. Facing a polynomial time problem, a competent researcher usually searches the literature and software for the appropriate polynomial time solution method.
   c. Relevant system design problems can be np-hard.
   d. All of the above are correct.
   e. All of the above are correct except (a) and (d).
6. Which is correct and most complete with reference to the problem in Eq. (19.7)?
   a. A global minimum for the problem has $x_i = 0$ for all i.
   b. The objective value in the problem can reach 2.0.
   c. GAs cannot be used for simulation optimization problems.
   d. All of the above are correct.
   e. All of the above are correct except (a) and (d).
7. Which is correct and most complete with reference to the problem in Eq. (19.8)?
   a. A global minimum problem has $x_i = 1$ for all i.
   b. The objective value in the problem can reach 25.0.
   c. Efficient solution methods can use variable sample sizes as they progress.
   d. All of the above are correct.
   e. All of the above are correct except (a) and (d).
8. Which is correct and most complete?
   a. Some GAs select fitter solutions for cross-over with higher probabilities.
   b. Elitist genetic algorithms involve copying no solution to the next generation.
   c. The algorithm in the appendix in this chapter is not an elitist GA.
   d. All of the above are correct.
   e. All of the above are correct except (a) and (d).
9. Suppose the constraint region in the optimization problem in Eq. (19.7) changed to $2.0 \leq x_i \leq 5.0$. How can the program in the appendix be adjusted to address the new problem?

References

Aizawa AN, Wah BW (1994) Scheduling of genetic algorithms in a noisy environment. Evol Comput 2:97–122

Allen TT, Bernshteyn M (2003) Supersaturated designs that maximize the probability of finding the active factors. Technometrics 45:1–8

Andradóttir S (1998) A review of simulation optimization techniques. In: Proceedings of the 1998 winter simulation conference, Washington, DC

Bernshteyn M (2001) Simulation optimization methods that combine multiple comparisons and genetic algorithms with applications in design for computer and supersaturated experiments. Ph.D. dissertation. The Ohio State University, Columbus, Ohio

Boesel J (1999) Search and selection for large-scale stochastic optimization. Ph.D. dissertation. Department of Industrial Engineering and Management Sciences, Northwestern University, Evanston, Illinois

Chelouah R, Siarry P (2000) A continuous genetic algorithm designed for the global optimization of multimodal functions. J Heuristics 6:191–213

De Jong KA (1975) An analysis of the behavior of a class of genetic adaptive systems. Ph.D. dissertation. The University of Michigan, Ann Arbor

Holland J (1975) Adaptation in natural and artificial systems. University of Michigan Press, Ann Arbor

Ittiwattana W (2002) A method for simulation optimization with applications in robust process design and locating supply chain operations. Ph.D. dissertation. The Ohio State University, Columbus, Ohio

Mühlenbein H, Schomish M, Born J (1991) The parallel genetic algorithm as function optimizer. In: Proceedings of the international conference on genetic algorithms, pp 271–278

Papadimitriou CH (1994) Computational complexity. Addison-Wesley, Boston, MA

Rechenberg I (1964) Kybernetische losungsansteuerung einer experimentellen forschungsaufgabe. Hermann-Fottinger-Institut fur Stromungstechnik der Technische Universitat, Berlin, Seminarvortrag

Rudolph G (1996) Convergence of evolutionary algorithms in general search spaces. In: Proceedings of the third IEEE conference on evolutionary computation. IEEE Press, Piscataway, NJ, pp 50–54

Swisher TR, Hyden PD, Jacobson SH, Schruben LW (2000) A survey of simulation optimization techniques and procedures. In: Joines JA, Barton RR, Kang K, Fishwick PA. Proceedings of the 2000 winter simulation conference, vol 1, pp 119–128

Yu L (2000) Expected modeling errors and low cost response surface methods. Ph.D. dissertation. The Ohio State University, Columbus, Ohio

Tolerance Design

T. T. Allen, *Introduction to Engineering Statistics and Lean Six Sigma*,
https://doi.org/10.1007/978-1-4471-7420-2_20

20.1 Introduction

"**Tolerance design**" refers to the selection of specifications for individual components using formal optimization. Specifications might relate to the acceptable length of a shaft, for example, or the acceptable resistance of a specific resistor in a printed circuit board. Choices about the specifications are important in part because conforming component parts can cause the entire engineered system to fail to conform to specifications. Also, sometimes the specification limits may be needlessly "tight" requiring expensive manufacturing equipment that does not benefit the customer.

"Statistical tolerancing" is the study of the properties of an ensemble of components using assumed properties of the individual components. Monte Carlo simulation from ► Chap. 18 is a powerful method in this study. Stochastic optimization from ► Chap. 19 can also be used to select the optimal combination of tolerances to achieve a variety of possible objectives.

"**Stackup analysis**" is statistical tolerancing when distances are associated with ensemble properties. Such analysis of specifications might involve all the complications of so-called geometric dimensioning and tolerancing (GD&T, e.g., see Krulikowski 1997). In some cases, Monte Carlo technology is built into the computer aided design (CAD) software for stackup analyses.

Example 20.1.1 Electronics Assembly Stackup

Question

Two resistors are in series and the resistance of each is assumed to be a normally distributed random variable with mean 10 Ω and standard deviation 0.5 Ω. What is the resistance distribution of the assembly for the two resistors in series and the chance that the entire component will conform to specifications, LSL = 17.1 Ω and USL = 22.1?

Answer

Monte Carlo simulation from ► Chap. 19 using Excel and Tools → Data Analysis → Random Number Generation derives that the series resistance has expected value equal to 20.0 Ω with standard deviation equal to 0.70. The chance of conformance can be similarly estimated to equal 0.997.

"**Optimal tolerancing**" involves using formal optimization to derive the nominal values and specification limits. The associated formulation may require the selection of specific processes which are associated with certain process capabilities, $6\sigma_0$, and thus feasible specification widths. Often, the specification limits derive from knowledge of what capabilities are feasible.

Example 20.1.2 Co-Packers' Problem

Question

A "co-packer" company inserts shampoo into bottles and sells the bottle to a well known brand for retail. The co-packer may have three potential equipment choices,

$i = 1, 2$, and 3. The equipment cost $c_1 = \$90K$, $c_2 = \$110K$, and $c_3 = \$115K$. The volumes of materials inserted are random variables whose distribution depends on the equipment choice and the nominal setting, μ. Past data confirm that the volumes are normally distributed to a good approximation and that the equipment is associated with standard deviations in ounces of $\sigma_1 = 0.35$, $\sigma_2 = 0.15$, and $\sigma_3 = 0.04$, respectively. Further, assume that the co-packer makes 10 million bottles a year with a material cost equal to \$0.01/ounce. Finally, assume that any units found below the lower specification limit of 16.0 ounce cost the company \$1 in penalty. Assume that USL − LSL = $10\sigma_i$ and USL + LSL = μ. By selecting USL and LSL, you are selecting specific equipment and the process mean. Which USL and LSL do you recommend?

Answer

One way to develop recommended "optimal tolerances" is to minimize the expected annual manufacturing cost using the following formulation in terms of the equipment choice, i, the nominal setting, μ, and the cumulative normal distribution function, Φ. This formulation assumes that the equipment fully depreciates in one year. It does not require stochastic optimization since Monte Carlo is not needed to evaluate the objective function.

Minimize: $c_i + \Phi(16.0, \mu, \sigma_i)(10{,}000{,}000)(\$1) + (10{,}000{,}000)(\$0.01)\mu$

$$\begin{aligned}&\text{By changing } i \text{ and } \mu\\ &\text{Subjectto}: c_1 = \$90\text{K}, c_2 = \$110\text{K}, \text{ and } c_3 = \$115\text{K}\\ &\qquad\qquad \sigma_1 = 0.35, \sigma_2 = 0.15, \text{ and } \sigma_3 = 0.04\end{aligned}$$

which has the solution $i = 3$ and $\mu = 16.15$. This can be found using a genetic algorithm or through brute force "enumeration" of alternatives using a spreadsheet. Therefore, the recommended USL = 16.15 + 5(0.04) = 16.35 and LSL = 16.15 − 5(0.04) = 15.95. Further, it is expected that the most expensive equipment is associated with a cost reduction of \$77K mainly in material cost savings.

20.2 Chapter Summary

This chapter has described an important type of simulation optimization application called statistical tolerance design. Statistical tolerance design helps engineers determine part specification limits that deliver desirable overall system performance. The phrase "co-packers problem" refers to decision-making on the part of companies charged with filling up bottles or glasses of processed good which can usefully be viewed as an application of tolerance design.

20.3 Problems

1. Consider two resistors in series each with resistance uniformly distributed with mean 10 Ω and uniform limits $a=8.5$ Ω and $b=11.5$ Ω. Which is correct and most complete?
   a. Pr{series resistance is between 19.0 and 21.0 Ω is} > 0.95.
   b. The series resistance is not uniformly distributed.
   c. The mean or average series resistance is 20.0 Ω .
   d. All of the above are correct.
   e. All of the above are correct except (a) and (d).
2. Consider three resistors in series each with resistance uniformly distributed with mean 10 Ω and uniform limits $a=8.5$ Ω and $b=11.5$ Ω. Which is correct and most complete?
   a. The series resistance is not uniformly distributed.
   b. The mean or average series resistance is 29.0 Ω.
   c. The chance that the series resistance is between 29.0 and 31.0 Ω is greater than 0.95.
   d. All of the above are correct.
   e. All of the above are correct except (c) and (d).
3. Resolve the co-packers's problem assuming $\sigma_1=0.4$, $\sigma_2=0.20$, and $\sigma_3=0.04$.

Reference

Krulikowski A (1997) Geometric dimensioning and tolerancing. Thomson Delmar Learning, Albany, NY

Design for Six Sigma

T. T. Allen, *Introduction to Engineering Statistics and Lean Six Sigma*,
https://doi.org/10.1007/978-1-4471-7420-2_21

21.1 Introduction

Design for six sigma (DFSS) methods can be viewed as part of six sigma or an alternative method as described in ▶ Chap. 1. These methods generally involve teams that have control over the design nominals and specifications. Having this "design control" often means that the teams have relatively great power to improve the system quality. It has been said that 80+% of product quality is determined by the product design specifications meaning that manufacturing and delivery can play only a relatively small role. Design teams, therefore, have great responsibility to ensure that new products or processes are put in place smoothly and foster high quality levels.

This chapter describes DFSS methods which are generally intended to offer deliberate and principled procedures that structure the processes of design. The examples here focus on DFSS in manufacturing contexts but DFSS methods have added value in other sectors including in government.

21.2 Design for Six Sigma (DFSS)

As described previously, six sigma is associated with an objective that relates to the idea that process variation causes the majority of quality problems (▶ Sect. 1.3.1). Specifically, it is assumed that the quality characteristic values of products produced in manufacturing vary with a characteristic standard deviation: "sigma" or σ_0. Sigma measures the standard deviation of quality characteristic values and the six sigma method attempts to reduce that standard deviation by removing its sources and/or making the process more robust. Therefore, six sigma focuses on making processes more consistent or "routinized" in the words of Welch and Welch (2005). Those authors also stated that, for the design of new products or radical re-designs, the role of six sigma is to deliver these products without a "hitch" or with minimal problems.

More generally, the phrase "design for six sigma" (DFSS) has been defined in at least two ways. First, DFSS has been defined as six sigma applied to the development and deployment of new systems or process which can be major re-designs of pre-existing systems or process. This definition, given in ▶ Sect. 1.3.1, makes it a special case of six sigma. Second, other authors have defined DFSS six sigma strategy which emphasizes the utilizations of today's most powerful tools for developing optimal designs (e.g., see El-Haik and Roy 2005). This emphasis can radically change the starting quality of the design when it enters the market place at the so-called "Job 1" time of introduction.

One other distinction that can be made is that design for six sigma (DFSS) is not so much a method as an approach or perspective. This view emphasizes the highly unpredictable nature of design processes and is reflected in the wide variety of phases described in the related literature. While there is a consensus (apparently) that DMAIC is appropriate for process improvements, ◘ Table 11.1 shows that many structures are being applied for design or radical re-design projects. Apparently also, DFSS has no universal roadmap and some have suggested that each company or major division should choose to implement their own version based on the needs of their domain and their internal culture (El-Haik and Roy 2005).

An important aspect of DFSS relates to the stage in the design process on which the method operates (El-Haik and Roy 2005). Essentially, DFSS is six sigma moved "upstream" to operations far in advance of manufacturing. The six sigma goal of routinizing relates most directly to manufacturing related quality characteristics and pre-existing processes. This means all systems are already in place and everything is already functioning. DFSS is "upstream" from six sigma's usual implementation area. As mentioned above, moving upstream gives six sigma potentially much more leverage because problems are identified sooner before they cause major expenses. One study in Allen et al. (2011) found that DFSS saved significantly more than other six sigma projects in the portfolio of projects at a Midwest manufacturer. The development cost could be lower, with a more robust product you have happier customers which could lead to a larger market share. For the same reasons this makes DFSS harder to quantify. This occurs because DFSS is used in the creation stage complicating comparisons.

The origins of DFSS can be traced back to the United States Department of Defense and the National Air and Space Agency (NASA). In their documentation, the common out line is as follows: create a charter, asses needs of customer, functional analysis, identify critical quality characteristics, concept selection, detailed design of product and creation processes, and control plans. Although each company may approach DFSS differently they all tend to use the same tool box including: benchmarking (▶ Chap. 3), Failure Modes and Effects Analysis (▶ Chap. 5), error proofing (▶ Chap. 5), Quality Function Deployment (QFD from ▶ Chap. 6), Design of Experiments (Part II), simulation, statistical optimization (▶ Chaps. 6 and 19), and Robust Design (▶ Chap. 15).

In general, the initial focus is on the creation of a clear list of the customer's needs and wants from the product to be used as system outputs or responses for optimization. This information generally drives the design creation. For example, a customer wants a fast, fuel efficient car. What does this really mean? They want a small 0–60 time? They want a great top speed? They want good city miles per gallon? They want good highway miles per gallon? They want it go around turns fast? DFSS helps participants in the design process address these ambiguities in a deliberate and principled manner.

As an example, consider that Samsung implemented DFSS in 2000, as a result they greatly increased sales and became a market leader (Park and Youngjoon 2006). In Samsung's implementation of DFSS they used a common DFSS tool called TRIZ (Theory of Inventive Problem Solving). TRIZ is an algorithmic method for inventing a new system (Oroloff 2006). TRIZ is based on the concept that inventions very generically spring from adapting technologies to meet the demands of new areas. The adaptations relate to new trade-offs in critical characteristics that are appropriate for the new circumstance. The generic case of competing output measure or responses is called a "contradiction" and finding a new balance is called "resolving" the contradiction. For example, to address a new market with cost conscious customers Toyota may want invent a car without air bags but with a slightly stronger frame. Then, Toyota would be positioning their car at a different balancing point in the cost, safety, and weight output space.

Because it is focused on the deployment of high quality new systems, DFSS emphasizes the designed system capabilities and performance measures or quality

Table 21.1 Six sigma versus DFSS

Six sigma versus DFSS	
Six sigma	**DFSS**
• DMAIC (▶ Sect. 1.3.1)	• IDOV (Identify, Design, Optimize, Validate) • DMEDI (Define, Measure, Explore, Develop, Implement) • DMADV (▶ Sect. 1.3.1) • DMADOV (Define, Measure, Analysis, Design, Optimize, Verify) • Etc.
Fixes pre-existing processes	Focus on up-front design to reduce variability and unpredictability
Reactive	Proactive
Easy to quantify	Difficult to qualify

characteristics. When applying DFSS, using a systematic, phased approach generally means that communication between team members is enhanced and Job 1 occurs at the scheduled time.

21.3 Implementation

In this section, an example of a DFSS procedure is defined. Somewhat arbitrarily we focus on Identify, Design, Optimize, and Validate (IDOV). This particular example is perhaps the most different from DMAIC of all those in Table 21.1 because it greatly deemphasizes the use of quantitative measurement early in the design cycle. It might be the most relevant, perhaps, for projects in which initial prototyping is difficult or impossible (Fig. 21.1).

Algorithm 21.1 IDOV Defined

Identify customer design needs, wants, and specifications
Design using systems engineering and operations research to design the process to create the product
Optimize improve initial designs for robustness
Validate test to be sure your process for creating the product is sigma level 4.5 or higher

The following phase descriptions paraphrase the work in El-Haik and Roy, IDOV.

Identify Phase: Create one or more ideas or design concepts. Once the ideas are created they should be clearly defined and documented. Also, ideas about performance measures should be documented. Related documentation might include items such as customers concerns for related products (e.g., see benchmarking in ▶ Chap. 3) and the voice of the customer. The entity being designed could be a business unit requiring a

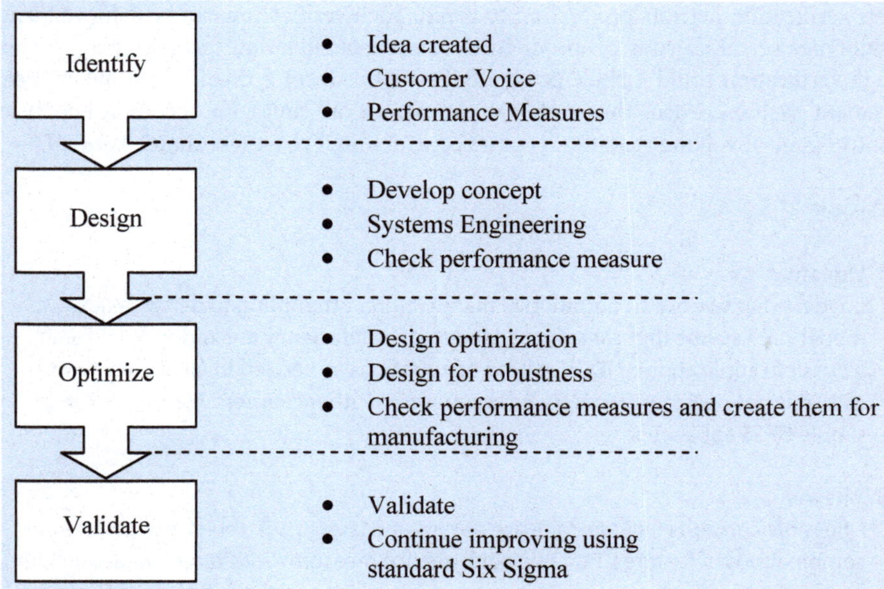

Fig. 21.1 IDOV roadmap example

business plans. Included in the list of performance measures are ways to quantify basic issues. The team should consider how the end result should look including questions like: What will it cost? How large is it? What does it do? Revelent techniques could include benchmarking, error proofing existing systems, and the development of measurement standard operating procedures.

Design Phase: The team should produce multiple solutions or design options, perhaps one designed to maximize each performance measure singly. Included in this documentation is generally diagrams and/or drawings of viable options. Next, the team should go through all the options and pick the ones that they are going to test. A potentially important aspect of this phase is the parameterization (▶ Chap. 12) of the design concepts to facilitate optimization. This involves defining factors or input variables whose settings define design spaces. Relevant techniques for this phase could include QFD and drafting of standard operating procedures (SOPs) described in ▶ Chap. 2.

Optimize Phase: The team needs to run real or virtual experiments to identify system settings fostering higher performance. It can sometimes be important to target settings that are robust to noise factor variation (▶ Sect. 14.1). Often, design of experiments methods from Part II are relevant in this phase. Alternatively, if the experiments are computationally inexpensive, the formal optimization methods from ▶ Chap. 20 might be relevant. In any case, such testing and optimization is often performed using computer experiments such as finite element analyses (▶ Chap. 17) or statistical simulation (▶ Chap. 19).

Validate Phase: In this phase, there is verification that all the performance measures meet targets and the customers and company needs (likely) will be satisfied.

This verification permits production to begin. Such verification can be achieved using prototypes or initial runs on production equipment. Relevant technologies for thorough verification could include gauge R&R methods from ▸ Chap. 4 to evaluate measurement systems. Also, thorough system evaluation could be accomplished using control plans involving statistical process control (SPC) charts (▸ Chaps. 4 and 7).

Example 21.2 IDOV

Question

Imagine that you are an automobile manufacturer attempting to design a new model car. Assume that car safety, speed and fuel efficiency are important. Sketch a possible application of IDOV stating assumptions as needed to fill in data. In this case, it is not necessary to apply techniques from other chapters. Merely, sketch a simple DFSS application.

Answer

I: Possible concepts include: mid-size sedans and large trunk vehicles with different combinations of features. Possible performance measures with target values include 30 mpg highway, 22 mpg city, ipod compatible radio, and the inclusion of features such as tilt and scoping steering wheels, comfortable seats, cup holders, power lock, power windows, curtain, side, front and knee airbags, rear safety best in class, 0–60 in under 6 s, have lost of reserve power that would not be used for everyday driving etc. Through benchmarking, the characteristics could be rated and correlated with design options. The business model might be to create a business unit that produces green, flex-fuel vehicles with reasonable power. This could constitute a new set of trade-offs for the company and thus a new resolution of the related design contractions.

D: Using formally designed meetings of various stakeholders (Chap. 3) consensus can be reached about what concepts on which to base initial drawings. The basic concepts can also be documented in augmented benchmarking matrices to study how customers might hypothetically respond to new designs in existing marketplaces. Parameterization could include the internal dimensions of the car related to the inclusion or exclusion of various features.

O: Virtual reality models of various concepts can be developed and testing both using finite element models to estimate performance values and with test subjects. Subjects can use goggles and other equipment to help imagine their experiences in the new car and rate their likely satisfaction level compared with alterative. Both screening and response surface experiments from Part II can be applied using various medium to tune the parameterized levels attempting to meet or exceed performance goals from the identify phase. Regression models of historical data from similar cars can be used to provide estimations that other models are unable to make. Also, historical data can provide calibration for the computer models using, e.g., comparison with a standard gauge R&R estimation (Chap. 4).

V: Once the virtual car meets system specifications, real prototypes can be developed and evaluated. These are generally based only on the one or two level combinations derived from the optimize phase. Short run production can often provide the information for control charts and the initial sigma levels. Such levels can be used to prioritize equipment improvement efforts and budget for rework and scrap during production. Finally, the car can be built and release for production (Job 1).

21.4 Problems

1. According to the chapter, which of the following is NOT commonly used in DFSS?
   a. IDOV.
   b. DMEDI.
   c. DMADV.
   d. DMAIC.
   e. DMADOV.
2. DFSS has been reported to be?
   a. A competing process with six sigma.
   b. A process within six sigma.
   c. A process that uses six sigma.
   d. Answers in part "a" and "c" are both correct.
   e. Answers in part "b" and "c" are both correct.
3. According to the chapter, why does DFSS more varied process phases than process improvement using six sigma and DMAIC?
   a. Different design domains are associated with differing types of contradictions and demands on design teams.
   b. New design projects generally involve cases in which initial quality measurements are NOT difficult.
   c. DFSS helps routinize pre-existing systems primarily.
   d. DFSS is primarily concerned with control planning which is creative.
   e. All answers are incorrect.
4. TRIZ is an algorithmic method with what primary objective?
   a. It helps select the phases of six sigma.
   b. It is primarily concerned with verification that designs conform to specifications.
   c. It helps deploy the correct shipments to your customers.
   d. TRIZ primarily helps the invention of new products using the concept of contradiction.
   e. Aids in benchmarking designs with alternatives.
5. What two United States of America agencies help to create DFSS?
   a. Federal Bureau of Investigation (FBI) and Central Intelligence Agency (CIA).
   b. Department of Homeland Security and Department of Agriculture.
   c. Department of Defense and Federal Bureau of Investigation (FBI).
   d. National Air and Space Agency (NASA) and Central Intelligence Agency (CIA).
   e. National Air and Space Agency (NASA) and Department of Defense (DOD).

6. A DFSS sigma level of 4.5 equates to (assume 1.5 sigma shift) approximately (under standard assumptions)?
   a. One defect per ten units.
   b. One defect per hundred units.
   c. One defect per thousand units.
   d. One defect per ten thousand units.
   e. One defect per million units.
7. Help General Motors come up an IDOV roadmap DFSS plan for new car called the Chevrolet Savor.
8. Do the same as number 7 except use DMADV roadmap for DFSS.
9. Compare your results from number 7 and number 8.

References

Allen TT, Brady JE, Schenk J (2011) Six sigma projects and personal experiences, In: A. Coskun (Ed.) Business management and economics 79–90

El-Haik BS, Roy DM (2005) Service design for six sigma: a roadmap for excellence. Wiley

Orloff MA (2006) Inventive thinking through TRIZ: a practical guide, 2nd ed. Springer Verlag, London

Park SG, Youngjoon (2006) How Samsung transformed its corporate R&D center. Res Technol Manag

Welch J, Welch S (2005) Winning. HarperBusiness, New York

Lean Sigma Project Design

T. T. Allen, *Introduction to Engineering Statistics and Lean Six Sigma*,
https://doi.org/10.1007/978-1-4471-7420-2_22

22.1 Introduction

The purposes of this chapter are: (1) to describe six sigma strategy and (2) to propose opportunities for additional research and evolution of six sigma. Part I of this book describes several methods that can structure activities within a project. Part II focuses on design of experiment (DOE) methods that can be used inside six sigma projects. DOE methods are complicated to the extent that decision-making about them might seem roughly comparable to decision-making about an entire project.

The extension of DOE theory from ▶ Chap. 18 and optimization methods from ▶ Chap. 19 to the design of projects constitutes perhaps the primary suggestion for future research. In ▶ Chap. 19, "strategy" is defined as decision-making about projects, focusing on the selection of methods to be used in the different phases. Brady (2005) proposed the following definitions:

Micro—dealing with individual statistical methods.
Meso—supervisor level decision-making about method selection and timing.
Macro—related to overall quality programs and stock performance.

Brady (2005) argued that the primary contributions associated with six sigma relate to the meso-level because the definition of six sigma relates to meso-level issues. These meso-level contributions and possible future meso-level analyses are explored in this chapter.

▶ Section 22.2 reviews the academic literature on six sigma based on Brady (2005). ▶ Section 22.3 explores the concept of "reverse engineering" six sigma, i.e., hypotheses about why six sigma works. The further hypothesis is suggested that understanding six sigma success is critical for additional contributions. ▶ Section 22.4 describes possible relationships of the sub-methods to decision problems that underly six sigma projects. ▶ Section 22.6 describes meso-level analysis of projects at one company reviewing results in Brady (2005). Section 22.7 concludes with suggestions for possible future research.

22.2 Literature Review

Brady (2005) analyzed over 200 articles from the academic literature. Trends identified include increasing levels of participation in six sigma-related research by academics, despite the fact that six sigma originated in industry among non-researchers. Brady divided the literature into those focusing on macro-, micro-, or meso-level issues.

First, Goh et al. (2003) and others analyzed the macro-level effects of the adoption of six sigma on corporate stock performance and found hints of short-lived benefits while their long term analysis was largely inconclusive. Their research, while important, investigated what might be considered a small sampling of relevant companies. Also on the macro-level, many articles include potentially helpful but ultimately subjective opinions about how to improve project return on investment through high level activities such as securing management commitment.

On the micro-level, a significant thread in the literature related to proposals for novel quantitative statistical methods having some relevance to six sigma.

For example, Ribardo and Allen (2003) proposed an optimization objective relevant to many formal optimizations in six sigma projects. Such micro-level research shares much in common with otherwise common statistical research about individual methods.

Related to meso-level, the most common type of article of any type was a case study describing an entire project. These articles included evidence that the project described achieved a good return on its associated investment. Also, much research related to how six sigma project leaders or "black belts" should be trained. Hoerl (2001) and the related discussion constitutes perhaps the most influential article on the topic of training. Yet, little quantitative research has been done in the literature despite the quantitative and evidence-based nature of six sigma, and many authors have suggested that such research might provide a rigorous foundation for six sigma and related instruction. Also, new valuable methods could be generated.

Example 22.2.1 Six Sigma Literature

Question

Which is correct and most complete?

a. Much literature has focused on what subjects should be taught to black belts.
b. The relationship between stock performance and six sigma adoption has been thoroughly studied.
c. A major theme in the literature focuses on subjective opinions about which strategies aid in achieving results.
d. Much literature has established a rigorous foundation for project design.
e. All of the above are correct.

Answer

Several articles in the highly regarded *Journal of Quality Technology* have focused on experts' opinions about what subjects are appropriate for inclusion in a curriculum. A series of articles has also addressed evidence that adoption helps stock performance, but results are somewhat inconclusive. Part (c) is also true. However, the selection of methods to apply in a project has received relatively little attention, in part because data relating to project actions and results has been largely unavailable. Also, modeling the performance of a series of statistical method applications is generally beyond the scope of statistical researchers.

22.3 Reverse Engineering Six Sigma

"**Reverse engineering**" can be defined as the study of successful products or services to understand how they achieve success. Montgomery (2001) and others have predicted that six sigma is likely to play an increasingly important role in business practices world wide for the foreseeable future. This section focuses on an admittedly subjective reverse engineering six sigma with the goal of clarifying its value proposition for industry to support understanding and, therefore, future research.

Six sigma projects are investments. Logically, for any project to achieve a desirable return on investment (ROI):

22

1. Some level of management support is needed, or else participants could not participate.
2. The project team involved must generate *desirable settings* for system input variables and then either implement (or have implemented) these settings.
3. Often, implementation is performed by others who must be convinced that the recommended settings result in a *safe* and profitable choices.

Table 22.1 revisits the definition of six sigma from ► Chap. 1 and suggests possible associated contributions of six sigma in the work place. The primary benefits of each aspect of the definition are considered only. For example, using an organized systematic problem-solving method might encourage management support or but that contributions is considered to be secondary.

Table 22.1 suggests that an associated, primary benefit of the entire six sigma method relates to slowing down team members and avoiding "**premature closure**" or termination of decision processes with associated key output variable (KOV) settings far from optimal. As indicated in Table 22.1, it is perhaps true that the primary benefit of the confirmation or verify phases relates to thorough confirmation. Through the use of the charting technology from ► Chap. 4 and the control planning technology from ► Chap. 8, settings are tested through at least 25 periods of sampling to make sure that performance is consistent and long lasting.

Table 22.1 Definition of six sigma related to requirements for return on investments

Definition or principle	Requirement		
	Management support	Desirable settings	Convincing evidence
Six sigma is an organized … problem-solving method … based on statistical methods …	–	No premature closure, based on data	Display of respect for the problem
Six sigma … includes as phases either … control … or … verify	–	–	Guarantees thorough confirmation
The six sigma method only fully commences … after establishing adequate monetary justification	Six sigma programs pay for themselves	Motivation of quantitative feedback	–
Practitioners applying six sigma can and should benefit … without … statistical experts	–	Motivation of reduced credit sharing	–

Many authors have commented that the attractiveness of six sigma to managers relates to its financial self-evaluation. It is suggested here, however, that the primary benefits of the two principles of six sigma relate to motivating project participants. First, quantitative financial evaluation of results offers those involved "**bragging rights**" about how much they are adding to their organizations. Further, by not necessarily involving outside experts, credit for successes do not need to be shared with non-engineers or non-scientists. The issue of "**credit assignment**" or sharing glory for successes seems to underly the success of six sigma. Credit assignment related benefits likely constitute the main reason why Welch and Welch (2005) wrote that the most "unheralded benefit of six sigma is its capacity to create a cadre of great leaders."

Yet the principle that statistics and optimization experts are not needed implies constraints on the methods used and substantial training costs. Clearly the methods developed by operations researchers and statisticians for use in six sigma projects must be "robust to user" to the extent that people with little training can benefit from them. Also, without statistical experts, many (if not all) personnel in companies must receive some level of training. Since many do not have the same level of technical knowledge as engineers or scientists, the technical level of training is generally lower than this book.

Example 22.3.1 Reverse Engineering Six Sigma

Question

Which is correct and most complete according to the text?

a. Scientists and engineers are generally eager to share credit with statisticians.
b. The only six sigma aspect strongly motivating management is cost justification.
c. Intense verification associated with control charting is often needed for acceptance of project recommendations.
d. The control or verify phases are not primarily related to improving settings.
e. All of the above are correct.
f. All of the above are correct except (a) and (e).

Answer

The text implies that engineers and scientists might generally sacrifice solution quality to avoid sharing credit with statisticians. Also, Table 22.1 implies that none of the other aspects of the six sigma definition provides important motivation for managers to adopt six sigma. The confirmation in the control or verify phases aids in acceptance but not improvement. Therefore, the most complete, correct answer is (f).

22.4 Uncovering and Solving Optimization Problems

In this section, assumptions are explored that could permit theoretical exploration of six sigma strategy. This theoretical exploration could permit comparison of strategies similar to the comparisons in ▶ Chaps. 13 and 18 showing that one response surface method design likely leads to greater prediction accuracy than another. In the broader six sigma context, such theory could indicate that one strategy likely fosters higher profits than another and/or provide a quantitative foundation for six sigma.

22

▶ Chapter 19 describes the assumption that a single optimization problem underlies each six sigma project. Under this assumption, specific strategies constitute heuristic approaches for uncovering and proposing a possible solution to the underlying problem. The following notation is helpful for the discussion:

1. $\mathbf{x}$ is a vector whose entries are supposed to be determined by the project team.
2. $\mathbf{x}_0$ contains the initial settings for $\mathbf{x}$, representing best guesses at the project start.
3. $\mathbf{x}_{op}$ is a vector whose entries are the settings recommended at the project end.
4. g quantifies the single true system objective of the project.
5. g_2 quantifies the single true system objective if uncertainty is removed.
6. g_3 quantifies the single true system objective accounting for project costs.
7. $\mathbf{Z}$ are uncontrollable factors causing variability during usual operations.
8. $\mathbf{Z}_2$ are uncontrollable factors causing method problems during the project.
9. M_1 are constraints on $\mathbf{x}$ known immediately by team members.
10. M_2 are constraints on $\mathbf{x}$ uncovered during the project.
11. M_3 are constraints on the strategy such as amount of time and total cost.

In terms of this notation, an optimization problem that might underly a six sigma project can be written

$$\begin{aligned} &\underset{\mathbf{x}}{\text{Maximize:}}\ g(\mathbf{x}) = \underset{\mathbf{Z}}{\mathrm{E}}\left[g_2(\mathbf{x},\mathbf{Z})\right] \\ &\text{Subject to:}\quad g_2(\mathbf{x},\mathbf{Z}) = y_{\text{est}}(\mathbf{x},\mathbf{Z}), \\ &\qquad\qquad\quad \mathbf{x} \in M_1,\ \text{and}\ \mathbf{x} \in M_2. \end{aligned} \tag{22.1}$$

Hazelrigg (1996) and others have remarked upon the fact that teams almost always fail to have a single quantifiable objective because of the mathematically provable irrational nature of groups. While this fact can have important consequences, the simplicity of a single objective suggests that it offers a natural starting point for quantitative analyses. For example, team members might agree that their primary goal is to increase expected profits while handling ethical and other considerations as constraints.

Note that many users of six sigma know little about formal optimization and would find Eq. (22.1) puzzling. This follows perhaps because the challenge in uncovering the objective function and constraint sets can be much more difficult than solving the problem once formulated. For example, once design of experiments has been applied, apparently desirable settings can come from inspection of main effects plots (▶ Chap. 12), 3D surface plots (▶ Chap. 13), or marginal plots (▶ Chap. 14). These practitioners are therefore attempting to solve an optimization problem without being aware of the associated formal vocabulary.

Table 22.2 reviews selected methods from previous chapters and their possible roles in solving a formulation of the form in Eq. (22.1). Some of the methods such as Pareto charting and cause and effects matrix construction result in subtle adjustments in the perceptions of objective or beliefs about which factors have no effects on the key output variables. Other methods such as regression, response surface methods, and formal optimization can play direct roles in exposing and solving the underlying formulation.

Table 22.2 Questions answered by selected methods and roles in optimization

Method	Question	Outcomes
Pareto charting	Which key output variables (KOVs) and subsystems should be focused on for improvement?	Qualitative information about $g(\mathbf{x})$ and $g_2(\mathbf{x})$
Xbar & R charting	What is the quality associated with current KIV level settings and is something unusual happening?	Quantitative information about $g(\mathbf{x}_0)$, $g(\mathbf{x}_{op})$, and $\mathbf{Z}$
Regression	Based on the data from the field, what key input variables (KOVs) likely affect the KOVs?	Potentially inaccurate quantitative knowledge of $g_2(\mathbf{x})$
Cause and effects matrices	Which factors likely do not affect the KOVs?	Potentially inaccurate quantitative knowledge of $g_2(\mathbf{x})$
Response surface methods	Provide quantitative predictions of KOV average values as a function of specific KIV level settings	Potentially relatively accurate quantitative knowledge of $g_2(\mathbf{x})$
Formal optimization	Given quantitative models, what settings of KIVs give the most desirable possible KOV values?	Potentially nearly optimal settings represented by $\mathbf{x}_{op}$

Researchers in optimal design of experiments (DOE) have long based their theories on the view that, once the DOE array or strategy is decided, many important consequences governing the prediction accuracy and decisions made later inevitably follow. This is the point of view underlying the evaluation of criteria described in ► Chap. 18. Extending such concepts to six sigma project and meso-analysis implies that certain strategies are likely to foster good outcomes in certain cases.

These considerations give rise to a second optimization problem focused on strategy and method selection

$$\begin{aligned} &\underset{\text{strategy}}{\text{Maximize:}} \quad E\{g_3[\mathbf{x}_{op}(\text{strategy}, \mathbf{Z}_2)]\} \\ &\mathbf{Z}_2 \\ &\text{strategy} \in M_3 \end{aligned} \tag{22.2}$$

In this formulation, possible strategies that could result might include choices to apply Pareto charting, cause and effect matrices, and statistical process control charting twice. These choices have implications for both the quality of the recommended settings that result, $\mathbf{x}_{op}$, and the cost of the project. Both types of considerations would, in general, be included in the function g_3 or reflected by the constraint set $\boldsymbol{M}_3$.

The next example describes the practically useful application of a formulation of the form in Eq. (22.2) to generate design of experiment (DOE) arrays. However, formulations are generally lacking from which to derive useful inferences about which combinations of distinct methods should be used in projects.

Example 22.4.1 Optimal Strategy Generation

22

Question

Provide one example of generating a strategy using optimization.

Answer

One strategic decision relates to which design of experiments (DOE) array should be used, i.e., the strategy. The EIMSE optimal DOE arrays in ▶ Chap. 13 were generated by minimizing the squared error objective (g_3). The EIMSE criterion taking into account unknown random errors, prediction points, and true model contained within by, $\mathbf{Z}_2$, in ▶ Chap. 18 using a genetic algorithm similar to the one described in ▶ Chap. 19. The act of generating the DOE constituted the use of formal optimization to create a method for project teams optimize their systems.

It is conceivable that the underlying system optimization in Eq. (22.1) might not correspond to a stochastic optimization problem as described in ▶ Chap. 18. However, it seems relatively likely that formulations of the type in Eq. (22.2) generally require stochastic optimization methods for their solution. This follows because most of the methods involve data collection and therefore uncertainty. Usually if the data could be accurately predicted in advance, it would not need to be collected. Also, data often enters in a non-linear way into decision-making.

Note also that Eq. (22.2) implies the "method-centered" view that, once the strategy determines which methods are to be used (their strategy), the project team essentially enters a relatively confining track. This "**method rollercoaster**" eventually terminates with recommendations, $\mathbf{x}_{op}$. Figure 22.1 illustrates a method rollercoaster which includes the strategy of creating a charter, using Pareto charting, fitting a preliminary regression model, and using a cause and effects matrix, response surface methods, and formal optimization. This specific strategy in this case results in recommended settings $\mathbf{x}_{op,1}$. Presumably if a different strategy were used or even the same strategy were repeated, different recommended setting would result. Therefore, the recommendations, $\mathbf{x}_{op}$, are a random variable, the expected value of its associated performance, $g(\mathbf{x}_{op})$, can be used to judge any given strategy.

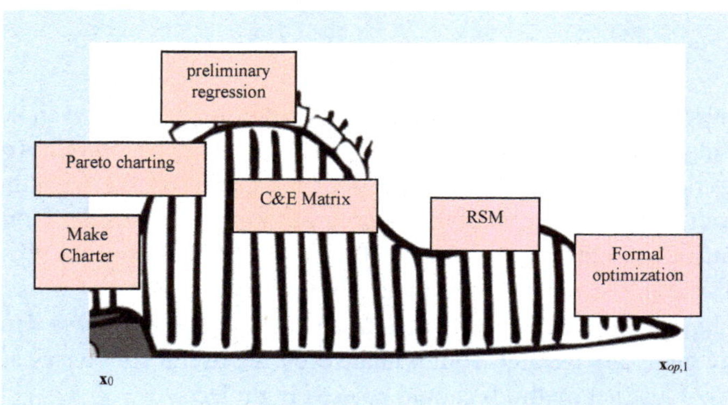

Fig. 22.1 "Method rollercoaster" is a strategy that starts with $\mathbf{x}_0$ and ends with $\mathbf{x}_{op,1}$

The method rollercoaster view of six sigma de-emphasizes the opportunistic nature of many projects. It also de-emphasizes the characteristics of individual participants and the role of science. As a result, the entire concept of "method rollercoaster" runs counter to both the six sigma and statistics cultures to a great extent. However, Fig. 22.1 and formulations similar to Eq. (22.2) are possibly unavoidable in the development of a scientific foundation for six sigma and new method development.

In fact, decision-making about strategy including method selection and the choice to perform a project has already been modeled usefully by several researchers. For example, Bisgaard and Freiesleben (2000) developed a quantitative method for determining whether to perform a six sigma project or not. Even though their approach ignores the skill of the team and many aspects of the problem under study, it is offers practically useful guidance in many cases.

Further, by studying actual historical case studies, Brady (2005) was able to quantify the effects of certain strategies on average profits, $g(\mathbf{x}_{op})$. For example, at a specific company the effects of training on profits were quantified. Also, Brady (2005) was able to demonstrate circumstances under which the use of design of experiments methods increased or decreased expected profits.

22.5 Future Research Opportunities

Six sigma was originally development by consultants, managers, and practitioners. Yet, it was built upon sub-methods such as Xbar & R charting (Shewhart 1931) and regular fractional factorials (Box and Hunter 1961) proposed by researchers. Therefore, it seems reasonable to expect researchers to continue to make useful contributions to six sigma and future methods of similar scope. This section describes several possible areas for future research.

Table 22.3 overviews an admittedly arbitrary sampling of proposed areas for future research. The first two rows represent continuing on-going threads of research likely to be received gratefully by practitioners in the short run. Continued quantitative research on the value of six sigma programs will likely be of interest to stock holders and management partly because past results are somewhat inconclusive.

Also, perceptions and performance continually change, and while over 40 papers have already published survey data, many questions remain largely unanswered. These include perceived needs for research and perceptions about effective training practices. Unfortunately, opinion based surveys might not produce results of long-run relevance.

22.5.1 New Methods from Stochastic Optimization

Research on new statistical micro-level methods for general uses can be highly valuable. Further, advances in computational speed and optimization heuristics provide unprecedented opportunities for new method development. Through applying optimization, it is possible that many if not all criteria used to evaluate methods can be used to generate new methods. For example, methods can be designed tailored to specific problems to maximize the chance of getting correct information or the expected profit

22

Table 22.3 Overview of possible topics for future research

Proposed area	Level	Possible outcomes
Quantitative analyses of management practices	Macro	Improved adoption and management guidelines
Opinion surveys	All	Improved understanding of needs for research and improvements
New statistics methods	Micro (new sub-methods)	User friendly software offering additional method options
Meso-analyses of project databases	Meso and macro	Improved training materials and strategies, expert system software
Test bed development for strategy evaluation	Meso	Criteria for theoretical evaluation of six sigma and other strategies
Optimal design of project strategies	Meso	Improved training materials and strategies, expert system software

from performing the study. However, studies of historical successes in micro-level method development such as Xbar & R charting from Shewhart (1931) suggest that practitioners may be slow to perceive value.

Six sigma explicitly creates different classes of practitioners interested in methods: green-belts, black belts, and master black belts. The reverse engineering of six sigma suggested that all of these classes are interested in using these practices without the aid of statistical methods. However, the requirement that methods be robust to the user does not mean that the methods cannot have complicated derivations and justifications. For example, many users of fractional factorials understand the process described in Box and Hunter (1961) for the derivation of the arrays used.

Possibilities for valuable new micro level-methods exist for cases in which current similar methods are not in widespread use, e.g., supersaturated designs from stochastic optimization in Allen and Bernshteyn (2003). Further, "data mining" or analysis of very large data sets using novel regression type or other methods continues to be an important area of investigation. Much of the related technology has not reached the level of maturity in which the methods are robust to user.

Also, new methods can potentially dominate many or all performance criteria relative to time-tested methods. For example, the EIMSE optimal design of experiments (DOE) arrays from Allen et al. (2003) offer methods with both fewer runs and lower expected prediction errors than either central composite or Box Behnken designs. Other possible areas under this category include new DOE arrays for problems involving categorical variables, custom arrays associated with reduced costs, and fully sequential response surface methods (RSM) deriving better results using reduced experimental costs. Also, multi-variate analysis and monitoring technologies tailored to specific problems can be developed.

In general, Bayesian decision analysis based methods (e.g., Degroot 1970) have not been fully exploited with respect to their abilities to generate practically useful methods. Also, it is possible that many (if not all) of the quantitative methods in this book could be improved in their ability to foster desirable outcomes and user robust methods using optimization and/or Bayesian analyses.

Example 22.5.1 New Micro-Level Methods

Question
According to the text, which is correct and most complete?

a. Robustness to user can imply that black belts and green belts can both benefit.
b. Practitioners have established a strong market for novel statistical methods.
c. There is no reason why researchers can improve on time-tested methods.
d. All of the above are correct.

Answer
Being conscious of user levels of knowledge is important, but it is possible to design robust methods that all can benefit by using, e.g., Xbar & R charts. In general, practitioners are not offering many grants for the development of new general purpose methods. Modern computers offer unprecedented opportunities for improved methods. Therefore, the answer is (a).

22.5.2 Meso-Analyses of Project Databases

With its emphases on monetary justification and documentation, six sigma has spawned the creation of many corporate databases containing the methods used in projects and the associated financial results. While the majority of these databases are confidential, practitioners at specific companies generally have access to their own organization's database. Brady (2005) proposed several approaches for analyzing such databases and the possible benefits of related research.

Results and benefits were illustrated using an example database describing 39 projects. A sampling of that project database is shown in Table 22.4. "#" refers to the project number. "M/I" indicates management (M) or individuals (I) identified the preliminary project charter. "A/P" indicates whether team members were assigned (A) or elected to participate (P). "#P" is the number of people on the team. "EC" is the number of economic analyses performed. The actual table in Brady (2005) also contained the number of SPC, DOE, and other quality methods used. Profits were estimated assuming a two-year payback period that addresses the fact that affected products typically have a two-year life cycle.

Brady (2005) showed how EWMA control charting (▸ Chap. 8) of the profits from individual six sigma projects can provide quantitative, statistical evidence of the monetary benefits associated with training programs. Also, the analysis indicated that design for six sigma (DFSS) can offer far higher profits than improvement projects. The control charting resulted in a point removed corresponding to the DFSS project found by charting to be not representative (following the approach in ▸ Chap. 4). The

Table 22.4 Sample from an open source six sigma project database

#	Expected savings	Expected time	M/I	A/P	#P	EC	…	Cost	Savings	Profit
1	$35,000	L	M	A	7	0	…	$48,700	$36,000	$−12,700
2	$70,000	L	M	A	1	1	…	$7,590	$0	$−7,590
3	$81,315	M	M	A	2	1	…	$35,300	$31,500	$−3,800
4	$40,000	M	M	A	1	0	…	$2,900	$0	$−2,900
5	$250,000	L	I	P	6	1	…	$325,500	$4E+06	$3,874,500
6	$150,000	L	M	P	4	0	…	$76,000	$170,000	$94,000
⋮	⋮	⋮	⋮	⋮	⋮	⋮	⋮	⋮	⋮	⋮

remaining data were analyzed using regression (▶ Chap. 15) and Markov Decision Processes (e.g., see Puterman 1994). Results included prescriptive recommendations about which sub-methods can be expected (at the related company) to achieve the highest profits in which situations.

Much of the information from the analyses in Brady (2005) based on the databases could be viewed as commonsensical. However, possible uses of information from such activities include feedback at the meso-level about needs for six sigma program management adjustments, improvements in training methods, and improvements to project budget planning. In general, modeling of effects of method applications on profits can provide many intangible benefits to statistical researchers including quantitative confirmation of specifics of the methods.

Additional research can consider larger databases and, possibly, achieve stronger inferences with larger scope than a single manufacturing unit. Also, a wider variety of possible analysis methods can be considered including neural nets, logistic regression, and many other techniques associated with data mining. Each of these methods might offer advantages in specific contexts and answer new types of questions. Overall, it seems that analyses of project databases is largely unexplored.

Example 22.5.2 Project Meso-Analyses

Question

According to the text, which is correct and most complete?

a. Confidentiality issues can offer a challenge for six sigma project analyses.
b. Meso-analyses could conclude that certain methods reduce average profits.
c. Some project profits might not be representative of usual operations.
d. All of the above are correct.
e. All of the above are correct except (a) and (d).

Answer

Yes, confidentiality makes many corporate databases unavailable. Also, hypothetically, results could conclude that a given set of quality systems are unable to apply certain methods with positive expected profits without changes. Finally, certain projects might not be representative. For example, project #6 in ◻ Table 21.4 corresponded to the only DFFS application because there was an unusual ability to change the design specifications. Therefore, the correct answer is (d).

22.5.3 Test Beds and Optimal Strategies

Quantitative evaluation of project strategies is possible similar to the evaluations of statistical quality control (SQC) and design of experiments methods in ▶ Chaps. 10 and 18. "**Test beds**" are problems with known solutions that can be used to compare methods. For example, a problem in a test bed could involve specific functions g and g_2 inserted into Eq. (21.2). Here, the methods could compare the hypothetical profit performance of alternative strategies on the same problem. Such test beds could facilitate training and provide the assumptions needed for the evaluation of criteria as described in ▶ Chap. 18.

One concern in the development of test beds relates to the realization that the underlying objective functions related to six sigma (g and g_2). For example, six sigma teams are rarely tasked with problems offering the opportunity for large improvements through the adjustment of a single factor. If such adjustments were possible, individuals associated with the system would likely have made them already. Special assumptions could be formulated to reflect this realization.

Finally, the same criteria and test beds associated with method evaluation and comparison can be used for generating optimal strategies. For example, Brady (2005) used an optimization method called stochastic dynamic programming to solve for the optimal sub-methods that maximize profits pertinent to many realistic situations. Yet the assumptions in Brady (2005) were not sufficiently realistic to provide valuable feedback to engineers. Much more research is possible to increase the scope and relevance of efforts to develop optimal improvement strategies.

22.6 Problems

1. Which is correct and most complete according to the text?
   a. Documenting a project case study is generally a micro-level contribution.
   b. Future opportunities for survey based contributions are generally not possible.
   c. Models predicting the value of specific strategies are entirely lacking.
   d. The subject of what should be taught to black belts has been addressed in the literature.
   e. All of the above are correct.
   f. All of the above are correct except (a) and (e).
2. Which is correct and most complete according to the text?
   a. Six sigma emphasizes credit-sharing between statistical experts and engineers.
   b. The control phase is important primarily because it motivates management.
   c. Encouraging adoption of project settings is critical for achieving project ROI.
   d. The quality of recommended settings is important but not required for ROI.
   e. All of the above are correct.
   f. All of the above are correct except (a) and (e).
3. Which is correct and most complete using the notation from the text?
   a. In general, $g(\mathbf{x})$ corresponds to the primary KOV.
   b. Control charting likely results in improved recommendations for settings, $\mathbf{x}_{op}$.
   c. Regression models cannot be used to specify constraint sets.
   d. Formal optimization is generally used by all teams in six sigma projects.
   e. All of the above are correct.
   f. All of the above are correct except (a) and (e).
4. Which is correct and most complete according to the text?
   a. The concept of a method rollercoaster emphasizes individual expertise levels.
   b. Optimization has not yet been used to produce methods or strategies.
   c. The nature of six sigma projects places no restriction on relevant test beds.
   d. Six sigma databases can help predict the costs in future projects.
   e. All of the above are correct.
   f. All of the above are correct except (a) and (e).

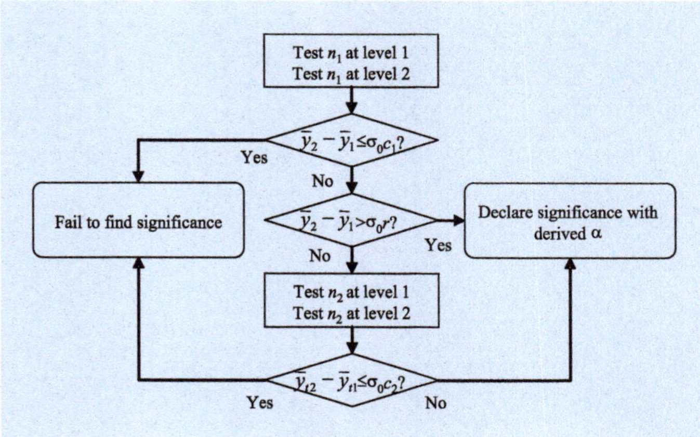

Fig. 22.2 Flowchart of proposed two-stage z-test procedure

5. Adapt the genetic algorithm in ► Chap. 19 to select n_1, n_2, c_1, r, and c_2 to minimize a weighted sum of the expected number of tests and the chance of errors for a t-testing procedure of the type shown in Fig. 22.2. Assume that σ_0 is known to equal 1.0. (Hypothetically, this procedure might be preferable to the certain well-known *z*-testing procedures.)

References

Allen TT, Bernshteyn M (2003) Supersaturated designs that maximize the probability of finding the active factors. Technometrics 45:1–8

Allen TT, Yu L, Schmitz J (2003) The expected integrated mean squared error experimental design criterion applied to die casting machine design. J R Stat Soc Ser C: Appl Stat 52:1–15

Bisgaard S, Freiesleben J (2000) Quality quandaries: economics of six sigma program. Qual Eng 13:325–331

Box GEP, Hunter JS (1961) The 2^{k-p} fractional factorial designs, part I. Technometrics 3:311–351

Brady JE (2005) Six sigma and the university: research, teaching, and meso-analysis. Ph.D dissertation, Industrial & Systems Engineering, The Ohio State University, Columbus

Brady JE, Allen TT (accepted) Six sigma: a literature review and suggestions for future research. In: Montgomery D (ed) Quality and reliability engineering international (special issue)

DeGroot MH (1970) Optimal statistical decisions. McGraw-Hill, New York

Goh TN, Low PC, Tsui KL, Xie M (2003) Impact of six sigma implementation on stock price performance. Total Qual Manag Bus Excell 14:753–763

Hoerl RW (2001) Six sigma black belts: what do they need to know? J Qual Technol 33:391–406

Hazelrigg G (1996) System engineering: an approach to information-based design. Prentice Hall, Upper Saddle River, NJ

Montgomery D (ed) (2001) Beyond six sigma. Qual Reliab Eng Int 17(4):iii–iv

Puterman ML (1994) Markov decision processes: discrete stochastic dynamic programming. Wiley, New York

Ribardo C, Allen TT (2003) An alternative desirability function for achieving six sigma quality. Qual Reliab Eng Int 19:1–14

Shewhart WA (1931) Economic quality control of manufactured product. ASQ, Milwaukee, WI (reprinted 1980)

Welch J, Welch S (2005) Winning. Harper Business, New York

Motivation and Change Management

T. T. Allen, *Introduction to Engineering Statistics and Lean Six Sigma*,
https://doi.org/10.1007/978-1-4471-7420-2_23

23.1 Introduction

The purposes of this chapter are (1) to describe some of the main contributions of Edwards Deming, (2) to clarify the motivating potential score and the implications for job design, (3) to explain the relevance of the field of human factors engineering, and (4) to offer brief highlights from the change management literature. Each of these subjects relates to topics on the ASQ Certified Quality Engineer (CQE) exam. Also, these topics are relevant to lean six sigma projects and general discussions of business management.

Note that Wikipedia is a helpful source on many of these issues. Some of the material here is paraphrased from Wikipedia pages.

23.2 Deming

In the 1980s in the United States, there was a major concern that Japan was emerging as the dominant economic power potentially unleashing massive unemployment in other countries. With this backdrop, a consultant named Edwards Deming, who had spent years in Japan absorbing and sharing ideas for improved productivity, offered important ideas that shaped the total quality management (TQM) movement.

Deming was responding to elements of U.S. business management that sought cost reductions at the expense of quality and efficiency at the expense of worker dignity. Several other key authors and consultants expounded related ideas. These included Crosby (1980) who highlighted the importance of viewing quality improvements as part of cost reduction activities. Also, Juran (1961), Juran and Godfrey (1999) had already assembled ideas based on Shewhart and others, which Deming largely championed and codified.

Deming's tone in his *Out of the Crisis* was partly alarmist, jobs are being lost, and partly commonsensical, e.g., slogans on the walls likely have minimal effects (Deming 1986). In this book, he identified his famous 14 points which are:

1. Create constancy of purpose toward improvement of product and service, with the aim to become competitive, to stay in business, and to provide jobs.
2. Adopt the new philosophy. We are in a new economic age. Western management must awaken to the challenge, must learn their responsibilities, and take on leadership for change.
3. Cease dependence on inspection to achieve quality. Eliminate the need for massive inspection by building quality into the product in the first place.
4. End the practice of awarding business on the basis of a price tag. Instead, minimize total cost. Move toward a single supplier for any one item, on a long-term relationship of loyalty and trust.
5. Improve constantly and forever the system of production and service, to improve quality and productivity, and thus constantly decrease costs.
6. Institute training on the job.
7. Institute leadership. The aim of supervision should be to help people and machines and gadgets do a better job. Supervision of management is in need of overhaul, as well as supervision of production workers.

8. Drive out fear, so that everyone may work effectively for the company.
9. Break down barriers between departments. People in research, design, sales, and production must work as a team, to foresee problems of production and usage that may be encountered with the product or service.
10. Eliminate slogans, exhortations, and targets for the workforce asking for zero defects and new levels of productivity. Such exhortations only create adversarial relationships, as the bulk of the causes of low quality and low productivity belong to the system and thus lie beyond the power of the workforce.
    a. Eliminate work standards (quotas) on the factory floor. Substitute with leadership.
    b. Eliminate management by objective. Eliminate management by numbers and numerical goals. Instead substitute with leadership.
11. Remove barriers that rob the hourly worker of his right to pride of workmanship. The responsibility of supervisors must be changed from sheer numbers to quality.
12. Remove barriers that rob people in management and in engineering of their right to pride of workmanship. This means, inter alia, abolishment of the annual or merit rating and of management by objectives.
13. Institute a vigorous program of education and self-improvement.
14. Put everybody in the company to work to accomplish the transformation. The transformation is everybody's job.

Clearly, many of Deming's points are controversial. For example, one still sees slogans relating to zero defects on walls in many profitable manufacturing operations (point 10). Also, listening to some instructors of lean six sigma and reading materials from several organizations, one sometimes feels that the intent is to "drive in fear" instead of out (point 8). Yet, Deming was surely responding to the reality we continue to observe in successful Japanese companies. Under Japanese management, workers are treated with respect and greater equality. As a result, workers under Japanese management generally feel more engaged in quality and system improvement.

Further, in adapting Shewhart's "do not react to noise" philosophy, Deming proposed that yearly personnel evaluations should be abolished. Some of his related philosophy dealt with the general realization that local authority should not be held accountable for causes beyond its control (Shewhart and Deming 1986). Deming estimated that, typically, 95% of the success associated with an individual is related to the environment of that individual. Therefore, evaluating the individual each year, with such a noisy system, is likely to be counterproductive and misleading.

Note that lean six sigma offers a partial solution to this situation. As noted by Jack Welch (Welch et al. 2005), lean six sigma offers the ability of individuals to be closely associated with major and measurable organizational gains. Therefore, lean six sigma has become an indispensable tool for human resource management and for identifying effective leaders. Without lean six sigma, the 95% rule of Deming perhaps applies. Yet, with lean six sigma, individuals have a greater chance, perhaps, to distinguish themselves legitimately.

23.3 Job Design: The Motivating Potential Score

Many organizations suffer high rates of absenteeism, turnover, and lack of worker engagement. These negative phenomena almost surely impact the quality of the products produced since unmotivate workers cause nonconformities through neglect or intention. Also, worker sabotage and exfiltration are serious concerns and some organizations have quality control and also quality defense (against worker) programs.

Some of the causes of poor motivation likely relate to the lack of respect employers display for their employees, which Deming observed. Yet, others have followed pioneering ideas of Herzberg (2017, first edition 1959) to create a practical system with which to score individual jobs for "motivating potential" and then to redesign the jobs to increase motivation.

Hachman and Oldham (1980) refined and extended Herzberg's two-factor model into a method that is still in common use. The refined system is based on the following definitions which relate to characteristics of specific jobs.

Skill Variety is the degree to which a job requires various activities, demanding that the worker develop a variety of skills and talents.

Task Identity is the degree to which the job requires jobholders to identify and complete a workpiece with a visible relevance to the final product or service.

Task Significance is the degree to which the job affects the lives of other people who are often the customer of the product or service.

Autonomy is the degree to which the job provides the employee with significant freedom, independence, and discretion to plan out the work and determine their job activities and the associated methods.

Feedback is the degree to which the worker has knowledge of results with actionable and objective information.

The derived system for job evaluation is based on worksheets and numbers. The Motivating Potential Score (MPS) is:

$$MPS = \frac{Skill\ \widehat{Variety} + Task\ \widehat{Identity} + Task\ \widehat{Significance}}{3} \times \widehat{Autonomy} \times \widehat{Feedback} \quad \text{(23.1)}$$

where the "^" indicates the average score value over the population. Typically, each issue is given three questions probing the related motivating potential on a discrete scale, e.g., on a given question, not-at-all motivating might be 1 point and extremely motiving might be 7 points.

The analyst typically uses a questionnaire, perhaps following materials from Hackman and Oldham (1980) to evaluate the perceptions of workers on the specific job. Hypothetically, the analyst could also observe the job and discuss the issues briefly with job holders. In either case, scores are entered and tabulated.

Table 23.1 Survey responses from a hypothetical manufacturing operation

Issue	Respondent 1	Respondent 2	Respondent 3	Average
Skill variety	2	1	2	1.67
Task identity	2	2	2	2.00
Task significance	2	1	1	1.33
Autonomy	1	2	1	1.33
Feedback	4	5	4	4.33

Table 23.2 Hypothetical data after a change to self-managing teams

Issue	Respondent 1	Respondent 2	Respondent 3	Average
Skill variety	2	1	2	1.67
Task identity	4	5	4	4.33
Task significance	4	4	4	4.00
Autonomy	7	5	5	5.67
Feedback	4	5	4	4.33
MPS				81.85

Example 24.1 Measuring and Increasing the MPS for Soup Manufacturing

Question

Consider the hypothetical data in Table 23.1 from a survey of workers in a manufacturing job with one question for each issue (instead of averaging over three). What is the MPS and what are the likely implications for the organization and how might the motivation be improved?

Answer

The MPS is $(1.67 + 2.00 + 1.33)/3 \times 1.33 \times 4.33 = 9.63$. This score is extremely low. The organization is likely to experience extremely high turnover with many hidden costs including quality losses. The organization might reorganize the work cell into a self-managing team. Then, the scores might likely become something like the hypothetical data in Table 23.2.

23.4 Elementary Human Factors Engineering

Human Factors is a major area of study which extends psychology ideas to engineering applications. The related research and practice may be viewed as an extension of the original work on "scientific management" by Taylor (1914). The subject can be divided into physical ergonomics, which focuses on avoiding physical injury, and cognitive engineering, which focuses on improving performance in general and avoiding errors.

There are many methods in cognitive engineering, e.g., see Woods et al. (2017). Here, only a selected sample of methods is briefly reviewed in part to prepare for the ASQ CQE exam. In general, a major theme in error analysis of human factors engineering is essentially the same as Deming's view that individuals have limited impacts on outcomes. Instead, the system or environment effects dominate.

Cognitive walkthrough is a usability inspection method in which a user point of view is employed to create task scenarios and identify design problems. Compared to other techniques considered here, this is a relatively theoretical design activity.

Ethnographic analysis involves observing the uses of technology in a real-world environment. It is a qualitative and observational method which involves taking extensive notes and then abstracting the patterns to support design decision-making.

Kansei method transforms consumer's responses, perhaps from surveys or interviews, to new products nominal values and specification limits.

Task analysis involves quantitatively describing human interaction, using a system or process to understand how to match the demands of the system or process to capabilities.

Usability analysis is a well-defined process for evaluating software and/or methods using cases and checklists of potential issues. Attributes of the users are considered systematically and testing, sometimes involving users, can identify likely issues early. Then, software and methods can hopefully be redesigned quickly so as to avoid losses. Unusable software and methods can cause quality issues, major frustration, and even cost lives. Good usability analysis can help the system avoid unintended consequences or side effects.

Work sampling is a method in which the job is tested at random intervals to determine the proportion of total time spent on a particular task. It provides insight into how often workers are performing tasks that might cause strain on their bodies.

23.5 Change Management

Lean six sigma methods generally have the scope of small to medium projects. Total quality management (TQM) from Deming and other management methods have a larger scope. As scope increases, one expects that the associated organization changes have more profound effects and can meet with more resistance. A major literature has developed dealing with preparing organizations for changes and making sure that the changes are enduring and positive.

In his book, Kotter (2008) offered a framework for managing effective organizational change. This involves an eight-step process:

1. *Create a Sense of Urgency*—Try to share the reasons that changes are needed widely.
2. *Build a Guiding Coalition*—Involve representatives of people who will be affected.
3. *Form a Strategic Vision and Initiatives*—Challenge the group to articulate the vision and steps needed.

4. *Enlist a Volunteer Army*—Once the changes are clear, then deploy personnel to follow the plan.
5. *Enable Action by Removing Barriers*—Management plays a key role in enabling change.
6. *Generate Short-Term Wins*—Instead of focusing only on the big picture, assure that positive events occur to build consensus that the change is positive and the movement is effective.
7. *Sustain Acceleration*—Management and others need to maintain a feeling of urgency.
8. *Institute Change*—When the vision and plans come into effect, the changes can become institutionalized in Standard Operating Procedures or design guides.

23.6 Problems

1. Relating to Deming, Crosby, and Juran.
   a. Deming major works came first with Juran and Crosby following.
   b. The three generally disagreed with Shewhart and his legacy.
   c. Deming argued that lean six sigma was critical for good evaluation.
   d. Deming, Crosby, and Juran generally supported Shewhart's legacy.
   e. Welch agreed with Juran that annual evaluations are not helpful.
   f. Welch took major issue with Shewhart's argument that noise is critical.
2. Which is correct and most complete according to the text?
   a. Deming argued that annual evaluations are critical for accountability.
   b. Deming argued that annual evaluations are counterproductive because 95% of outcomes are beyond the control of the individual.
   c. Crosby and Juran both advocated for the importance of quality.
   d. Welch argued that lean six sigma is critical for human resources.
   e. All of the above are correct.
   f. All of the above are correct except (a) and (e).
3. Which is correct and most complete using the notation from the text?
   a. In the MPS formula, task variety and autonomy have equal importance.
   b. In the MPS formula, feedback and autonomy have equal importance.
   c. In the Hackman and Oldham methods, surveys are often used.
   d. In MPS theory, increasing feedback is expected to increase motivation.
   e. All of the above are correct.
   f. All of the above are correct except (a) and (e).
4. Which is correct and most complete using the notation from the text?
   a. In ▶ Sect. 23.2, the example involves decreasing MPS from autonomy.
   b. In the MPS formula, feedback and task identity have equal importance.
   c. In the Hackman and Oldham methods, results are not quantitative.
   d. In ▶ Sect. 23.2, the example involves increasing MSP with autonomy.
   e. All of the above are correct.
   f. All of the above are correct except (a) and (e).

5. Which is correct and most complete according to the text?
   a. Usability engineering relates to evaluating systems to avoid problems.
   b. Kansei relates to translating user needs into specifications.
   c. Ethnography involves studying systems and abstracting design needs.
   d. Cognitive walkthroughs can involve analyses to improve design.
   e. All of the above are correct.
   f. All of the above are correct except (a) and (e).
6. Study an actual job (from internships, relatives, or local retailers) and estimate the motivating potential score. What steps does the MPS theory suggest might make the job more interesting and motivating?

References

Crosby PB (1980) Quality is free: the art of making quality certain. Signet

Deming WE (1986) Out of the crisis. MIT press

Hackman JR, Oldham GR (1980) Work redesign, vol 72. Reading, Addison-Wesley

Herzberg F (2017) Motivation to work. Routledge

Juran JM (1961) Japan revisited. Industrial quality control 17(9):32

Juran J, Godfrey AB (1999) Quality handbook, 5th edn. Republished McGraw-Hill, pp 173–178

Kotter JP (2008) Force for change: how leadership differs from management. Simon and Schuster

Shewhart WA, Deming WE (1986) Statistical method from the viewpoint of quality control. Courier Corporation

Taylor FW (1914) The principles of scientific management. Harper

Welch J, Welch S, Primus B, Winkelmann H, Grawe S, Szymczyk M (2005) Winning. HarperCollins, New York

Woods DD, Dekker S, Cook R, Johannesen L, Sarter N (2017) Behind human error. CRC Press

Software Overview and Methods Review: Minitab

T. T. Allen, *Introduction to Engineering Statistics and Lean Six Sigma*,
https://doi.org/10.1007/978-1-4471-7420-2_24

24.1 Introduction

The purposes of this chapter are to review many of the most powerful statistical techniques from the previous chapters and to illustrate their application with Minitab® software. To focus the review, ten questions about simple experimental systems are asked and answered using Minitab and previously mentioned techniques.

The first questions relate to a simple system which is the folding of towels, e.g., in a laundry processing operation or hotel. The initial or starting standard operating procedure is shown in ◘ Fig. 24.1 from Allen and Tseng (2018). In statistical process control and Part I of this book, an SOP such as this can be evaluated for its possible improvement and sources of variation can be identified or mitigated. Then, an improved standard operating procedure can be generated.

The other questions relate to the design of paper helicopters and Part II of this book. The methods of design of experiments permit the data-driven parameter design of the system indicated in ◘ Fig. 24.2. Even a simple paper helicopter offers many relevant design decision options and opportunities for performance improvements.

Overall, the chapter objectives include reviewing *t*-testing, ANOVA, multiple comparisons, fractional factorials, quality tools including control charts, and acceptance sampling planning. The reader should understand the related vocabulary and concepts as well as how to perform related tasks supported by Minitab.

Minitab is available on many university campus computers, but if you want to work from home, you can download a student trial version. Copy and paste the link (▶ http://www.minitab.com/en-us/) into your browser and follow the steps provided

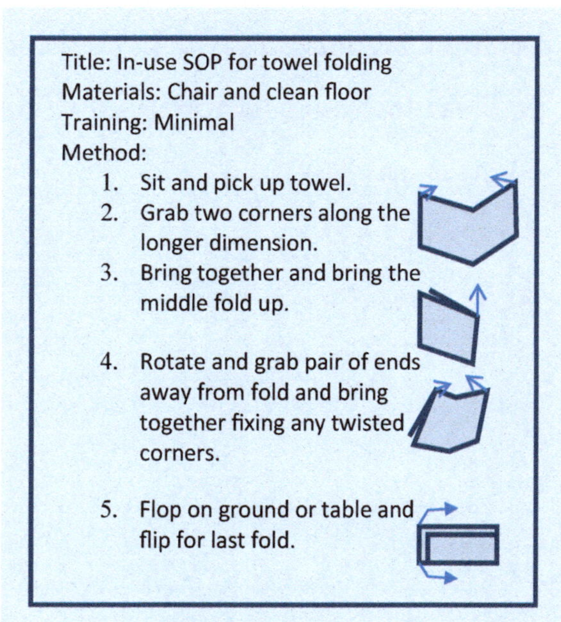

◘ **Fig. 24.1** An initial standard operating procedure (SOP) for towel folding

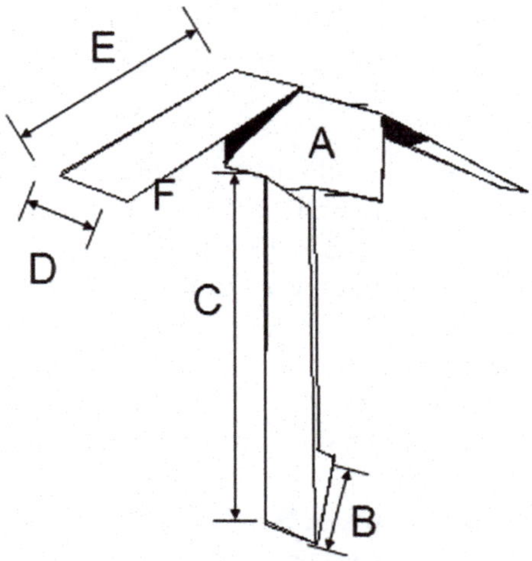

Fig. 24.2 Paper helicopter parameter design problem with selected factors

under a free trial. Make sure to use your university email so you get the full trial version for free in many cases. The versions of Minitab have not changed greatly so illustrations here apply generally to versions 16 and 17.

▶ Section 24.2 and ▶ Sect. 24.3 illustrate data visualization and Pareto charting from ▶ Chap. 2 and ▶ Chap. 3, respectively. In Sects. 24.3 and 24.4, continuous and attribute control charts from ▶ Chap. 4 are described. ▶ Section 24.5 illustrates methods from acceptance sampling in ▶ Chap. 7. Next, t-testing, ANOVA, and multiple comparisons from ▶ Chap. 11 are illustrated in ▶ Sect. 24.5, ▶ Sect. 24.6, and ▶ Sect. 24.7 respectively. ▶ Section 24.10 explores fractional factorial methods from ▶ Chap. 12. In ▶ Sect. 24.11, regression methods are described from ▶ Chap. 15.

24.2 Quality Data Visualization

In this section, we focus on the main "Stat" menu of Minitab. The items under the other relevant "Graph" menu can generally be accessed from the Stat menu as well since there is some redundancy in the interface. Minitab covers material that is taught in multiple courses at this university including our SQC course, design of experiments, and reliability courses. These topics can help with preparing for machine learning which is also addressed in part in Minitab and further in JMP.

Overall, the next three sections will be broken down into specific questions related to the towel-folding project. These questions will address the key issues associated with developing consistently high quality and efficient towel folds. Each laboratory of the following laboratories will show how various methods contribute to an improved process. While Minitab has materials relating to organizing lean six sigma projects, the intent here is to focus on methods primarily relevant to the measure and analyze phases. Yet, methods are potentially relevant in multiple phases of a lean six sigma project.

Users of Minitab often have a Microsoft Excel sheet open to process the Minitab inputs. This relationship between Minitab and Excel is becoming more formalized including through "Minitab Express" software available on the Minitab website (▶ www.minitab.com).

Table 24.1 and Fig. 24.3 show the data pasted from the towel-folding project write-up. We use the symbols X_1, X_2, and X_3 to describe the measured cycle times for the three towel-folding processes evaluated in each subgroup. These characterize the SOP in Fig. 24.1. We use this notation even though there are outputs (Ys) to follow the usual convention for Xbar and R charting. The counts of nonconformities are on the right-hand side. These data are relevant for the first three laboratories. As is often the case, the data must be preprocessed (tabulated and rearranged) to be ready for Minitab to analyze. Figure 24.3 also shows the sums of the columns associated with each nonconformity type.

The summary statistics are shown in Fig. 24.4 and a matrix plot of the X values are displayed in Fig. 24.5. The summary statistics are reasonably consistent with towel-folding times independently identically distributed (IID) from the normal distribution. Yet, the matrix plot shows cross-correlations which invalidate the IID assumption. These cross-correlations are ignored in the control charting applications.

24.3 Pareto Charting

In this section, Minitab is applied to address the following question which relates to exploratory data analysis of the available quality-related data:

- What are the most common nonconformities and is there an associated Pareto rule which might make improving the system relatively easy?

To answer this question, we will apply Pareto charting by selecting Stat → Quality Tools → Pareto Chart.

Using Paste Special… and selecting the transpose option of the G18:M18 cell data give the result in Fig. 24.6 in cells E3:E9. Note that a new worksheet in the workbook has been created to hold the transposed Excel data. Also, the number, type, and nonconformity have been included from the towel-folding project write-up.

Now, the data can be entered into Minitab. This can be done with a simple copy and paste approach putting the titles "Number"…"Counts" in the top row. Minitab reserves the top row for variable titles. With the data entered into Minitab, select Stat → Quality Tools → Pareto Chart. Next, select "Defect or attribute data in" and double-click on "Nonconformity" and then select "Frequencies in" and double click on Counts. Keep all the defaults and select OK. The result is in Fig. 24.7.

Note that right-clicking on the figure in Minitab and selecting "Edit Figure Region…" allow for changing the fill pattern and border and fill lines to white or (effectively) colorless. In this context, colorless is easier on printers and conforms more to the way textbooks and journal articles generally appear. Click on Background Color (under Fill Pattern) and Color (under Borders and Fill Lines) to select white for both the fill pattern and the borders and fill lines.

Table 24.1 Data from a towel-folding study

Stage	Subgp.	X_1	X_2	X_3	1							2		
					1	2	3	4	5	6	7	1	2	3
1	1	24.6	21.2	26.6		1								
1	2	25.0	16.9	29.6		1		1				1	1	1
1	3	20.7	29.2	21.2	1								1	
1	4	21.7	26.2	21.3	1	1						1		
1	5	23.8	20.4	24.5		1						1		
1	6	22.3	26.1	23.2	1	1						1		
1	7	24.3	23.7	24.0	1			1					1	
1	8	21.5	21.9	23.2	1	1		1				1		
2	9	26.4	19.6	25.9										
2	10	25.4	30.0	25.2				1					1	
2	11	22.9	19.5	24.3										
2	12	25.7	19.5	24.3				1						
2	13	24.8	22.5	23.7										

(continued)

Table 24.1 (continued)

2				3							Major	Minor	Demerits
4	5	6	7	1	2	3	4	5	6	7			
1				1							2	1	11
1				1	1						5	3	28
1	1				1						3	2	17
1				1			1				4	2	22
									1		2	1	11
1				1			1				4	2	22
					1						3	1	16
1							1				3	3	18
1											0	1	1
							1				1	2	7
							1				0	1	1
1											0	2	2
1											0	1	1

	A	B	C	D	E	F	G	H	I	J	K	L	M	N	O	P	Q	R	S	T	U	V	W	X	Y	Z	AA	AB	AC	AD	AE	AF	AG
1																															=AVERAGE(D4:F4)		
2										1							2							3								=MAX(D4:F4)-MIN(D4	
3		Stage	Subgp.	X_1	X_2	X_3	1	2	3	4	5	6	7	1	2	3	4	5	6	7	1	2	3	4	5	6	7	Major	Minor	Demerits	Xbar	R	s
4		1	1	25	21	27		1									1				1							2	1	11	24.13	5.40	2.730079
5		1	2	25	17	30		1		1				1	1	1	1				1	1						5	3	28	23.83	12.70	6.429878
6		1	3	21	29	21	1								1		1	1				1						3	2	17	23.70	8.50	4.769696
7		1	4	22	26	21	1	1						1			1				1			1				4	2	22	23.07	4.90	2.720907
8		1	5	24	20	25		1						1												1		2	1	11	22.90	4.10	2.193171
9		1	6	22	26	23	1	1						1			1				1			1				4	2	22	23.87	3.80	1.985783
10		1	7	24	24	24	1			1					1							1						3	1	16	24.00	0.60	0.3
11		1	8	22	22	23	1	1		1				1			1							1				3	3	18	22.20	1.70	0.888819
12		2	9	26	20	26											1											0	1	1	23.97	6.80	3.789899
13		2	10	25	30	25				1					1									1				1	2	7	26.87	4.80	2.715388
14		2	11	23	20	24																		1				0	1	1	22.23	4.80	2.468468
15		2	12	26	20	24				1							1											0	2	2	23.17	6.20	3.251666
16		2	13	25	23	24											1											0	1	1	23.67	2.30	1.150362
17							5	6	0	5	0	0	0	5	4	1	9	1	0	0	4	3	0	5	0	1	0						
18			=SUM(G4:G16)				14	13	1	19	1	1	0																				
19			=G17+N17+U17																														

Fig. 24.3 Towel-folding data with excel formulas for calculation

Descriptive Statistics: X1, X2, X3

Variable	N	N*	Mean	SE Mean	StDev	Minimum	Q1	Median	Q3	Maximum
X1	13	0	23.777	0.498	1.794	20.700	22.000	24.300	25.200	26.400
X2	13	0	22.82	1.11	4.00	16.90	19.55	21.90	26.15	30.00
X3	13	0	24.385	0.611	2.202	21.200	23.200	24.300	25.550	29.600

Fig. 24.4 Summary statistics for the time data

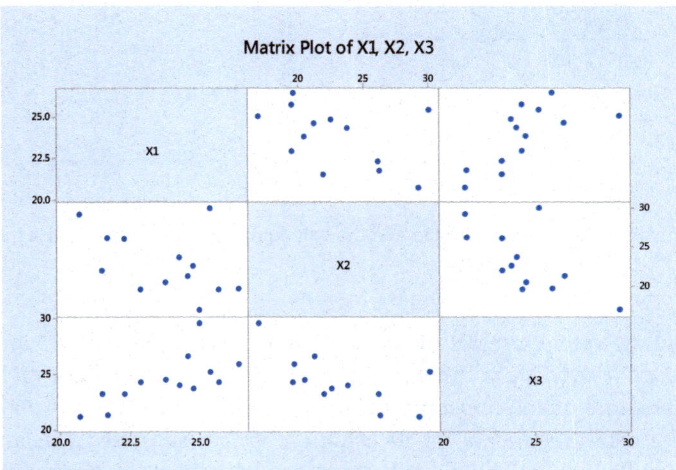

Fig. 24.5 Matrix plot showing the cross-correlations of the time data

For the Pareto rule to apply, at least 20% of the nonconformity clusters must account for 80% or more of the observed nonconformities. In the folding towel example, the news to be gleaned is negative in the sense that the Pareto rule does not apply. The data in Fig. 24.6 provide direct insight but Fig. 24.7 may help to improve intuition.

Number	Type	Nonconformity	Counts
1	Major	Corners not straight	14
2	Major	Top not flat	13
3	Minor	Not square	1
4	Minor	Corners not together	19
5	Minor	Poor placement	1
6	Minor	Corner twisted	1
7	Minor	Edges not straight	0

raw data + sums and avgs

Fig. 24.6 The inputs for a Pareto chart based on the towel-folding data

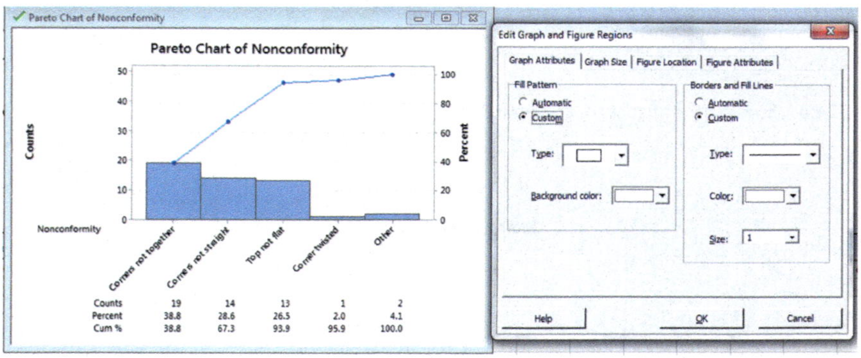

Fig. 24.7 The Pareto chart from Minitab and the edit figure region… Menu to turn the fill and borders white

In the folding towel example, the news to be gleaned is negative in the sense that the Pareto rule does not apply. In no instance do 20% of the nonconformity types constitute 80% or more of the nonconformities. Sometimes, the Pareto rule does apply and it offers a data-driven case for focusing on certain subsystems in improvement efforts. The results here, however, imply that we need to focus on the corners and also the flatness of the top of the folded towel.

	A	B	C	D	E	F	G	H	I	J
1										
2		Stage	Subgp.	X_1	X_2	X_3	Xbar	R	s	
3		1	1	24.6	21.2	26.6	24.13	5.40	2.730079	
4		1	2	25	16.9	29.6	23.83	12.70	6.429878	
5		1	3	20.7	29.2	21.2	23.70	8.50	4.769696	
6		1	4	21.7	26.2	21.3	23.07	4.90	2.720907	
7		1	5	23.8	20.4	24.5	22.90	4.10	2.193171	
8		1	6	22.3	26.1	23.2	23.87	3.80	1.985783	
9		1	7	24.3	23.7	24	24.00	0.60	0.3	
10		1	8	21.5	21.9	23.2	22.20	1.70	0.888819	
11										
12						Xbarbar	23.46	5.21		
13						=AVERAGE(G3:G10)		Rbar	=AVERAGE(H3:H10)	
14										
15						d2(3) = 1.693 Table B, Besterfield's textbook				
16		=G12+3*(H12/1.693)/SQRT(3)				UCL=	28.80			

Xbar and R chart / Xbar and R with 2 stages / Parame

Fig. 24.8 Stage 1 or startup phase data for subgroup-based Xbar and R charting

24.4 Control Charts: Variables

This section explores standard variables control charting with data in subgroups. It addresses the questions:

- How do I measure and monitor processes with variable (continuous) characteristics?
- How do I establish the quality level thoroughly before and after a change?

The data from the startup phase (Stage I) for setting up the control chart are shown in Fig. 24.8. Typically, one has 25 subgroups each of size i with $n > 3$. Yet, for illustration and for student projects, we have a smaller number. The Microsoft Excel worksheet also includes calculations to derive the estimated standard deviation and the control limits.

Pasting the data into Minitab, use Stat → Control Charts → Variables Charts with Subgroups → Xbar-R to open the dialog shown in Fig. 24.9. With only the data shown in Fig. 24.8, the selected inputs and OK produce the charts shown in Fig. 24.10.

If the values were input together with the Stage 2 values shown in Fig. 24.11, we would need to filter the Stage 2 values out. This can be achieved using Data Options… in the dialog in Fig. 24.9. The result gives the dialog in Fig. 24.11 on the left. Clicking on Condition… gives the value on the right-hand-side of Fig. 24.11. In the Condition area, we include Stage = 1 (Fig. 24.12).

Often, in the context of a lean six sigma project, a staged control chart is required. This permits the success or failure of the project to be thoroughly evaluated. To produce this evaluation, select Stages under Xbar-R Chart options. Figure 24.14 shows the result in two stages. In the example in Fig. 24.13, the project's effect on the average time is not readily apparent.

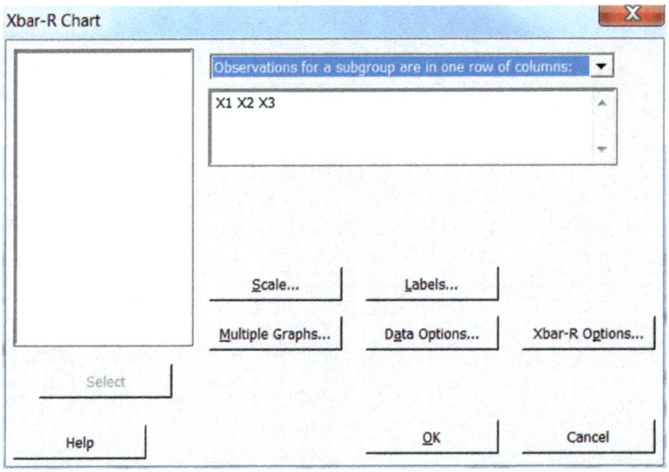

Fig. 24.9 Minitab dialog to set up initial Xbar and R charts

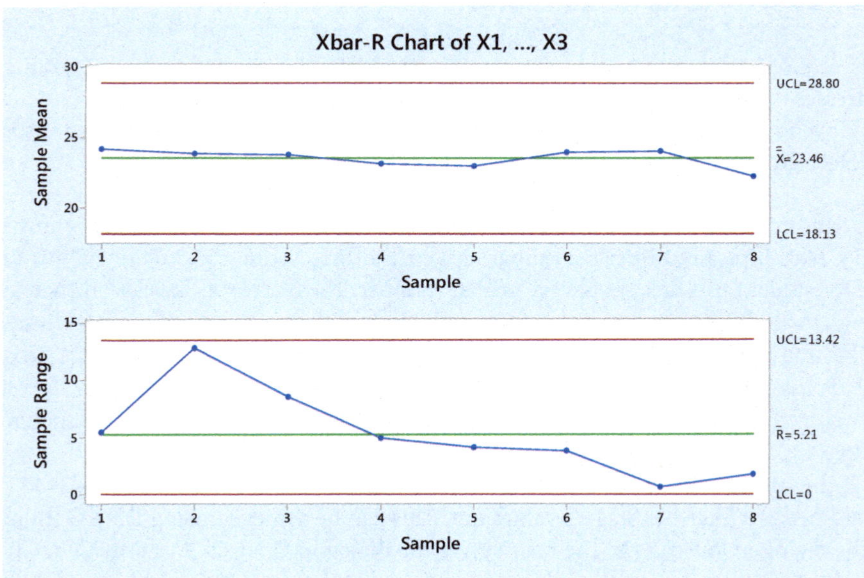

Fig. 24.10 Initial Xbar and R chart outputs for the time system output, i.e., the times required for towel folding

24.5 Control Charts: Attributes

This section addresses the questions:

- How do I measure and monitor processes with attribute (discrete count) data?
- How do I establish the quality level thoroughly before and after a change?

Minitab does not currently support demerit charting but we can chart the counts of nonconformities on a u-chart. Figure 24.14 shows the data and outputs for a staged

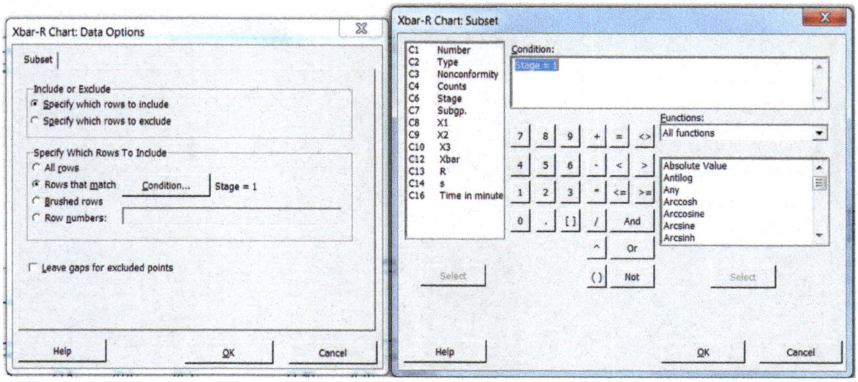

Fig. 24.11 Details for selecting only the Stage 1 data for the case in which both Stage 1 and Stage 2 are in the same columns

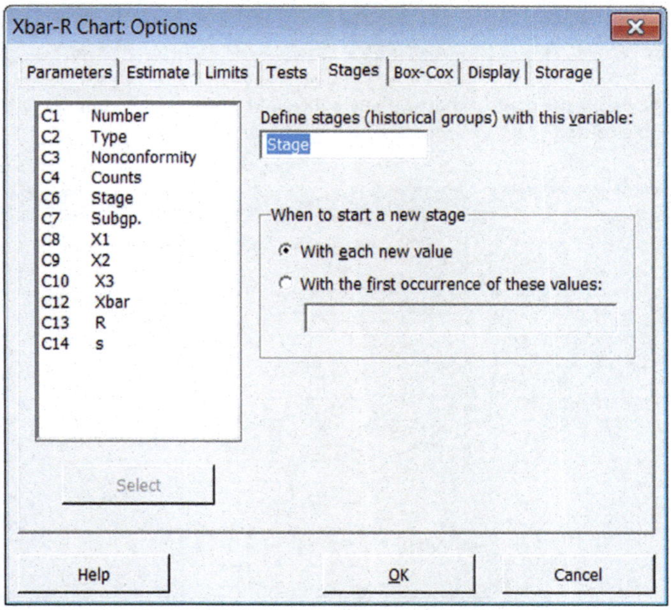

Fig. 24.12 Details for selecting only the Stage 1 data for the case in which both Stage 1 and Stage 2 are in the same columns

u-chart (Stat → Control Charts → Attribute Charts → U Charts… → Options → Stages). As for variables data, the stage option permits an improvement project to be thoroughly evaluated. In this case, the effect of the project is readily apparent. This provides strong evidence that the changes did significantly (both statistically and practically) reduce the counts of nonconformities and improve the quality.

A common issue with attribute charts is the appearance of large numbers of effectively false positives. This can occur because the sample size is large enough that any

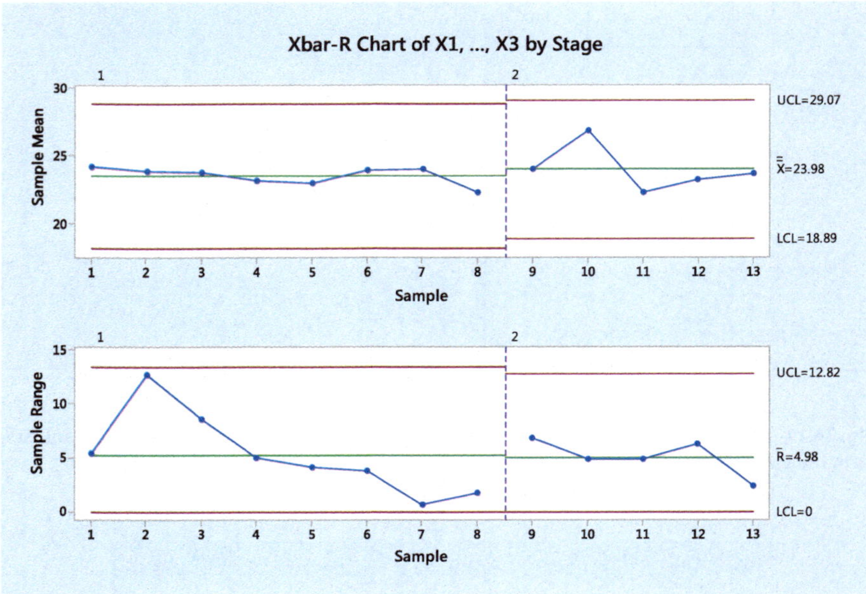

Fig. 24.13 The two-stage control chart with evidence that the project recommendations were a desirable assignable cause

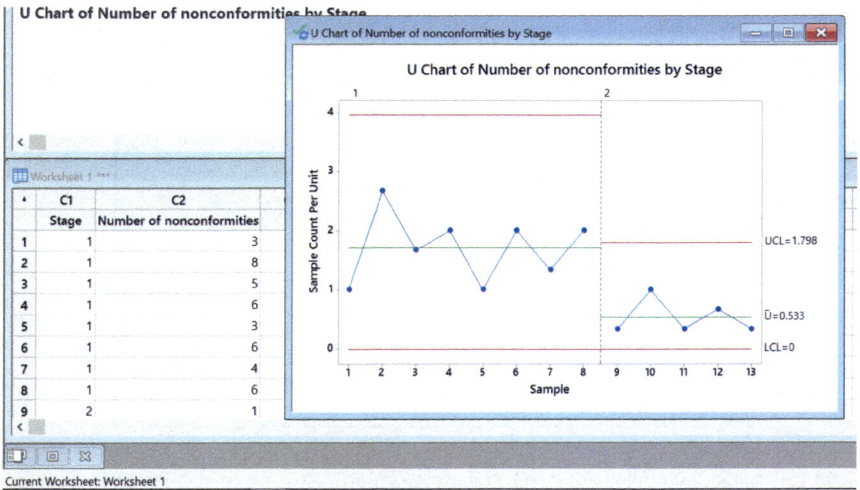

Fig. 24.14 The two-stage attribute control chart with evidence that the project recommendations were a desirable assignable cause

small shift in the process parameters can be detected. This can be called "over dispersion" and the Laney chart option mitigates the issue to some extent. Laney charts are more cautious by focusing on different sources of variation but are not clearly pertinent if the root cause is simply extremely large sample sizes. A potentially preferable option to Laney charts is simply adjusting the limits to 4, 5, or higher sigma levels instead of the usual 3. You can display additional limits under Options… → Limits then enter 4 standard deviations. If you use these wider limits for monitoring, the false alarm rate will be substantially reduced.

24.6 Acceptance Sampling

This section addresses the question:

- How do I analyze and recommend specific acceptance sampling plans?

Relevant support is under the Stat → Quality Tools → Acceptance Sampling by Attributes… option.

In the dialog, select "proportion defective" (might be better as "nonconforming") for the units. The dialog asks for an acceptable quality level (AQL). Lots at this level would be judged acceptable and the small fraction of nonconforming units would not cause substantial harm. In the default example, this is set at 0.02. The rejectable quality level (RQL) is the fraction such that the lot would be considered unacceptable. If you leave the risks at their default value as done in Fig. 24.15, you obtain $n = 306$ and $c = 10$ as the recommended plan in the session window. It might be desirable to iterate this process or consider a double sampling plan for more efficiency.

24.7 Two Sample Comparisons

This section and the remaining sections in this chapter relate to specific questions about the system design and the paper helicopter. These questions will address how flight times of a paper helicopter are affected by different factors and their levels. Each section will show how various methods are used to examine the effects of these factors. And they will also demonstrate how to use Minitab as a supplement to class materials to determine the statistical significance of the factors.

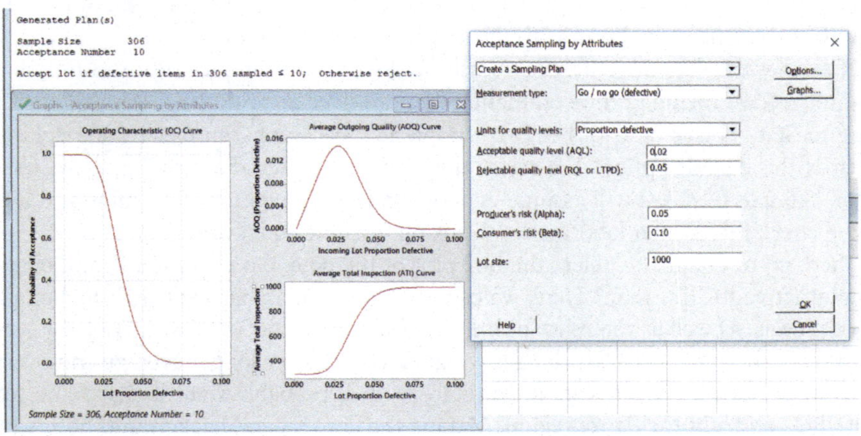

Fig. 24.15 Inputs and outputs relevant to deriving a single sampling plan

Table 24.2 Data from testing paper helicopters

Run	Factor A	Time
1	Level 2	2.1
2	Level 1	1.6
3	Level 1	2.1
4	Level 2	2.2
5	Level 1	1.8
6	Level 2	1.9
7	Level 2	2.1
8	Level 1	1.8

Specifically, in this section we seek to answer:

- How do I show that level 1 of factor A results in a significantly better engineering design than level 2 with all other factors held constant?

For our paper helicopter example, let us say we are trying to learn that having short wing length significantly affected the time the helicopter was in the air. To do this, we will use two-sample t-testing and create a box-and-whisker plot.

It may be easiest to first enter the data into Excel to keep track of the numbers and calculations and then transfer that data into Minitab. Here, the data for factor A, wing length are entered into Excel. As you can see there are four data points for each level. See Table 24.2 for the data. Also, the run order is randomized so that proof is possible to derive from the analysis. Level 1 is the short wing length and level 2 is the long wing length. It should be noted that Excel itself supports t-testing based on the Analysis ToolPak Add-In, which is likely already installed on your version of Excel and available under the Data ribbon.

Next copy and paste these data into the Minitab worksheet. Minitab can analyze the data either with separate columns for each level, or with response data in a single column and a separate column with the labels as we have done.

Now that the data are entered, we will perform a two-sample t-test in Minitab. In the.pptx, I have included the commands that are needed to do this. We will start by going to Stat → Basic Statistics → Two-Sample t… We want "samples in one column" which is the default option. We select the response under "Samples" and the factor under "Sample IDs". We will assume equal variances throughout the course and learn how to check this in later laboratories. Click on "assume equal variances."

Then, go to Graphs…, select the box plot of the data. Go to "options" and change the alternative to "less than". Here, we can see that Minitab has provided the box plot of both levels, as well as the relevant information needed for the t-test. Looking at the plot, it is fairly evident that the longer wings (level 2) foster longer average and overall flight times. As with most plots in Minitab, we will probably want to select the plot, right-click, and edit the figure region. Making the borders and background both white makes the plot less busy and generally looks better in reports (Fig. 24.16). Next, we examine the numerical results.

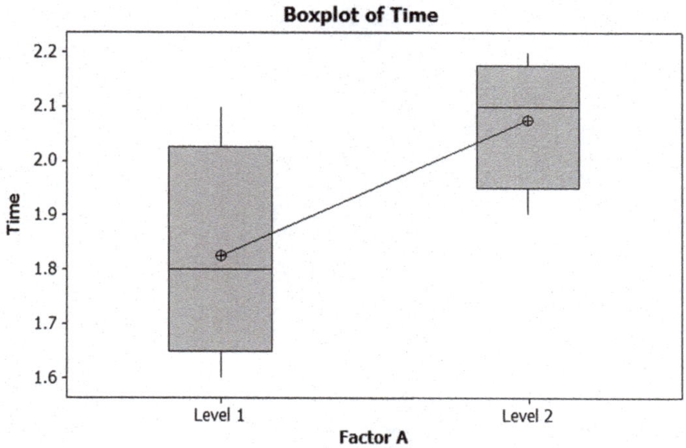

Fig. 24.16 Box-and-whisker plot of times for the levels of wing length

To determine statistical significance, the relevant data are the t-value and the p-value. We are looking for p-value less than 0.05, and in this case, it is 0.042 so we can declare significance. Considering that the t-value is greater than t-critical or, equivalently, that the p-value is less than 0.05 and a randomized run order was applied, the results prove that using level 2 (long wing length) results in significantly longer average flight times than level 1 (short wing length) with alpha equal to 0.05.

24.8 ANOVA

In this section, the objective is to show how Minitab can be used to answer:

- If we have more than 1 factor and/or greater than 2 levels, how do we show significance associated with factors?

In terms of our paper helicopter example, assume we have three different kinds of papers, and we want to know if the factor "paper type" significantly affects average flight times.

We will use Minitab to perform single factor ANOVA. Start again by entering the data into Excel reflecting the randomized run order that was presumably used in the data collection (see Table 24.3). Then, we transfer the data into Minitab. On the right is our raw response data for the three different paper types using a randomized run order for the output Y time in the air in second listed for each run. Now that the data are pasted in, we can have Minitab perform ANOVA and select the first option which is "One way". At the end of this lab, we will show how the General Linear Model option can give the same results with different options. In the .pptx, I have listed the Minitab command that we will need to do this. Start off by going to STAT, and to ANOVA. Then, go to 'one way, select each level, click "store residuals" and "store fits". Also select box plot under Graph…

Table 24.3 Data for Laboratory 2

Factor A	Time
Level 2	1.4
Level 3	2.1
Level 1	1.5
Level 2	1.7
Level 1	1.6
Level 3	2.2

Click O.K. Here, we see Minitab has given us the information for ANOVA as well as the information we need for our multiple comparison tests. From the result, we can locate several numbers needed to show that paper type significantly affects average flight times. The mean squared error of the residuals will give us an estimate of the response variance, sigma-squared, and the F-values and P-values will be what we will use to determine significance. To prove significance, we need F-value greater than F-critical, and a P-value less than 0.05.

To determine F-critical, we can look it up in a table using a confidence level of 0.05, the degrees of freedom of the estimated effects, 2, and the degrees of freedom of the residuals 3. Or we can use the same data and find the F-critical on excel using the F-inverse function. From Minitab, the F-value we are given is 13.09, and the F-critical is 9.55. The p-value given to us from Minitab is 0.033. Since the F-value is greater than F-critical, and, equivalently, the p-value is less than 0.05 and randomization was applied in experimentation, we can declare that paper type significantly affects average flight times with alpha equal to 0.05.

If we have more than a single factor, a general way to fit Analysis of Variance is through the General Linear Model option. This can be applied for the single factor case also resulting in identical findings. The differences relate to options about graphs and multiple comparisons. We see that the p-value is again 0.033. The models derived are identical.

24.9 Multiple Comparisons

In this section, we seek to answer:

- For the significant factors in question 2, how do I show which levels are significantly better than others?

In terms of our paper helicopter example, say we have three different kinds of papers, and we already proved that at least two levels are associated with significantly different average values of flight times using Analysis of Variance (ANOVA).

The method that we will use is based on "Fishers Least Significant Difference" and it is only supposed to be applied after ANOVA has found a factor to have a significant effect. This is because it is based on the so-called "individual error rate". This means that if we directly applied Fisher's LSD, we would likely make a lot of Type I errors because we would be doing, in general, a large number of comparisons simultaneously.

Table 24.4 Output from Minitab multiple comparisons via the ANOVA feature

Factor A	N	Mean	Grouping	
Level 3	2	2.15	A	
Level 1	2	1.55		B
Level 2	2	1.55		B

This explains why Fisher's LSD is buried underneath ANOVA. Note that other multiple comparison techniques like Tukey's, Hsu's, or simultaneous interval based methods do not need the "ANOVA" protection and offer benefits in specific circumstances that we will not describe here.

We have the same data as for ANOVA in lab 2. Next, we go to "One way…" under the ANOVA menu. Then, we go to "Comparisons…" and select Fisher's individual error rate. Generally speaking, this approach offers a desirable balance between Type I and Type II errors while being reasonably simple and transparent. It is similar to doing multiple t-tests with a pooled standard deviation estimate after having the go-ahead from ANOVA. We also select "box plot" under graphs. My customers of the analysis will mainly want to see the box-and-whisker plot. As with most plots in Minitab, we will probably want to select the plot, right-click, and edit the figure region. We would like to make the borders and background both white. This makes the plot less busy and generally looks better in our reports. We next examine the numerical results.

We click "OK" and look in the session window for the results. We see the repeat of the ANOVA table showing significance. We also see a table of the means and the grouping. This indicates that level 3 (Paper Type 3) fosters significantly higher average flight times than levels 1 or 2. Table 24.4 shows the Minitab output. Note that moving B over to the right from A does not mean that levels 1 and 2 have higher means, just different. This can be a little confusing.

24.10 Regular Fractional Factorials

This section addresses screening experiments using regular fractional factorials or Plackett–Burman designs. Therefore, this lab will focus on answering:

- How do I, with a small number of tests, find the significant factors from the long list of a large number of factors?

To do this, we will plan the experiment using one of the A-optimal or D-optimal designs and a randomized run ordering. Included in this section are rules of thumb to follow in deciding when to use each design. For regular fractional factorials, the number of runs available is equal to $2i$, where i is 3, 4, 5, …. For Plackett–Burman, designs are available for more cases including when $n = 4 * i$, where $i = 2, 3, 4, 5, \ldots$

As a rule of thumb, we often want to start using the pre-tabulated D- and A-optimal design with the first few runs. This helps with experimental costs by not locking all our budget into one experiment with a single selection of levels. If we have m fac-

Table 24.5 Pick the smallest with enough columns for the factors rule of thumb

Preferred	n	# factors	Design	# columns
*	4	3	Reg.	3
	4		PB	3
*	8	4–7	Reg.	7
	8		PB	7
	16	8–11	Reg.	15
*	12		PB	11
*	16	12–16	Reg.	15
	16		PB	15

tors, we need to pick a design with at least $n = m + 1$ runs so as to fit our model with the constant term equivalent to the grand mean in ANOVA.

As another rule of thumb, regular fractional factorials are chosen as a tiebreaker over Plackett–Burman designs although some prefer Plackett–Burman designs relating to the so-called partial confounding of interactions. The results of these two rules of thumb are shown in Table 24.5.

For our Minitab example, we will use the factors listed in the corresponding high and low level values to see what factors affect average flight time.

Since we have seven factors, we will need eight columns, and so will pick a regular fractional factorial with 8 runs. Before running our experiment, we will need to create the design in Minitab to show which levels to test for each factor on each run.

To prepare, consult the instructions in the PowerPoint which include the commands needed to create the desired regular fractional factorial or Plackett–Burman design.

To create a regular design in Minitab, start by going to Start → DOE → Factorial → Create Factorial Design, and "2-level default generators." Go to Design, which will show that we need 8 runs for a so-called "Resolution III" design.

Going to Options…, make sure that the randomized run order box is selected. Enter 2 for the random seed so that all the experimental plans will look the same. Going next to Factors…, select −1 and 1 for our lower and higher levels. Note also that we could use Minitab to generate the experimental plan in engineering units instead of 1 and −1 units. To do this, we could select "Factors…" in the dialog box and type in the factor names. Clicking O.K. will display our design. The results, later shown, in Table 24.6 and Table 24.7, respectively, depend on whether you typed in the factor ranges or names.

Now that our design is created we see that each of our 8 runs has different settings level for each factor. We can ignore the run order and block columns. The fact that we have only one block means that all runs are performed in a single batch, which is usually the case for such a small experimental plan. The level settings might appear random and the run order is randomized, but they represent a pattern that is the solution to an optimization problem. It is easy to check that the dot product of any two columns is 0 so that the columns are orthogonal.

Table 24.6 A regular fractional factorial experimental design in coded −1, 1 units

A	B	C	D	E	F	G	Time
1	−1	−1	−1	−1	1	1	0.25
1	1	−1	1	−1	−1	−1	0.6
−1	−1	−1	1	1	1	−1	1.74
−1	−1	1	1	−1	−1	1	1.17
−1	1	−1	−1	1	−1	1	1.3
1	−1	1	−1	1	−1	−1	0.93
1	1	1	1	1	1	1	1.32
−1	1	1	−1	−1	1	−1	0.7

Table 24.7 The same regular fractional factorial design in engineering units

Body height	Fold-over length	Tail length	Wing width	Wing length	Pitch angle	Paper type	Time
4.0	0.5	2	2	3	20	Regular	0.25
4.0	1.5	2	4	3	0	Lined	0.60
1.5	0.5	2	4	6	20	Lined	1.74
1.5	0.5	5	4	3	0	Regular	1.17
1.5	1.5	2	2	6	0	Regular	1.30
4.0	0.5	5	2	6	0	Lined	0.93
4.0	1.5	5	4	6	20	Regular	1.32
1.5	1.5	5	2	3	20	Lined	0.70

Next, we imagine running the experiment building 8 paper helicopters according to the specifications. We build the first helicopter with body height 4 cm, fold-over length 0.5 cm, tail length 2 cm, …, down to paper type regular. Testing this helicopter, the drop time is estimated at only 0.25 s. We continue running our experiment to create a Y or "Time" column to store the output for each run.

Once we have run the experiment following the experimental planning output for each run in Minitab under the Time column, we can create a plot to analyze the effect of the factors. Start by creating a normal probability plot of estimated effects. To do this, we will go to Stat → DOE → Factorial → Analyze Factorial Design….

Select Y, and go to Graphs…. Here we ask for the normal probability plots with an alpha value of 0.05. Next, go to Storage…, ask to store Effects…, and click O.K. We prefer using the normal probability plots of estimated effects instead of ANOVA for regular fractional factorials (only) because there are so few degrees of freedom and the coefficients should all be approximately normally distributed. In other analysis situations, we might generally prefer ANOVA.

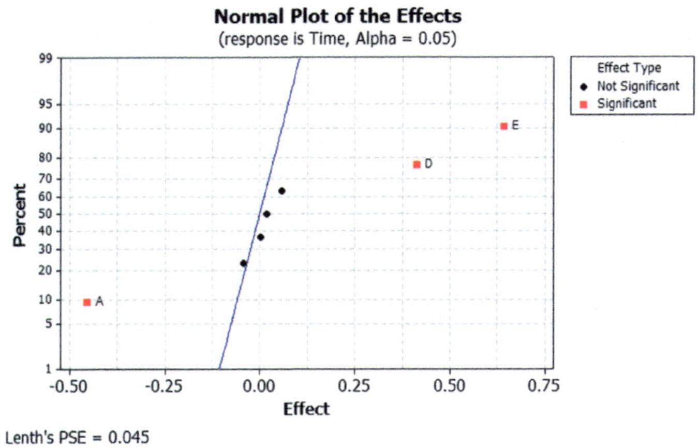

Fig. 24.17 Normal probability plot of estimated effects for the paper helicopter two-level regular fractional factorial analysis

Looking at our normal probability plots, we see that Minitab has flagged factors A, D, and E as significant. The remaining factors should approximately line up. This plot lesson normalizes the effect of the factors. Factors that don't seem to come from the normal distribution must have an unusual effect on the output (Fig. 24.17).

To further investigate the effects of the factors, we can create a main effects plot. To do this, we will start by going to Stat… → DOE → Factorial → Factorial Plots… and check the 'main effect plot box' Next, go to setup, and make sure Time is selected for the 'responses'. You can then select all the factors that you are interested in examining under 'selected'. Click O.K.

Our main effect plot shows the relative effect of each factor, indicating responses for each level of the factors. The steeper the slope of the line, the greater is the effect on the output of a level change in that factor. As with most plots in Minitab, we will probably want to select the plot, right-click, and edit the figure region. Making the borders and background both white makes the plot less busy and generally looks better in reports. We next examine the numerical results.

Looking at factor A, we see that it has the steepest slope and that the negative level greatly increases the output mean. In conclusion, we have declared that factor A, Body Height significantly affects flight time, and that the low level of 1.5 cm should be used to optimize flight time. The normal probability plot proves that factors A, D, and E are associated with significant effects. We can then look at the main effects plot for system design decision-making. A− (1.5 cm in body height), D+ (4 cm wing widths), and E+ (6 cm wing lengths) seem particularly promising (Fig. 24.18).

As a final check, we also consider planning an experiment with 12 factors and 16 runs. Again, we use a regular fractional factorial or "2-level fractional factorial (default generators)". Here, we make up the response data. We select Graphs… Effects Plots → Normal. We see that there are 15 points on the line even though we only have 12 factors. This is because the columns of the full design are used to create quantities that would be independent identically distributed under the hypoth-

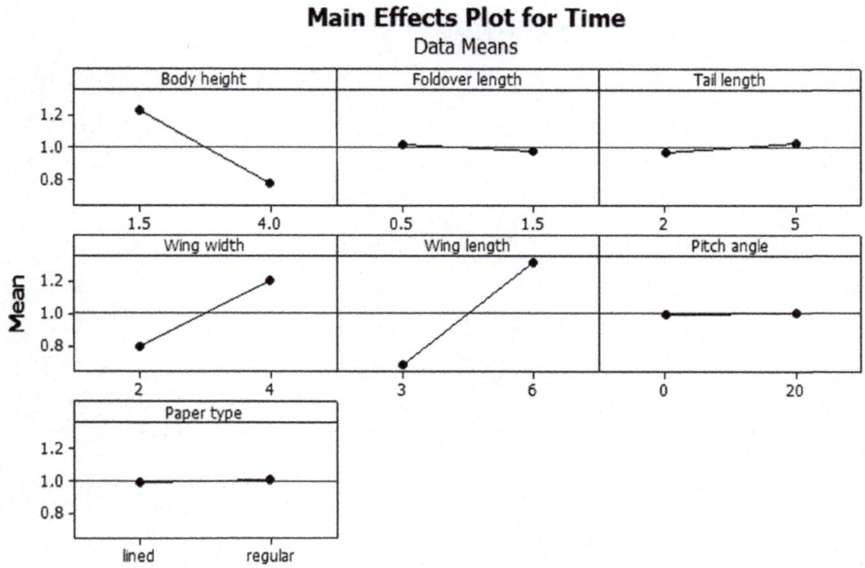

Fig. 24.18 Main effects plot for the flight time responses for the regular fractional factorial example

esis that the data are normally distributed deviates. In other words, we are looking for quantities that would be normally distributed if all the effects are zero to establish the line.

24.11 Regression

This section seeks to use Minitab to answer:

- How do I develop a relatively accurate prediction model with a small number of factors?

The results will be relevant to both on-hand data analysis not from a randomized experiment and analysis following a response surface experiment. Technically, we could also use a first-order model to fit the outputs from a two-level regular fractional factorial. However, we prefer using the normal probability plot of estimated effects for regular fractional factorials (only) because there are so few degrees of freedom and the coefficients should all be approximately normally distributed for that case only. We will be using normal probability plotting again but not of the coefficients. Instead, we will use it to check model quality by plotting the residuals.

The data for this example are given in Table 24.8. In this example, they are meant to represent the results of studying several helicopter makers both the controllable body heights and the measured flight times. Body height was not varied using the arbitrary ability as in a randomized experiment but was assigned a single measured value. As a

Table 24.8 Data for the regression example

Data Point	Body height	X_1	X_1^2	Flight time
1	1.5	−0.44	0.19	1.91
2	1.4	−0.52	0.27	1.83
3	2.7	0.52	0.27	0.86
4	1.1	−0.76	0.58	1.72
5	0.9	−0.92	0.85	1.28
6	0.8	−1.00	1.00	1.09
7	2.9	0.68	0.46	0.79
8	2.2	0.12	0.01	1.10
9	3.3	1.00	1.00	0.81
10	1.8	−0.20	0.04	1.67
Minimum	0.8			
Maximum	3.3			

result, we will not be able to prove that it was body height that affected average flight time. Still, we can develop a defensible and suggestive model from the on-hand data analysis.

Before we paste the data into Minitab, it is generally a good idea to create a scaled version of the inputs using Excel, as is shown in Table 24.5. We calculate the minimum and maximum of the inputs variables using the=MIN() and=MAX() functions. Then, we create the coded version of the inputs using "=−1+2 * (A2 − A12)/(A13 −A12)" and copying down this function. We also create the X12 term using "=C2^2". Creating coded inputs helps in avoiding round-off errors and also with the interpretation of the results. Is there a significant squared term? The answer depends on the scaling of the inputs and so it is natural to use the [−1, 1] scaling. We can type or paste these data into Minitab which will treat X12 as if it were any other input.

Paste the data into Minitab and select Stat → Regression → Regression. Change the name of X12 to X1^2 to be clear about its interpretation. It is often a good practice to start with a first-order model, particularly with on-hand data. With on-hand data, the inclusion of squared terms or interaction terms generally produces variance inflation factor problems. Being able to safely model with these terms is one of the benefits of using a response surface method (RSM) experimental design. We select Flight Time as the response and X1 as the input. Then, we select Graphs… and Normal plot under residual plots. Again, a normal probability plot of effects is not appropriate for any data other than after a regular fractional factorial experiment or Plackett–Burman design. In older versions of Minitab, under Options… we select variance inflation factors (VIF). In newer versions, VIFs are standard in the output in the session window.

We click ok, to see if our simple model is defensible. Looking into the session window, we see that the adjusted R-squared is 34.0% which is a reasonable value for on-hand data. The reasonably high R-squared is related to the significant value of F0 (they are proportional). The ANOVA p-value of 0.045 means that the model is associated with a level of significance and that we can look at the p-values of individual coefficients. Checking ANOVA first and, upon significance, looking at individual p-values for coefficients are very similar to checking ANOVA before using Fisher's individual error rate or least significant difference (LSD) approach.

The VIF is 1.0 which is passable and the normal probability plot of residuals looks acceptable. We do not see any outliers in the lower left or upper right or a strong pattern. We conclude that the first-order model is defensible. By looking at the −0.40 coefficient, we conclude that the effect of changing body height from the lowest setting (0.8 cm) to the highest setting (3.3 cm) is approximately $2 \times 0.4\ \text{s} = 0.8\ \text{s}$. The minus sign on the coefficient means that the low level is expected to foster the higher average flight time. We regard these results with some degree of skepticism because we know that other factors were also varied and we did not apply a randomized experimental design. In other words, we did not systematically change body height so we are less than sure that body height is what drove the 0.8 s effect. Still, body height is the likely factor.

If some part of the regression checklist had failed, putting the X1^2 term into the model would have been a possible corrective action. This could potentially have improved the residual plot and/or made the ANOVA significant if it was not. An alternative possibly useful approach for cases involving two or more factors would relate to pretending that the data derived from a Response Surface Design and using the Define Custom Design… option. If this approach were taken, Minitab would actually create the X1 and X1^2 coded unit inputs. This would save some Excel work although there are slightly fewer options on that route, e.g., relating to Variance Inflation Factors.

24.12 Problems

1. True or False. Minitab can be easily obtained in many university computer rooms?
   a. True, Minitab is expensive commercial software only available to students at high prices.
   b. False, they even have a product called Minitab express which includes an Excel-type spreadsheet.
2. True or False, Minitab seems unaware that users often use their product together with Microsoft Excel?
   a. True, they almost never do observational studies or interviews.
   b. False, they even have a product called Minitab express which includes an Excel-type spreadsheet.
3. What is the count of all nonconformities with the highest count in ► Sect. 24.3?
   a. 2.
   b. 5.
   c. 6.
   d. 14.
   e. 19.
4. In Pareto charting, one is often hoping that?
   a. The nonconformities are concentrated into a few types so that it is clear where to concentrate improvement efforts.

   b. Nonconformities are evenly divided among many types so that it is difficult to apportion blame.
   c. The Pareto rule does NOT hold so that prioritization of issues is easy.
   d. All of the above.
5. In the Pareto chart of nonconformities (in ▶ Sect. 24.3), if one grouped the corner nonconformities (2 types) together, would the 80–20 rule be followed?
   a. Yes, there would be four categories and the resulting count would be greater than 80% of the total nonconformities.
   b. No, the combine count would be under 40% of the total nonconformities.
6. What is a typical number of subgroups in the startup or Phase I of a control chart process?
   a. 2.
   b. 5.
   c. 16.
   d. 17.
   e. 19.
   f. 25.
7. What is the button to open the dialog to select the stages tab for staged control charting?
   a. Xbar R Chart: Options…
   b. Responses…
   c. Stages…
   d. Machine…
   e. All of the above.
8. What would evidence of an efficiency improvement look like in a staged control chart?
   a. The mean would shift lower, then higher, then lower on the Xbar chart.
   b. The range would increase (shift up) on a sustained basis.
   c. The range would decrease on a sustained basis.
   d. The range and sample mean would fluctuate.
   e. All of the above.
9. Which of the following might be a cause for excessive false alarms?
   a. The selection of a small sample size.
   b. The applicability of the normal distribution.
   c. The use of a large sample.
   d. A lack of autocorrelation in the data.
   e. All of the above.
10. Does Minitab support demerit charting in the current version?
   a. Yes, under individuals charting.
   b. Yes, under DOE → Factorial.
   c. Yes, under time series modeling.
   d. No, there is no support.
   e. All of the above.
11. Based on the u-chart example in ▶ Sect. 24.5, your estimate for the capability improvement is?
   a. Approximately 0.02/item.

b. Approximately 1.2 nonconformities/item.
c. The range would decrease on a sustained basis.
d. The range and sample mean fluctuate.
e. All of the above.

12. Consider the example in ▶ Sect. 24.7. If unequal variances are not assumed, the p-value is?
a. 0.001.
b. 0.032.
c. 0.049.
d. 0.054.
e. 0.198.

13. Consider the example in ▶ Sect. 24.7. The degrees of freedom assuming equal variances are?
a. 0.001.
b. 4.
c. 6.
d. 8.
e. 12.

14. Consider the example discussed in ▶ Sect. 24.8. The MSE in the output for the example is (from the session window)?
a. 0.001.
b. 0.018.
c. 0.049.
d. 0.085.
e. 0.091.

15. Consider the example discussed in ▶ Sect. 24.8. Fisher's multiple comparison method results in which result?
a. All differences are not significant.
b. One or more mean differences are identified.
c. The ANOVA result is nullified.
d. The p-value is 0.291.
e. None of the above.

16. Consider the example discussed in ▶ Sect. 24.8. The MSF in the output for the example is (from the session window)?
a. 0.001.
b. 0.050.
c. 0.043.
d. 0.085.
e. 0.240.

17. Consider the example discussed in ▶ Sect. 24.9. Under generalized linear models, there is no option for Fisher's least significant difference multiple comparisons, true or false?
a. True.
b. False.

18. Consider the example discussed in ▶ Sect. 24.10. The estimated effect for factor A is?
a. −0.59120.

b. −0.4525.
c. 0.59120.
d. 0.4525.
e. None of the above.

19. Consider the example discussed in ▶ Sect. 24.10. The value in the lower left-hand corner of the normal probability plot of estimated effects is?
a. Standard error.
b. Lenth's PSE.
c. The coefficient of determination.
d. The process sigma.
e. The Plackett–Burman design.

20. In Minitab, the background of a plot can be changed by altering the color of the?
a. Amplitude.
b. Fill pattern.
c. Solid.
d. Frequency.
e. None of the above.

21. Consider the example in ▶ Sect. 24.11. Which applies?
a. The fitted model is a first-order polynomial.
b. The fitted model is a second-order polynomial.
c. The fitted model is an artificial neural net.
d. The R-squared value is spuriously high.
e. None of the above.

22. Identify data with controllable inputs and at least one important continuous output that can be modeled using ordinary least squares regression (or possibly logistic regression). Perform the analysis and generate recommendations that at least one imagined (or real) decision maker might gratefully consider and apply.

Reference

Allen TT, Afful-Dadzie A, Tseng SH (2018) Case study implementation of an intervention to improve instruction of lean six sigma in a statistical quality control course. 中原企管評論 1;16(1):1–4

Supplementary Information

T. T. Allen, *Introduction to Engineering Statistics and Lean Six Sigma*,
https://doi.org/10.1007/978-1-4471-7420-2

Problem Solutions

▶ Chapter 1

1. c.
2. e.
3. b.
4. d.
5. KIVs—study time, number of study mates KOVs—math GPA, english GPA.
6. KIVs—origin account type, expediting fee KOVs—time until funds are available, number of hours spent.
7. d.
8. b.
9. Six sigma training is case based. It is also vocational and not theory based.
10. TQM might be too vague for workers to act on. It also might not be profit focused enough to make many managers embrace it.
11. b.
12. e.
13. Having only a small part of a complex job in mass production, the workers cannot easily perceive the relationship between their actions and quality.
14. Shewhart wanted skilled workers to not need to carefully monitor a large number of processes. He also wanted a thorough evaluation of process quality.
15. By causing workers to follow a part through many operations in lean production, they can understand better how their actions affect results. Also, with greatly reduced inventory and one piece flows, problems are discovered downstream much faster.
16. A book being written is an engineered system. Applying benchmarking and engaging proof readers together constitute part of an improvement system.
17. In grading exams, I would be a mass producer if I graded all problem 1s then all problem 2s and so on. I would be a lean producer if I graded entire exams one after another.
18. c.
19. d.
20. e.
21. a.
22. b.
23. e.
24. Green belts should know terminology and how to apply the methods competently. Black belts should know what green belts know and have enough understanding of theory to critique and suggest which methods should be used in a project.
25. Knowledge: Physics refers to attempt to predict occurrences involving few entities with high accuracy.
    Comprehension: One table in physics might show the time it takes for various planets to orbit the sun.
    Application: An example application of physics is predicting the time it takes for the earth around the sun.
    Analysis: An analysis question relates to identifying whether a given type of theory can achieved the desired level of accuracy.
    Synthesis: An issue for synthesis is how to connect mathematical technology with a need to forecast blast trajectories.
26. a.
27. b.
28. c.
29. c.
30. d.

31. b.
32. a.
33. b.
34. c.

▶ Chapter 2

1. b.
2. e.
3. a.
4. d.
5. a.
6. c.
7. c.
8. a.
9. Acceptance Sampling, Process Mapping, Regression.
10. Acceptance Sampling, Control Planning, FMEA, Gauge R&R, SPC Charting.
11. d.
12. d.
13. e.
14. a.
15. a.
16. d.
17. See the examples in the chapter.
18. An additional quality characteristic might be the groove width with USL = 0.60 mm and LSL = 0.50 mm.
19. b.
20. a.
21. e.
22. c.
23. See the example in ▶ Chap. 4.
24. See the examples in ▶ Chap. 9.
25. In Step 5 of the method, which order should the base be folded. Fold up first from the bottom or from the sides?
26. It is a little difficult to tell which are cuts and which are folds. For example, is there a cut on the sides or a fold only?

▶ Chapter 3

1. e.
2. b.
3. e.
4. d.
5. b
6. a.
7. d.
8. e.
9. a.
10. b.
11. See the examples in ▶ Chap. 9.
12. e.
13. a.
14. a.
15. b.
16. Writing this book slowed progress on many other projects including course preparation, writing grant proposals, and writing research papers.
17. e.
18. b.
19. c.
20. d.
21. b.
22. a.
23. a.
24. d.
25. b.
26. e.
27. c.
28. c.
29. b.
30. a.
31. e.
32. In decision-making, there is a tendency not to focus on the failures that affect the most people in the most serious way. Instead, we often focus on problems about which someone is bothering us the most. Pareto charting can bring perspective that can facilitate alignment of project scopes with the most important needs.
33. For many important issues such as voting methods, there is a tendency to abandon discipline and make decisions based on what we personalty like. Through a careful benchmarking exercise, one can discover what

make customers happy, drawing on inspiration from competitor systems. This can be important ethically because it can help make more people happy.

34. b.
35. e.
36. e.
37. d
38. b.
39. c.
40. a.
41. d.
42. b.

▶ Chapter 4

1. b.
2. d.
3. a.
4. b.
5. b.
6. c.
7. c.
8. a.
9. b.
10. c.
11. d.
12. Under standard assumptions, the measurement system is not gauge capable. The measurement system cannot reliably distinguish between parts whose differences are comparable to the ones used in the study.
13. c.
14. a.
15. b.
16. b.
17. b.
18. c.
19. The following is from Minitab (Fig. PS.1).
20. e.
21. d.
22. b.
23. d.
24. The following is from Microsoft Excel (Fig. PS.2).
25. e.
26. a.
27. c.
28. a.
29. f.
30. c.
31. d.
32. The following was generated using Minitab with subgroups 18 through 22 removed (Fig. PS.3).
33. e

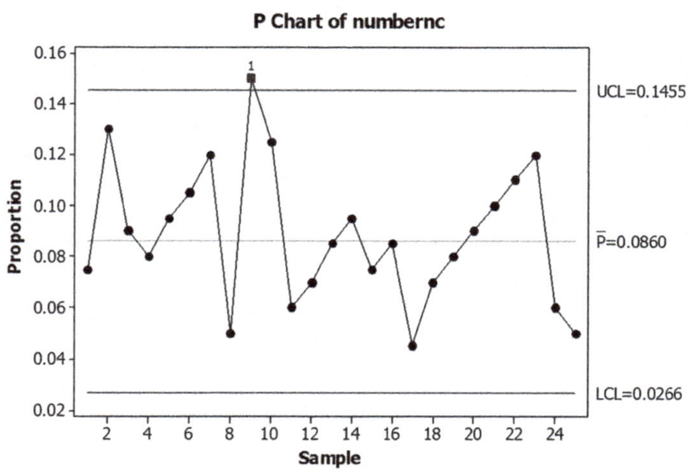

Fig. PS.1 P chart showing the fraction of nonconforming units

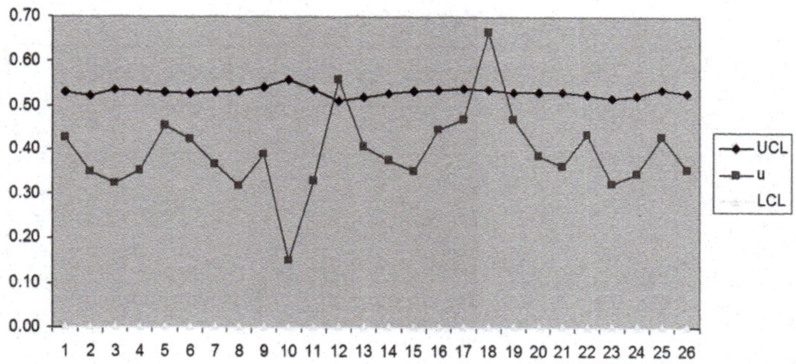

Fig. PS.2 U chart showing the number of conformitities per item

34. Crossed gauge R&R could help quantify the gauge capability of the system including both inspectors. Also, the inspectors could meet and document a joint measurement SOP. Finally, it might be helpful to create standard units using an external expert. Then, comparison with standards could be used which would be most definitive.
35. Assignable causes might include non-regular, major job performance evaluations or major life decisions. In general, the charting could aid in the development of a relatively rational perspective and acceptance that life has peaks and valleys. Also, the chart could help in rational approaches to determine that a major life decision has been a success or a failure.
36. The majority of quality problems are caused by variability in the process. Otherwise, all of the units would be conforming or nonconforming. The width of the limits on R-charts quantifies the amount of variability in the process with only common causes operating. Since major quality projects are often designed to reduce the common cause variation, a pair of charting activities before and after the project can quantify the effects.
37. Suppose that we are playing doubles tennis and coming up to the net at every opportunity. Unfortunately, we have a string of pathetic volleying attempts causing lost points. It still is probably advisable to continue coming to net since the strategy still makes sense.
38. Variation generally causes quality problems or else the faction of nonconforming units would be 0% or 100%. Therefore, reducing the variation as measured by the R chart on average would be a desirable outcome.
39. Measurement error is defined here as the difference between the output variables derived and the standard values. This means that if you measure a standard value one 1 plug to be 1.02 in. you measurement error is 0.02 in.
40. The data can be put into Minitab standard format as follows.

Part#	Op.	Measurement
1	1	0.92
2	1	1.07
3	1	1.05
4	1	1.08

Then, using Minitab and the Xbar & R option, the estimated gauge R&R% is 9.5, so the gauge is capable and measurement error can often be ignored.

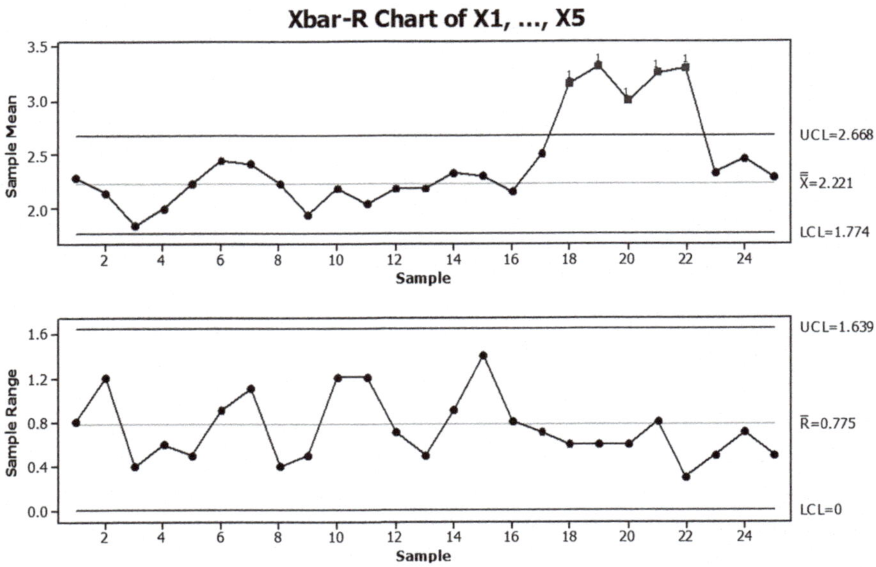

Fig. PS.3 Minitab Xbar and R chart

41. $n = 200$

    $p_0 =$ (total number nonconforming) $\div$ (total number inspected) $= 210/5000 = 0.042$,

    $UCL_{trial} = 0.042 + 3.0 \times \text{sqrt}\,[(0.042) \times (1 - 0.042) \div 200] = 0.085$,

    $CL_{trial} = 0.042$, and

    $LCL_{trial} = \text{Max}\,\{0.042 - 3.0 \times \text{sqrt}\,[(0.050) \times (1 - 0.050) \div 200], 0.0\} = 0.0$.

    Figure PS.4 shows a p-chart of the startup period. Clearly, the p value in week 5 and 6 subgroups constitute out-of-control signals. These signals were likely caused by an assignable cause (power outages).

    The revised limits are:

    $p_0 =$ (total number nonconforming) $\div$ (total number inspected) $= 128/4600 = 0.028$,

    $UCL = 0.028 + 3.0 \times \text{sqrt}\,[(0.028) \times (1 - 0.028) \div 200] = 0.063$,

    $CL = 0.028$, and

    $LCL = \text{Max}\,\{0.028 - 3.0 \times \text{sqrt}\,[(0.028) \times (1 - 0.028) \div 200], 0.0\} = 0.000$.

Chapter 5

1. d.
2. e.
3. b.
4. a.
5. b.
6. c.
7. b.
8. a.
9. C&E matrices help engineers communicate about their beliefs related to causality. The process of creating these matrices forces engineers to think about all of the customer issues. Also, C&E matrices are needed for the construction of the house of quality which many people feel helps them make system design decisions.
10. e.
11. a.
12. d.
13. c.
14. b.
15. d.

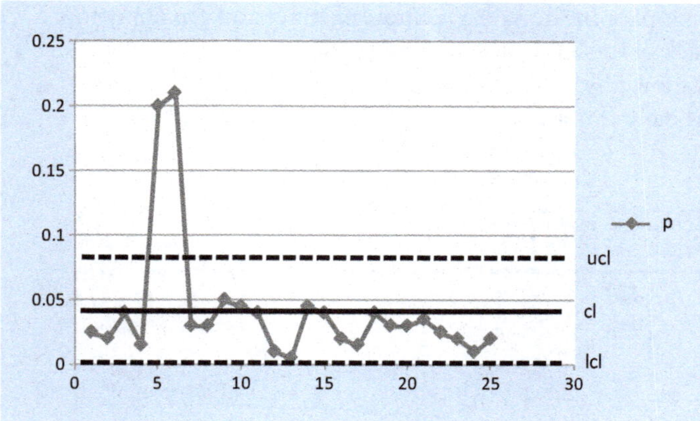

Fig. PS.4 Car Dealer *p*-chart during the startup period

16. It is not clear how parents could fail to detect their child's pinching their fingers in doors. If the failures were not detected because the baby sitters failed to report the problem, that needs to be clarified. Similarly, it is not clear how parents could fail to detect too much TV watching unless they were previously unaware of the issue or there were baby sitters involved.
17. e.
18. Lean relates to objectives for the system. Six sigma relates to the problem solving method which could help to reach lean objectives and benefit from lean-related sub-methods.
19. See the example at the end of ► Chap. 9.
20. The central principle is to avoid waste, i.e., any actions that the practitioner would not willingly pay for.
21. Value stream mapping, Poka-yoke, Kanban, 5S
22. Sort any items laying around for possible donations or trash. Then, the items are straightened to permit sweeping. Then, the key items are assigned specific clear places. Sustaining means returning each item after use.
23. It would be ideal, perhaps, to eliminate all operations besides report and homework writing, preparing for exams, exercise, and socializing.
24. Batch production is eliminated so that each sub-order is handled one-at-a-time.
25. Generally in the U.S. the primary constraints for engineering professors relate to research funding. The way to elevate that constraint is to focus on partnering with successful other professors and organizations which takes time. According to the TOC, other activities such as meetings for university service are generally not related to the bottleneck and should be reduced when possible. After funding has reached an acceptable level, the key issue generally becomes writing seminal research papers and/or teaching well.
26. These diagrams aid in the reduction of parts travelled.
27. My car offered many problems including cost and parking issues. Now, I ride my bike. Also, I often used to forget where my car was when I parked it. The poke-yoke

solution was to always park in the same place at the back of the lot.

28. Spaghetti diagrams show physical flows while VSM daigrams are more abstract and can apply to information.

29. The diagram is below.

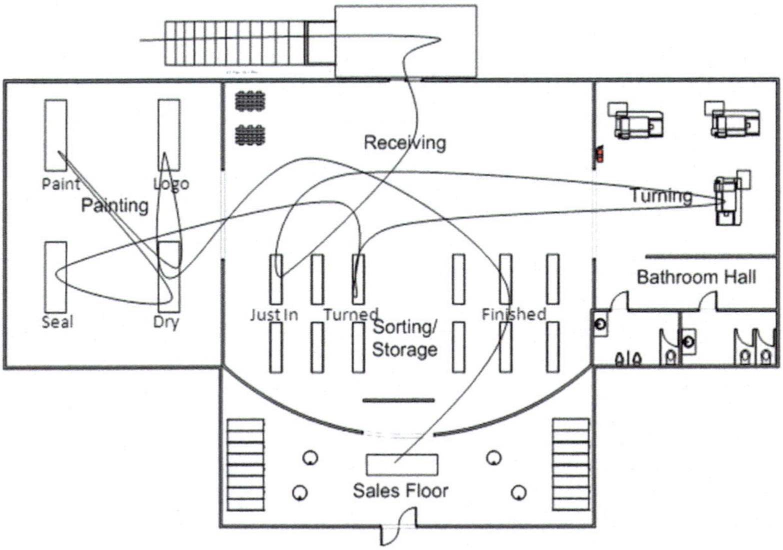

30. Possible recommendations include:
    a. Rearrange the storage shelf area so that the "turned" sub-area is closer to turning area,
    b. Move the finished parts area closer to the painting area, and
    c. Move paint booths closer together, move all of storage up and shrink overly large receiving area.

31. Receive to storage $n_1 = 1$ time.
    Storage to Cut $n_2 = 2$ times.
    Cut to Storage $n_3 = 1$ time.
    Storage to turning $n_4 = 1$ time.
    Turing to storage $n_5 = 1$ time.
    Cutting to paint $n_6 = 2$ times.
    Paint to print $n_7 = 1$ times.
    Paint to cut $n_8 = 1$ time.
    Paint to storage $n_9 = 1$ time.
    Storage to sales $n_{10} = 1$ times.
    TTD = (1)(100) + (2)(75) + (1)(75) + (1)(150) + (1)(150) + (2)(115) + (1)(50) + (1)(115) + (1)(115) + (1)(74) = 1209 feet.

32. A maid service operates in private homes. The workers could generate spaghetti diagrams for a few generic house plans to communicate best cleaning practices. Such diagrams could help in planning sweeping and mopping activities to make sure that materials are transported minimum distances and that areas of the house are always near and available for cleaning during drying periods. Hypothetically, efficiency improvements could permit the firm to serve more houses with fewer crews.

▸ Chapter 6

1. b.
2. c.
3. c.
4. b.
5. c.
6. a.
7. d.
8. QFD might be better able to address more customer considerations and input and output variable related issues. Also, QFD forces the decision-makers to study at least some of their competitors which can provide invaluable insights.
9. Targets can function as effective customer ratings, changing the ranking of the companies. For example, one company could dominate the customer ratings but how key input variable or key output variable settings far from the targets. Then, that company's approach might not be worthy of emulation.
10. See Table 6.3.
11. c.
12. c.
13. d.
14. See the snap tab example in the chapter.
15. d.

▸ Chapter 7

1. e.
2. d.
3. a.
4. c.
5. a.
6. a.
7. a.
8. d.
9. Monitoring using control charts requires the greatest on-going expense.
10. c.
11. c.
12. b.
13. Acceptance sampling can be used when the sampling is destructive. Acceptance sampling results generally results in reduced inspection costs.
14. e.
15. a.
16. $N = 1000$, $n_1 = 100$, $n_2 = 350$, $c_1 = 3$, $c_2 = 7$, and $r = 6$.
17. 450.
18. 100.
19. Because the company is applying some form of sampling, perhaps control charting or acceptance sampling. Acceptance sampling could be in place to potentially make firing decisions for a company supplying services.
20. e.
21. *Define* project goals including the current process. *Measure* current relevant process and collect data. *Analyze* the data and verify cause-and-effect relationships. Determine areas for improvement. *Design* or optimize the process by the use of design of experiments etc. *Verify* ensure everything is correct and new process setup has been tested.
22. Double sampling with $n_1 = 35$, $c_1 = 1$, $n_2 = 45$, $c_2=3$.
23. This is single sampling with $n = 15$ and $c = 0$.

▸ Chapter 8

1. e.
2. d.
3. a.
4. d.
5. Monitoring the several major index funds simultaneously is possible. Monitoring the results customer satisfaction surveys for all questions simultaneously is possible.

▶ Chapter 9

1. d.
2. b.
3. c.
4. I did in fact recommend performing design of experiments using even more factors. It seemed that they were lucky to find the source of the majority of the variation using only four factors.
5. e.
6. d.
7. c.
8. c.
9. c.
10. b.
11. d.
12. c.
13. Examples of criticisms are found in Chap.9.
14. For the air wing project, one might propose a goal to increase the air flight time to 2.0 s.
15. See the example in ▶ Chap. 9.

▶ Chapter 10

1. d.
2. e.
3. a.
4. 0.0013.
5. Triangular[$a = \$2.2M, b = \$3.0M, c = \$2.7M$] (others are possible).
6. 0.4.
7. a.
8. Yes.
9. b.
10. 0.81.
11. a.
12. d.
13. c.
14. a.

15a. 0.94.

15b. 0.94.

16. The following is from Excel (◘ Fig. PS.5).
17. The following is from Excel (◘ Fig. PS.6).
18. The policy in problem 17 is always more likely to accept lots because the OC curve is always above the preceding OC curve.
19. The ideal OC curve would look like ◘ Fig. PS.7. This follows because the policy would function like complete and perfect inspection but at potentially reduced cost.

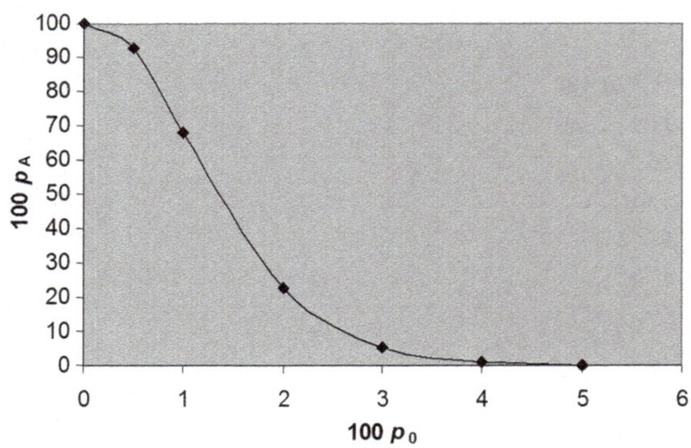

Fig. PS.5 Operating Characteristic curve plots percentage accepted vs. true percentage nonconforming

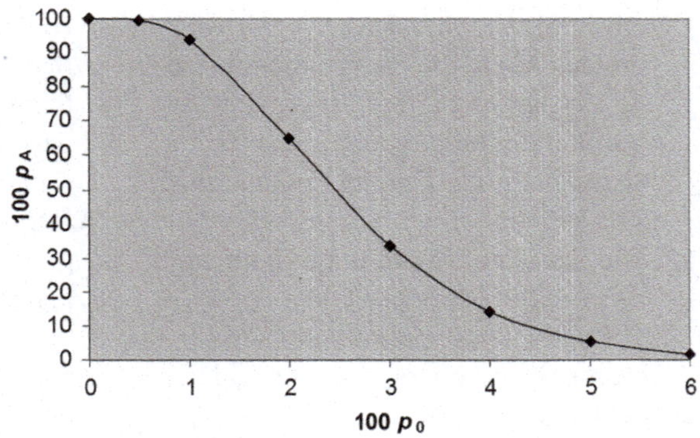

Fig. PS.6 Ideal Operating Characteristic curve

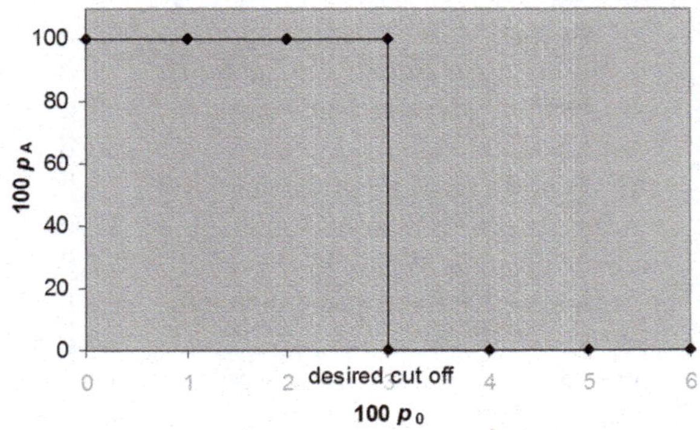

Fig. PS.7 Operating Characteristic curve

▶ Chapter 11

1. f.
2. b.
3. a.
4. d.
5. c.
6. $df = 4$ using the rounding formula.
7. f.
8. b.
9. f.
10. b.
11. a.
12. e.
13. c.
14. b.
15. b.
16. b.
17. a.
18. c.
19. A paired t-test can test non-homogeneous products and needs no assumptions, this is done because you are comparing each product to itself in different circumstances.
20. The author has assisted an observational study recently (with no randomization) and helped them to provide evidence (not proof) that software assistance significantly reduces the average number of errors in certain types of intelligence analyses.
21. Consider the hospital discharge case study in which I tested the average discharge times before and after

our intervention in a lean six sigma project: Allen TT, Tseng SH, Swanson K, McClay, MA (2009) Improving the hospital discharge process with Six Sigma methods. Quality Engineering 22(1):13–20.

22. The size is not found to be significant because the p-value is 0.89.
23. The author has applied ANOVA on numerous occasions to provide evidence in regression analyses that model predictions can be trusted (see ▶ Chap. 15).
24. In all the case studies in ▶ Chapter 17 the ANOVA p-values were significant.
25. ANOVA is a set of methods that include methods relevant to the following purpose. When there is a single factor varied over more than two levels or multiple factors, ANOVA often provides a standard way of evaluating whether the specific factors are associated with significant average response difference. It does this while permitting the user to specify the desired chance of making one or more Type I errors (wrongly finding significance). In ▶ Chap. 12, an alternative method to ANOVA called Lenth's relevant to fractional factorial data analysis is presented. In this specific context, Lenth's method likely offers a superior combination of Type I and Type II error rates to ANOVA and, potentially, ANOVA may be simply nonapplicable since the degrees of freedom for the residuals might be zero.

▶ Chapter 12

1. g.
2. b.
3. c.
4. a.
5. d.
6. c.
7. a.
8. d.
9. a.
10. d.
11. b.
12. b.
13. d.
14. f.
15. c.
16. e.
17. See the examples in ▶ Chap. 18.
18. Naming the factors A through E, $E = AD$.
19. b.
20. e.
21. c.

Short Answer Section:

1. D it the entire matrix below:

The $n = 16$ run regular fractional factorial array

Run	x_1	x_2	x_3	x_4	x_5	x_6	x_7	x_8	x_9	x_{10}	x_{11}	x_{12}	x_{13}	x_{14}	x_{15}
1	−1	1	1	−1	1	−1	1	1	−1	−1	−1	1	1	−1	−1
2	1	1	1	1	1	1	1	1	1	1	1	1	1	1	1
3	1	1	−1	−1	1	−1	−1	1	1	1	−1	−1	−1	−1	1
4	1	−1	1	1	−1	−1	−1	1	−1	−1	1	1	−1	−1	1
5	−1	−1	−1	1	−1	−1	1	1	1	1	1	−1	1	−1	−1
6	1	−1	−1	1	1	1	−1	−1	1	−1	−1	1	1	−1	−1
7	1	1	−1	1	−1	−1	1	−1	−1	1	−1	1	−1	1	−1
8	1	1	1	−1	−1	1	−1	−1	−1	1	1	−1	1	−1	−1
9	1	−1	1	−1	1	−1	1	−1	1	−1	1	−1	−1	1	−1
10	−1	−1	−1	−1	1	−1	−1	−1	−1	1	1	1	1	1	1
11	−1	1	1	1	−1	−1	−1	−1	1	−1	−1	−1	1	1	1
12	−1	−1	1	−1	−1	1	−1	1	1	1	−1	1	−1	1	−1
13	1	−1	−1	−1	−1	1	1	1	−1	−1	−1	−1	1	1	1
14	−1	1	−1	−1	−1	1	1	−1	1	−1	1	1	−1	−1	1
15	−1	−1	1	1	1	1	1	−1	−1	1	−1	−1	−1	−1	1
16	−1	1	−1	1	1	1	−1	1	−1	−1	1	−1	−1	1	−1

Using $X = (1|D)\ x$ is below

$$\mathbf{X} = \begin{bmatrix}
1 & 1 & 1 & 1 & 1 & 1 & 1 & 1 & 1 & 1 & 1 & 1 & 1 & 1 & 1 & 1 \\
1 & 1 & 1 & -1 & -1 & 1 & -1 & -1 & 1 & 1 & 1 & -1 & -1 & -1 & -1 & 1 \\
1 & 1 & -1 & 1 & 1 & -1 & -1 & -1 & 1 & -1 & -1 & 1 & 1 & -1 & -1 & 1 \\
1 & -1 & -1 & -1 & 1 & -1 & -1 & 1 & 1 & 1 & 1 & 1 & -1 & 1 & -1 & -1 \\
1 & 1 & -1 & -1 & 1 & 1 & 1 & -1 & -1 & 1 & -1 & -1 & 1 & 1 & -1 & -1 \\
1 & 1 & 1 & -1 & 1 & -1 & -1 & 1 & -1 & -1 & 1 & -1 & 1 & -1 & 1 & -1 \\
1 & 1 & 1 & 1 & -1 & -1 & 1 & -1 & -1 & -1 & 1 & 1 & -1 & 1 & -1 & -1 \\
1 & 1 & -1 & 1 & -1 & 1 & -1 & 1 & -1 & 1 & -1 & 1 & -1 & -1 & 1 & -1 \\
1 & -1 & -1 & -1 & -1 & 1 & -1 & -1 & -1 & -1 & 1 & 1 & 1 & 1 & 1 & 1 \\
1 & -1 & 1 & 1 & 1 & -1 & -1 & -1 & -1 & 1 & -1 & -1 & -1 & 1 & 1 & 1 \\
1 & -1 & -1 & 1 & -1 & -1 & 1 & -1 & 1 & 1 & 1 & -1 & 1 & -1 & 1 & -1 \\
1 & 1 & -1 & -1 & -1 & -1 & 1 & 1 & 1 & -1 & -1 & -1 & -1 & 1 & 1 & 1 \\
1 & -1 & 1 & -1 & -1 & -1 & 1 & 1 & -1 & 1 & -1 & 1 & 1 & -1 & -1 & 1 \\
1 & -1 & -1 & 1 & 1 & 1 & 1 & 1 & -1 & -1 & 1 & -1 & -1 & -1 & -1 & 1 \\
1 & -1 & 1 & -1 & 1 & 1 & 1 & -1 & 1 & -1 & -1 & 1 & -1 & -1 & 1 & -1
\end{bmatrix}$$

$X'X = n \times \mathbf{I}$ there for this DOE is orthogonal and A is given below:

$\mathbf{A} = (\mathbf{X}'\mathbf{X})^{-1}\mathbf{X}'$ for the 16 run plan.

$$\mathbf{A} = 0.0625\begin{bmatrix} 1 & 1 & 1 & 1 & 1 & 1 & 1 & 1 & 1 & 1 & 1 & 1 & 1 & 1 & 1 & 1 \\ -1 & 1 & 1 & 1 & -1 & 1 & 1 & 1 & 1 & -1 & -1 & -1 & 1 & -1 & -1 & -1 \\ 1 & 1 & 1 & -1 & -1 & -1 & 1 & 1 & -1 & -1 & 1 & -1 & -1 & 1 & -1 & 1 \\ 1 & 1 & -1 & 1 & -1 & -1 & -1 & 1 & 1 & -1 & 1 & 1 & -1 & -1 & 1 & -1 \\ -1 & 1 & -1 & 1 & 1 & 1 & 1 & -1 & -1 & -1 & 1 & -1 & -1 & -1 & 1 & 1 \\ 1 & 1 & 1 & -1 & -1 & 1 & -1 & -1 & 1 & 1 & -1 & -1 & -1 & -1 & 1 & 1 \\ -1 & 1 & -1 & -1 & -1 & 1 & -1 & 1 & -1 & -1 & -1 & 1 & 1 & 1 & 1 & 1 \\ 1 & 1 & -1 & -1 & 1 & -1 & 1 & -1 & 1 & -1 & -1 & -1 & 1 & 1 & 1 & -1 \\ 1 & 1 & 1 & 1 & 1 & -1 & -1 & -1 & -1 & -1 & -1 & 1 & 1 & -1 & -1 & 1 \\ -1 & 1 & 1 & -1 & 1 & 1 & -1 & -1 & 1 & -1 & 1 & 1 & -1 & 1 & -1 & -1 \\ -1 & 1 & 1 & -1 & 1 & -1 & 1 & 1 & -1 & 1 & -1 & 1 & -1 & -1 & 1 & -1 \\ -1 & 1 & -1 & 1 & 1 & -1 & -1 & 1 & 1 & 1 & -1 & -1 & -1 & 1 & -1 & 1 \\ 1 & 1 & -1 & 1 & -1 & 1 & 1 & -1 & -1 & 1 & -1 & 1 & -1 & 1 & -1 & -1 \\ 1 & 1 & -1 & -1 & 1 & 1 & -1 & 1 & -1 & 1 & 1 & -1 & 1 & -1 & -1 & -1 \\ -1 & 1 & -1 & -1 & -1 & -1 & 1 & -1 & 1 & 1 & 1 & 1 & 1 & -1 & -1 & 1 \\ -1 & 1 & 1 & 1 & -1 & -1 & -1 & -1 & -1 & 1 & 1 & -1 & 1 & 1 & 1 & -1 \end{bmatrix}$$

2. Type one error is a false postive. Answers may very, anything where an experiment said something was accepted but in reality it should have been rejected. Type two error is a false negative. Answers will very and this is the opposite of a type one error so you rejected something because of an experiment but really it should have been accepted.

▶ Chapter 13

1. c.
2. a.
3. b.
4. a.
5. e.
6. a.
7. The number of factors is three and the number of levels is three.
8. e.
9. d.
10. a.
11. f.
12. b.
13. b.
14. c.
15. b.
16. c.
17. e.
18. The lack of fit is marginally significant.
19. Response surface methods generally generate more accurate prediction models than screening methods using fractional factorials, often resulting in recommended settings with more desirable engineered system average response values.
20. Response surface methods are experimentation and analysis procedures to explore continuous variable input-output relationships. Two classes could refer to one-shot or sequential or central composites or Box Behnken designs.

21. Box Behnken designs (BBDs) were argued to generate relatively accurate predictions because they minimize so-called "bias" errors under certain reasonable assumptions. Central composite designs (CCDs) were argued to offer the advantage that they permit more flexibility in level adjustments. Also, CCDs can be used in two-step sequential response surface methods. In these methods, there is a chance that the experimental will stop with relatively few runs and decide his or her prediction model is satisfactory.

Chapter 14

1. b.
2. f.
3. a.
4. RDPM is designed to elevate the bottleneck subsystem identified using a TOC approach. Dependencies are modeled and settings are chosen to maximize throughput subject to appropriate fractions of nonconforming units.
5. c.
6. c.
7. a.
8. d.
9. $\sigma_1^2(5x_1 + 2x_2 + 2)^2 + \sigma_2^2(2x_1 + 8x_2 - 1)^2$.
10. (i) Derives the settings that directly maximize the profits and not an obscure measure of quality. (ii) Build upon standard RSM which is taught in many universities and might result in improved recommendations.
11. (i) Generally, Taguchi Methods are easier for practitioners to use and understand without the benefit of software. (ii) If all noise factors or all control factors are *ETC*, the cost of experimentation might be far less using Taguchi product arrays compared with standard RSM based experimentation.
12. See Allen et al. (2001).

Chapter15

1. c.
2. a.
3. c.
4. a.
5. e.
6. a.
7. b.
8. f.
9. a.
10. b.
11. a. $y_{est}(x_1, x_2) = 0.443 + 0.0361\ x_1 + 0.00151\ x_2$.
    b. $y_{est}(x_1, x_2) = 0.643 + 0.0352\ x_1 + 0.000564\ x_2 - 0.000231\ x_1^2 + 8.11066\text{e-}007\ x_2^2 + 1.12021\text{e}{-}005\ x_1x_2$.
    c. 0.778.
    d. 1.3.
    e. 0.678.
    f. 4.15.
12. f.
13. It seems likely that genre plays an important role in movie profitability. However, it is not clear that all of the levels and contrasts are needed to quantify that effect. Therefore, the analysis here focuses only on the action contrast. Starting with scaled inputs and a first order fit model, the following table can be derived. It indicates that stars likely play a far greater role in non-action movie profits than they do in action movie profits. Considering the potentially cancelling effects of video sales and marketing costs, it seems reasonable that fifth week revenues roughly correspond to profits. Then, stars are worth roughly \$8M a piece on average, and that value depends greatly on the type of movie. Note that this analysis effectively assumes that critic rating is controllable through the hiring of additional writing talent.

Scaled factor	Coefficients	Standard error	t Stat	p-value	VIF
Constant	109,691,377	19,447,351	5.64	0.000	–
Action (0 to 1 → −1 to 1)	23,853,177	18,149,094	1.31	0.207	2.05
#Stars (0 to 2 → −1 to 1)	8,539,237	24,084,687	0.35	0.728	1.28
Sequel (0 to 1 → −1 to 1)	25,435,916	17,204,916	1.48	0.159	1.57
CritNo. (68 to 92→ −1 to 1)	65,602,653	28,438,046	2.31	0.035	1.43
Action × #Stars	−29,689,954	24,026,140	−1.24	0.234	1.45
Action × Sequel	−16,057,519	15,687,510	−1.02	0.321	1.30

(Note also that leaving stars out of the model might make sense but it would result in no help for setting their pay.)

14. Many possible models can be considered. However, models including interactions between the city and the numbers of bedrooms and baths do not make physical sense. After exploring models derived in coded –1 and 1 units, the following model was fitted in the original units: y_{est} = Offering Price (in \$) = –103678 – 83766.2 × City_Columbus + 214823 × #Bedrooms – 2671.63 × #Baths – 48753.9 × #Bedrooms^2 – 30911.9 × #Baths^2 + 58820.9 × #Bedrooms × #Baths

This model can be used to develop reasonable prices in line with other offering prices in the market. This model gives an acceptable normal probability plot of residuals shown in Fig. PS.8, seems decent from the VIF and PRESS standpoints, and makes intuitive sense .

Chapter 16

1. a.
2. f.
3. c.
4. f.
5. a.
6. e.

Chapter 17

1. b.
2. f.
3. e.
4. c.
5. f.
6. d.
7. b.
8. See the first example in Sect. 13.4.

Chapter 18

1. f.
2. a.
3. d.
4. b.
5. b.
6. b.
7. a.
8. e.
9. c.
10. d.
11. a.
12. 1.7.

Chapter19

1. c.
2. b.
3. e.
4. c.
5. d.
6. a.
7. c.
8. a.

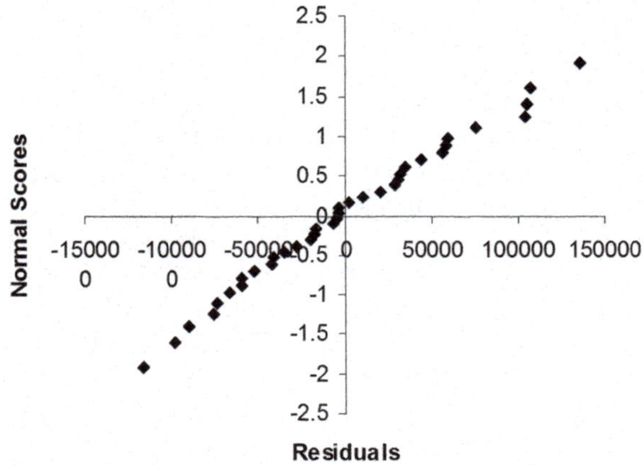

Fig. PS.8 Residual plot for a specific offering price regression model

9. /* This change de-codes the chromosome to address the new constraint. */ for (j = 0; j < SIZE; j++) vect[j]=2.0 +(x -> vector[j]) * 3.0;

▸ Chapter 20

1. e.
2. a.
3. The solutions is unchanged. The selected system is the most expensive.

▸ Chapter 21

1. d.
2. e.
3. a.
4. d.
5. e.
6. c.
7. Identify—Customer surveys & focus groups suggest, e.g., that low cost and highly reliable vehicles are needed.
   Design—A design using a tight cost and reliability constraint is developed.
   Optimize—Additional Optimizations are Performed using experimental design, QFD, or some other activity.
   Validate—The design is prototyped, built, and tested.
8. Define—The problem is defined with rough needs guessed for a new market using an existing product to be improved.
   Measure—Customers and others are exposed and feedback is generated.
   Analyze—Analysis and additional experiments possibly using design of experiments or QFD are performed.
   Design—The vehicle is designed and prototyped.
   Verify—Tests confirm that the vehicle hopefully meets the needs or iteration is performed.
9. IDOV is more relevant when there is no existing product and tries to gain feedback early for the new design. DMADV is more relevant for product improvement and gives more time to study leveraging the existing design.

▸ Chapter 22

1. d.
2. c.
3. a.
4. d.
5. // ** translate solution vector to method design.

```
c1crit = x->vector[0]*NLITTLE;
rcrit = c1crit + x->vector[1]*(NLITTLE   -c1crit);
c2crit = x->vector[2]*NLITTLE;
Ntotal = (long) ((NBIG  - MINRUNS) * (x ->vector[3])) + MINRUNS;
n1 = (long) ((Ntotal) * (x->vector[4]));
n2 = Ntotal - n1;

// ** generate truth
temp = ranS(&seed); difference = 0.0;
if(temp < trueProb) {difference = diffParm;}
else {difference = 0.0;}

// ** generate data
for(i=0;i<n1;i++) //first round
{         y11i[i]=gasdev(&seed);
                    y12i[i]=gasdev(&seed)+difference; }
for(i=i;i<Ntotal;i++) //second round (might be wasted)
{         y21i[i]=gasdev(&seed);
                    y22i[i]=gasdev(& seed)+difference; }

// ** do test
for(i=0;i<n1;i++)
{ y11avg+=y11i[i]; y12avg+=y12i[i]; }
 y11avg=y11avg/((double) n1); y12avg=y12avg/((double) n1);
for(i=i;i<Ntotal;i++)
{ y21avg+=y21i[i]; y22avg+=y22i[i]; }
 y21avg=y21avg/((double) n2); y22avg=y22avg/((double) n2);
if(y12avg - y11avg < c1crit)


{declareD = 0; nUsed=((double)n1);} //1st stage says no diff
else {if(y12avg  - y11avg > rcrit)
          {declareD = 1; nUsed=((double)n1);} //1st stage says diff
          else { nUsed=((double)Ntotal);
temp = (y12avg-y11avg)*((double)n1)–(y22avg-y21avg)*((double)n2);
                    if(temp > c1crit * ((double)Ntotal) && n2 > 0)
                    {declareD = 1;} else {declareD = 0;}
          } //finish 2nd stage
} //finish testing

// ** evaluate results
if((difference>0.00000001) && (declareD < 1)) {errorP   = 1.0;}
if((difference<0.00000001) && (declareD > 0)) {errorP = 1.0;}
return nUsed + weight*errorP;
```

Chapter 23

1. d.
2. b.
3. f.
4. d.
5. e.
6. Increase skill variety, task identity, task significance, autonomy, and/or feedback.

Chapter 24

1. a.
2. b.
3. e.
4. a.
5. a.
6. b.
7. f.
8. a.
9. c.
10. d (as of Version 18).
11. b.
12. d.
13. c.
14. b.
15. b.
16. e.
17. a.
18. d.
19. b.
20. b.
21. b.
22. See the examples in Chap. 17.

Glossary

5S a lean production tool meaning, sort, straighten, sweep, standardize, sustain, which is used when organizing a work area to be lean.

AdjustedR-squaredof prediction a regression diagnostic that estimates the fraction of the observed variation explained by the model based on a cross-validation based estimate of the SSE.

Anaylsis of Variance (ANOVA) set of methods related to the statistical comparison of variances. Often, these methods use variances to evaluate the effects of factors on mean response values while regulating the rate of making at least a single Type I error.

Assignable cause variation a set of changes in a given quality characteristic caused by factors changing over which local authority can control because they are attributable to something "fixable".

Bayes formula define $\Pr(B|A_i)$ as the probability that the event$B$ will happen given that the event A_i has happened. Also, let $A_1, A_2, \ldots, A_\infty$ be an infinite sequence of disjoint events that exhaust the sample space. Then, Bayes' formula, provable using Venn diagrams and the definition of conditional probability, states

$$\Pr(A_i|B) = \frac{\Pr(B|A_i)\Pr(A_i)}{S_{j=1\ldots\infty}\Pr(B|A_j)\Pr(A_j)}.$$

Benchmarking an activity in which people systematically inspect the relative performance of their product or service compared with alternatives.

Bonferonni inequality establishes that the chance of making no mistake in q trials having chances $\alpha_1, \ldots, \alpha_q$ of making mistakes is greater than $1 - \Sigma_{i=1,\ldots,q}\,\alpha_i$.

Categorical variable system input that can assume only a finite number of levels and these levels have no natural ordering.

Common cause variation a set of changes in a given quality characteristic caused by factors changing over which local authority has minimal ability to control.

Confounding generating an input pattern in which system inputs correlate such that accurate empirical attribution of causes is impossible because of multicollinearity

Continuous factor input variable that can assume, theoretically, any of an infinite number of levels (with levels having a natural ordering).

Control factor a system input that can be controlled during normal system operation and during experimentation.

Control group people in a study who receive the current system level settings and are used to generate response data (related to counteracting any Hawthorne effects).

Control limit when the charted quantity is outside these numbers, we say that evidence for a possible assignable cause is strong and investigation is warranted.

Decision space the set of solutions for system design that the decision-maker is choosing from.

Defining relation a set of generators that are sufficiently complete as to identify uniquely the fractional factorial.

Design for six sigma (DFSS) a systematic method for the radical redesign of existing systems or the design of new systems. It can be viewed as part of six sigma or an alternative method to six sigma because it necessarily includes design control within its scope and varies more greatly in its implementation across business units than DMAIC-based improvement strategies.

Double blind attribute of an experimental plan such that test subjects and organizers in contact with them do not know which input factor combination is being tested.

Easy-to-change factors (*ETC*) inputs with the property that if only their settings are changed, the marginal cost of each additional experimental run is small.

Engineered system an entity with controllable inputs (or "factors") and outputs (or "responses") for which the payoff to stakeholders is direct.

Event set of values that a random variable can assume.

Expected value the theoretical mean of a random variable derivable from the distribution or probability mass function, written *E*[ ].

Failure mode refers to an inability to meet engineering specifications expressed using a causal vocabulary.

Fidelity level the degree of difference, evaluated either quantitatively or subjectively, of the prototype system to the relevant engineered system.

Flat file database of entries that are formatted well enough to facilitate easy analysis with standard software.

Fractional factorial(FF) designs input patterns or arrays in which some possible combinations of factors and levels are omitted (standard FFs have the property that all columns can be obtained by multiplying together other columns).

Full factorialdesigns input patterns or arrays in which all possible combinations of factors and levels are used.

Functional form the relationship that constrains and defines the fitted model global.

Generator a property of a specific regular fractional factorial array showing how one columns may be obtained by multiplying together other columns, e.g., AB = C

Hard-to-change factors(HTC) inputs with the property that if any of their settings are changed, the marginal cost of each additional experimental run is large.

Hawthorne effect a difference in average responses caused by the attention associated with a study (often they occur because studying improves performance).

Heteroscedasticity situation in which the variance is nonconstant, e.g., in regression this can be detected by normal probability plotting the residuals and seeing tails.

Heuristic solution method that is not guaranteed to find a global optimum solution.

Improvement system any set of activities resulting in a recommended set of engineered system inputs.

Interaction terms in polynomials involving products, e.g., $2.0x_1^2x_3$, are called interaction terms.

Job 1 the time in the design cycle in which the first product for mass market is made

Kanban a lean production tool which uses an indicator, typically a card of color, to let the employees know the extent to which inventory buffers are full.

Kriging modeling an approach for fitting a nonlinear "spline" function to data usually using maximum likelihood estimation which can be viewed as a generalization of polynomial regression.

Lean manufacturing a set of techniques as well as the principle of seeking zero waste. The techniques were generally invented by Toyota to make a manufacturing system work more efficiently and associated with the elements of the Toyota Production System.

Lean sigma of lean six sigma a principled approach to problem solving that combines the elements of six sigma, the theory if constraints, and lean manufacturing including their tool boxes with the goals of zero waste including zero (or nearly zero) nonconformities produced.

Least squares estimation a process of deriving the parameters of a fitted meta model through formal optimization minimizing the sum of squares error.

Leverage in regression the issue of input patterns causing the potentially strong influence of a small number of observations on the fitted model attributes.

Liebniz rule the formula that allows calculation of derivatives of integrals is

$$\frac{d}{dx}\int_{a(x)}^{b(x)} f(x,u)du = \int_{a(x)}^{b(x)} [\frac{d}{dx}f(x,u)]du$$
$$+[\frac{d}{dx}b(x)]f(b(x),x)] - [\frac{d}{dx}a(x)]f(a(x),x)]$$

Linear regression curve fitting in which the model form can be written $\mathbf{f}(\mathbf{x})'\boldsymbol{\beta}_{est}$.

Local optimum the best solution to a formal optimization problem in which the optimization is constrained to a subspace not containing the global optimum.

Metamodel a functional relationship between inputs and outputs that could be a regression, kriging, neural net, or other empirical model.

Multicollinearity situation in which the input pattern or design of experiments array make fitting a model form of interest prone to high errors and/or misinterpretation.

Neural nets a curve fitting approach useful for either classifying outputs or as an alternative to linear regression sharing some characteristics with the human mind

Noise factor an input to the system that can be controlled during experimentation but which cannot be controlled during normal system operation.

Normally distributed random variables associated with probabilities of taking on certain values given by a distribution function of the following form: $f(x) = (2\pi\sigma^2)^{-1}\exp[(x-\mu)^2/(2\sigma^2)]$, where μ is the distribution mean and σ is the standard deviation. In layperson's terms, if X is normally distributed it is probably within a number σ of the mean μ.

Np-hard the property of optimization problems that no known procedures can guarantee finding a global optimum solution in a time bounded by a polynomial in the problem size parameters, e.g., m the number of factors (a heuristic is needed).

Null hypothesis the belief that the factors being studied have no effects, e.g., on the mean response value.

Observational study data collection and analysis for cases in which randomization has not been used.

On-hand data response numbers associated with an observational study.

Parameterization one way to turn a system design problem into the selection of settings on well-defined "dials".

Optimum the true best solution to a formal optimization problem constrained to lie in a decision space.

p-value the value of alpha in hypothesis testing such that the test statistic equals the critical value. Small values of alpha can be used as evidence that the effect is "statistically significant". Under standard assumptions it is the probability that a larger value of the test statistic would occur if the factor in question had zero effect on the mean or variance in question.

Poka-yoke a lean manufacturing tool used to "error proof" a system.

Posterior probabilities the estimated probabilities of events calculated after data has been collected using Bayes' formula to update the prior probabilities.

Process flow the movement of material or information through a system.

Quality characteristic a system output or response that can be used to define whether the system is acceptable or not.

Random effects models fitted curves to data and associated tests for cases in which some of the factor levels are relevant only for their ability to be representative of populations (e.g., people in a drug study or parts in manufacturing).

Random errors the difference between the true average for a given set of outputs and the actual values of those outputs (caused by uncontrolled factors varying).

Randomization the allocation of blocking factor levels to runs in a random or unpatterned way in experimental planning.

Reductionist tendency an instinct to, when confronted with complexity, abandon a disciplined approach to decision making and/or become hostile to ideas.

Region of interest the set of solutions for system design that the decision-maker feels is particularly likely to contain desirable solutions.

Regression model an equation derived from curve fitting response data useful for predicting mean outputs for a given set of inputs.

Residual difference between the metamodel prediction and the actual data value

rigorous method (in the context of optimization)—an algorithm associated with a rigorously proven claim about the objective function of the derived solution.

Responses (throughout) variables that characterize the outputs of a system

Response surface model a fitted model with a model form involving quadratic terms.

Sample standard deviation(s) one measure of the "dispersion" of a collection of n numbers, $y_1, \ldots, y_n$ which has the formula $s^2 = S_{i=1,\ldots,n}(y_i - -\bar{y})^2/(n - -1)$.

Sample variance (s^2) the sample standard deviation squared.

Screening experimentation in which the primary goal is to identify the "important factors" which aids in the elimination of some factors from consideration.

Significance level synonym for p-value.

Six sigma an organized and systematic problem-solving method for strategic system improvement and new product and service development that relies on statistical methods and the scientific method to make dramatic reductions in customer defined defect rates and/or improvements in key output variables.

Spaghetti diagrams a diagram that can be used to visualize and quantify the flow through a facility, often used as a tool for eliminating waste.

Specification limit these are numbers that determine the acceptability of a part. If the critical characteristic is inside the limits, the part is acceptable.

Statistical power probability of *not* making a Type II error, i.e., finding significance for a case when there is a nonzero difference.

Stochastic optimization the study of problems in which the decision-maker believes that it is necessary to estimate the objective function, $g(\mathbf{X})$, using some form of numerical integration.

Test for normality an evaluation of the extent to which a decision-making can feel comfortable believing that responses or averages are normally distributed.

Trace of a matrix the sum of the diagonal elements of a matrix, written Tr[].

Theory of Constraints (ToC) a system for system improvement based on focusing and alleviating bottleneck subsystems.

Theory of Constraints Lean Six Sigma (TLS) a system for process improvement that includes six sigma, ToC, and lean engineering.

Total Travel Distance (TTD) the total distance traveled through a facility as shown on a spaghetti diagram.

Type I error the event that a hypothesis testing procedure such as t-testing results in the declaration that a factor or term *is significant* when, in the true engineered system, that factor or term has *no effect* on system performance.

Type II error the event that a hypothesis testing procedure such as t-testing results in a *failure* to declare that a factor or term is significant when, in the true engineered system, that factor or term has a *nonzero* effect on system performance.

Within and between subject designs experimental plans involving within and between subject variables (relates to the allocation of levels or runs to subjects).

Variance Inflation Factors(VIFs) numbers that permit the assessment of whether there is any chance reliable predictions and inferences can be derived from the combination of model form and input pattern (VIFs < 10.0).

Index

F

G

H

I

J

K

T

U

V

W

Y

Printed by Printforce, the Netherlands